AF775109

High-Throughput Mass Spectrometry in Drug Discovery

High-Throughput Mass Spectrometry in Drug Discovery

Edited by

Chang Liu
SCIEX
Concord, Canada

Hui Zhang
Entos Inc.
La Jolla
USA

WILEY

This edition first published 2023
© 2023 John Wiley & Sons, Inc.

All rights reserved. No part of this publication may be reproduced, stored in a retrieval system, or transmitted, in any form or by any means, electronic, mechanical, photocopying, recording or otherwise, except as permitted by law. Advice on how to obtain permission to reuse material from this title is available at http://www.wiley.com/go/permissions.

The right of Chang Liu and Hui Zhang to be identified as the authors of the editorial material in this work has been asserted in accordance with law.

Registered Office
John Wiley & Sons, Inc., 111 River Street, Hoboken, NJ 07030, USA

Editorial Office
111 River Street, Hoboken, NJ 07030, USA

For details of our global editorial offices, customer services, and more information about Wiley products visit us at www.wiley.com.

Wiley also publishes its books in a variety of electronic formats and by print-on-demand. Some content that appears in standard print versions of this book may not be available in other formats.

Trademarks: Wiley and the Wiley logo are trademarks or registered trademarks of John Wiley & Sons, Inc. and/or its affiliates in the United States and other countries and may not be used without written permission. All other trademarks are the property of their respective owners. John Wiley & Sons, Inc. is not associated with any product or vendor mentioned in this book.

Limit of Liability/Disclaimer of Warranty
In view of ongoing research, equipment modifications, changes in governmental regulations, and the constant flow of information relating to the use of experimental reagents, equipment, and devices, the reader is urged to review and evaluate the information provided in the package insert or instructions for each chemical, piece of equipment, reagent, or device for, among other things, any changes in the instructions or indication of usage and for added warnings and precautions. While the publisher and authors have used their best efforts in preparing this work, they make no representations or warranties with respect to the accuracy or completeness of the contents of this work and specifically disclaim all warranties, including without limitation any implied warranties of merchantability or fitness for a particular purpose. No warranty may be created or extended by sales representatives, written sales materials or promotional statements for this work. The fact that an organization, website, or product is referred to in this work as a citation and/or potential source of further information does not mean that the publisher and authors endorse the information or services the organization, website, or product may provide or recommendations it may make. This work is sold with the understanding that the publisher is not engaged in rendering professional services. The advice and strategies contained herein may not be suitable for your situation. You should consult with a specialist where appropriate. Further, readers should be aware that websites listed in this work may have changed or disappeared between when this work was written and when it is read. Neither the publisher nor authors shall be liable for any loss of profit or any other commercial damages, including but not limited to special, incidental, consequential, or other damages.

Library of Congress Cataloging-in-Publication Data applied for
Hardback ISBN: 9781119678434

Cover Design: Wiley
Cover Images: Back cover image courtesy of Thomas R. Covey and Ella Potyrala
Cover Image: Courtesy of Thomas R. Covey
Cover Design Concept: Courtesy of Ella Potyrala

Set in 9.5/12.5pt STIXTwoText by Straive, Pondicherry, India
Printed and bound by CPI Group (UK) Ltd, Croydon, CR0 4YY

C9781119678434_260623

Contents

Part 6 Conclusion

List of Contributors

Leigh Bedford
SCIEX, Concord, ON, Canada

Aivett Bilbao
Earth and Biological Sciences Directorate, Pacific Northwest National Laboratory
Richland, WA, USA

Thomas R. Covey
SCIEX, Concord, ON, Canada

Sammy Datwani
Beckman Coulter Life Sciences
San Jose, CA, USA

Sergei Dikler
Bruker Scientific, LLC
Billerica, MA, USA

Eva Duchoslav
SCIEX, Concord, ON, Canada

Lucien Ghislain
Beckman Coulter Life Sciences
San Jose, CA, USA

John Janiszewski
National Center for Advancing Translational Sciences (NCATS)
Rockville, MD, USA

Mengxuan Jia
Preclinical Development ADMET/BA
Merck & Co., Inc
South San Francisco, CA, USA

Yang Kang
SCIEX, Concord, ON, Canada

Brendon Kapinos
Pfizer Worldwide Research and Development, Groton, CT, USA

Adam Latawiec
SCIEX, Concord, ON, Canada

Sylvain Letarte
R&D Department, Phytronix Technologies Inc., Québec
QC, Canada

Chang Liu
SCIEX, Concord, ON, Canada

Juncai Meng
Discovery Technology and Molecular Pharmacology (DTMP)
Janssen Research & Development LLC, Spring House, PA, USA

Olivier Mozziconacci
Discovery Pharmaceutical Sciences
Merck & Co., Inc, South San Francisco CA, USA

Réal E. Paquin
Université Laval
Québec, QC, Canada

Pierre Picard
R&D Department, Phytronix Technologies Inc., Québec QC, Canada

Elizabeth Pierson
Analytical R&D, Merck & Co., Inc., Rahway, NJ, USA

Subhasish Purkayastha
SCIEX, Framingham, MA, USA

Jonathan Rochon
R&D Department, Phytronix Technologies Inc.,
Québec, QC, Canada

Dylan H. Ross
Earth and Biological Sciences Directorate, Pacific Northwest National Laboratory
Richland, WA, USA

Bradley B. Schneider
SCIEX, Concord, ON, Canada

Wilson Z. Shou
Lead Discovery and Optimization
Bristol-Myers Squibb Company
Princeton, NJ, USA

Richard D. Smith
Earth and Biological Sciences Directorate, Pacific Northwest National Laboratory
Richland, WA, USA

Aaron Stella
SCIEX, Framingham, MA, USA

Christopher F. Stratton
Discovery Technology and Molecular Pharmacology (DTMP), Janssen Research & Development, LLC
Spring House, PA, USA

Lawrence M. Szewczuk
Discovery Technology and Molecular Pharmacology (DTMP), Janssen Research & Development, LLC
Spring House, PA, USA

Matthew D. Troutman
Hit Discovery and Optimization
Pfizer, Inc., Groton, CT, USA

Andrew D. Wagner
Lead Discovery and Optimization
Bristol-Myers Squibb Company
Princeton, NJ, USA

Jianzhong Wen
Preclinical Development ADMET/BA
Merck & Co., Inc
South San Francisco, CA, USA

Jonathan Wingfield
Mechanistic and Structural Biophysics
Discovery Sciences, R&D
AstraZeneca, Cambridge, UK

Hui Zhang
Entos Inc.
Department of Analytical
Technologies, Entos
San Diego, CA, USA

Wenpeng Zhang
State Key Laboratory of Precision
Measurement Technology and
Instruments, Department of Precision
Instrument, Tsinghua University
Beijing, China

Xueyun Zheng
Earth and Biological Sciences
Directorate, Pacific Northwest National
Laboratory, Richland, WA, USA

Preface

The automated and integrated high-throughput sample analysis is critical to the drug discovery process. Traditional high-throughput bioanalytical technologies such as colorimetric microplate-based readers are often constrained by linear dynamic range. In addition, they need label attachment schemes with the propensity to modify equilibrium and kinetic analysis. On the other hand, mass spectrometry (MS) based methods can achieve label-free, universal mass detection of a wide arrange of analytes with exceptional sensitivity, selectivity, and specificity. However, these techniques are limited by the speed of sample introduction. In recent decades, there have been a lot of efforts to improve the throughput of MS-based analysis for drug discovery. Along with those developments, a dedicated book would be helpful to introduce the fundamentals, experimental details, and applications of a wide variety of technologies that enabled high-throughput mass spectrometry-based screens in supporting broad drug discovery applications. The key research areas include hit discovery by label-free screen, synthetic reaction optimization, lead optimization and SAR support, ADME (absorption, distribution, metabolism, and excretion), toxicology screening, etc.

This book starts with an overview of the 40 years of efforts to improve the analytical throughput of MS-based approaches (Chapter 1). Then, technologies with high-speed sequential and parallel chromatographic sample introduction, high repetition rate lasers, ion mobility, and low-volume MS samplings were summarized.

Due to its high specificity and high sensitivity, the LC-MS technology has been widely used in various steps of the drug discovery workflow. In Part 2 (Chapter 2–3), the efforts to improve the LC-MS analytical throughput are introduced. The development of the high-speed sample introduction for LC-MS and its application on ADME and HTS applications is described in Chapter 2. Another approach for throughput improvement utilizing paralleled multiplexing LC is described in Chapter 3.

Following the conventional LC-MS-based technologies, other electrospray ionization (ESI)MS-based high-throughput platforms without chromatographic separation are summarized in Part 3 (Chapter 4–7). Direct online solid-phase extraction (SPE) MS and its application in ADME and HTS workflows are described in Chapter 4. The utilization of the acoustic energy for non-contact transfer samples from microplates to MS for high-throughput analysis, including the acoustic mist ionization (AMI) and through the open-port interface (OPI), is summarized in Chapter 5. By skipping the chromatographic separation process, these approaches demonstrated higher analytical throughput than the conventional LC-MS approach. However, there would be the risk of potential isomeric/isobaric interference. Ion mobility spectrometry (IMS) and differential mobility spectrometry (DMS), described in Chapters 6 and 7, respectively, provide the additional dimension of the selectively, potentially solving the specificity issues of these high-throughput technologies for some drug discovery assays.

Part 4 (Chapters 8–10) summarized the MS-based high-throughput hit identification technologies based on the drug-target interaction. Affinity-selection mass spectrometry (ASMS) is a rapidly developing technology for high-throughput hit identification. The off-line and in-line ASMS approaches are introduced in Chapters 8 and 9. In addition, as a direct confirmation tool for the protein-drug binding, native MS has been rapidly developed in the past decade, which is described in Chapter 10.

Part 5 (Chapter 11–13) introduces developments of ambient ionization technologies other than the conventional ESI and their applications in the high-throughput drug discovery workflows, such as Laser Diode Thermal Desorption (LDTD, Chapter 11), Matrix-Assisted Laser Desorption/Ionization (MALDI, Chapter 12), and Desorption Electrospray Ionization (DESI, Chapter 13).

The last chapter (Chapter 14) provides perspectives for future development opportunities after a brief reflection of the realized impacts of high-throughput MS on drug discovery and the pharmaceutical industry.

We believe our goal in this book is accomplished through the extensive coverage of fundamentals, experimental details, and applications of state-of-art technologies that enable high-throughput MS-based screens in supporting drug discovery. We hope it could benefit scientists in pharmaceutical/biopharmaceutical companies and CROs who design and perform the studies and provide analytical support throughout drug discovery processes. We would like to acknowledge the commitment and contributions of all authors of the book chapters and the support and valuable discussions with colleagues and collaborators in the SCIEX research team and Pfizer Discovery Science department. In addition, we sincerely thank the editorial team at John Wiley & Sons, especially Adalfin Jayasingh, Stacey Woods, Jonathan Rose, Andreas Sendtko, and Sabeen Aziz, for their generous support of this book. Finally, we are grateful to our family members for their understanding and support for our editing work in the evening and on weekends.

Chang Liu and Hui Zhang

List of Abbreviations

%-RBA	relative binding affinity percentage
μFLC	microflow liquid chromatography
2d	two-dimensional
2-HG	2-hydroxyglutarate
3CLpro	3-chymotrypsin-like cysteine protease
4EBP1	Eukaryotic translation initiation factor 4E-binding protein 1
A	pre-exponential factor constant
ACE50	affinity competition experiment 50% inhibitory concentration
AChE	acetylcholinesterase
ADC	antibody–drug conjugate
ADE-OPI-MS	acoustic droplet ejection-open port interface-mass spectrometry
ADME	adsorption, distribution, metabolism, and excretion
AEMS	acoustic ejection mass spectrometry
AMI-MS	acoustic mist ionization-mass spectrometry
AMS	affinity mass spectrometry
ANSI	American National Standards Institute
APCI	atmospheric pressure chemical ionization
API	atmospheric pressure ionization
APIs	active pharmaceutical ingredients
APPI	atmospheric pressure photo ionization
ASAP	atmospheric solids analysis probe
ASMS	affinity selection mass spectrometry
ASMS	American society mass spectrometry
Asp	aspartic acid
ATD	arrival time distribution
ATP	adenosine triphosphate

AUC	analytical ultracentrifugation
AUC	area under the curve
BACC	bacterial acetyl coenzyme-A carboxylase
BACE	beta-site APP cleaving enzyme
BAMS	bead assisted mass spectrometry
Bcl-xL	B-cell lymphoma-extra large protein
bdf	batch data file
BE	buffer exchange
Bead-GPS	bead-based global proteomic screening
BFA	bound fraction analysis
BKM120	Buparlisib
BSA	bovine serum albumin
BTE	Boltzmann transport equation
C18	octadecyl stationary phase
C8	octyl stationary phase
CCS	collision cross section
CD	circular dichroism
CDMS	charge detection mass spectrometry
CEM	chain ejection model
cGAMP	cyclic GMP-ATP
cGAS	cyclic GMP-AMP synthase
CHCA	α-cyano-4-hyroxycinnamic acid
CHK1	checkpoint kinase
CID	collision induced dissociation
CIU	collision induced unfolding
CN	cyano stationary phase
CoV	compensation voltage
CPATI	cytosolic proteome and affinity-based target identification
CRIMP	Compression Ratio Ion Mobility Programming
CRM	charged residue model
CV	coefficient of variation
CYP	cytochrome P450
Da	Dalton, measurement unit used in mass spectrometry
DAR	drug-to-antibody ratio
DART	direct analysis in real time
DDI	drug–drug interaction
DEC	desorption enhancing coating
DEL	DNA-encoded library
DESI	desorption electrospray ionization
DHAP	2,5-dihydroxyacetophenone
DHFR	dihydrofolate reductase

diCQA	dicaffeoylquinic acid
DI-GCE/MS/MS	direct injection/on-line guard cartridge extraction/tandem mass spectrometry
DIMS	differential IMS
DIOS	desorption ionization on silicon
DLS	dynamic light scattering
DMA	differential mobility analyzer
DMS	differential mobility spectrometry
DP	declustering potential
DQ	DiscoveryQuant
DSF	differential scanning fluorimetry
DT	drift time
DTIMS	drift tube IMS
DUB	deubiquitilase
E_a	energy of activation
ebox	electronics box
E_d	bound dissociation energy
EDTA	ethylenediaminetetraacetic acid
EHDI	electrohydrodynamic ionization
EI	electron impact
EM	electron microscopy
ERK1/ERK2	extracellular signal-regulated kinase 1 and 2
ESI	electrospray ionization
ESI-MS	electrospray ionization mass spectrometry
E_λ	energy associated with the vibrational wavelength
FAB	fast atom bombardment
FAIMS	high field asymmetric waveform ion mobility spectrometry
FAK	focal adhesion kinase
FASN	fatty acid synthase
FIA	flow injection analysis
FLD	fluorescence detector
FP	fluorescence polarization
FTE	full-time equivalent
FTICR	Fourier-transform ion cyclotron resonance
FWHM	full width at half maximum
GABA	γ-aminobutyric
GC	gas chromatography
GLP	good laboratory practice
GPC	gel permeation chromatography
GST	glutathione S-transferase
GWAS	genome-wide association studies

HBSS	Hank's buffered salt solution
HCV	hepatitis C virus
HDMA	high-density micropatterned array
HEPES	4-(2-hydroxyethyl)-1-piperazineethanesulfonic acid
HIC	hydrophobic interaction chromatography
HLM	human liver microsomes
HMW	high molecular weight species
HPLC	high-performance liquid chromatography
HRMS	high-resolution mass spectrometry
HT-ADME	high-throughput absorption, distribution, metabolism, excretion
HTE	high-throughput experimentation
HT-LC/MS/MS	high-throughput mass spectrometry
HT-MALDI	high-throughput matrix-assisted laser desorption/ionization
HT-MS	high-throughput mass spectrometry
HTRF	homogenous time-resolved fluorescence
HTS	high-throughput screening
IC_{50}	half maximal inhibitory concentration
ID	internal diameter
IDH1	isocitrate dehydrogenase 1
IEX	ion exchange chromatography
IM	ion mobility
IMAC	immobilized metal ion affinity chromatography
iMALDI	immuno-matrix-assisted laser desorption/ionization
IMS	ion mobility spectrometry
IR-MALDESI	infrared matrix-assisted desorption electrospray ionization
IS	internal standards
isoAsp	isoaspartic acid
ITC	isothermal titration calorimetry
ITO	indium tin oxide
IVIVC	in vitro to in vivo correlations
k	rate constant
LC	liquid chromatography
LC/MS/MS	liquid chromatography tandem mass spectrometry
LC-MALDI	liquid chromatography-matrix-assisted laser desorption/ionization
LC-MS	liquid chromatography mass spectrometry
LDLR	low-density lipoprotein receptor
LDTD	laser diode thermal desorption
LESA	liquid extraction surface analysis
LLE	liquid–liquid extraction

LOD	limits of detection
LogD	distribution coefficient
LOQ	limit of quantitation
LPS	lipopolysaccharides
M3	microfabricated monolithic multinozzle
mAbs	monoclonal antibodies
MagMASS	magnetic microbead affinity selection screen
MALDI	matrix-assisted laser desorption ionization
MALDI-2	laser-induced postionization
MALDI-FTICR MS	matrix-assisted laser desorption/ionization Fourier-transform ion cyclotron resonance mass spectrometry
MALDI-TOF MS	matrix-assisted laser desorption/ionization time-of-flight mass spectrometry
MetAP2	methionyl aminopeptidase 2
MnESI	microflow-nanospray electrospray ionization
MPS	mesoporous silica
MRM	multiple reaction monitoring
MRO	medical review officer
MS	mass spectrometer
MS/MS	tandem mass spectrometry
MSI	mass spectrometry imaging
MTBE	methyl tert-butyl ether
MTP	microtiter plate
MuRF	muscle RING-finger protein
NADPH	nicotinamide adenine dinucleotide phosphate
NALDI	nanostructure-assisted laser desorption/ionization
Nano-DESI	nanospray desorption electrospray ionization
NAPA-LDI	nanopost array-laser desorption/ionization
NDM-1	New Delhi metallo-lactamase1
NDX	native-denatured exchange
nESI	nano electrospray ionization
NHS	*N*-hydroxysuccinimide
NIMS	nanostructure-initiator mass spectrometry
nL	nanoliter
NMR	nuclear magnetic resonance
nMS	native mass spectrometry
NSAID	nonsteroidal anti-inflammatory drugs
NSP14	nonstructural protein 14
OATP2B1	organic anion transporting polypeptide 2B1
OIMS	overtone IMS
OPSI	open port sampling interface

PAH	polycyclic aromatic hydrocarbon
PBED	polybrominated diphenyl ether
PBS	phosphate-buffered saline, buffer solution about pH 7.4
PCB	polychlorinated biphenyl
PCB	printed circuit board
PC-mass-tags	photocleavable mass-tags
PFAS	per- and polyfluoroalkyl substances
PK	pharmacokinetic
pK_a	acid dissociation constant
PKCα	protein kinase C-α
PMF	peptide mass fingerprinting
PoC	percentage of control
POE	percent of enrichment
PPT	protein precipitation technique
PROTAC	proteolysis targeting chimera
PTP1B	tyrosine phosphatase 1B
PUF-MS	pulsed ultrafiltration-mass spectrometry
PVDF	polyvinylidene difluoride
QA/QC	quality assurance and quality control
qPCR	quantitative polymerase chain reaction
qTOF	quadrupole time-of-flight
QuEChERS	quick easy cheap effective rugged and safe
R	universal gas constant
R^2	coefficient of determination
RAM	restricted access media, usually a type of filtering or extraction media
RAM	restricted access medium
RF-MS	RapidFire – mass spectrometry
ROI	return on investment
RXRa	retinoid X receptor-a
S/N	signal-to-noise ratio
SALLE	salt assisted liquid–liquid extraction
SAM	*S*-adenosyl-L-methionine
SAMDI	self-assembled monolayers and matrix-assisted laser desorption ionization
SAR	structure-activity relationship
SEC	size-exclusion chromatography
SEC-TID	size-exclusion chromatography for target identification
SEM	scanning electron microscope
SESI	secondary electrospray ionization

SEZ	staggered elution zone chromatography
SIK2	salt-inducible kinase 2
SIMS	secondary ion mass spectrometry
Sirt3	Sirtuin 3
SISCAPA	stable isotope standards and capture by anti-peptide antibodies
SLIM	structures for lossless ion manipulations
SLS	static light scattering
SME	small molecular entity
SmyD3	SmyD3 histone methyltransferase
SNP	single-nucleotide polymorphism
SPE	solid phase extraction
SPE-MS	solid-phase extraction mass spectrometry
SPME	solid-phase microextraction
SPR	surface plasmon resonance
SRM	selected reaction monitoring
SSP	surface sampling probe
SUPER	Serpentine Ultralong Path with Extended Routing
SV	separation voltage
SWATH	sequential window acquisition of all theoretical mass spectra
T	absolute temperature in Kelvin
TCP	tumor cell percentage
THC	tetrahydrocannabinol
TIMS	trapped ion mobility
TLC	layer chromatography
TMA-lyase	trimethylamine-lyase
TM-DESI	transmission mode DESI
TM-IMS	transversal modulation IMS
TOF	time-of-flight
TR-FRET	time-resolved fluorescence energy transfer
TRIS	Tris (hydroxymethyl) aminomethane
TWIMS	traveling wave ion mobility
UFA	unbound fraction analysis
UHPLC	ultrahigh-performance (or pressure) liquid chromatography
UHPLC/MS	ultrahigh-performance liquid chromatography-mass spectrometry
uHT-MALDI	ultrahigh-throughput matrix-assisted laser desorption/ ionization
uHTS	ultrahigh-throughput screening
UPLC	ultra performance liquid chromatography

UV	ultraviolet, usually meant to describe absorbances between 190 and 400 nm
UVPD	ultraviolet photodissociation
WBA	whole-body autoradiography
XRD	X-ray diffraction
Δ^9-THCC	Carboxylic Δ^9-tetrahydrocannabinol
λ	phonon wavelength

Section 1

Introduction

1

Forty-Year Evolution of High-Throughput Mass Spectrometry: A Perspective

Thomas R. Covey

SCIEX, Concord, Ontario, Canada

1.1 Introduction

The field of drug discovery has been the primary driver behind the development of quantitative high-throughput mass spectrometry (HT-MS) over the past several decades. Hypothesis-driven science guides the search for effective chemical interventions in disease increasingly supplemented with stochastic methods implemented to broaden the range of chemistries to be tested for efficacy. This later approach has generated the need to be able to make quantitative measurements on tens to hundreds of thousands of drug candidates per day in multiple in vitro and in vivo experimental scenarios. For any method of measurement such as mass spectrometry that is serial in nature, addressing daily numbers of that magnitude would require an analysis to be completed in approximately one second (1 Hz). It is fair to say that it has been only within the past few years that mass spectrometry-based throughputs at this rate have been shown to be possible in a way that can be practically implemented into the drug discovery process. It is the purpose of this manuscript to attempt to explain how this came about. For a more thorough discussion of the role of high-throughput experimentation (HTE) and the impact HT-MS has on pharmaceutical R&D, see Chapter 14 of this book [1], review articles [2–7], and two earlier books on the topic of mass spectrometry in drug discovery [8, 9].

This chapter is not an exhaustive review of the literature but rather attempts to define what the trends in the field of HT-MS were over the past 40 years using specific examples to illustrate its evolution. It is a personal perspective where many of the examples cited are technologies the author was in some way involved

High-Throughput Mass Spectrometry in Drug Discovery, First Edition. Edited by Chang Liu and Hui Zhang.
© 2023 John Wiley & Sons, Inc. Published 2023 by John Wiley & Sons, Inc.

in either their development or early testing. This perspective attests to the importance of addressing as many of the bottlenecks in the overall HT-MS workflow as possible, so a wide variety of technologies are shown to be contributors to the overall solution above and beyond just the speed of sample introduction. Historical context is provided for advances in all of the areas of consideration so that the gains in the field of HT-MS since the first indications of its possibility nearly 40 years ago can be fully appreciated. Failures as well as successes are included in this perspective as they are shown to have provided to all working in the field valuable clues regarding what new directions to pursue eventually leading to where the industry is at today.

The areas to be covered, in more or less chronological order, are summarized in this chapter. Section 1.2 begins with a brief historical perspective on the development of ionization technologies and interfaces to mass spectrometry for LC/MS which laid the foundations for HT-MS today. This culminated in the domination of atmospheric pressure ionization (API) for online LC/MS in the late 1980s utilizing both electrospray ionization (ESI) and atmospheric pressure chemical ionization (APCI). The later technique was the first to demonstrate the feasibility of HT-MS in 1986 but ESI now serves as the basis for most systems today because it provides the broadest compound coverage, has been developed to operate reliably at high fluid linear velocities similar to APCI, and like APCI, has been demonstrated to be sufficiently resistant to contamination to sustain 24/7 operation under high sample loads.

Included in Section 1.2 are the origins of alternative sampling techniques that utilize ESI and APCI, popularly referred to as "ambient ionization," as part of a mission to bypass chromatography to gain speed. Some are beginning to indicate possible utility for HT-MS. Also in this section is the development of methods for the direct and indirect measurement of molecular affinities to biological targets which have begun to be incorporated into industrialized drug discovery programs with some of their earliest origins 20–30 years ago.

As described in Sections 1.3 and 1.4 of this chapter, approximately a decade after API took hold in the early 1990s, there was a proliferation of means to improve the speed of liquid chromatography and chromatographic systems which have evolved into the most widely used approach to HT-MS today. These developments are more thoroughly covered in Chapters 2 [10], 3 [11], and 4 [12] of this book.

Section 1.5 of this chapter describes how the 1 Hz sampling rate barrier was finally broken using matrix-assisted laser desorption ionization (MALDI) around 2005 with high repetition rate lasers. This was in response to the realization that high performance liquid chromatography (HPLC) needed to be bypassed to achieve this goal. MALDI originated in the 1980s as a means to obtain molecular weight information on large biomolecules and has emerged recently as the HT-MS method of choice for solid sample introduction thoroughly covered in Chapter 12 of this book [13].

Section 1.6 describes the understanding that some form of high-speed chemically based separation to replace HPLC would remain a requirement to keep the versatility of HT-MS on par with LC/MS. This initiated more concerted efforts to find a high-speed substitute in ion mobility technology considered in detail in Chapter 6 of this book [14]. In 2008, differential ion mobility spectrometry (DMS) was adapted to the mass spectrometer (MS) for HT-MS because of its unique separation mechanism that, like HPLC, separates based on the chemical properties of molecules. Further discussion of DMS is provided in Chapter 7 of this book [15] as well as Section 1.6 of this chapter.

By the second decade of the twenty-first century, the cumulative gains in API MS sensitivity over the past 40 years approached one million-fold and are described in Section 1.7 of this chapter. This development has played a major role in the evolution of HT-MS as it has made possible the reduction in the volume of sample required to achieve biologically relevant limits of quantitation (LOQ) by a similar amount, approximately 6 orders of magnitude. This relaxes the requirements for sample preparation, reduces system contamination, and enables the use of high-speed low-volume dispensers in the picoliter to nanoliter range.

The final topic in Section 1.8 describes the emergence of low-volume sample introduction techniques into the MS representing an important culmination of the past 40 years of HT-MS evolution. These approaches have the potential to streamline several bottlenecks in the high-throughput workflow such as the elimination or minimization of sample preparation, the elimination of the time wasting and ambiguous results generated from sample cross contamination, and simplification of automation by leveraging the microtiter plate format. Chapter 5 of this book [16] also elaborates on some applications of this approach to many aspects of the drug discovery process in addition to Section 1.8 of this chapter.

As throughputs increased over this time, new software and robotic systems designed to improve sample handling, management, data acquisition, and data analysis advanced in lock step. The role they play is vital to enable these HT-MS developments to reach their throughput potentials. This aspect is conspicuously omitted from this chapter for no other reason than its scope is of sufficient magnitude to be the topic of another book. Initial access to information regarding these aspects is provided in references [17–21].

1.2 Ionization Foundations of High-Throughput Mass Spectrometry

During the late 1960s through the 1980s, two different approaches to ionization emerged in the field of mass spectrometry that established the foundation for today's HT-MS technologies, API [22–27], and MALDI [28]. The efforts leading to MALDI were primarily driven by the quest to find a means to create intact gas-phase ions from biopolymers. Until then, obtaining molecular ions from molecules with molecular weights above 2000 amu was considered a heroic effort, a single mass spectrum the centerpiece of an entire PhD thesis. API became popular primarily from the efforts to find a viable means to interface the HPLC to the MS although its capacity for high mass measurement contributed to its success as well [29, 30].

In the 1980s, the vast majority of commercial instruments were based on ionization inside the MS vacuum system [31]. The difficulties faced interfacing HPLC and MS prior to API were expressed in a famous icon created by Patrick Arpino in 1982 shown in the insert in Figure 1.1. Titled "*A Difficult Courtship*," it is referred to the fundamental incompatibility between the vacuum-based gas-phase world

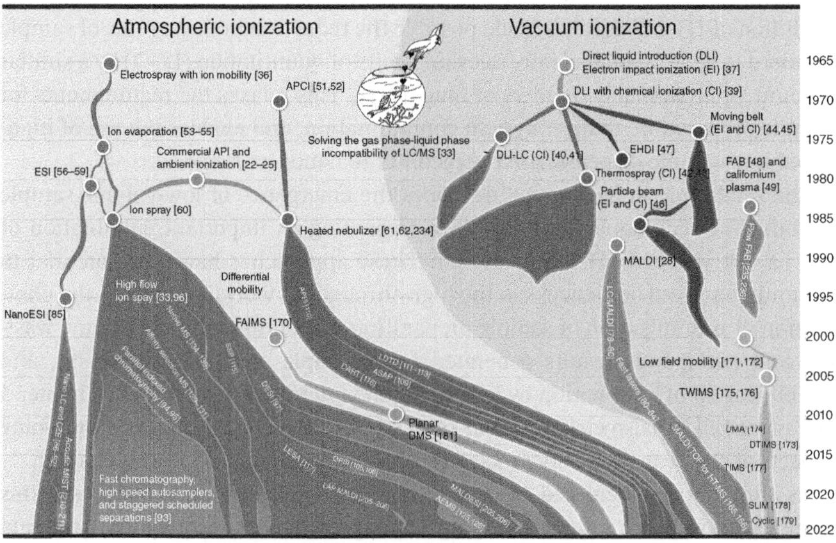

Figure 1.1 Histomap of LC/MS interfaces from 1967 to present and some emerging HT-MS technologies superimposed. The width of the region (*x*-axis) that each technique occupies illustrates its growth from initial invention and early publications (thin line) to an approximation of its commercial proliferation and incorporation into industrialized processes relative to the other techniques during that period of time. The map is bifurcated down the middle into two halves. On the left is shown atmospheric pressure ionization LC/MS interfaces and on the right are those based on ionization under vacuum conditions. The famous 1980s bird/fish icon represents the incompatibility of the gas-phase world of MS and liquid-phase world of LC as viewed at that time. *Source:* Reproduced with permission from Arpino [32]. All acronyms are defined in the chapter text. Ion mobility is included in the Histomap even though it is not an LC/MS interface or sample introduction system because of its potential as a high-speed substitute for some functions of HPLC in emerging HT-MS systems. LAP MALDI and MALDESI are included in the electrospray area because they both utilize the Ion Evaporation ionization mechanism of ESI at atmospheric pressure. This Histomap was adapted from a similar Histomap in references [27, 33, 34]. The Histomap concept was developed by John Sparks and published by Rand McNally to track the course of world cultural history since the dawn of civilization in a 6-ft long Histomap [35].

of MS, the bird, and the liquid-phase world of LC, the fish [32]. A line from the Broadway play "Fiddler on the Roof" summarizes the difficulties of such a relationship when Tevye says to his daughter intent on marrying against his wishes: "a bird may love a fish, but where would they build a home together?" The home would eventually reside in an atmospheric pressure ion source when it was finally realized that spraying liquids into vacuum systems or through vacuum locks and stages was not such a good idea after all.

1.2.1 Historical Context of the Development of LC/MS. Ionization in Vacuum or at Atmospheric Pressure?

The Histomap in Figure 1.1 provides a perspective on the extent to which the LC/MS interfacing problem was being addressed in research groups worldwide and the diversity of approaches that were developed, commercialized, and proliferated. From its earliest beginnings in the 1960s, the LC/MS field was divided into two camps, those generating ions created at atmospheric pressure initiated by Malcom Dole in Chicago [36] (ESI on a mobility analyzer at atmospheric pressure), and those creating ions under vacuum conditions initiated by Victor Talroze in Moscow [37] (low-volume liquids introduced into an electron impact ionization source in a vacuum). The atmosphere versus vacuum ionization divide would extend for two more decades and as seen in the Histomap, the vacuum-based techniques for LC/MS interfacing overwhelmingly dominating the commercial landscape through the 1980s. This was largely because commercial API systems initially did not exist and when one emerged in 1978, it was designed primarily for direct air pollution monitoring only relegating it to the commercial fringes of the analytical mass spectrometry field. Modifications to main stream commercial mass spectrometers for API operation were extensive whereby it was seemingly more pragmatic to develop LC interfaces that maintained vacuum-based ionization. After all mass spectrometers were designed to analyze ions created in a vacuum since the earliest beginnings of its use for chemical analysis in 1912 [38] and there had been little thought to do otherwise for much of the twentieth century.

LC/MS interfaces with ionization in a vacuum followed two paths. The route proposed by McLafferty [39], Henion [40, 41], and Vestal [42, 43] involved chemical ionization using the vaporized LC mobile phase as the reagent gas with either the direct liquid introduction or Thermospray interfaces.

The other route championed by McFadden [44], Games [45], and Willoughby [46] upheld the importance of maintaining the option of electron impact ionization capability with the moving belt and particle beam interfaces. These interfaces removed all of the mobile phase using various types of vacuum locks and stages so only dried sample material would enter vacuum on either a Kapton belt or as a dried gas-phase particle. Thermal desorption of the dried sample inside vacuum enabled electron impact ionization or chemical ionization when reagent gas was piped in.

Kelsey Cook was exploring an early form of ESI under vacuum called Electrohydrodynamic Ionization (EHDI) [47]. With this approach, low flows of liquid charged by a high voltage on the capillary transporting it were sprayed directly inside the vacuum system. With the exception of the work of Thomson discussed later in this Section 1.2.1, the attributes of atmospheric ESI were not yet realized, its extraordinarily mild ionization and multiple charging properties. These capabilities were not observed with EHDI because of the poor evaporation rates of droplets in a vacuum due to inefficient thermal energy transfer at these pressures. The droplets could not evaporate to the Rayleigh limit and undergo the coulomb explosions required to achieve ion emission diameters. Also, the creation of energetic electrons by the Townsend discharge process occurring at these pressures using the voltages required to form a spray caused fragmentation of whatever molecules made it to the gas phase. Had the sprayer been moved from inside the vacuum system to atmosphere, requiring substantial modifications to the MS as mentioned earlier, history may have been rewritten. For the reasons described above, it was not commercialized as seen in its dead ended trace in the Histomap.

During this time, attempts to ionize higher molecular weight compounds were done under vacuum with the high-energy particle impact techniques of Fast Atom Bombardment [48] (FAB) and Californium Plasma Desorption [49], both of which were replaced by MALDI which could produce intact gas-phase molecular ions, as could ESI, from much larger molecules.

References to the initial publications of these ionization and LC/MS interfacing techniques are provided in Figure 1.1 with a more detailed description of the various interfaces displayed provided in earlier published versions of this Histomap [27, 34, 50].

In 1978, an important event occurred that provided a base upon which to develop HT-MS systems in the future. As noted in the Histomap, an API MS was commercialized as a component of a van-based mobile laboratory for environmental and regulatory applications as described in more detail later in this chapter. Although samples were introduced as gasses or solids by a variety of techniques, not in the liquid form, this technology drew the attention of a few experts in the LC/MS field as a potential solution to the LC/MS interfacing problem particularly because some earlier liquid introduction work at atmospheric pressure by the Horning's with APCI [51, 52] and Thomson with Ion Evaporation [53–55] provided proof-of-principle.

The Thomson work in the late 1970s described the theory behind the ESI mechanism which remains as the primary explanation of the electrospray process today, referred to as the Ion Evaporation Theory for ion production. He also showed its unique ability to produce multiply charged ions when interfaced to a MS [55]. By the mid-1980s, John Fenn brought attention to electrospray to a broader audience in the West [56, 57] while Gall independently developed it in the Soviet Union at the same time [58, 59]. A means to couple ESI with conventional flow rate liquid chromatography followed in 1986 borrowing elements of the Thomson Ion Evaporation and the Fenn/Gall Electrospray, given the moniker Ion Spray [60], because it was the combination of the two which made it practical. From ion evaporation was borrowed the formation of droplets by high-velocity gas shear forces. The ion evaporation interface charged the droplets remotely with an induction electrode. From electrospray was borrowed direct electrical charging of the liquid to replace the inductive charging of the ion evaporation interface which increased the charge density of the droplets and improved sensitivity. Ion Spray greatly increased the sensitivity of the Ion Evaporation interface and greatly increased the practicality for coupling to liquid chromatography Fenn and Gall's Electrospray. Figure 1.2a–d shows the first published drawings and photographs of these early developments and how they evolved into the most commonly used interface today.

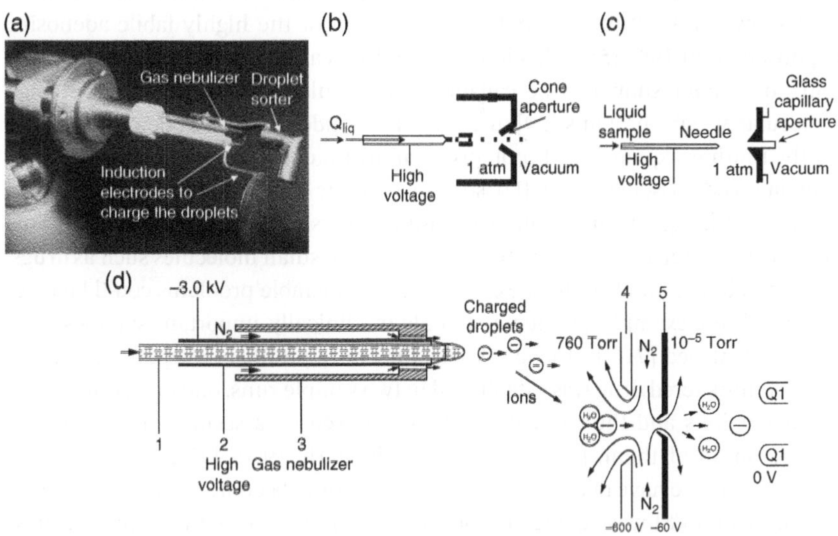

Figure 1.2 The evolution of electrospray LC/MS interfaces. (a) The ion evaporation interface of Thomson showing pneumatic nebulization and inductive charging of the droplets. (b) Depiction of the Gall electrospray device showing direct electrical charging of the fluid. *Source:* Adapted from Alexandrov et al. [59]. (c) Depiction of the Fenn electrospray device showing direct electrical charging of the fluid. *Source:* Adapted from Whitehouse et al. [57]. (d) The ion spray interface of bruins which combined the ion evaporation nebulization with the direct electrical charging of electrospray. *Source:* Reproduced with permission from Bruins et al. [60]/American Chemical Society.

The earliest compelling publication of the potential of API for HT-MS was not with ESI but rather APCI. Based on the earlier APCI LC/MS work of the Horning's [52] and the APCI LC/MS/MS work of Henion and Thomson [61], the heated nebulizer APCI LC/MS interface was redesigned to operate at higher flow rates by increasing the desolvation chamber power and prototyped in the mid-1980s and is shown in Figure 1.3c, d. Demonstrated in 1986 was a successful quantitative pharmacokinetic study by LC/MS/MS at the rate of one sample per minute, monitoring a drug and three of its metabolites for extended periods of time as shown in Figure 1.3a, b [62]. It is important to put this data in the context of the time when the successful completion of 1–10 LC/MS chromatograms per day before a catastrophic system failure occurred was considered an accomplishment. A common ritual, upon entering a laboratory exploring the utility of the various LC/MS interfaces in the 1980s, was to Spray N′ Pray.

Increasing amounts of ESI data from labs with early pre-commercialization versions manifested in the testing of a large swath of chemical space from different application areas. A rapid realization was emerging from this data that the ionization process was uniquely mild among all other forms of ionization considered to that date. Clues that this would be true were present in the earlier data of Thomson who showed a multiply charged mass spectrum of the highly labile adenosine triphosphate in 1982 [55]. This indicated ESI was a much more versatile form of ionization, later shown to be capable of molecular ion generation from high molecular weight proteins [29] and oligonucleotides [30] and opening the door for the routine sequencing of proteins from the uniquely simple to interpret collision induced dissociation (CID) spectra of their doubly charged tryptic peptides [63–70]. Equally important was gaining access to a mass spectrometry-based approach to determine the structures of very labile small molecules such as drugs, their metabolites, and conjugates. Previously intractable problems could be readily solved, for example unknown steroids in clinically important samples were readily identified [71]. In the late 1980s, the entire profile of over 20 metabolites of an administered drug was established in two sample runs, one to identify all the molecular ions and their retention times followed by a second to acquire CID spectra on all of them, published later in the early 1990s [72], a problem that would otherwise have taken years to solve, and would become even more straightforward when ESI was adapted to routine tandem high-resolution accurate mass systems like QqTOF's [73] and Orbitraps. Previously unresolved chemical mysteries, that presented life or death consequences, could suddenly be solved in a matter of minutes or hours [74, 75].

During this time it was also observed that ionization at atmospheric pressure was far more resistant to performance degradation due to sample contamination compared to the vacuum-based ionization techniques. This derives from the fact that focusing electric fields play a very minor role in determining an ion's

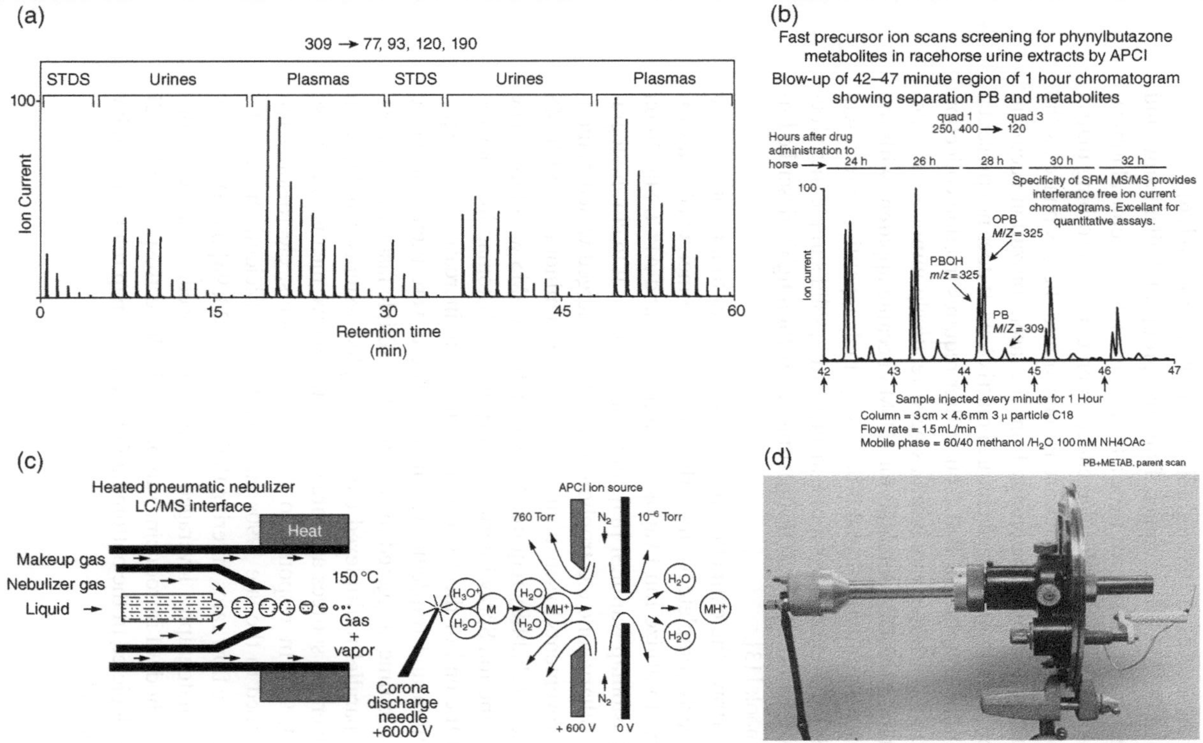

Figure 1.3 First demonstration of the feasibility of HT-MS using the heated nebulizer with APCI circa 1986. (a) MRM ion current from the pharmacokinetic study of phenylbutazone (PB) and its metabolites after administered to a horse demonstrating high-quality quantitative data at a rate of one sample per minute for one hour. Urine and plasma samples taken from 0 to 48 hours post dose and analyzed in that order at one-minute intervals from 0 to 30 minutes. Sample set repeated from 30 to 60 minutes to demonstrate long-term reproducibility. (b) MRM ion current trace from the urine samples monitoring two metabolites and parent drug. *Source:* (a and b) Reproduced with permission from Covey et al. [31]/American Chemical Society. (c and d) Heated nebulizer APCI prototype interface [234]. Corona discharge created reagent ions from the vaporized mobile phase flowing at 1 mL/min which transfers the charge to the gas-phase analyte by the chemical ionization process. The nebulized droplets (not charged) impacted the hot quartz walls of the desolvation chamber (~400 °C) to vaporize both the mobile phase and the analyte as a neutral entity. The resulting gas containing analyte was ~150 °C. Column 3 cm × 3 mm i.d. with 3 μm C18 spherical particles developed by Ken Ogan at Perkin Elmer in the 1980s. *Source:* (c) Reproduced with permission from Covey et al. [31]; Covey et al. [31, 62]/Reproduced with permission from American Chemical Society.

trajectory at atmospheric pressure. Gas flows that are not perturbed by distortions in electric fields caused by the contamination of lenses dominate an ion's fate at atmospheric pressure. The evidence was accumulating that throughputs in far excess of the historical norms were possible. All of these features considered, broad compound coverage, sensitivity, compatibility with high fluid flow sample introduction, and stability under heavy sample loads contributed to the decision to launch the first dedicated API LC/MS/MS system with the Ion Spray and Heated Nebulizer ion sources in 1989 as indicated in the Histomap [76].

In parallel with API, MALDI emerged from the group of Hillenkamp and Karas [28] and is generally considered to be a form of gas-phase chemical ionization [77]. With MALDI, solid samples embedded in an energy-absorbing matrix present in great excess over the analytes are launched into the gas phase as the matrix vaporizes upon absorbing the energy from the laser. Ionization occurs in the gas phase by the radiation-damaged charged matrix ions. It does not present a means for direct online coupling to HPLC but rather requires fraction collection and co-crystallization [78–80], so its role has been relatively minor as an integrated LC/MS system. MALDI's potential for HT-MS became apparent with the demonstration of its quantitative capabilities using high repetition rate lasers around 2005 [81–84] and today plays a role incorporating solid sample introduction into the HT-MS workflow on TOF instruments as thoroughly described in Chapter 12 of this book [13].

The Histomap portrays the extinction of LC-MS interfaces to ion sources in a vacuum as being nearly complete by the early 1990s and the emergence of a variety of different ways to utilize Ion Evaporation Ionization and APCI. Shown on the far left of the Histomap NanoESI [85], and adaptations of it to chromatography [86–89] and capillary zone electrophoresis [90–92], emerged which demonstrated marked improvements to sensitivity but has played a minor role in HT-MS largely due to a lack of reliability required for demanding HT-MS workflows. The expansion of use of the Ion Spray interface shown in the Histomap includes the emergence toward the end of the 1990s improvements to increase speed beginning with fast column and trap/elute technology. These were then married to methods that offset in time-staggered injections into multiple fast columns [93]. At the same time, parallel multiplexed chromatographic systems coupled with indexed multi-sprayer ion sources appeared as an alternative means to increase the throughput [94, 95]. In response to these developments, electrospray ion sources were developed in the late 1990s that could more efficiently handle the increased mobile-phase flow rates entering the ion source [33, 96] that these new approaches to chromatography leveraged to increase speed. And new non-chromatographic methods of introducing samples for both electrospray and APCI began to emerge as depicted in the bottom half of the Histomap referred to today as "Ambient Ionization."

1.2.2 Ambient Sample Introduction Methods (Ambient Ionization) into an API Ion Source Without LC and Their HT-MS Potential

Approaches to sample introduction for ESI and APCI that bypass chromatographic separations in recent years have been somewhat misrepresented by the generalized term "ambient ionization" [97–104] because they are not new forms of ionization but rather are new or modified forms of sample introduction. They all produce ions by either chemical ionization at atmospheric pressure (APCI) which is a gas-phase ionization process or by the Ion Evaporation mechanism that launches ions preformed in the liquid phase by field ion emission from charged droplets.

The first commercial API MS in 1978 utilized a variety of ambient sample introduction techniques, that is to say sample introduction methods and devices that bypass chromatography. API single and triple quadrupole mass spectrometers were mounted in vans and airplanes for the purpose of real-time tracking of fugitive environmental emissions in the atmosphere and deployed at airports and military installations for the detection of contraband, explosives, and chemical warfare agents [22, 25]. As with many ambient ionization techniques today, the motivation back then was not necessarily the high-speed execution of large numbers of batched samples, but rather fast turn-around time from when an individual sample is taken and results are presented, i.e. real-time analysis. Inlet systems were developed for the direct analysis of components in air such as pollutants, volatiles such as pesticides on food, pheromones from animals, and breath for diagnostic purposes. Photographs of this first mobile "ambient ionization" system and some of its applications that include its deployment to infamous chemical disaster sites are shown in Figure 1.4. Several devices were developed and commercialized to sample particles from cargo and clothing for drugs, explosives, and other contraband using handheld samplers that directly transferred the samples into a thermal desorption APCI ion source as seen in Figure 1.5. The photographs in Figures 1.4 and 1.5 indicate the degree of sophistication and level of advanced development and deployment these original ambient ionization systems achieved, which current ambient ionization techniques are striving to emulate. There was a poignant urgency for these capabilities in the 1980s as chemical dump sites from the earlier unregulated industrial age were being discovered in communities such as the Love Canal in Buffalo, NY, and terrorism was on the rise in airports as exemplified by a series of bombings at Heathrow and Lockerbie airports during the time of the Irish "Troubles."

Some of the recent versions of ambient sampling (aka ambient ionization) have begun to show possibilities for high sample introduction speeds and are shown in the Histomap in Figure 1.1. Some introduce samples in a flow injection-like manner to bypass chromatography like the Open Port Sampling Interface

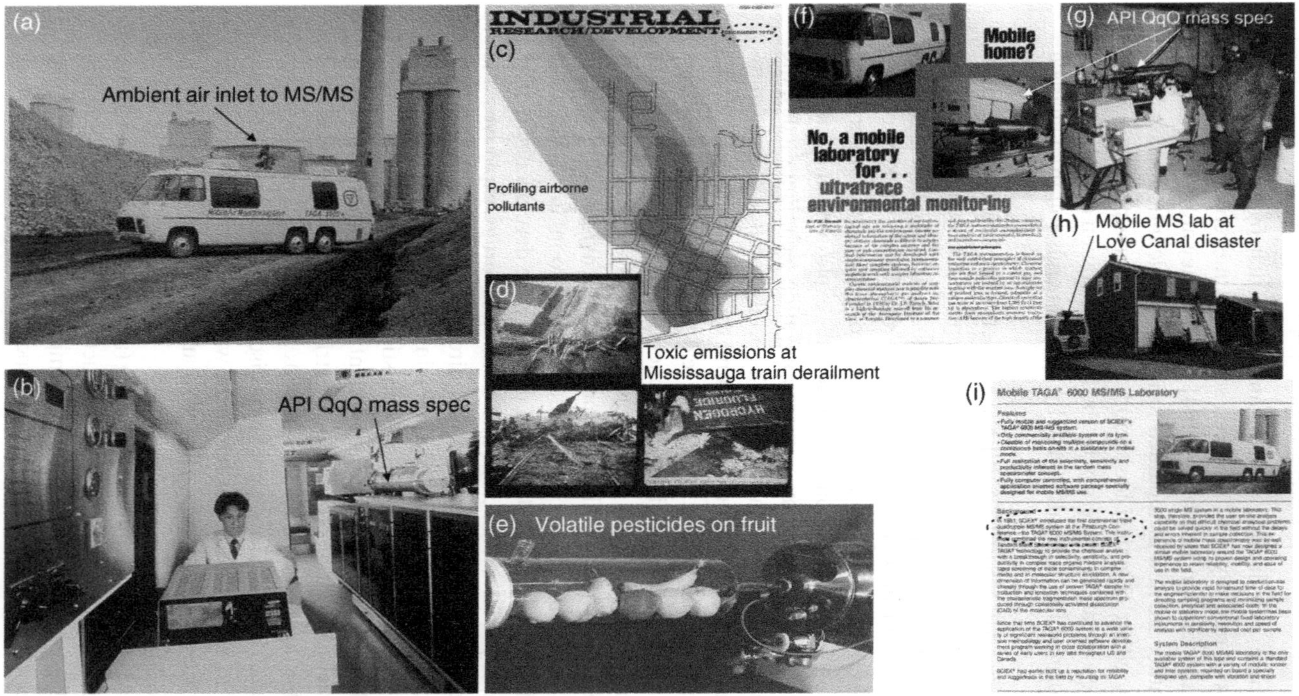

Figure 1.4 Images and applications of the first ambient sampling (aka ambient ionization) MS and MS/MS systems used for "real-time" analysis. (a) Mobile lab containing the TAGA (trace atmospheric gas analyzer) and API mass spectrometer with scientist John Fulford extending a sampling periscope above the roof to track fugitive atmospheric emissions in real time at an industrial site. *Source:* John Fulford. (b) Scientist Bori Shushan operating the TAGA MS/MS system in the mobile lab. *Source:* Bori Shushan. (c) Example data output from real-time emissions tracking in 1978. (d) The TAGA van tracked the emissions from the Mississauga train derailment in 1979 (photographs). The city of 200,000 people was evacuated as a result. (e) "Ambient sampling" of pesticides from fruit. (f) System description in 1978 of the API mobile lab using a single quadrupole MS. (g) Direct monitoring of chemical warfare agents at a Canadian military base in the early 1980s. (h) TAGA van monitoring airborne dioxins at the 1980 Love Canal toxic waste dump disaster in Buffalo, NY, which led to the abandonment and razing of the community. (i) TAGA tandem MS/MS system description identifying March 1981 as the launch date of the first commercial triple quadrupole mass spectrometer.

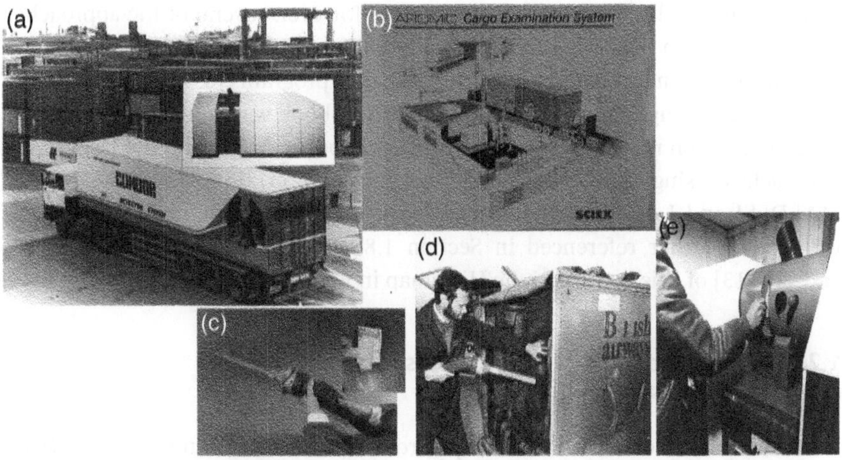

Figure 1.5 Images and applications of the second ambient sampling (aka ambient ionization) system circa 1985. (a) The British Aerospace mobile lab Condor contained among other instruments the Aromic API triple quadrupole MS (insert) whose purpose was to detect contraband and explosives at shipping ports and airports. (b) Drawing of system in operation using sampling devices that utilized vacuum to collect particles on filters for thermal desorption APCI mass spectrometry. (c) Modeled after the Black & Decker Dust Buster, the "drug buster" was a handheld vacuum with a removable filter cartridge that inserted directly into the ion source for thermal desorption APCI after sampling. (d) Scientist Bill Stott sampling luggage with the Drug Buster in a British Airways terminal. *Source:* Bill Stott. (e) The Drug Buster cartridge being inserted into the APCI ion source by scientist Rupert Van Veen for thermal desorption. *Source:* Rupert Van Veen.

(OPSI) [105, 106] which can be used with either ESI or APCI ionization. An example of its use can be found in Chapters 5 [16] and 7 [15] of this book as well as Section 1.8 of this chapter. Others introduce samples on solid surfaces and thermally desorb them followed by gas-phase chemical ionization like Direct Analysis Real Time (DART) [107, 108], Atmospheric Solids Analysis Probe (ASAP) [109], Atmospheric Pressure Photo Ionization (APPI) [110], and Laser Diode Thermal Desorption (LDTD) [111, 112], the later also covered in Chapter 11 of this book [113]. Solid samples can also be analyzed by ESI extracting them with a liquid as done with the Surface Sampling Probe (SSP) [114–116] commercialized by Advion as Liquid Extraction Surface Analysis (LESA) [117, 118]. An approach to extracting surfaces using a flowing stream of liquid was commercialized by Prosalia as the FlowProbe [119]. Extracting surfaces by impacting them with high-velocity charged liquid droplets, as is done with Desorption Electrospray Ionization (DESI) [97, 120, 121], is covered in Chapter 13 of this book [122]. The relatively minimal impact of the ambient sampling techniques in the LC/MS Histomap of Figure 1.1 is because, as with MALDI, their utilization as an interface to chromatography has been marginal to none. However, several of the approaches cited above are in the early stages of evaluation for HT-MS and to varying degrees are being tested in industrial-scale drug discovery programs [6, 7]. Some frontrunners are beginning to emerge in 2020, when this book was being written such as acoustic ejection mass spectrometry (AEMS), acoustic MIST, and the laser-based approaches using electrospray-like (ion evaporation) ionization of infrared MALDESI and LAP MALDI (ultra violet liquid atmospheric pressure MALDI) which are further referenced in Section 1.8 of this chapter, Chapters 5 [16] and 11 [113] of this book, and the Histomap in Figure 1.1.

1.2.3 Direct and Indirect Affinity Measurements with ESI/MS for HTS

Both Affinity Selection Mass Spectrometry (ASMS) and Native Mass Spectrometry have become important adjuncts to the drug discovery process in recent times but their origins date back 20–30 years as indicated in Figure 1.1. Although the relevance of molecular affinity to biological activity has been understood since the beginnings of molecular biology, activity-based assays have provided the bulk of the information for drug discovery decisions partially because they can be adapted for high-throughout measurements using optical-based detectors. But over the ensuing years, the need for additional information than that provided by HTS regarding a candidate's potential has emerged from this grand experiment. Around 1998, the ASMS technique showed that although each affinity-based assay took several minutes, throughput was gained by measuring hundreds of compounds in each assay [129–131]. In only the past few years this technique has been revived, improved, and has been implemented widely in drug discovery programs as summarized in Chapters 8 [132] and 9 [133] of this book.

Native MS has its roots in the field of non-covalent interactions, which requires that some degree of a protein's 3-D structure be maintained as it transitions from the liquid to the gas phase. An indication of this possibility was observed in the early 1990s (see Figure 1.1) when the first gas-phase cross sections of protein ions were measured, and these cross sections varied depending on the degree of solution-phase denaturation [134]. In parallel, studies proceeded to show for the first time using an important biological target (RAS protein) that the affinity for its small molecule activators and inhibitors (FK 506 and Rapamycin) could be observed in the electrospray mass spectrum [135, 136]. One year earlier this was observed with non-covalent interactions between enzymes and their substrates and other model proteins having specific affinities for certain small molecules [137, 138]. Although the study of non-covalent interactions with mass spectrometry produced many inspiring publications in the scientific literature since, it did not emerge as an industrialized tool in the drug discovery process until nearly 30 years after this was first proposed. Currently it is not a high-throughput technique, but the importance of adding affinity information to the activity-based data to triage potential drug candidates has been recognized resulting in its incorporation on an industrial scale in some drug discovery programs as detailed in Chapter 10 of this book [139].

It took two to three decades for both affinity MS techniques to rise from initial proof-of-principle to industrial utilization. A similar trend is visible in the Histomap for several technologies having relevance to HT-MS that had long periods of dormancy after their initial demonstration emerging in the commercial scene nearly a decade later. Three different reasons why this occurred are illustrated in the stories behind some of these technologies.

The first reason is the status quo can be sufficiently entrenched to inhibit the rate at which new concepts emerge. This was experienced in the early days of LC/MS interface development when vacuum ionization was utilized in the vast majority of commercial instruments for nearly a century, which diverted efforts toward developing a solution that incorporated vacuum ionization instead of API. Eventually, when compelling data continued to emerge, the status quo was breached and the dominant LC/MS vacuum-based ionization interfaces went extinct, but it took nearly two decades.

The second reason is the vicissitudes of markets that are largely financed by government funding, regulations, and geopolitical events. These types of markets for instrumentation can go from boom to bust over the course of a single change in political administrations. This was the case in the late 1970s and 1980s with the discovery of cancer-inducing toxic waste dumps throughout the United States that resulted in the formation of the EPA Super Fund to correct the damage. At the same time, terrorism in airports such as Heathrow and Lockerbie appeared. The development of ambient MS sampling methods emerged responding to the need and commercial opportunity as depicted in Figures 1.4 and 1.5. Priorities were changed by shifts in US political control and a temporary reduction in the terrorist threat in the United Kingdom, both of which dried up funding and quenched the market. Despite the technical sophistication and the advanced degree of validation in real-world situations for those early ambient MS technologies, development and investment in the field soon collapsed and for the most part they vanished from the market. Like a phoenix, ambient sampling would re-emerge 20 years later as environmental and security issues returned to the headlines.

The third reason explains why it took until the late 1990s for the emergence of high-speed injection technologies when the possibility of high-throughputs was first clearly demonstrated in 1986. In this instance, the technology simply preceded the need. In the 1980s, there were no situations where the ability to analyze 100,000 samples per day were compelling, because that many samples did not exist on a daily basis. The approach that emerged during the 1990s to screen the vast pharmaceutical compound libraries with fluorescent plate readers changed that situation rekindling the efforts in the late 1990s to try to address this application with mass spectrometry using fast chromatography, the topic of the next section of this chapter. This also explains the lag in adaptation of the affinity MS techniques. The results of the grand experiment to discover drug leads based on HTS activity-based measurements indicated many years later that activity-based assays alone were not sufficient and the orthogonal affinity-based measurements from years past could finally prove useful.

To summarize this section, and as captured in the Histomap, the era defined by the early research in the development of LC/MS brought us atmospheric pressure ion sources which provided the foundation for the majority of industrialized approaches to HT-MS today and the opportunity to explore many different methods of sample introduction to further advance this field. MALDI emerged at the same time as part of an attempt to extend the mass range of analytes that could be analyzed by mass spectrometry and with the introduction of high repetition rate lasers has demonstrated utility for HT-MS as well. The first concerted efforts toward deploying HT-MS with ESI occurred in the late 1990s using two different approaches to increase chromatographic throughput. As shown in the Histomap, these two approaches emerged simultaneously and are the topics of the next two sections on high-speed serial and parallel chromatographic technologies.

1.3 High-Speed Serial Chromatographic Sample Introduction

At the American Society of Mass Spectrometry Conference held in Chicago in 1999, the workshop on LC/MS technologies [140] drew an audience of over 2000 people indicating the keen interest in the topic "Strategies for Achieving UltraHigh-Throughput LC/MS: Parallel versus Fast Serial Chromatography." A heated debate ensued between the proponents of both sides. Mark Cole from Pfizer argued that the essence of the speed advantage with the fast serial approach was fast chromatography, enabled by high flow rate ion sources, which were sometimes simplified to basic trap and elute-type separations to minimize analysis time. The injection times were staggered between two or more columns to improve the duty cycle and the analytes shunted though a valve into a single sprayer ion source one at a time as they eluted in sequence. Dan Kassel from Glaxo argued that the essence of the speed advantage with the parallel chromatography approach was up to eight chromatographic channels to be operated in parallel such that the injection and elution from each column occurred simultaneously requiring separate ionizing sprayers for each channel to be "indexed" to the MS detector. Only one of these two approaches advanced to today as a viable solution to HT-MS. The following sections describe activities in both camps at that time, which one survived and why, along with the development of high flow rate ESI ion sources and fast, high flow rate LC technology, both of which contributed to the speed of this new approach to liquid chromatography as the sample inlet to mass spectrometers.

1.3.1 High Flow Rate Ion Sources

High mobile-phase flow rates are important to high-speed analysis for several practical and fundamental chromatographic reasons. The high fluid linear velocities that are achieved at high flow rates (100–2000 μL/min) and needed for rapid transport of the sample through the column and plumbing can also be achieved in the microflow (<100 μL/min) or nanoflow (<1 μL/min) regimes with the appropriate choice of column and tubing diameters. However, column diameters drop from several mm in diameter to a few hundred microns or less across the high flow to nanoflow range and tubing i.d.'s decrease from ≥100 to <25 μm to prevent peak broadening. Practical issues arise with these reductions such as an increased frequency of clogging and increasingly difficult to manage dead volume effects. From a chromatography perspective, sample loading capacity decreases proportional to the square of the column diameter and the negative influence of column wall effects on separations increases (ratio of wall surface to bed area). Add to this the fact that system back pressures and surface tension effects scale with the 4th power of tubing i.d., the balance in favor of higher flow rate systems to reliably execute large sample number assays becomes apparent.

As mentioned earlier, the APCI heated nebulizer interface, operating at 1 mL/min, successfully demonstrated the feasibility of HT-MS. This led to its early implementation for quantitative analytical methods in support of pharmaceutical drug development and clinical trials reaching throughput levels of 200–400 samples per day [141, 142]. One notable example, which highlights an improvement in 2 orders of magnitude in throughput, reliability, and LOQ over previous mass spec-based approaches cites the quantitation of over 40,000 human serum samples for cotinine, a metabolite of nicotine, accomplished over several months. This was part of a survey by the Center of Disease Control in Atlanta, Georgia, of the North American population for second-hand exposure to cigarette smoke [143].

Drug development assays, particularly those in support of clinical trials, tend to be validated for one or a few targeted analytes per assay. For those APCI assays cited above, it was determined *a priori* that the drugs and their metabolites of interest were stable to the thermal vaporization and chemical ionization process used in the heated nebulizer interface. In the drug discovery process, the more common situation is an assay is conducted with many different types of analytes at the same time making the versatility of the ionization process a crucial consideration to avoid failed measurements. Estimates have been made that in excess of 30% of biochemically relevant small molecule chemical space would result in fragmentation with APCI whereas >95% is covered by ESI [144]. This drove the efforts to develop ESI-based ion sources toward higher flow rates making them more useful for HT-MS applications.

Ion Spray raised the flow rate limitations of Fenn's and Gall's ESI from ~1 μL/min or less to the microflow regime of ~100 μL/min which made chromatographic coupling a practical reality. Improvements to this ion source in the early 1990s provided sensitivity gains, which involved increasing desolvation rates with heat [96] but sensitivity levels remained significantly lower than APCI at flow rates approaching 1 mL/min. It became apparent that more improvements could be made with a more careful consideration of the ionization process from first principles. The pneumatically generated charged droplets of ~10 μm diameter can be evaporated to their Rayleigh limits, within their flight time in the ion source (~1 ms), at which point coulomb explosions rapidly complete their final droplet size reduction to ion emission diameters of 10 nm, if they are deeply entrained in gas at temperatures as high as 800 °C [33].

Figure 1.6 is a photograph of an ion source that achieves this condition. Temperatures of this magnitude seem alarming considering the possibility of

Figure 1.6 High flow rate ESI ion source operating at 800 °C. Evaporative cooling keeps the droplet temperature to ~<50 °C at all times. The thermal gradient provides a cool region for ion emission from the 10 nm diameter droplets in the vicinity of the curtain gas aperture (center hole). This approach matched the high flow rate capabilities of the APCI heated nebulizer and enabled high-speed chromatographic systems with flow rates in the range of 0.1 to >2 mL/min to be used with ESI. *Source:* Covey [33]/Reproduced with permission from Elsevier.

thermal degradation. However, the temperature of the liquid in the evaporating droplet reaches a maximum of only ~50 °C due to the evaporative cooling effect [145] and only increases when the droplet is reduced to a dry residue at which time it would spontaneously ignite if it remained at this temperature. The volume of the 800 °C injection gas is approximately 10 times the volume of the nebulizer gas as dictated by the law of conservation of momentum. The rapid decrease in the nebulizer gas velocity is compensated for by an uptake or entrainment of surrounding lower velocity gas at approximately this volumetric ratio. As a result, initial desolvation occurs rapidly at high temperatures while ions clustered with solvent molecules, that absorb additional vibrational energy from the surrounding gas, are emitted from the 10 nm diameter droplets in a cooler region near the vacuum aperture [33]. The residence time of the ion-solvent cluster in this region is a few tens of microseconds as it is swept out by the vacuum drag of the MS. Thermal degradation is primarily avoided and declustering is achieved by collisions with background gas in a vacuum in a controlled fashion with a DC potential and the use of a dry counter current flow of nitrogen sometimes referred to as a curtain gas [22].

This development in 2001 provided sensitivity gains up to 20–30-fold above the previous generation of heated ESI ion sources [27, 33] at high fluid flow rates bringing the sensitivity of ESI at par with APCI in this higher flow rate regime. However, the compound coverage is much greater with ESI making it the preferred approach for HT-MS systems. As described in Thomson's Ion Evaporation mechanism, the local electric field strength around these drops due to the concentration of charge within them is sufficient to overcome the surface tension of the liquid, emitting ions directly from the liquid to the gas phase using energies well below typical molecular bond strengths. It is a very different situation than when molecules are thermally vaporized into the gas phase from a solid or liquid as with the heated nebulizer or other thermal desorption APCI techniques. In these situations, the sample molecules must be transferred to the gas phase from the solid or liquid state by heat alone. Weaker molecular bonds are ruptured by the excess vibrational energy required to exceed the vapor pressure of the molecule to drive it into the gas phase with heat. The gas phase ion-molecule reactions with the reagent ions present in high excess are also capable of rupturing covalent bonds if the difference in the proton affinities of the analyte and reagent ions exceeds the weakest bond strength in the analyte molecule.

The dramatic improvements in the performance of pneumatically assisted electrospray ion sources introduced around the turn of the twentieth century was largely motivated by the increasing trend in the HT-MS field toward high-speed, high-flow rate chromatographic systems being developed at that time as discussed below in Section 1.3.2.

1.3.2 Fast Serial Scheduled, Staggered Chromatographic Separations with Fast Autosamplers

During the 1990s, sample run times rapidly decreased from several minutes down to approximately one minute per sample by taking advantage of the specificity of mass detection which allowed separation requirements to be relaxed. In the late 1990s, a technique called staggered elution zone chromatography (SEZ) emerged from Pfizer initiating the concept of staggering the injection times from more than one column such that the analyte eluting from each chromatographic channel could be serially shunted into the ion source using a post column selector valve [93]. Using 1 cm columns packed with 40 μm bed particle diameters operating at flows in excess of 1 mL/min, high speed trap and elute separations were achieved. Originally designed using the large deck capacity Gilson autosampler, the duty cycle improvement provided by this approach soon outpaced the speed of commercial autosamplers. A custom-made high-speed dual arm autosampler, the DAAS shown in Figure 1.7a, was designed that enabled speeds up to 10 seconds per sample to be achieved and fed by a large 16 μL plate capacity deck [146–148] robotically loaded. Two chromatographic channels had their injection times slightly staggered so the eluting analytes could be individually shunted to a single sprayer ion source using the plumbing and selector valve arrangement shown in Figure 1.7b. Eliminating the valve with a non-indexed two sprayer ion source, shown in Figure 1.7c, improved the speed to seven seconds [149, 150]. This speed gain was not sufficient to offset the added complexities involved in tuning the electric fields and gas flows of two physically spaced independent sprayers to make their responses equal, so the dual sprayer modification was not commercialized. This system also introduced autotuning MS conditions for large numbers of compounds and data basing the parameters for enterprise-wide utilization [151]. A four-column staggered serial sample introduction system based on commercially available components was demonstrated at the same time [152] as well as a system based on 8-injector valves feeding a single column [153].

With the path forward established by the SEZ concept, its supporting software, and the DAAS autosampler, various companies developed commercial systems to emulate this approach. These became a new generation of high-speed autosamplers with similar valving arrangements for either fast trap and elute or chromatography, and software to schedule and shunt the multiple channels to a single MS sprayer at the appropriate time. Apricot used the DAAS platform to develop their dual arm autosampler shown in Figure 1.7e. Sound Analytics, who were involved in developing the original SEZ and DAAS prototypes, advanced their Discovery Quant and LeadScape software and designed a new high-speed autosampler, the LS1 [154] as further described in Chapter 2 of this book [10] and shown in Figure 1.7g. BioTrove, who had access to the Pfizer SEZ and DAAS system through a joint MIT/Pfizer

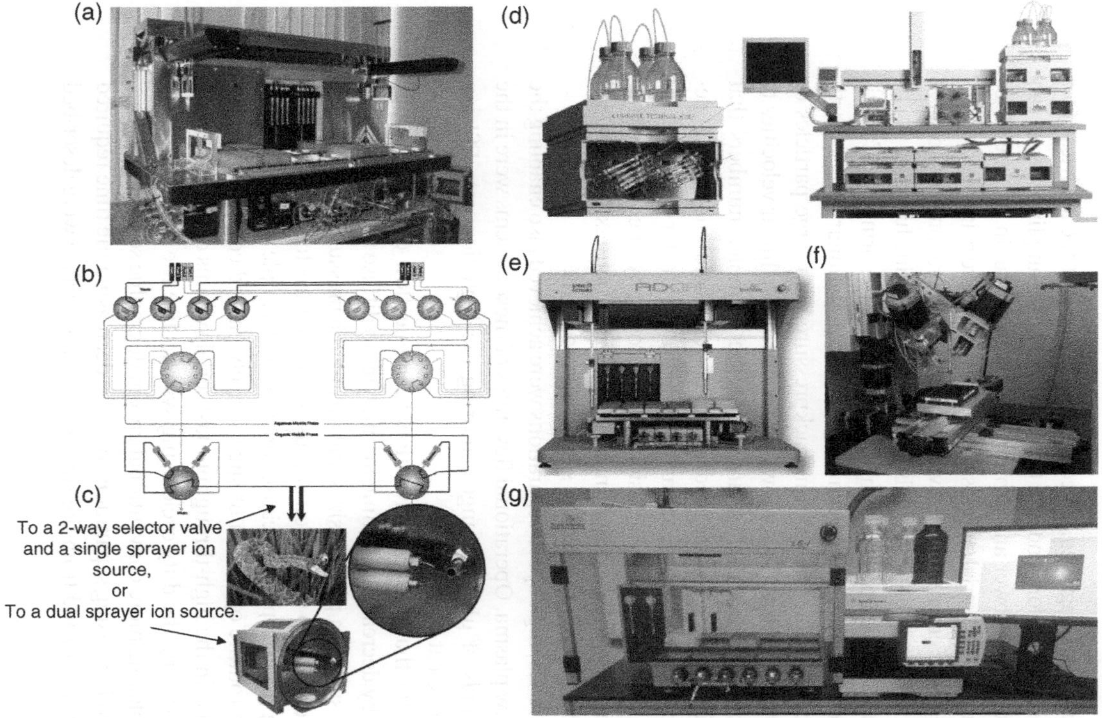

Figure 1.7 Fast serial, staggered chromatography systems and autosamplers to emerge in late 1990s and early 2000s. (a) The Pfizer dual arm autosampler (DAAS), grandfather of the fast autosamplers and systems to be developed for scheduled staggered chromatography was capable of 500 samples per hour. A Cavro RSP 9000 dual-arm robot comprises the core component, each arm configured with four aspiration needles. The deck accommodated 16, 96-, or 384-well plates. A total of eight 6-port 2-position injection valves, two 10-port, 2-position column switching valves, and two 10-port multi-position valves and 8 injection ports are integrated into the unit. (b) Schematic representation of the valves and flow path integrated into the sample delivery system. (c) A non-indexed dual sprayer ion source could replace the stream selector valve to increase speed. It was called the Cobra ion source because the sprayers resemble the fangs of a viper about to strike. (d) The Cohesive scheduled, staggered chromatography system utilizing a more conventional CTC autosampler. (e) The Apricot ADDA dual arm autosampler, a first-generation descendant of the Pfizer DAAS. (f) The rapid-fire system, a first-generation descendant of the Pfizer DAAS. (g) The LS-1 a second-generation descendant of the DAAS.

contract, developed from the Pfizer concept a dedicated fast trap and elute system, the Rapid-Fire [155], shown in Figure 1.7f now provided by Agilent and described in detail in Chapter 4 of this book [12]. Cohesive developed the Aria system [156] using conventional autosamplers and is now provided by Thermo as the Transcend 4-channel system shown in Figure 1.7d. Additional examples of software designed to stagger the introduction of analyte from multiple chromatographic systems, such as the Sciex MPX, are provided in Chapter 3 of this book [11].

1.3.3 High-Speed Column Stationary Phases

A common theme with the staggered, scheduled systems is that the analytes enter the ion source in a serial fashion, one at a time. This implies sampling speeds can increase as the peaks get narrower and shorter in retention time. In the late 1990s and early 2000s, when the fast serial systems described above were being prototyped, column technologies were emerging to provide high-speed separations and to serve a larger purpose than trap and elute which is largely just a desalting function.

The stationary phases developed to do this were based on the principle of maximizing turbulent flow conditions between and within the stationary phase particles. This is done by reducing the column backpressure with large particles or highly porous stationary phases. This allows for mobile-phase linear velocities to reach turbulent flow conditions lying far beyond the Reynolds numbers that define optimal laminar flow velocities used in conventional columns with single-digit micron-sized particles. Polystyrene particles of 40 μm diameter were used for the Cohesive and the Perceptive Perfusion chromatography media. Another approach used stationary phases casted into monolithic structures from sol–gel materials pioneered by Merck-Darmstadt, so-called speed rods.

As shown in Figure 1.8, high-speed separations were achieved also enabling the direct injection of raw plasma. Operational flow rates of these systems were in the hundreds to thousands of μL/min range requiring the high flow rate ESI ion sources being developed at this time.

UPLC emerged later that utilizes conventional laminar flow chromatography but increases speed by decreasing the time for analytes to diffuse to the stationary phase surface using sub 2 μm particle sizes and raising the pressure limitations of the pumps. Improvements to separation speeds of two to threefold over conventional chromatography are typical. Because some form of sample pre-purification is important for the routine operation of such densely packed columns, it is not found in common use in the high-throughput fast serial systems described above, but plays a role improving the duty cycle using systems like the Aria/Transcend equipped for multiplexing more conventional chromatographic separations.

Two approaches to increasing separation speed by decreasing the time required for the analyte to diffuse to and from the stationary phase surface were described

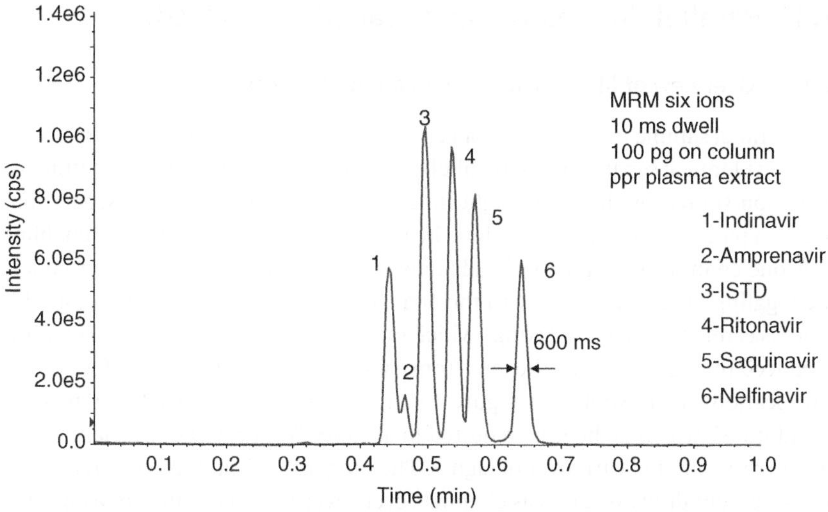

Figure 1.8 MRM ion current trace using a monolithic stationary phase casted from sol–gel material introduced around 2000. Speed Rods were the first LC columns to obtain sub-one second wide peaks while maintaining chromatographic separations and were tolerant to complex sample matrices. Developed by Merck-Darmstadt, Chromolith Speed Rods of 4.6 mm × 5 cm dimensions operated at flow rates of up to 6 mL/min. Three mL/min shown in the figure.

above. The use of sub 2-μm bed particles operating at high pressures is one. The second involved creating turbulent flow between and within large diameter and highly porous particles. A third way to increase speed by decreasing diffusion times is to use supercritical fluids for the mobile phase, which have diffusion rates close to those obtained with gas chromatography. The selectivity of the separations with packed column SFC can be similar to those of HPLC particularly if the CO_2 mobile phase is doped with alcohols or water, but the speed is greater producing peak widths on the order of a few hundred milliseconds and it can be readily interfaced to either electrospray or APCI [157, 158]. Its use for HT-MS has not proliferated because of a lack of commercial analytical SFC pumping systems although SFC chromatography is commonly used for preparative work. This approach remains as an opportunity to increase the throughput of staggered fast serial sample introduction although it will still be rate limited by autosampler considerations.

For HT-MS applications today that use this fast serial sample introduction approach, a wide variety of column technologies, including those mentioned above, are used depending on the application requirements and are discussed in the aforementioned Chapters 2–4 of this book [10–12].

1.4 Parallel Chromatographic Sample Introduction

1.4.1 Overview of Multichannel Indexed Ion Sources

The parallel sample introduction concept enables four to eight columns to be conducting identical separations in parallel without offsetting their injection times by using ion sources with four to eight indexed sprayers. Although the separations occur in parallel, the detection is done in a serial manner by sequentially enabling only one channel at a time to provide ions to the MS analyzer and synchronizing the toggling of each channel with the arrival of the ions at the MS detector. This process is referred to as "indexing" which enables the recording of which channel produced the signal. Gating the ions from each channel can be done in four different locations as illustrated in Figure 1.9. The most critical consideration is the "dead time," which is the time required to advance from channel-to-channel and avoid cross-channel carryover of signal. The longer the dead time the greater the measurement duty cycle loss as channels cycle over the time course of a chromatographic peak. These approaches differ significantly with regards to their dead times, which became a major criterion for whether or not to pursue a particular

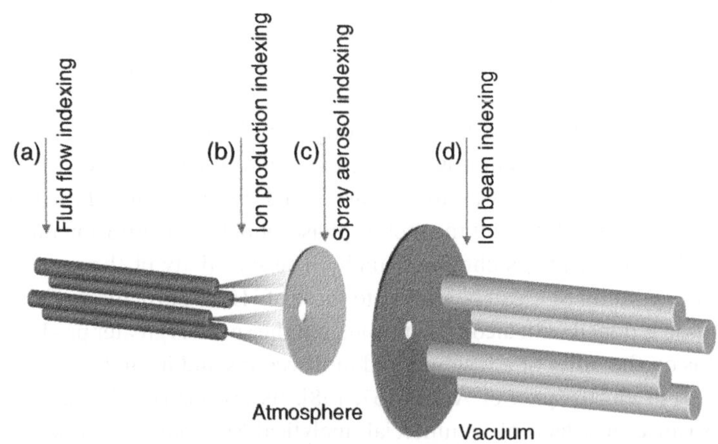

Figure 1.9 Illustration of the four positions where signal can be toggled on and off for indexing multichannel LC systems. (a) Fluid flow indexing uses a fast-rotating selector valve requiring only one sprayer. (b) Ion production indexing turns on and off either the voltage or fluid delivery at the tip of each sprayer. (c) Spray aerosol indexing uses a rotating spray chopper with an iris in a disc to interrupt the droplets in flight to the vacuum entrance aperture. (d) Ion beam indexing requires multiple sprayers in front of multiple atmosphere-to-vacuum apertures. The multiple ion beams in a vacuum can then be either deflected or focused into the ion optics quickly with a lens.

approach. For a comprehensive overview of the topic of multichannel and multiplexed ion sources, see reference [159].

1.4.2 Fluid Indexing

A means of indexing was prototyped and tested using a modified post column selector valve to shunt one chromatographic channel at a time to a single sprayer ion source. As illustrated in Figure 1.10a, the 4-port selector valve (Rheodyne) was modified to be continuously rotating at a rate of one revolution per second (1 Hz) with encoders registering which channel was being sampled during rotation.

The dead time for this approach was approximately 200 ms per channel. The valve could advance between channels faster than this, but the issue became one of the effects of fluid flows on peak shapes. During rotation, channels would be

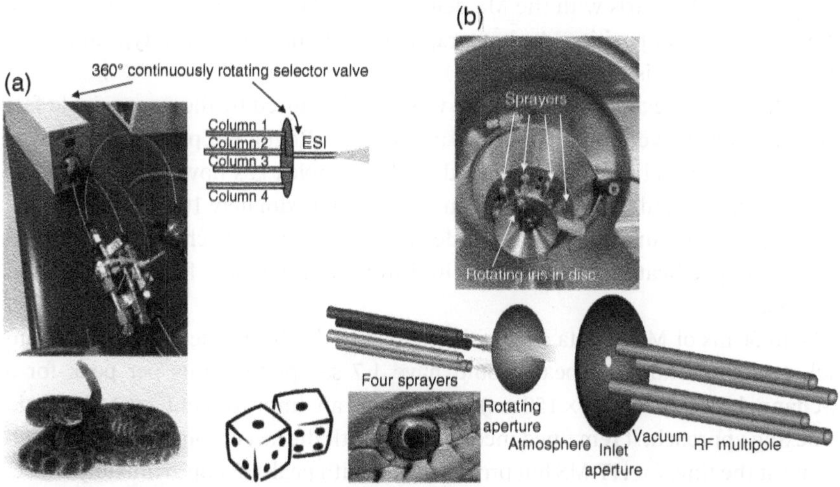

Figure 1.10 Two examples of indexing systems prototyped and tested. (a) Fluid flow indexing. A photograph and schematic of the "Rattler,", a four chromatographic channel indexed parallel sample introduction system based on a high-speed rotating selector valve (Rheodyne). The project was code named the Rattler because, when at speed, the valve would enter a state of resonance and sound like the northern most rattlesnake, the Mississauga rattler, about to strike. (b) Spray aerosol indexing. A photograph of an early prototype and schematic of a 4-channel device. The first prototype had two channels indexed with a rotating iris to pass the spray from each of two channels resulting in the project name Snake Eyes. The project was also given this moniker because it was initially considered to be a roll of the dice. It was eventually expanded to four chromatographic channels with four sprayers as shown. A 4-channel device was first developed and commercialized as MUX by MicroMass. *Source:* Schneider and Covey [159]/Reproduced with permission from ELSEVIER.

momentarily blocked building up pressure which, upon release, would create a momentary high velocity flow in the immediate downstream tubing. Peaks had severe frontal tailing due to the parabolic flow created in the resulting velocity gradients created across the diameter of the tube (Poisseile flow), which was a large contributor to this relatively long dead time. With 50 ms of data acquisition per channel, a 4-channel system would be limited to one data point per one second wide peak. Because of this limitation fluid, indexing was abandoned.

1.4.3 Spray Aerosol Indexing

This approach emerged near simultaneously from the University of Alberta [160], Glaxo Smith Kline in Chapel Hill [94], and MicroMass in Manchester [95] who ultimately commercialized it as MUX with four channels [161, 162]. An iris in a rotating disc allowed the charged aerosol from one sprayer at a time to reach the entrance to the mass analyzer with indexing being achieved by registering the position of the iris with the MS detector. The Alberta approach was to rotate the sprayers out of position. A photograph and schematic of a prototype of the iris type of ion source is shown in Figure 1.10b.

With spray indexing, the dead time was largely caused by the unavoidable fact that all sprayers were producing ions at all times. Rogue ion producing droplets from each channel remain suspended in the complex gas flows encountered in front of and behind the blocking rotating iris. Approximately 100 ms per channel was required to purge this area in order to achieve the 0.1% channel-to-channel carryover specification typically required from autosamplers for most quantitative assays.

With 50 ms of MRM data acquisition added to the 100 ms dead time per channel, a one second wide peak would have 1.7 sampling points per peak for a 4-channel system (1 s/(4 × 150 ms)). Although an improvement over the fluid indexing approach, it remained incompatible with the fast chromatography developing at the time for HT-MS but proved useful with peaks of approximately 10 seconds width typical of conventional chromatography doing qualitative analyses.

1.4.4 Ion Beam Indexing

Ion beam indexing requires multiple sprayers to be situated in front of a corresponding atmosphere to vacuum aperture. The resulting multiple ion beams inside the vacuum system can be turned on and off with a lens in <1 ms. However, each atmosphere-to-vacuum aperture must be proportionally reduced in area so as not to overwhelm the vacuum system resulting in a sensitivity loss for each channel. This approach was never commercially pursued for this reason but was patented [163, 164].

1.4.5 Ionization Indexing

Shutting off the production of ions from the sprayers not being sampled provides a means to eliminate the residual lingering ion producing droplets of the spray indexing approach thus minimizing the dead time required to clear the ion source of analyte between measurements. Turning off the ion production by interrupting the fluid flow to the sprayers, without disrupting the peak shape as the rotating valve did, also presents a way to increase the number of channels to eight without overflowing the ion source with excessive amounts of liquid and aerosols.

The approach taken to do this involved using a solenoid valve-controlled fluid divert line situated close to each of the eight separate ESI emitters that were arranged in Gatling-like array as shown in Figure 1.11b [159, 165]. With this arrangement, all sprayers share a common nebulizer gas and high voltage connection resulting in the same response per channel, which was difficult to achieve

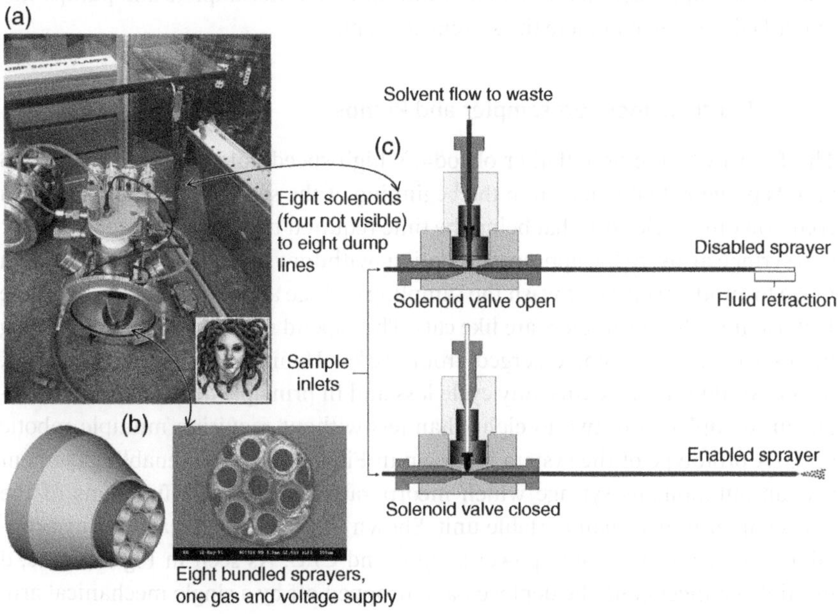

Figure 1.11 Ion production indexing. (a) Photograph of the Medusa ion source. (b) Eight sprayers arranged in a bundle which enabled a single tune for all eight (one voltage, gas flow, and position for all eight simultaneously). The ion production from each was gated by retracting the solvent from the sprayer tip a few millimeters using solenoid-actuated dump lines located close to the sprayer tip as shown in (c). This approach proved to be faster with less dead time between channels than the other indexing approaches, but it was not commercialized as the specter of the complexity and reliability of an 8-channel HPLC system became as terrifying as the project name suggests. *Source:* Schneider and Covey [159]/Reproduced with permission from Elsevier.

with the other independent sprayer approaches. A slight negative pressure applied to the divert lines draws the fluid back from the ESI tip ~3 mm when the solenoid valve on the divert line is opened as illustrated in Figure 1.11c. This effectively shuts off the production of ions. With this approach, the dead time between channels was reduced to 10 ms enabling eight channels to be cycled in ~500 ms using 50 ms of MRM data acquisition per sample. Chromatographic peaks 5 seconds wide for an 8-channel system or 2.5 seconds wide for a 4-channel system would be required to obtain 10 sampling points per peak for reproducible peak area calculations.

The 8-channel ion source is shown in Figure 1.11a with only four of the eight solenoid valves being visible. The system was code named Medusa because the management of the plumbing for eight columns and an 8-channel ion source was an altogether frightening concept. Even more daunting was the concept of eight autosamplers and eight gradient HPLC pumps per system. Despite these trepidations, a new approach to the design of multiplexed autosamplers and pumps was embarked upon to complete the system concept.

1.4.6 Multichannel Autosampler and Pumps

The DAAS was the grandfather of today's high-speed autosamplers. The DAAS and its progeny that emerged in the beginning of the second millennium share a common duty cycle cost, that being the time required to move the syringe arm (or two syringe arms) to the appropriate station, withdraw and dispense sample, then move to another station to wash the sampling syringe and injector valve. To quote Jack Henion, "Autosamplers are like cats. They spend most of their time cleaning themselves!". A concept emerged from Parker-Hannifin, supported by Sciex, which would minimize this duty cycle loss and in principle be able to scale a single autosampler from two to eight channels without requiring multiple robotic arms. A prototype of the system is shown in Figure 1.12a. The enabling element was an autonomous syringe which incorporated most of the functions of the autosampler into a small portable unit. Shown in Figure 1.12b each syringe contained its own drive motor, power supply, and CPU. As seen in Figure 1.12c, d multiple syringes could be deployed and dropped off by a single mechanical arm to retrieve samples, dock in wash stations, or inject, carrying out these functions independently and simultaneously for all channels. The project was named Checkmate because the autonomous syringes were strategically deployed to different positions across the deck of samples, wash stations, and injector valves giving the visual impression of a high-paced chess match [166].

To complete the concept, an 8-channel gradient pumping system was designed by Eksigent using 16 pneumatically driven pressure amplifiers capable of delivering rapid, steep mobile-phase gradients to maximize separation speed for this and

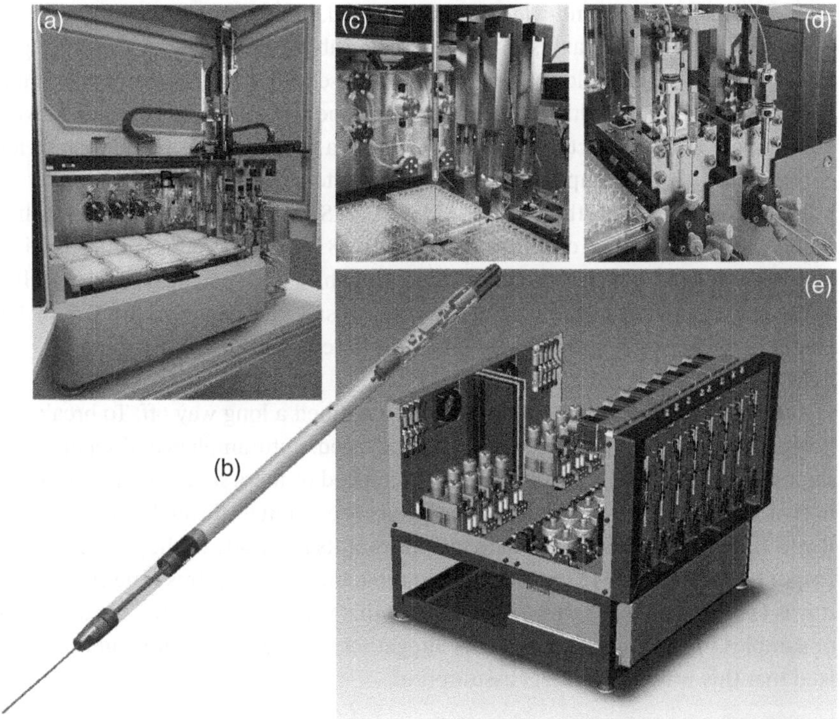

Figure 1.12 Components of the multichannel autosampler (Parker-Hannifin) and HPLC pumping system (Eksigent) being prototyped in 2006–2008 in sponsorship with Sciex. (a) A single multipurpose arm with a mechanical gripper was used to deploy multiple autonomous syringes to retrieve sample from a well plate on the 10-plate deck. The syringes would be dropped off by the arm in an injector, wash station, or sample well while the arm proceeded to move another syringe to perform another task thereby improving the duty cycle of the autosampler. Ultimately, an 8-channel system was envisioned having 8-injector valves and wash stations. (b) The autonomous syringe contains its own drive motors, power supply, and CPU. Various numbers of syringes could be deployed over the deck all involved in various stages of aspiration, delivery, and washing simultaneously. (c) Photo of one syringe aspirating a sample while others in the background are washing. (d) Two injectors shown in this photograph with one of them receiving the sample. With this design concept, the number of injectors and wash stations could be scaled to the desired number of channels. (e) Prototype eight gradient channel HPLC pumping system by Eksigent. Instead of reciprocating piston pumps simplified pneumatic pressure amplifiers capable of UPLC pressures and delivery of instantaneous gradients were used.

other applications like the development of chiral separations. A drawing of this system is shown in Figure 1.12e.

Prototypes of Medusa, Checkmate, and the 8-channel pump were developed and tested. Although the deadtime in the Medusa ion source was minimized, the requirement for 2.5-second-wide peaks remained. Although speed was gained by operating in parallel, it was also lost by not being able to take advantage of the high-speed separation technology being effectively used with the fast serial sampling methods where peak widths were approaching one second or less. A major consideration was also the practicality of maintaining a multichannel HPLC system for continuous high-throughput operation, so the system was not commercialized.

The debate held at the 1999 ASMS LC/MS workshop mentioned earlier had come to a definitive conclusion around 2008. The serial sample introduction approach turned out to be faster, more reliable, and practical. The parallel approach for the most part left the commercial stage as shown in the Figure 1.1 Histomap. Fast serial systems proliferated and remain a mainstay of the HT-MS field today.

Achieving the 1 Hz sampling target still remained a long way off. To break the throughput barrier, a harsh reality had to be faced: autosamplers and separations imposed fundamental limits to the analysis speed of mass spectrometry. In order to rival the speeds of the optical plate readers used in high-throughput screening, it was time for the MS to be emancipated from its servitude to the HPLC and no longer be held hostage to the chromatographic time scale. At this time circa 2005, it was assumed the only way to accomplish this was to abandon the introduction of samples in flowing liquid streams altogether. Ten years later it would be realized that this was not a correct assumption.

1.5 High Repetition Rate Lasers

By 2005, MALDI had been long established as a method for determining molecular weights of biopolymers using TOF mass analyzers. Typical systems up to that time used nitrogen lasers which provided relatively high energy pulses (300 μJ/pulse) at low frequencies of ~40 Hz to produce transient signals of variable shot-to-shot intensity as the laser rastered the surface and the MS spectra were averaged to improve the S/N. Its utility for quantitation using targeted acquisition modes like MRM on a triple quad was not obvious. In early 2005, a prototype source was bolted onto a triple quad to explore these possibilities. See photographs of the prototype instrument used for initial exploratory work in Figure 1.13a–c.

It was observed with this instrument that reproducible peak-like signals could be obtained by keeping the laser in a stationary position and drilling down to the metal target plate, as long as the crystallization of the sample was uniform as shown in Figure 1.14c. By increasing the laser repetition rate into the kHz range and lowering the energy to a few μJ/pulse, this method became much faster. As shown in Figure 1.14a, b, the improvement in peak widths and intensity were over 50-fold [81, 82]. This data showed that the 1 Hz barrier was no longer

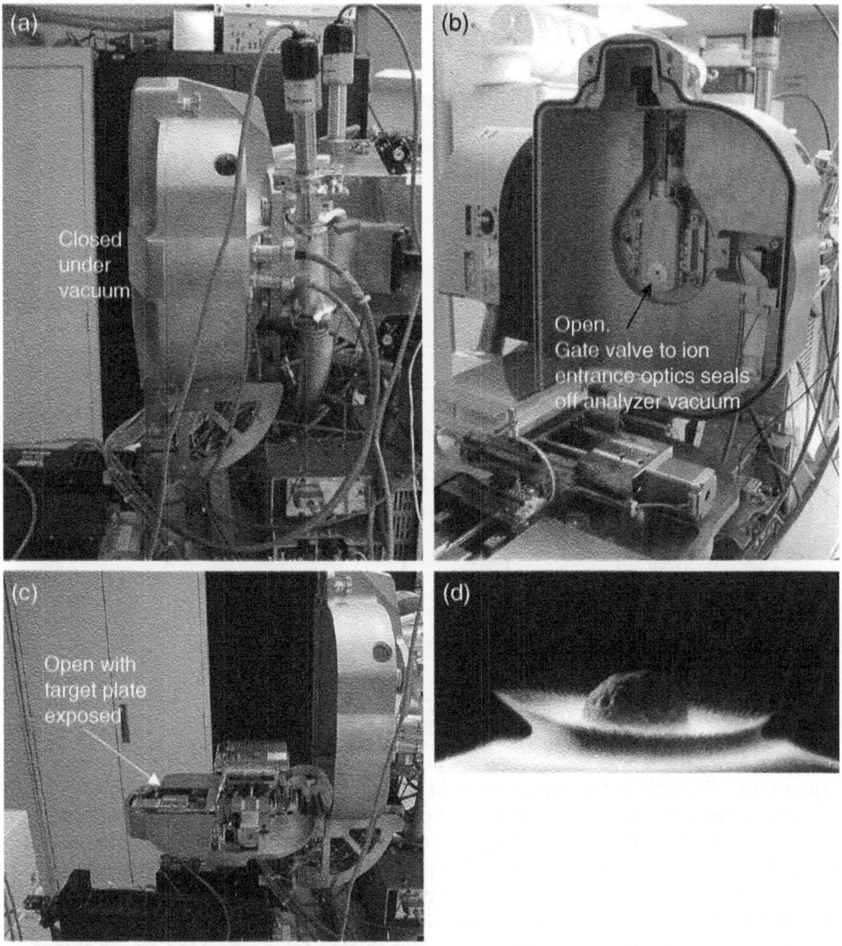

Figure 1.13 The prototype MALDI ion source adapted to a triple quadrupole instrument with a high repetition rate laser. (a) Closed and under vacuum for operation. (b and c) Open for plate loading. This research grade instrument was the first to break the 1 Hz HT-MS sample speed barrier which gave it the name Asteroid. (d) Because of its potential to initiate an evolutionary change in the HT-MS field, or so it was thought during its early development. Its commercial progeny [81, 82], FlashQuan, would not live up to these expectations for reasons described in the text.

insurmountable, and a MALDI triple quadruple system for HT-MS was commercialized called Flash Quan.

For in vitro ADME assays, purifying samples with solid-phase extraction, and using internal standards to reduce variabilities in crystallization and spot morphology, provided good quantitative results. CVs < 10% and LLOQ's for drug-like

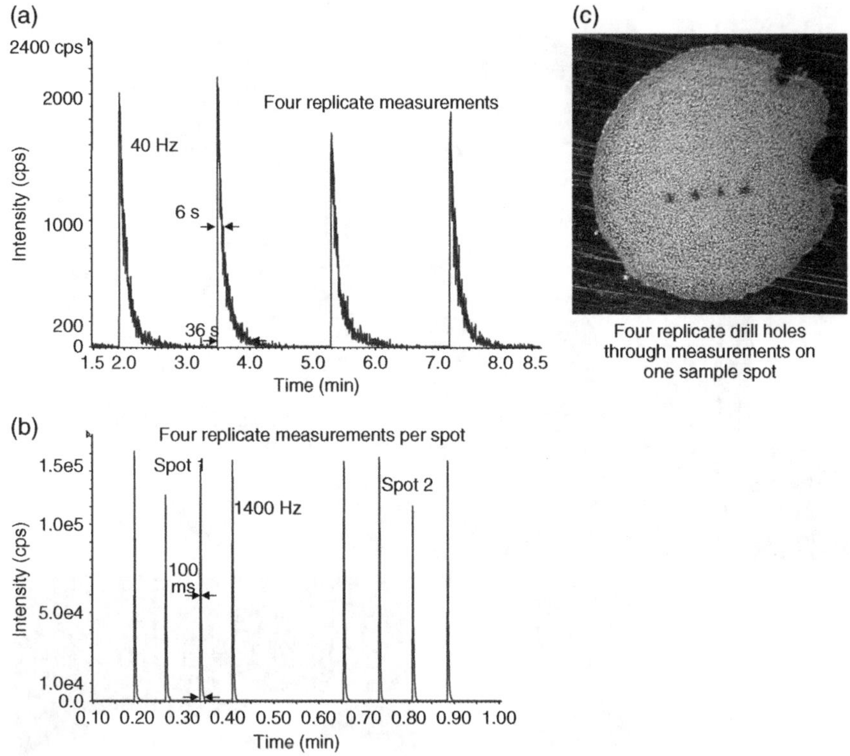

Figure 1.14 Example of the improvement in quantitative data when using a high repetition rate laser and preferred method of data acquisition. (a) Conventional N_2 laser signal transients produced from drilling sample to target plate with a 40 Hz repetition rate and 300 μJ/pulse. (b) High rep rate diode pumped laser signal transients. Repetition rate = 1400 Hz with 5 μJ/pulse. Same amount of analyte crystalized in the spot in a and b. (c) Preferred method for producing quantifiable peaks by targeting a single area and desorbing the sample to the target plate. The four ablated spots reflect the groups of four replicates in a and b. *Source:* (c) Corr et al. [81] / Reproduced with permission from ELSEVIER.

compounds in the low ng/mL, nM concentration range were achieved. For in vivo serum or plasma samples, additional sample preparation was required to obtain quantitative results as both ionization suppression and crystallization nonuniformities occurred [82, 84].

In 2007, Flash Quan was abandoned largely because MALDI, being a relatively mild form of thermal desorption chemical ionization [77], introduces more internal energy into the analytes during ionization than ESI. The energy transfer to the analyte during both the desorption and charge transfer process resulted in the most labile molecules fragmentating [82, 83]. Across small molecule chemical space, the sensitivity and molecular ion integrity produced from MALDI and heated nebulizer APCI proved to be similar falling short of ESI [144]. Since compound coverage is an important consideration for many HTS and ADME assays, in 2007, the balance was tipped in favor of putting resources back into developing an approach compatible with ESI.

MALDI has had significant success on TOF machines in recent years for HT-MS, as depicted in the Figure 1.1 Histomap, demonstrating the ability to break the 1 Hz barrier using high rep rate lasers for many HTS screening assays. It remains today as an important option for HT-MS, is developing a track record of successes [167, 168], and is thoroughly covered in other reviews [4–7] and in Chapter 12 of this book [13]. In recent years, the use of high repetition rate lasers to vaporize the liquid sample and ionize it using different implementations of electrospray have begun to emerge and are further referenced in Section 1.8 covering "High-Speed Sub-Microliter Volume Sampling."

At this time around 2008, four bottlenecks to throughput using ESI remained to be solved: isobar separation speed, injection speed, pre-MS speed limitations imposed by fluid transfer rates, and sample preparation. A potential solution to the first of these, the topic of the next section, was found in an unlikely institution, a Russian military research site developing detectors for landmines during their 1980s war in Afghanistan. Solutions to the remaining three bottlenecks would come after 2012.

1.6 Ion Mobility for High-Speed Gas-Phase Separations

1.6.1 Motivation and Commercial Options

There are many situations encountered in drug discovery and other fields where compounds exist in samples that have the same elemental compositions and are sufficiently close in structure to have similar, if not identical, CID mass spectra. In such a situation, a separation based on the differences in their chemistries is required to measure them independently, such as HPLC provides. It became increasingly clear during the first decade of the 2000s that abandoning chemically based separations prior to mass filtering as a component of an HT-MS system would be, in the long term, a shortsighted approach.

Ion mobility prior to the mass separation stage has been suggested because it is considerably faster than HPLC and several different forms of mobility have been coupled to mass spectrometry. As shown in the Figure 1.1 Histomap, two different forms of combined high field/low field differential ion mobility analyzers that operate at atmospheric pressure, Differential Mobility Spectrometry (DMS) [169] and High Field Asymmetric Waveform Ion Mobility Spectrometry (FAIMS) [170], emerged coupled to mass spectrometers. For the most part, low field mobility analyzers coupled to mass spectrometers operate under vacuum as shown in the lower right of the Histomap. Commercial options are drift tube ion mobility (DTIMS) [171–173], differential mobility analyzer (DMA) [174], traveling wave ion mobility (TWIMS) [175, 176], trapped ion mobility (TIMS) [177], structures for lossless ion manipulations (SLIM) [178], and CYCLIC IMS [179]. DMA is an exception as it operates at atmospheric pressure but is shown in the Histomap grouped with the vacuum-based analyzers because its separation is based on low field principles.

Another reason for mobility pre-separation is to improve LOQ's by reducing the contribution of background molecules and clusters to baseline chemical noise. Ion mobility provides the possibility for both isobar separation and noise reduction, but the harsh reality is that ion mobility is lacking in both the selectivity and peak capacity of HPLC. The separation versatility of HPLC cannot be matched by any form of ion mobility largely because of the intimacy of the chemical interactions that occur in the condensed phase and the wide variety of stationary and mobile phases that can be accessed or designed to address particular chemistries.

Low field mobility separates molecules based on their physical size or collision cross section. The separation of two isomers of a drug-like molecule differing by the position of a single hydroxylation would be considered a herculean task by low field ion mobility because of the similarity in their physical size, but relatively simple with HPLC that separates based on their unique chemical interactions with their immediate environment. Excessively long flight times incompatible with the emerging high-speed sample introduction technologies for HT-MS would be required. In addition, most low field mobility devices (except DMA) require the pulsing of the ion beam into the mobility cell introducing losses of ion current from continuous beam ion sources like ESI and scan modes like MRM used for quantitation.

For these reasons, a form of ion mobility that increases separation selectivity based on chemical interactions of the ions with the transport gas, performs separations in 1–10 ms and operates without requiring the segmenting and pulsing of a continuous ion beam, would be an ideal candidate within the ion mobility family to replace many of the functions of HPLC in the quantitative HT-MS drug discovery application space. Differential ion mobility/mass spectrometry (DMS/MS) has these properties and is the remaining focus of this section, Chapter 7 of this book [15], and is reviewed in reference [169]. A review of the other forms of ion mobility/MS is provided in Chapter 6 of this book [14].

1.6.2 Origins of DMS

In the 1980s, a new physical principle called DMS was discovered by Soviet scientists in their national laboratory network that showed promise for application as a handheld chemical detector in the battlefield theater. Planar and cylindrical analyzer geometries emerged from this effort [180]. The cylindrical geometry migrated to Mine Safety Associates (MSA) in Pennsylvania to be developed as a field-based toxic gas sensor. Referred to as FAIMS by MSA, the cylindrical device was adapted as the inlet to a MS by Ionalytics in the early 2000s, eventually to be acquired by Thermo. The planar device, referred to as a Differential Ion Mobility Spectrometer (DMS), demonstrated both a speed and a selectivity advantage over the cylindrical device using polar modifiers in the transport gas and was designed as a general-purpose handheld chemical detector by Sionex, a spin-out of Draper Labs [181, 182]. Sciex acquired the rights to this technology in 2006. Details of this historical context can be found in reference [169].

1.6.3 Chemically Based Selectivity with DMS to Mimic Chromatography

DMS devices designed to serve as an inlet to a MS were prototyped using some of the design principles in the handheld DMS detectors developed by Sionex described above [183]. Figure 1.15a shows that it is sealed to the atmospheric side of the atmosphere-to-vacuum aperture of an API ion source and shows the region where vaporized polar solvents are added to the N_2 transport gas. Figure 1.15b illustrates the principles by which ions are separated in a DMS cell where flight times can be <5 ms. Of particular importance is Figure 1.15c which provides the physical explanation of the chemically based separation mechanism. Measuring the difference in mobility of an ion in its clustered versus unclustered state is the primary distinguishing attribute of DMS over other forms of ion mobility making it attractive as a substitute for HPLC in HT-MS situations that bypass chromatography [184–188]. Figure 1.15d provides a perspective of its size relative to common currency and Figure 1.15e an exploded view of the assembly.

Some of the initial data from the prototypes showing the separations of various types of isobaric species using the chemical clustering effect are shown in Figure 1.16a–d. Figure 1.16a,a' show the dramatic effect on separation selectivity that occurs when invoking the ion clustering mode for the separation of isobaric lipids. The separation of the stereoisomers of drugs and amino acids are shown in Figure 1.16b, c using isopropanol in the transport gas. Separation of the positional isomers of the glucuronide conjugates of propranolol extracted from the liver of a dosed rat is shown in Figure 1.16d [183]. Although HPLC can readily separate all of these examples, DMS does so on a millisecond time frame compatible with recently emerging high-speed injection systems. More examples of its recently demonstrated utility for HT-MS applications and detailed explanations of this approach are provided in Chapter 7 of this book [15].

DMS/MS is another example of a technology to be developed ahead of its time, more than a decade in advance of the appearance of low volume, sub one

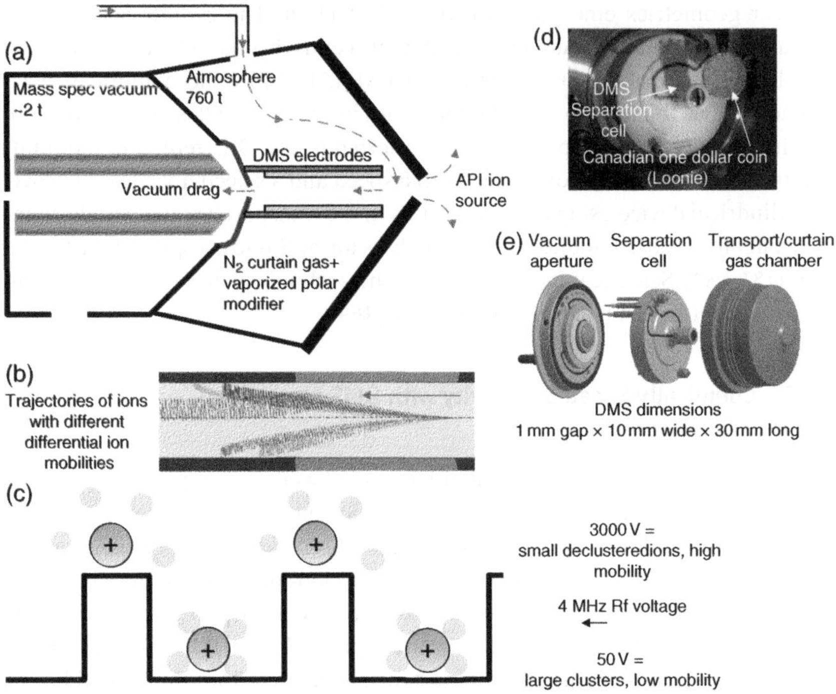

Figure 1.15 Illustrations of a differential ion mobility analyzer coupled to a mass spectrometer and its mechanism for separations. (a) The MS vacuum draws from the atmospheric pressure ion source the ions in the N_2 transport gas doped with polar modifiers. The DMS cell is 3 cm long. (b) Under the influence of an electric field oscillating from 0 to 4000 V at 4 MHz frequency the ions follow a saw-toothed trajectory and drift toward one or the other electrode depending on the difference between their low field (<100 V) and high field (>1000 V) mobility. If their high and low field mobilities were the same, they would travel on the central axis. Ions of a particular differential ion mobility (alpha) are steered back on axis into the MS with a compensating DC voltage (compensation voltage) producing a continuous beam of the selected ions entering the MS. (c) When the N_2 transport gas contains a few % vaporized polar solvents or water, the ions and solvent aggregate into clusters of specific number and size depending on the chemical affinity of the analyte ion with the polar solvent molecule. The cluster number and its orientation are largely a function of the unique dipole moment of a particular ion. The clusters are stripped off then reformed again during each cycle of the RF waveform thousands of times as they transit the cell. Since ions in the clustered low voltage state have a much lower mobility than when stripped of solvent at high voltage, the differential mobility is amplified as a function of this chemical affinity specific for that analyte. (d) Photograph of the entrance to the DMS cell in relationship to a Canadian one-dollar coin (Loonie) showing the 1 mm gap between the two planar surfaces defining the ion path. (e) Assembly of the mobility cell sandwiched between the vacuum aperture on the left and curtain gas confining plate on the right.

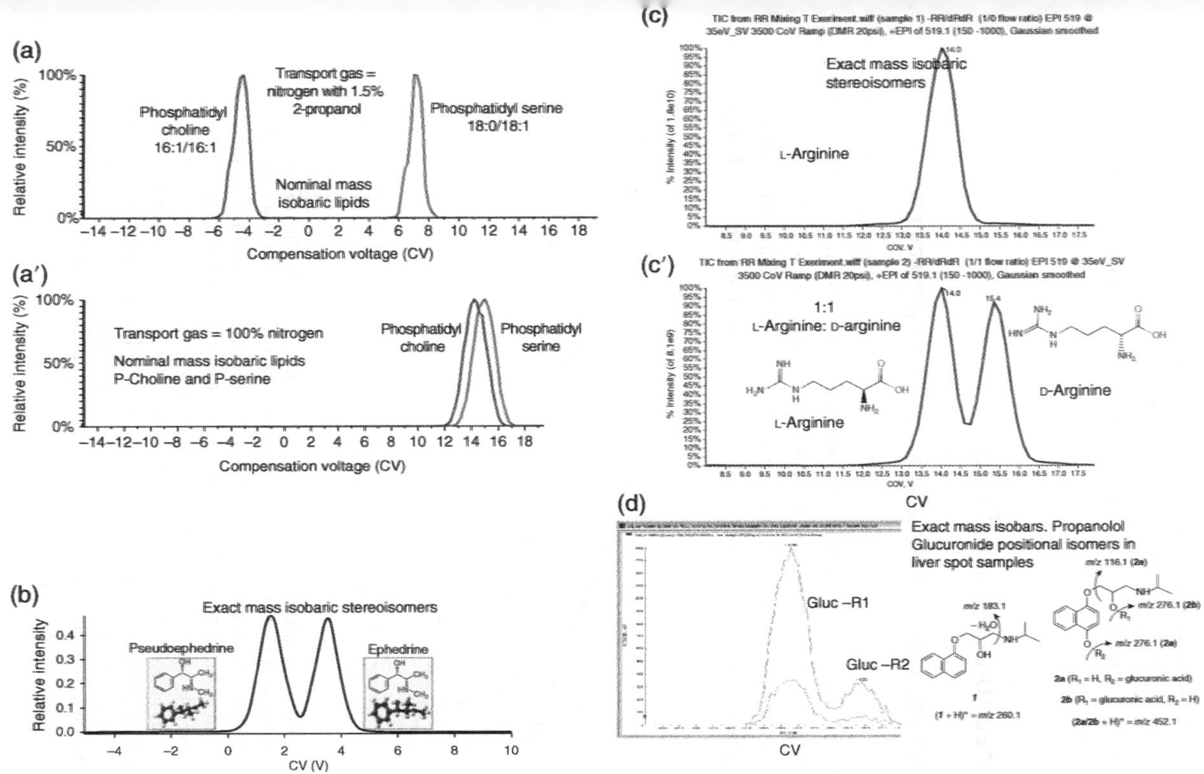

Figure 1.16 Ionograms created by scanning the CV voltage during a continuous infusion of a sample to demonstrate separation capability. Isopropanol modifier added to the N_2 transport gas used in all cases except a′. (a and a′) The separation of two nominal mass isobaric lipids with and without isopropanol modifier to demonstrate the selectivity of the separation when chemical interactions are utilized. (b) Separation of the stereoisomers ephedrine and pseudoephedrine, using isopropanol as the clustering modifier, that have the same elemental composition, identical masses, and identical product ion spectra as a demonstration of a separation that cannot be done by high-resolution tandem mass spectrometry. (c and c′) Separation of the d and l stereoisomers of arginine without (c) and with (c′) clustering modifier, for a similar demonstration as in b. (d) Separation of the glucuronide conjugate positional isomers using propranolol clustering modifier. The sample is from liver of a dosed rat using the surface sampling probe (LESA) as a demonstration of its utility for metabolism studies. *Source:* Figure adapted from data in Parson et al. [189].

second per sample API systems for HT-MS. High-speed sampling systems, where HPLC is not an option, provide the most compelling scenario for the utilization of high-speed gas-phase separations such as ion mobility provides.

Although ion mobility is faster than HPLC, it does not address one of its critical functions, that being the reduction of ionization suppression with either ESI or APCI, resulting from excessive amounts of sample matrix mixed with the sample during the ionization process. This is because separation occurs post ionization with ion mobility. Developing a means to address the suppression issue as well as tackle the remaining bottlenecks to HT-MS are components of the discussion in the final two sections.

1.7 Mass Spectrometer Sensitivity

1.7.1 Historical Gains and Motivation for Sensitivity Improvements

The sensitivity of API mass spectrometers has been improving over the past 40 years accruing ~6 orders of magnitude gains since the first systems of the late 1970s [190]. Figure 1.17 shows the accumulation of those gains from the first manufacturer of dedicated API-based mass spectrometers. It needs to be pointed out that similar graphs could be constructed by other manufacturers, although

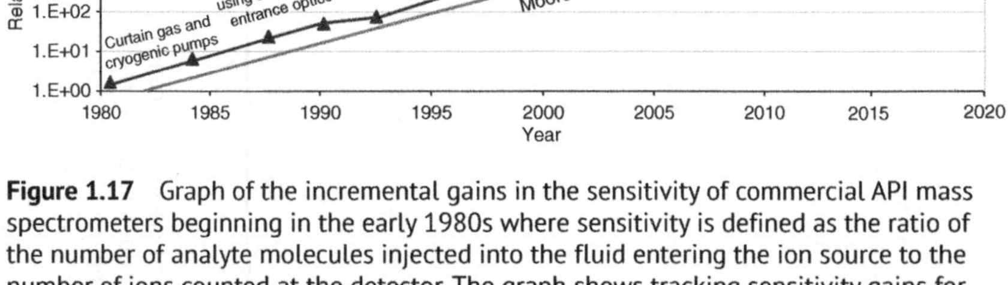

Figure 1.17 Graph of the incremental gains in the sensitivity of commercial API mass spectrometers beginning in the early 1980s where sensitivity is defined as the ratio of the number of analyte molecules injected into the fluid entering the ion source to the number of ions counted at the detector. The graph shows tracking sensitivity gains for fluid introduction flow rates relevant to conventional HPLC and HT-MS applications (>10 μL/min) and do not apply to NanoESI in all cases. How the sensitivity gain was achieved is indicated on the graph at particular points in time. The improvement ~10^5 to 10^6 has challenged the gains predicted by Moore's law for biannual increases in the speed of microelectronics. This development has been a key enabler of today's HT-MS systems. *Source:* Adapted from Covey [190].

not going as far back in time, but would share in common the areas and degrees to which these efficiencies were improved. The product of these technical advances qualifies as a major contributor to making HT-MS a reality today.

The motivation to improve sensitivity during this time has been as much about achieving improvements to throughput as it has been for lowering LOQ's. It improves throughput because the sample volumes required to achieve biologically relevant LOQ's have been proportionally reduced which simplifies and generalizes sample preparation, enables the use of low-volume high-speed sample dispensers, reduces MS contamination when subjected to high sample loads, and enables a means to control ionization suppression by applying large dilution factors to the sample. In the 1980s, one mL of plasma, suitably extracted and concentrated, was required to achieve LOQ's in the ng/mL or nM range. Today, approximately 10^5–10^6 less sample is required to do that, i.e. low nL quantities, which can be directly injected, provided sufficient dilution is used to manage ionization suppression.

Studies have indicated that dilution factors of 1000–10,000 will reduce suppression effects to manageable levels in biological fluids [191, 192]. Why dilution would work can be rationalized from the Ion Evaporation Theory for ion production by ESI. When all sites on the surface of the ion emitting droplets are occupied with matrix molecules, the surface field desorption of analyte ions is prevented. Dilution makes available to the analyte surface desorption sites [193]. Otherwise, to eliminate suppression, protein precipitation methods and other forms of time-consuming sample purification techniques coupled with separations prior to ionization are typically utilized [194–196].

Sensitivity is defined as the ratio of the number of analyte molecules injected into the fluid stream entering the ion source to the number of analyte ions detected. Advances in ESI efficiency, the efficiency of transporting ions from atmosphere-to-vacuum (transport efficiency), the confinement of the ion beam in the gas expansion immediately upon entering vacuum (focusing efficiency), and the confinement of the ion beam in the collision cell have been the major contributors to improvements to sensitivity and LOQ's. The specifics of the technologies responsible for improving the efficiencies indicated in Figure 1.17 have been described in detail elsewhere [26, 27] and most recently explained in a perspective paper titled "Where Have All the Ions Gone, Long Time Passing" in a special issue of Rapid Communications in Mass Spectrometry dedicated to the 40th anniversary of the triple quadrupole MS [190].

Today, from ~20 molecules (33 yoctomoles of reserpine) injected into the ion source, one ion will reach the detector with nanoESI and the same for 200 molecules (330 yoctomoles) injected at conventional LC flow rates bearing in mind that LOQ's will be much higher than this, in the low attomole range, due to chemical noise and ion statistics limitations. Nevertheless, the efficiency of ion production, transmission, and detection is currently approaching the practical maximum with only about a factor of 10-fold left to be mined in the future [190]. According to Avogadro what lies beyond yoctomoles is nomormoles (😉).

Of course, this only applies to compounds that have physical–chemical properties resulting in near 100% ionization efficiency (electrospores) which is known to occur with both Nano and conventional flow ESI [190]. Improvements to the ESI efficiency of poor ionizers (electronors) have largely come from altering their physical–chemical properties by derivatization to improve their surface activity and their ability to be charged in solution or adduct another charged ion such as ammonium, sodium, lithium, formate, or acetate [197–201]. An example of this is shown in Figure 1.18a, b where the ionization efficiency of a poor ionizer like testosterone can be improved 40-fold with derivatization [202]. A 1000-fold difference in LOQ's between an ideal electrospore and electronore is shown in Figure 1.18a, c. However, with regards to high-throughput techniques, derivatization complicates the method and is avoided in situations where high speed is desired.

Although the extraordinary gains in sensitivity for API mass spectrometers of the past 40 years will not be repeated in the future, they have made possible a reduction in the volume of sample required to achieve biologically relevant single-digit nM or ng/mL LOQ's by 10^5–10^6. Thus, for many compounds the use of nanoliter volume samples without the use of pre-concentration sample preparation methods is feasible today. As shown in Figure 1.18b, c, a 2.5 nL acoustic dispense into a stream of 450 μL/min of methanol will achieve this for compounds with high and intermediate ionization efficiency. As shown in Figure 1.18c, for poor ionizers, sub μM is achievable and an improvement of about 10× in LOQ's will be required to bring the worst ionizers into the lower ranges of biologically relevant concentrations (low nM) for HT-MS techniques that utilize small volume injections. The use of low-volume contact-free dispensers for HT-MS is the topic of the next and final section.

1.8 High-Speed Sub-Microliter Volume Sampling

1.8.1 Small Sample Size and Low Volume Dispensing HT-MS Technologies

The development of means to dispense small volumes of fluids in the picoliter to nanoliter range for various industrial, consumer, and biomedical research applications has been ongoing for the past 50 years [203]. For example, the use of acoustic energy pulses to dispense low nanoliter volume droplets at speeds in excess of 1 Hz has emerged as a useful tool for managing large pharmaceutical compound libraries [204]. But the utility of low volume dispensers as a means for

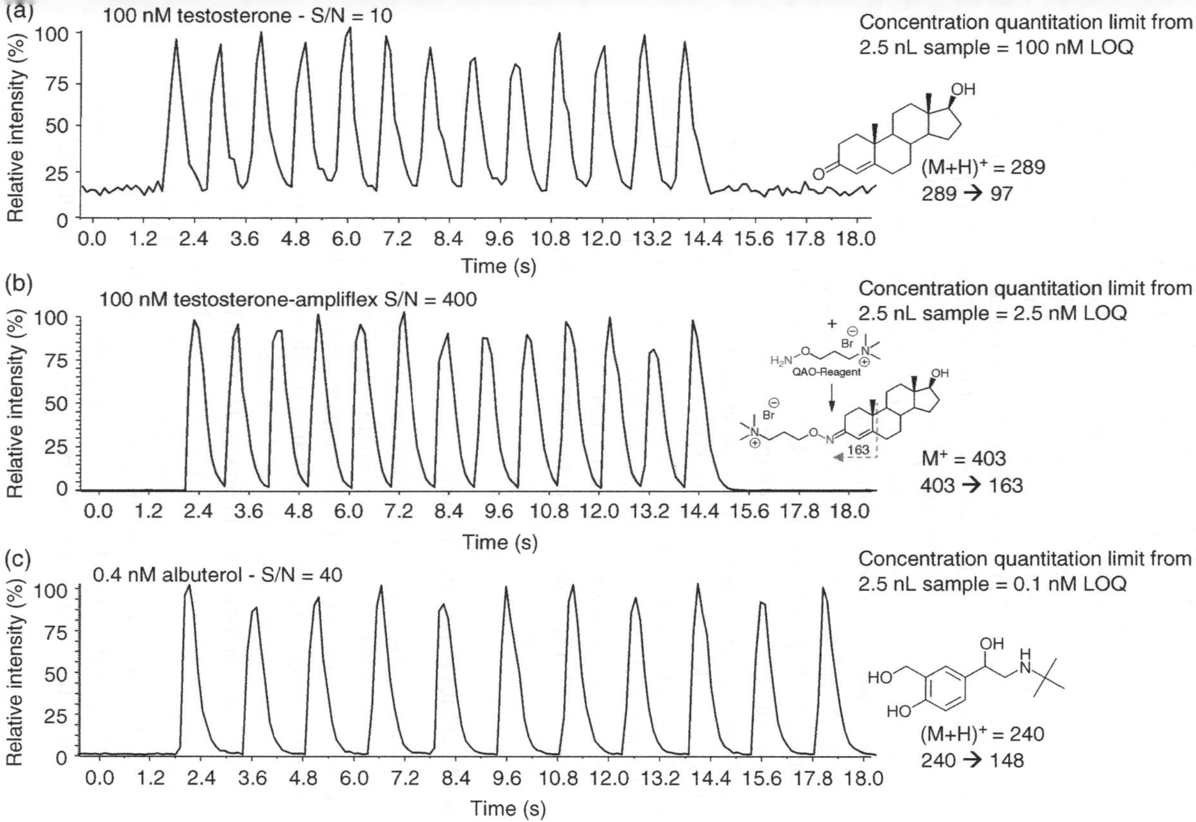

Figure 1.18 Data from three compounds of varying ionization efficiency using Acoustic Ejection Mass Spectrometry (AEMS) as the means for sample introduction into a standard ESI ion source. Each peak is from a 2.5 nL acoustic dispense at one sample per second into the OPI operating at 450 μL/min using 100% methanol as transport fluid. The limit of quantitation for the three compounds is compared and shown to cover a range of 10^3.
(a) Testosterone a poor ionizer. (b) Testosterone derivatized with a quaternary amine (Ampliflex). (c) Albuterol a known electrospore. The differences in LOQ shown are primarily due to the differences in their ionization efficiency because in all three cases the product ion monitored constituted >50% of the total precursor ion current and the background noise is similar. Relative ratio of LOQ's: Albuterol 1/Testosterone-Ampliflex 25/Testosterone 1000.

directly injecting biological fluids and assay buffers at high speed for mass spectrometry analysis has been hampered by a lack of sensitivity until recent years. In the past decade, a variety of high-throughput techniques achieving 1–30 Hz sample injection rates have emerged using lasers to ablate solid samples crystalized in a UV adsorbing matrix [13, 81, 82, 167, 168] (MALDI using chemical ionization), IR (IR MALDESI) [205, 206], and UV (liquid atmospheric pressure MALDI = LAP MALDI) [207, 208] lasers to vaporize liquid samples using electrospray-like (ion evaporation) ionization, microfluidic-based sample stacking strategies combined with electrospray [209], and utilizing acoustic energy to dispense low volumes of liquid samples. With regards to the later, two different approaches take advantage of acoustic sound waves to dispense samples. Acoustically created picoliter droplets in the form of a mist [210, 211] can be inductively charged sufficiently to create ions by the ion evaporation mechanism of electrospray. Chapter 5 of this book [16] provides more detail on the applications of the picoliter MIST approach. Single, discrete nanoliter volume droplets [125, 126] can also be acoustically dispensed and segregated in a flowing liquid for conventional pneumatically assisted electrospray also further covered in Chapter 5 of this book [16].

To date, only the MALDI and nanoliter acoustic dispensing approaches have been commercialized. Chapter 12 of this book [13] provides more detail on the MALDI approach. The remainder of this chapter focuses on the development of the commercial acoustically generated nanoliter droplet approach otherwise referred to as acoustic ejection MS (AEMS). With the sensitivity of modern-day triple quadrupole instruments, nanomolar LOQ's can now be achieved by directly injecting 2.5 nL volumes of drug discovery assays, with no sample preparation, at speeds exceeding 1 Hz. This has enabled routine daily sample throughputs on the order of 70,000 per day [212] of various pharmaceutical drug discovery HTS and ADME assays [6, 7, 213–220].

1.8.2 Shoot N' Dilute Nanoliter Droplets

Acoustic dispensers in commercial plate replicators typically have their energy and frequency calibrated to deliver low nanoliter volume droplets (~150–200 μm diameters) that are large in comparison to picoliter droplets, visible to the eye, have substantial momentum, straight trajectories, and are reproducible in volume. At the 2015 ASMS conference in St. Louis, a video of this process was shown by Labcyte. As a result, a concept was proposed at that time by Don Arnold, Chang Liu, and the author of this chapter to capture and transport these droplets to a standard Ion Spray ion source with the OPSI device developed by Gary Van Berkel [105, 106].

A diagram and photograph of the first breadboard of an acoustic ejection mass spectrometer (AEMS) is shown in Figure 1.19a, b, code named the Millenium Falcon echoing the words of Hans Solo "She may not look like much but she's got

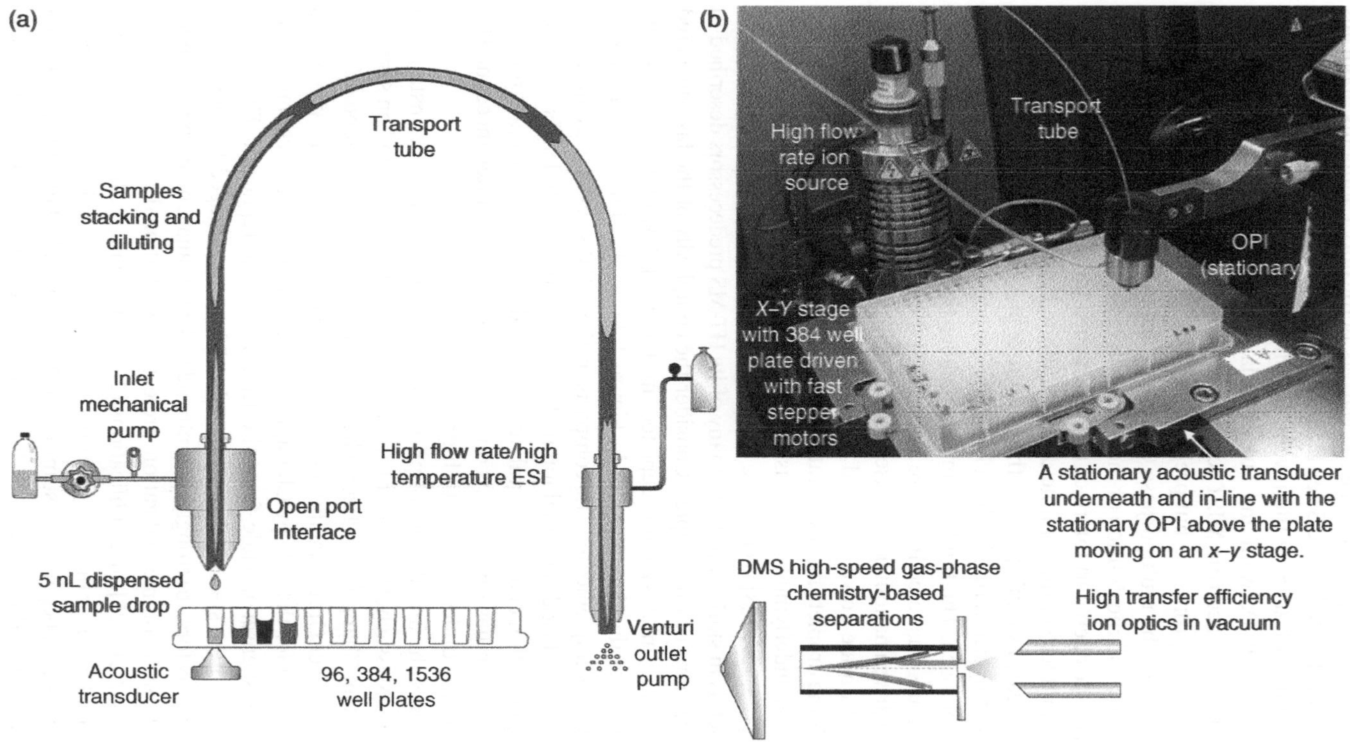

Figure 1.19 (a) Diagram of a prototype Acoustic Ejection Mass Spectrometry system (AEMS) designed for low nanoliter sample dispensing and a 1 Hz sampling speed equipped with differential ion mobility for isobar separations. (b) Photograph of research breadboard used to provide the initial proof-of-principle of the AEMS technology concept code named the "Millennium Falcon."

it where it counts, kid!" after Luke Skywalkers derogatory comments about his craft. It was constructed at the Sciex facilities in Concord and tested in Redwood City in December of 2015 by the inventors using a Labcyte acoustic transducer provided by Sammy Datwani and Luke Ghislain who were also involved in the development of the acoustic MIST technology of Astra Zeneca and Waters. The initial results were promising but preliminary.

Optimization of the multipurpose OPSI interface to increase both the speed of droplet transfer and the dilution of the sample to reduce suppression effects led to prototypes of the open port interface (OPI) designed for the expressed purpose of capturing nanoliter volume sample droplets, dilute them >1000-fold, and transport the segregated sample plugs to the ion source at 1 Hz and faster. This process of online diluting was termed "Shoot N' Dilute" as opposed to the classical "Dilute N' Shoot" approach. The fluid dynamic principles that were considered to make this possible are illustrated in Figure 1.20a [125]. Figure 1.20b is a diagram of an acoustically driven nanoliter dispenser used to deliver the sample to the OPI when it is oriented upside down. Focused acoustic sound waves deliver reproducible 1–10 nL volume droplets from separate wells at up to 6 Hz [126, 212].

The physics underlying both acoustic dispensing [204] and the fluid dynamics of the OPI [125, 127, 128] is complex. But the mechanical and fluidic simplicity of this approach is in stark contrast to many of its HT-MS predecessors described in the earlier sections of this chapter. Consider the complexity of the fast-staggered chromatographic inlet systems depicted in Figure 1.7 where the plumbing required to create the flow diagram in Figure 1.7b invokes images of a "snakes wedding" when implemented. In comparison the AEMS system, as shown in Figure 1.19a, with fast differential ion mobility separations included, is an exemplar of simplicity that enables several different modes of HT-MS operation [126]. This simplification has implications on ease of operation, maintenance, and stability particularly under continuous high sample loads.

One year after the initial proof-of-principle, a second prototype was installed in Framingham, Massachusetts, in November of 2016, which was tested on samples provided by Hui Zhang from Pfizer. The results compelled the breadboard to be moved to the Pfizer Groton facility in June 2017 where two additional prototypes were constructed for that site by Chang Liu and the author of this chapter using the acoustic dispenser module from the EDC ATS plate replicator. The results emerging from these systems during 2017 led Pfizer, encouraged by their scientific board of directors, to strongly recommend to Sciex management in September they embark on a commercialization project. The additional results, presentations [123, 124, 221–225], and publications [213, 214] to emerge from the multiple prototypes at Pfizer over the next year provided the evidence to initiate a commercialization project to begin in the summer of 2018 eventually launching two years later in May 2020. In the spring of 2019, both Labcyte and EDC were purchased by

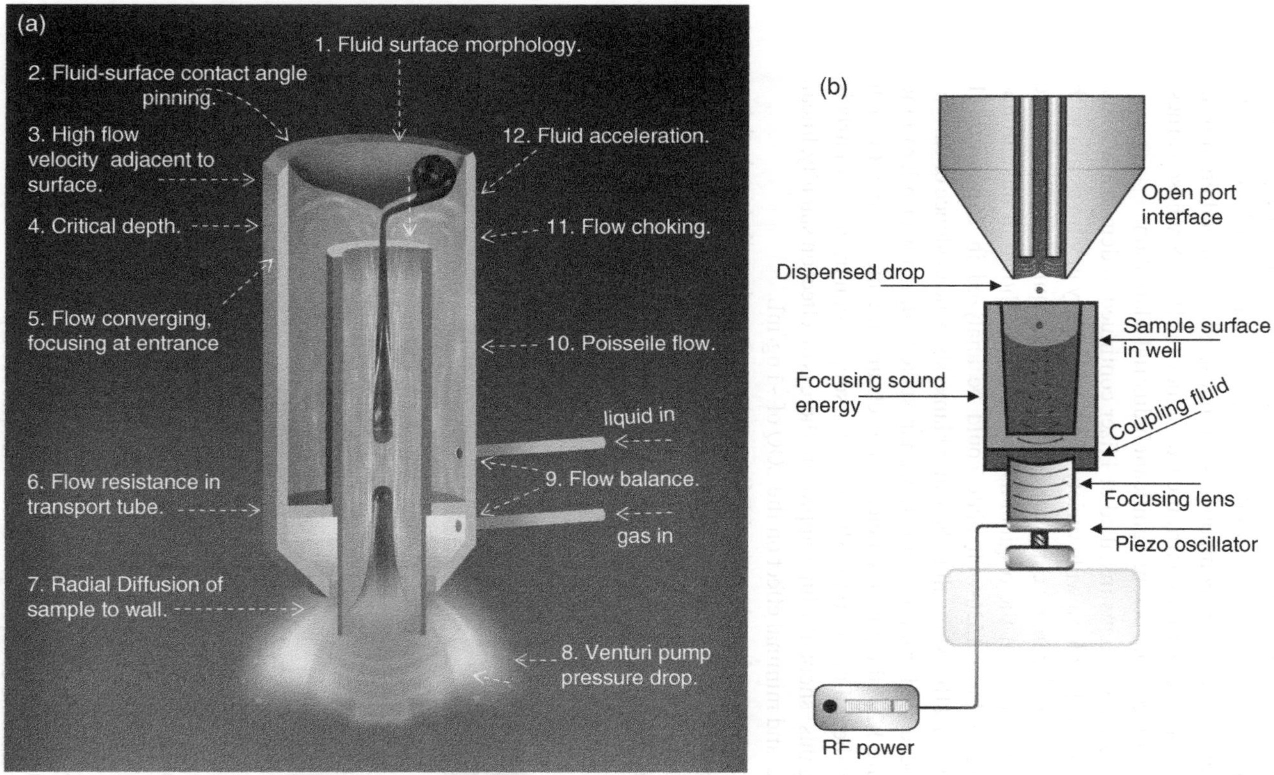

Figure 1.20 (a) Axial cross-section of the OPI interface operating under conditions for most effective capture, dilution, and transport of nanoliter volume droplets. Shown here in the "up position," it is rotated 180° to the "down position" for acoustic sampling that fires droplets upward against gravity. Twelve important elements of the fluid dynamics of the system that contribute to this are identified. The length of the transport tube is altered at position #6 to achieve different speeds. *Source:* Liu et al. [125] / Reproduced with permission from American Chemical Society. (b) Illustration of the acoustic transducer delivering sound energy through the bottom of a microtiter plate sample well focused on the surface meniscus to eject 2.5 nL droplets into the OPI. *Source:* Reproduced with permission from Liu et al. [126] / American Chemical Society.

Beckman, a subsidiary of Danaher Corporation, who now manufacture the acoustic component of the AEMS system.

Although it may seem that this technology moved quickly from concept to commercialization, the research leading to it began 20 years earlier for both the OPI fluidic and acoustic dispensing components. In that sense, it is no different from the previous technologies that contributed to achieving HT-MS whose earliest inceptions were 10–20 years before significant industrial adaptation.

The three prototype systems installed at Pfizer continued to demonstrate sustainable operation at 1 Hz for a wide variety of drug discovery in vitro assays with single-digit nM LOQ's for many compounds without the requirement for sample preparation or alteration of the original assay buffers. As shown in the chronogram in Figure 1.21a, a 1536-well plate could be analyzed in 26 minutes (1 sample/s). Figure 1.21b, c show that the water blanks show no evidence of carryover. The data generated from dextromethorphan-spiked human liver microsome matrix in Figure 1.22 illustrate the detection levels and the management of ion suppression by the online dilution. Although 30% suppression from this matrix is observed using this "shoot n' dilute" approach, there is no effect on stability, linear dynamic range, and minimal effect on the LOQ of ~1 ng/mL.

Additional prototypes based on the EDC transducer were installed in the laboratory of Wilson Shou at Bristol Meyers Squib, Princeton, NJ, and Andreas Luippold at Boehringer Ingelheim, Germany, in 2019, as well as prototypes based on the Labcyte transducer at Kevin Bateman and Dave McLaren's labs at Merck

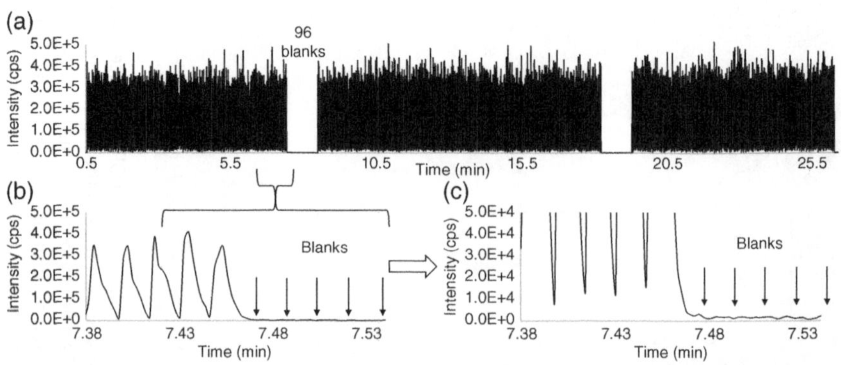

Figure 1.21 (a) Sonogram of 1536 samples dispensed as one 5 nL drop per sample at 1 Hz. A single concentration of 100 ng/mL of propranolol was used to calculate the CV for the absolute peak areas. CV = 7.8%. Gaps at 7.5 and 18 minutes were two rows of 96 water blanks to illustrate lack of carryover effects. (b) Expansion of *y*-axis between 7.38 and 7.54 minutes to illustrate peak shape and height reproducibility. (c) Expansion of *y*-axis and *x*-axis (10×) in the region between 7.38 and 7.54 minutes to illustrate the lack of carryover signal. All data is unsmoothed. MRM dwell time = 30 ms. *Source:* Reproduced with permission from Liu et al. [126]/American Chemical Society.

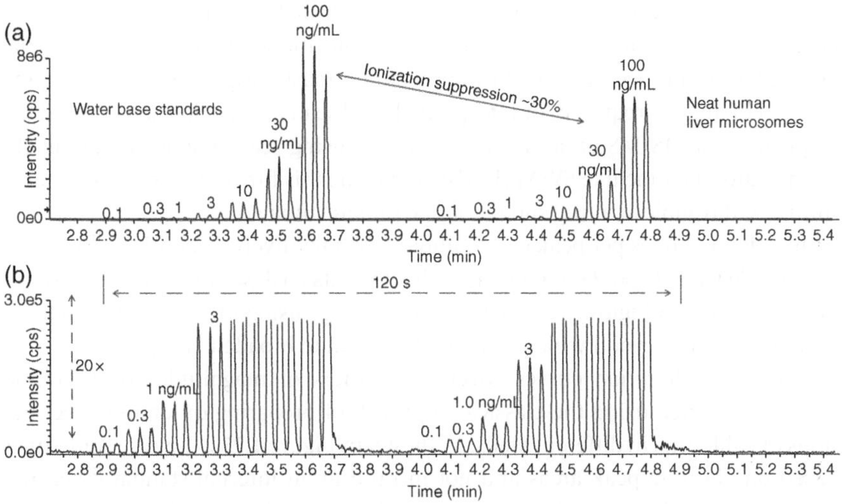

Figure 1.22 (a) Data illustrating limits of quantitation and suppression effects for dextromethorphan in a neat liver human microsome assay buffer system. (b) Expansion of "*y*-axis" of a to reveal S/N at the low-level injections. Although 30% suppression is observed, there is no effect on stability, linear dynamic range, and minimal effect on LOQ. Sampling rate = 1.2 seconds per sample. Assay buffer spiked with Dextromethophan (MRM 272 → 216, 20 ms dwell) using D3-Dextromethorphan as an internal standard (275 → 215). Transport flow = methanol at 475 μL/min.

the same year. These systems validated a variety of applications for AEMS in the HT-MS drug discovery setting including ADME and HTS [215, 216]. Acoustic dispensing samples from a 2-phase fluid as a means for direct liquid–liquid extraction was shown as a component of an HTS assay method [217]. Its use for quality control of libraries and nanoscale reaction synthesis optimization has been shown using high-resolution, accurate mass TOF analyzers instead of triple quads [218]. Nanoscale reaction optimization in 1536-well plates is of increasing importance in the drug discovery arena [226] and this technique appears to be well suited to address those needs and the needs of the synthetic biology industry, which require direct monitoring of large numbers of fermentation media for the production of the desired products [227, 228].

Outside the field of drug discovery, AEMS has been implemented in agriculture as the detector for an automated system designed to determine the sex of chicken eggs before they hatch at a rate of >50,000 per day [229]. This is part of a program to eliminate the culling of male chicks after birth.

New modes of operation have emerged from the work of these early adapters of the technologies [126]. As shown in Figure 1.23b, d, repetitive dispenses from the same well can be done at a frequency faster than the OPI purges, typically

10–20 Hz, to produce an on-demand and controllable continuous signal that consumes sample at a rate of ~1 μL/min. Applications of this mode are signal averaging to improve ion statistics for accurate mass assignments, lowering the amount of analyte required for interpretable full scan product ion spectra, and the acquisition of MS/MS on all ions in a sample using scan functions that benefit from additional time like SWATH. Also, the value this function has for automated tuning and troubleshooting should not be underestimated.

Injection volumes per peak can be altered by repetitively dispensing from a single sample well at 100 Hz or greater, so the droplets coalesce in the OPI as a single peak as shown in Figure 1.23a. The dynamic range using this method is approximately 10 drops beyond which peak widths begin to broaden [126].

Although calibration curves covering 3 orders of magnitude concentration range are created using the classical approach of preparing different concentrations of calibrants in different wells, as shown in Figure 1.23c, they are sometimes based on absolute peak areas and not in ratio to an internal standard, because peak area variations introduced from sample preparation have been eliminated. However, the big difference over the classical LC/MS approach is that now these calibration curves can be constructed in 10–20 seconds using sampling speeds of 1 Hz [126].

The use of ESI for monitoring reaction kinetics is another example of an application that was demonstrated 30 years ago but now presenting itself as an industrially useful method with this new technology. In 1989, the ability to continuously monitor chemical reactions and calculate K_m values was demonstrated [230]. And a level of deep understanding of the mechanism of inhibition of human leucocyte elastase was elucidated [231, 232].

With AEMS, reaction kinetics can be determined from a single reaction well without the need for quenching the reactions at different time points in different wells. This is because only a few nanoliters of the sample are required to be dispensed from the bulk reaction without sample preparation intervening. The reaction stops when it enters the OPI containing a transport fluid with a high concentration of methanol [126]. Large arrays of reactions in microtiter plates can be simultaneously monitored offering an element of HT-MS to the enzyme kineticist. Many of these applications and more are highlighted in Chapter 5 of this book [16].

The speed is not limited to one sample per second using the OPI approach for nanoliter droplet capture and transfer. An OPI interface designed to segregate samples delivered to it at 1 Hz can achieve 3 Hz using a familiar method in mass spectrometry referred to as multiplexing speeds can be increased. The multiplexing method requires that a different compound with different MRM transitions is being assayed in adjacent wells. For 3× multiplexing as shown in Figure 1.24a three different compounds having different MRM transitions are monitored, along with their

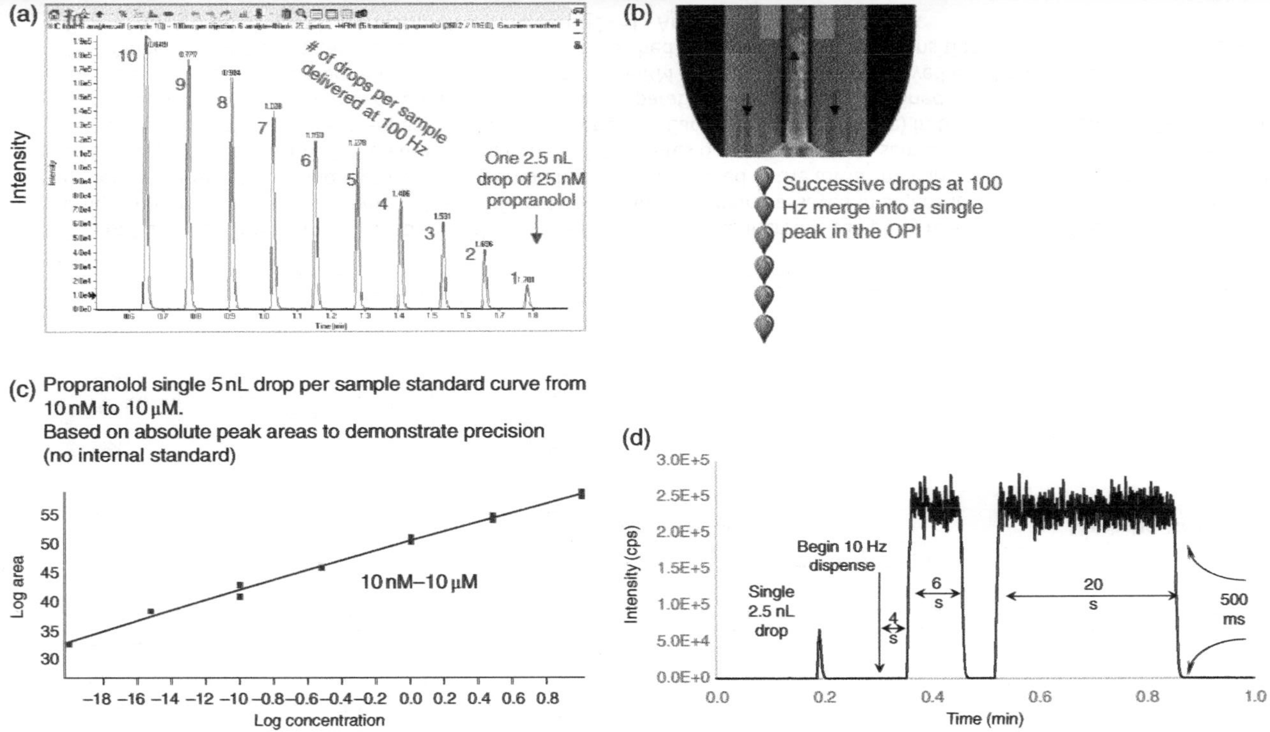

Figure 1.23 Different modes of operating AEMS. (a) Data illustrating the ability to adjust the injection volume 10-fold by injecting sequential drops at 100 Hz and maintain the same peak widths. At this speed, successive drops merge into a single peak in the OPI. The dynamic range of this method is approximately 10 drops after which peaks begin to broaden. (b) Drawing of the OPI capturing rapidly dispensed drops merging into a single peak. (c) Calibration curve for propranolol from single 5 nL drop injections of 10 nM–10 μM generated from absolute peak areas without internal standard correction. Each calibration point was created from a single drop injection from a different well having a different concentration as is typically done for conventional LC/MS methods. (d) Infusion mode of operation. Drops are injected at a rate of 10 drops per second from a single well to create a continuous stream of analyte in the OPI transfer line. Time from initiation of dispensing to appearance of signal = transit time through OPI = 4 seconds. Time for signal to rise and fall from baseline to maximum signal = 500 ms. Sample consumption rate based on a 2.5 nL droplet at 10 Hz = 25 nL/s = 1.5 μL/min. Data is unsmoothed. *Source:* Liu et al. [126]/Reproduced with permission from American Chemical Society.

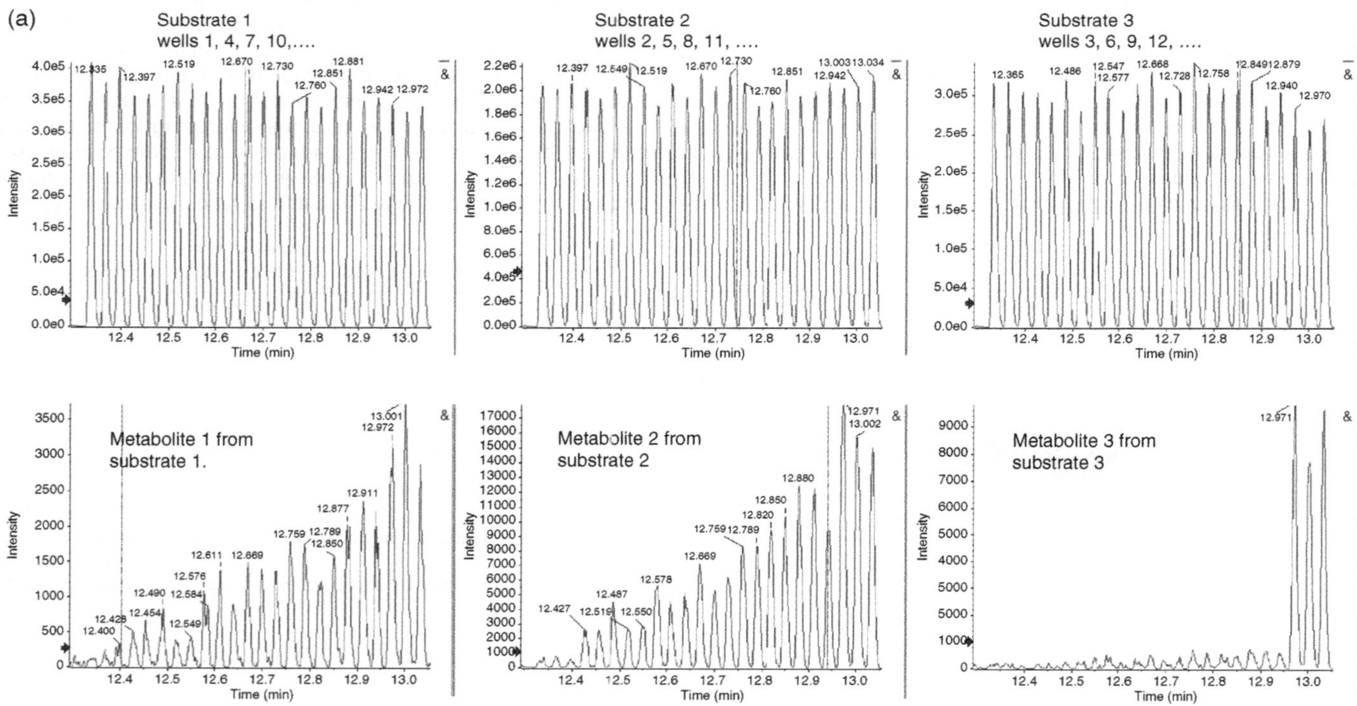

Figure 1.24 (a) Chronograms illustrating the multiplexing method to increase sampling rate from 1 to 3 Hz for three cytochrome P450 drug–drug interaction assays where three different compounds are being assayed in adjacent wells. Signal from the drug substrates is in the upper ion current traces and that from the metabolites being produced in the lower ion current traces. Data from one row of samples of a 384-well plate is shown with the entire plate done in 2.1 minutes operating at three samples per second. The parent drug concentration is 1 μM. The metabolites begin to appear at ~2 nM concentration. Each time point (sample) in the assay was obtained by quenching the P450 enzymatic reaction with methanol, but no other sample preparation was used. CVs obtained for this assay were 8%. (b) The multiplexing method requires that a different compound with different MRM transitions is being assayed in adjacent wells. 3× multiplexing is shown here where the peaks overlap, but the signals are distinguished by the different MRM transitions from the different compounds. *Source:* (b) Reproduced with permission from Liu et al. [126] / American Chemical Society.

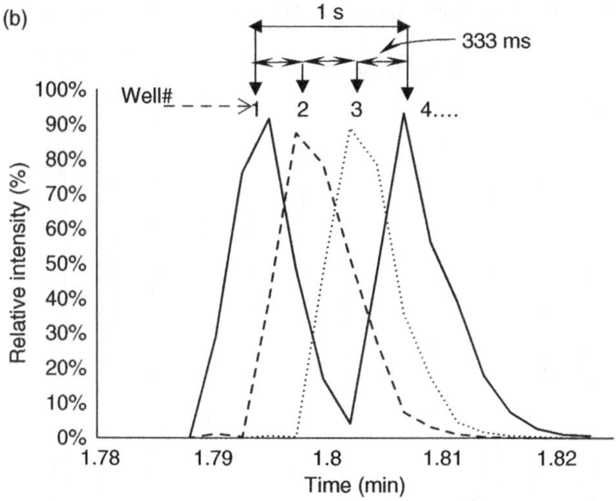

Figure 1.24 (Continued)

metabolites, in adjacent wells. With the stages moving at 0.33 seconds per well, the peaks overlap, but their different MRM transitions allow them to be independently integrated as shown in Figure 1.24b. For more details see reference [126].

It was soon demonstrated that, with some minor modification to the OPI transport tube dimensions, three samples per second could be achieved without multiplexing and 6 Hz with multiplexing [212]. Additional modifications to the OPI to increase speed were implemented based on the fluid dynamic principles described in Figure 1.20a. As shown in Figure 1.25, the peak widths at baseline are 68 ms which would accommodate ~15–20 samples per second, provided a droplet dispenser could keep pace [125]. A fast pneumatically driven nanoliter dispenser was used in this case operating at 10 Hz. Current manifestations of acoustic dispensers are limited to 3–6 Hz for reproducible dispensing, because each sample well needs to be calibrated for fluid depth before droplet launch [126]. When the acoustic dispense speed restrictions are overcome, at these speeds, the calibration curve constructed in 20 seconds at 1 Hz mentioned earlier could conceivably be constructed in little more than one second.

1.9 Conclusions and Future Prospects

In the 1980s, during the early development of LC/MS, the daily ritual upon entering the laboratory was to "Spray n' Pray" in desperate hope of achieving a single successful LC/MS run that day. One consequence of the explosion of research

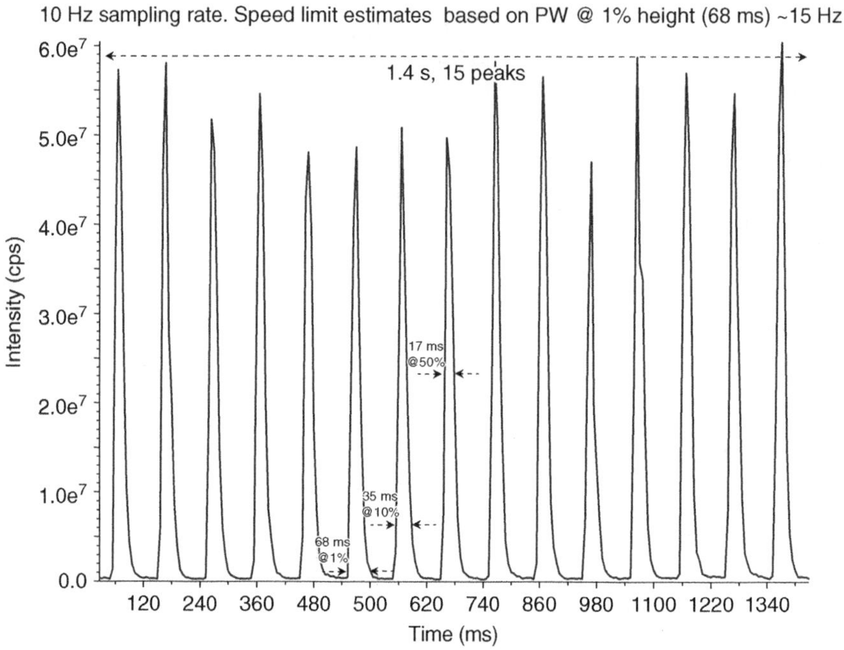

Figure 1.25 Chronogram using an OPI interface modified to obtain maximum speed and with 5 nL droplet introduction using a fast pneumatically driven nanoliter dispenser operating at 10 Hz. A low resistance transfer tube and low viscosity transport solvent (acetonitrile) were used to maximize the fluid linear velocity. Based on the separation of the peaks at baseline, this demonstrates the feasibility of HT-MS speeds of 15–20 Hz. Compound monitored by MRM was DEET with the transport flow rate 2.5 mL/min. *Source:* Reproduced with permission from Liu et al. [125] / American Chemical Society.

activities into developing LC/MS during that decade was the emergence of API that set the stage for daily sample throughputs to increase into the hundreds of thousands per day by 2022.

The improvements to API MS sensitivity have been of a similar magnitude to the throughput gains over these 40 years. The cumulative gains in MS sensitivity has enabled the reduction of sample volumes, eventually reaching the point in the second decade of the twenty-first century, where the analysis of nanoliter volume samples dispensed contact free at high speed became possible while maintaining biologically relevant LOQ's. This has important implications for addressing many of the bottlenecks in the overall HT-MS workflow including sample preparation, eliminating time-consuming cross-contamination issues, minimizing instrument contamination, enhancing reliability, and enabling the leveraging of the sophisticated laboratory robotics and automation equipment that has been developed around microtiter plate formats. A likely consequence could be a sea-change in the field of HTS toward the use of mass spectrometry as the field shifts toward the development of label-free assays.

The development of HT-MS technologies has been an evolutionary process where many attempts never made it to successful commercialization as their limitations were revealed during development or the requirements rapidly evolved beyond their initial capabilities. Both the successes and failures played a role in setting new directions from the lessons learned. Of those that have succeeded to survive to the present day, it is a common theme that it takes up to 10 years or longer to enter the industrial sphere from their first proof-of-principle. Innovation is as much about failing as it is about succeeding, and as much about persistence as it is about inspiration, as illustrated in the story behind the development of LC/MS with which this chapter began.

The HT-MS field was initiated by the development of high-speed column technology, construction of fast autosamplers, and methods and software for staggering fast separations in the late 1990s and early 2000s driving speeds well under the one minute per sample mark. The one sample per second target, established to meet the growing demands of the drug discovery process, was first achieved in 2005 with MALDI using high repetition rate lasers. At the same time, millisecond separation speeds using ion mobility technology were being explored as an alternative to HPLC and in some manifestations could mimic the chemically based separation mechanisms of chromatography that are orthogonal to separations based on mass.

The drug discovery process is largely responsible for catalyzing these developments which are now beginning to have impact in areas outside HTS and ADME. Parallel synthesis and reaction condition optimization for drug discovery is one example. Its potential extends far beyond that, for example synthetic biology, which utilizes the advances in genetic technologies to induce rapid evolutionary change in microorganisms toward the production of industrially useful materials, an area in dire need of the most recent advances in HT-MS. Agriculture and food production are being impacted by HT-MS with one example being the rapid sexing of poultry eggs having the goal of screening 1.3 trillion eggs per year for their sex by mass spectrometry. The implementation of HT-MS into the field of human disease diagnosis is only being gated by the validation of reliable biomarkers which is on the rise [233]. Analysis speeds are now in place to accommodate these needs.

It is thought provoking to consider the implications of the speeds being shown feasible for solving problems beyond just the analysis of more samples per day. Twenty samples per second is reasonably within reach which begs the question of how to effectively leverage these speeds in the design and control of the instrumentation itself. For example, the creation of a quantitation calibration curve covering 3 orders of magnitude in concentration, fully equipped with replicates and

QC's, would take less than a second enabling detailed self-checking and troubleshooting of both the instrumentation and the method triggering corrective action to be automatically implemented. One example would be the detection of excessive ionization suppression in samples of unexpected composition and the correction of the measurement by the immediate repeat of the analysis using a higher dilution ratio which could be achieved by a variety of automated online means.

There remain hurdles to overcome. Sensitivity improvements are key as they would expand the number of compounds measurable with HT-MS at biologically relevant concentrations in the sub to low nM range. As the practical limits to sensitivity and LOQ's are being approached, it is becoming increasingly difficult to achieve gains. The 10^6 gains experienced over the past 40 years will not be repeated but lowering chemical noise by implementing new techniques to improve selectivity and harnessing what sensitivity gains are left could yield further improvements in LOQ's by 10–30-fold.

Improvements to high-speed dispensers will also be important. Techniques to transfer samples into an ESI ion source, such as the OPI has demonstrated, have exceeded the speed of current-day acoustic dispensers by ~10-fold imposing limits on throughputs. Other techniques are in the works such as using electrostatic, pneumatic, piezoelectric, or laser-based liquid dispensing compatible with APCI and electrospray (ion evaporation) ionization.

As speeds reach 10 Hz or greater, the data acquisition rate of the MS will begin to be limiting. The fastest acquisition rate mass spectrometers today are QqQ's in the MRM mode and QqTOF's in the full scan mode which can acquire accurate quantitative data from peaks 100 ms wide but will begin to be challenged as they exceed these speeds. Present-day ion trapping analyzers are at least 10 times slower, so are currently challenged by the 1 Hz sampling speeds shown to be achievable today for quantitation work.

In the 1980s, 1–10 successful LC/MS samples per day was the norm. The one sample per second barrier was broken in 2005 with MALDI and in 2018 with ESI. The new goal is now to be able to construct entire 20-point calibration curves in one second or, to put it another way, in time frames similar to "the blink of an eye."

References

1 Troutman, M. (2022). The impact of HT-MS to date and it's potential to shape the future of metrics-based experimentation and analysis, Chapter 14. In: *High-Throughput Mass Spectrometry in Drug Discovery* (ed. C. Liu and H. Zhang), 455–465. Wiley.

2 Grainger, R. and Whibley, S. (2021). A perspective on the analytical challenges encountered in high-throughput experimentation. *Org. Process Res. Dev.* 25 (3): 354–364. https://doi.org/10.1021/acs.oprd.0c00463.

3 Wildey, M.J., Haunso, A., Tudor, M. et al. (2017). High-throughput screening. *Annu. Rep. Med. Chem.* 50: 149–195. https://doi.org/10.1016/bs.armc.2017.08.004.

4 Vervoort, N., Goossens, K., Baeten, M., and Chen, Q. (2021). Recent advances in analytical techniques for high-throughput experimentation. *Anal. Sci. Adv.* 2 (3, 4): 1–19.

5 Kempa, E., Hollywood, K.A., Smith, C.A., and Barran, P.E. (2019). High-throughput screening of complex biological samples with mass spectrometry – from bulk measurements to single cell analysis. *Analyst* 144: 872–891. https://doi.org/10.1039/c8an01448e.

6 McLaren, D.G., Shah, V., Wisniewski, T. et al. (2021). High-throughput mass spectrometry for hit identification: current landscape and future perspectives. *SLAS Discovery* 26 (2): 168–191. https://doi.org/10.1177/2472555220980696.

7 Shou, W.Z. (2020). Current status and future directions of high-throughput ADME screening in drug discovery. *J. Pharm. Anal.* 10 (3): 201–208. https://doi.org/10.1016/j.jpha.2020.05.004.

8 Lee, M.S. (2002). *LC/MS Application in Drug Development, Wiley-Interscience Series on Mass Spectrometry* (ed. D. Desiderio and N.M.M. Nibbering). New York: Wiley.

9 Lee, M.S. (ed.) (2005). *Integrated Strategies for Drug Discovery Using Mass Spectrometry*. Wiley. Chapters 1–18.

10 Kapinos, B. (2022). Development and application of a enterprise-level, high-capacity sample delivery system: lead sampler LS-1, Chapter 2. In: *High-Throughput Mass Spectrometry in Drug Discovery* (ed. C. Liu and H. Zhang), 77–102. Wiley.

11 Latawiec, A. (2022). Evolution of multiplexing technology for high-throughput LC/MS analyses, Chapter 3. In: *High-Throughput Mass Spectrometry in Drug Discovery* (ed. C. Liu and H. Zhang), 103–120. Wiley.

12 Wagner, A. (2022). Direct online SPE-MS for high-throughput analysis in drug discovery, Chapter 4. In: *High-Throughput Mass Spectrometry in Drug Discovery* (ed. C. Liu and H. Zhang), 123–142. Wiley.

13 Dikler, S. (2022). Accelerating drug discovery with ultrahigh-throughput MALDI-TOF MS, Chapter 12. In: *High-Throughput Mass Spectrometry in Drug Discovery*, 393–422. 2e. Wiley.

14 Zheng, X. (2022). Recent advances in ion mobility-mass spectrometry for drug discovery, Chapter 6. In: *High-Throughput Mass Spectrometry in Drug Discovery* (ed. C. Liu and H. Zhang), 183–214. Wiley.

15 Schneider, B. (2022). Differential mobility spectrometry and its application to high-throughput analysis, Chapter 7. In: *High-Throughput Mass Spectrometry in Drug Discovery* (ed. C. Liu and H. Zhang), 215–266. Wiley.

16 Liu, C. (2022). Acoustic ejection mass spectrometry for high-throughput drug discovery. In: *High-Throughput Mass Spectrometry in Drug Discovery*, Chapter 5 (ed. C. Liu and H. Zhang), 143–182. Wiley.

17 Elpa, D.P., Rakesh, G., Prabhu, D. et al. (2019). Automation of mass spectrometric detection of analytes and related workflows: a review. *Talanta* 208: 120304. https://doi.org/10.1016/j.talanta.2019.120304.

18 Luippold, A.H., Arnhold, T., Jörg, W. et al. (2011). Application of a rapid and integrated analysis system (RIAS)as a high-throughput processing tool for in vitro ADME samples by liquid chromatography/tandem mass spectrometry. *J. Biomol. Screening* 16 (3): 370–377.

19 Luippold, A.H., Arnhold, T., Jorg, W., and Sussmuth, R.D. (2010). An integrated platform for fully automated high-throughput LC-MS/MS analysis of in vitro metabolic stability assay samples. *Int. J. Mass Spectrom.* 296: 1–9. https://doi.org/10.1016/bs.armc.2017.08.004.

20 Janiszewski, J., Schneider, R., Kapinos, B. et al. (2012). An enterprise level high-speed, system for LC-MS/MS bioanalysis. *Bioanalysis* 4 (9): 1039–1056.

21 Shou, W.Z. and Zhang, J. (2012). Recent development in software and automation tools for high-throughput discovery bioanalysis. *Bioanalysis* 4: 1097–1109.

22 French, J.D., Reid, N.M., and Buckley, J.A. (1977). Apparatus for analyzing trace components in air. US Patent 4,023,398.

23 Carroll, D.I., Dzidic, I., Horning, E.C., and Stilwell, R.N. (1981). Atmospheric pressure ionization mass spectrometry. *Appl. Spectrosc. Rev.* 17 (3): 337–406.

24 Mitchum, R.K. and Korfmacher, W.A. (1983). Atmospheric pressure ionization mass spectrometry. *Anal. Chem.* 1485A–1499A.

25 French, J.B., Thomson, B.A., Davidson, W.R. et al. (1984). Atmospheric pressure chemical ionization mass spectrometry. In: *Mass Spectrometry in Environmental Sciences* (ed. F.W. Karasek, O. Hutzinger, and S.H. Safe), 101–121. New York: Plenum.

26 Bruins, A.P. (1991). Mass spectrometry with ion sources operating at atmospheric pressure. *Mass Spectrom. Rev.* 10: 53–77.

27 Covey, T.R., Thomson, B.A., and Schneider, B.B. (2009). Atmospheric pressure ion sources. *Mass Spectrom. Rev.* 28: 870–897. https://doi.org/10.1002/mas.20246.

28 Karas, M. and Hillenkamp, F. (1988). Laser desorption ionization of proteins with molecular masses exceeding 10,000 daltons. *Anal. Chem.* 60 (20): 2299–2301. https://doi.org/10.1021/ac00171a028.

29 Fenn, J.B., Mann, M., Meng, C.K. et al. (1989). Electrospray ionization for mass spectrometry of large biomolecules. *Science* 246: 64–71.

30 Covey, T.R., Bonner, R.F., Shushan, B.I., and Henion, J.D. (1988). The determination of protein, oligonucleotide, and peptide molecular weights. *Rapid Commun. Mass Spectrom.* 2 (11): 249–256.

31 Covey, T.R., Lee, E.D., Bruins, A.P., and Henion, J.D. (1986). Liquid chromatography mass spectrometry. *Anal. Chem.* 58 (14): 1451A–1461A.

32 Arpino, P. (1982). On-line liquid chromatography/mass spectrometry? An odd couple! *TrAC, Trends Anal. Chem.* 1: 154–158.

33 Covey, T.R. (2007). Pneumatically and thermally assisted electrospray ionization. In: *Encyclopedia of Mass Spectrometry*, vol. 6 (ed. M. Gross and R. Caproli), 426–434. Elsevier.

34 Covey, T.R. (2010). A tribute the scientific career of Andries Bruins. *Eur. J. Mass Spectrom.* 16: 1–10.

35 Sparks, J.B. (1990). *Histomap of World History*. Rand McNally. Available from the John Sparks Trust Fund by contacting his granddaughter Jackie Glanz at jacquieglanz@yahoo.com. Makes a great stocking stuffer for history buffs. Also Google "Histomap".

36 Dole, M., Mack, L.L., Hines, R.L. et al. (1968). Molecular beams of macroions. *J. Phys. Chem. Phys.* 49 (1): 2240–2249.

37 Talrose, V.L., Karpov, G.V., Gorodetskii, I.G., and Skurat, V.E. (1968). Capillary system for the introduction of liquid mixtures into an analytical mass-spectrometer. *Russ. J. Phys. Chem.* 42 (12): 1658.

38 Thomson, J.J. (1913). *Rays of Positive Electricity and Their Application to Chemical Analysis*. Longmas, Green and Co., republished by The American Society for Mass Spectrometry.

39 Baldwin, M.S. and McLafferty, F.W. (1973). The direct introduction of liquid solutions into a chemical ionization mass spectrometer. *Org. Mass Spectrom.* 7: 1111–1112.

40 Henion, J.D. (1978). Drug analysis by continuously monitored liquid chromatography/mass spectrometry with a quadrupole mass spectrometer. *Anal. Chem.* 50 (12): 1887.

41 Covey, T.R. and Henion, J.D. (1983). Direct liquid introduction interface for liquid chromatography/mass spectrometry. *Anal. Chem.* 55: 2275–2280.

42 Blakley, C.R., Carmody, J.J., and Vestal, M.L. (1980). Liquid-chromatograph-mass spectrometer for analysis of nonvolatile samples. *Anal. Chem.* 52: 1636.

43 Covey, T.R., Crowther, J.B., Dewey, E.A., and Henion, J.D. (1985). Thermospray mass spectrometry interface and application for determination of drugs and their metabolites in biological fluids. *Anal. Chem.* 57: 474–480.

44 McFadden, W.H., Schwartz, H.L., and Evans, S. (1976). Direct analysis of liquid chromatographic effluents. *J. Chromatogr.* 122: 386–296.

45 Games, D.E., Hirter, P., Kuhnz, W. et al. (1981). Studies of combined liquid chromatography-mass spectrometry with a moving belt interface. *J. Chromatogr.* 203: 131–138.

46 Willoughby, R.C. and Browner, R.F. (1984). Monodisperse aerosol generation interface for combining liquid chromatography with mass spectrometry. *Anal. Chem.* 56: 2626–2631.

47 Cook, K.D. (1986). Electrohydrodynamic mass spectrometry. *Mass Spectrom. Rev.* 5: 467–519.

48 Barber, M., Bordoli, R.S., Sedgwick, R.D., and Tyler, A.N. (1981). Fast atom bombardment of solids (FAB)-a new ion source for mass spectrometry. *J. Chem. Soc. Chem. Commun.* 325–327.

49 MacFarlane, R.D. (1983). Californium-252 plasma desorption mass spectrometry. *Anal. Chem.* 55: 1247A–1250A.

50 Covey, T.R. (2020). The Vouros Odyssey. Liquid chromatography/mass spectrometry from a chromatographic perspective and a Theseus Paradox. *Mass Spectrom. Rev.* 39 (1, 2): 15–34. https://doi.org/10.1002/mas.21564.

51 Carroll, C.N., Dzidic, R.N., Stillwell, R.N. et al. (1974). Subpicogram detection system for gas phase analysis based in atmospheric pressure ionization mass spectrometry. *Anal. Chem.* 46 (6): 706.

52 Carroll, D.I., Dzidic, I., Stilwell, R.N. et al. (1975). Atmospheric pressure ionization mass spectrometry: corona discharge ion source for use in liquid chromatograph-mass spectrometer-computer analytical system. *Anal. Chem.* 47 (14): 2369–2373.

53 Thomson, B.A. and Iribarne, J.V. (1978). The fate of electrical charges in evaporating cloud droplets. *Rev. Geophys. Space Phys.* 16 (3): 431–434.

54 Thomson, B.A. and Iribarne, J.V. (1979). Field induced ion evaporation from liquid surfaces at atmospheric pressure. *J. Chem. Phys.* 71 (11): 4451–4463.

55 Thomson, B.A., Iribarne, J.V., and Dziedzic, P.J. (1982). Liquid ion evaporation/ mass spectrometry/mass spectrometry for the detection of polar and labile molecules. *Anal. Chem.* 54: 2219–2224.

56 Yamashita, M. and Fenn, J.B. (1984). Electrospray ion source. Another variation on the free-jet theme. *J. Phys. Chem.* 88: 4451–4459.

57 Whitehouse, M., Dreyer, R.N., Yamashita, M., and Fenn, J.B. (1985). Electrospray interface for liquid chromatographs and mass spectrometers. *Anal. Chem.* 57: 675–679.

58 Alexandrov, M.L., Gall, L.N., Krasnov, N.V. et al. (1984). Extraction of ions from solution under atmospheric pressure as a method for mass spectrometric analysis of bioorganic compounds. *Dokl. Akad. Nauk SSSR* 277: 379–383.

59 Alexandrov, M.L., Gall, L.N., Krasnov, N.V. et al. (2008). Extraction of ions from solution under atmospheric pressure as a method for mass spectrometric analysis of bioorganic compounds. *Rapid Commun. Mass Spectrom.* 22: 267–270. English translation of reference 40.

60 Bruins, A.P., Covey, T.R., and Henion, J.D. (1987). Ion spray interface for combined liquid chromatography/atmospheric pressure ionization mass spectrometry. *Anal. Chem.* 59: 2642–2646.

61 Henion, J.D., Thomson, B.A., and Dawson, P.H. (1982). Determination of sulfa drugs in biological fluids by liquid chromatography/mass spectrometry/mass spectrometry. *Anal. Chem.* 54: 451–456.

62 Covey, T.R., Lee, E.D., and Henion, J.D. (1986). High-speed liquid chromatography/tandem mass spectrometry for the determination of drugs in biological fluids. *Anal. Chem.* 58: 2453–2460.

63 Covey, T.R. (1989). MS/MS of tryptic peptides. *37th ASMS Conference on Mass Spectrometry and Allied Topics*, ThPl 232, Miami Beach. Florida: American

Society of Mass Spectrometry (June 1989). https://www.asms.org/conferences/past-conferences.

64 Lee, E.D., Henion, J.D., and Covey, T.R. (1989). Microbore high performance liquid chromatography-ion spray mass spectrometry for the determination of peptides. *J. Microcolumn Sep.* 1 (1): 14–18.

65 Covey, T.R., Huang, E.C., and Henion, J.D. (1991). Structural characterization of protein tryptic peptides via liquid chromatography/mass spectrometry and collision-induced dissociation of their doubly charged molecular ions. *Anal. Chem.* 63: 1193–1200.

66 Hemling, M.E., Roberts, G.D., Johnson, W. et al. (1990). Analysis of proteins and glycoproteins at the picomole level by on-line coupling of microbore high performance liquid chromatography with flow fast atom bombardment and electrospray mass spectrometry: a comparative evaluation. *Biomed. Environ. Mass Spectrom.* 19: 677–691.

67 Ling, V., Guzzetta, A.W., Canova-Davis, E. et al. (1991). Characterization of the tryptic map of recombinant DNA derived tissue plasminogen activator by high-performance liquid chromatography-electrospray ionization mass spectrometry. *Anal. Chem.* 63 (24): 2909–2915.

68 Hess, D., Covey, T.R., Winz, R. et al. (1992). Analytical and micropreparative peptide mapping by high performance liquid chromatography/electrospray mass spectrometry of proteins purified from 2-D gel electrophoresis. *Protein Sci.* 2: 1342–1351.

69 Tsarbopoulos, A., Pramanik, B.N., Nagabhushan, T.I., and Covey, T.R. (1995). Structural analysis of the CHO-derived interleukin-4 by liquid chromatography/electrospray ionization mass spectrometry. *J. Mass Spectrom.* 30: 1752–1763.

70 Covey, T.R. (1996). Liquid chromatography/mass spectrometry for the analysis of protein digests, Chapter 7. In: *Methods in Molecular Biology Vol. 61: Protein and Peptide Analysis* (ed. J.R. Chapman), 83–99. Totowa, NJ: Humana Press.

71 Wong, T., Shackleton, C.H.L., Covey, T.R., and Ellis, G. (1992). Identification of the steroids in neonatal plasma that interfere with 17 alpha-hydroxyprogesterone radioimmunoassays. *Clin. Chem.* 38: 1830–1836.

72 Weidolf, L. and Covey, T.R. (1992). Studies on the metabolism of omeprazole in the rat using liquid chromatography/ionspray mass spectrometry and the isotope cluster technique with [^{34}S] omeprazole. *Rapid Commun. Mass Spectrom.* 6: 192–196.

73 Hopfgartner, G., Chernushevish, I.V., Covey, T.R. et al. (1999). Exact mass measurement of product ions for the structural elucidation of drug metabolites with a tandem quadrupole orthogonal-acceleration time-of-flight mass spectrometer. *J. Am. Soc. Mass Spectrom.* 10: 1305–1314.

74 Center for Disease Control (1990). Analysis of l-tryptophan for the etiology of eosinophilia-myalgia syndrome. *MMWR Morb. Mortal. Wkly. Rep.* 39 (34): 34–36.

75 Eckerlin, R.H., Ebel, J.G., Henion, J.D., and Covey, T.R. (1989). The case of the toxic dexamethazone. *Anal. Chem.* 61: 53A–57A.

76 Uden, P.C. (1989). Editorial report on the editorial. Conference report: 1989 Pittsburgh conference and exposition on analytical chemistry and applied spectroscopy. March 6th–10th 1989, Atlanta, Georgia, USA. *Analyst* 114: 757–758. https://doi.org/10.1039/AN9891400757.

77 Karas, M. and Kruger, R. (2003). Ion formation in MALDI: the cluster ionization mechanism. *Chem. Rev.* 103: 427–439.

78 Peters, E.C., Brock, A., Horn, D.M. et al. (2002). Automated LC–MALDI FT-ICR MS platform for high-throughput proteomics. *LC GC Eur.* 2–7.

79 Chen, H., Reitar, T., Andrev, V. et al. (2005). High-speed, high-resolution monolithic capillary LC-MALDI MS using an off-line continuous deposition interface for proteomic analysis. *Anal. Chem.* 77 (8): 2323–2331.

80 Kovarik, P., Hodgson, R.J., Covey, T.R. et al. (2005). Capillary-scale frontal affinity chromatography/MALDI tandem mass spectrometry using protein-doped monolithic silica columns. *Anal. Chem.* 77: 3340–3350.

81 Corr, J.J., Kovarik, P., Schneider, B.S. et al. (2006). Design considerations for high speed quantitative mass spectrometry with MALDI ionization. *J. Am. Soc. Mass Spectrom.* 17: 1129–1141.

82 Gobey, J., Cole, M., Janiszewski, J. et al. (2005). Characterization and performance of MALDI on a triple quadrupole mass spectrometer for analysis and quantification of small molecules. *Anal. Chem.* 77: 5643–5654.

83 Schneider, B.B., Lock, C., and Covey, T.R. (2005). AP and vacuum MALDI on a QqLIT instrument. *J. Am. Soc. Mass Spectrom.* 16: 176–182.

84 Volmer, D.A., Sleno, L., Bateman, K. et al. (2007). Comparison of MALDI to ESI on a triple quadrupole platform for pharmacokinetic analyses. *Anal. Chem.* 79: 9000–9006.

85 Wilm, M. and Mann, M. (1994). Electrospray and Taylor-Cone theory, Dole's beam of macromolecules at last? *Int. J. Mass Spectrom. Ion Process.* 136: 167–180.

86 Valaskovic, G.A. and McLafferty, F.W. (1996). Long-lived metallized tips for nanoliter electrospray mass spectrometry. *J. Am. Soc. Mass Spectrom.* 7: 1270–1272.

87 Schultz, G.A., Corso, T.N., Prosser, S.J., and Zhang, S. (2000). A fully integrated monolithic microchip electrospray device for mass spectrometry. *Anal. Chem.* 72: 4058–4063.

88 Schneider, B.B., Guo, X., Fell, L.M., and Covey, T.R. (2005). Stable gradient flow nanoLC-MS. *J. Am. Soc. Mass Spectrom.* 16: 1545–1551.

89 Covey, T.R. and Pinto, D. (2002). Nanospray electrospray ionization development. LC/MS and CE/MS application. In: *Applied Electrospray Mass Spectrometry*, Chapter 2 (ed. B.N. Pramanik, A.K. Ganguly, and M.L. Gross), 105–148. New York: Marcel Dekker, Inc.

90 Smith, R.D., Baranaga, C.J., and Udseth, H.R. (1989). Improved electrospray interface for capillary electrophoresis mass spectrometry. *Anal. Chem.* 60: 1948–1952.

91 Lee, E.D., Muck, W., Henion, J.D., and Covey, T.R. (1989). Liquid junction coupling of capillary zone electrophoresis/ion spray mass spectrometry. *Biomed. Environ. Mass Spectrom.* 18: 844–850.

92 Lee, E.D., Muck, W., Henion, J.D., and Covey, T.R. (1989). Capillary zone electrophoresis/tandem mass spectrometry for the determination of sulfonated azo dyes. *Biomed. Environ. Mass Spectrom.* 18: 253–257.

93 Janiszewski, J.S., Rogers, K.J., Whalen, K.M. et al. (2001). A high capacity LC/MS system for the bioanalysis of samples generated from plate-based metabolic screening. *Anal. Chem.* 73 (7): 1495–1501.

94 Wang, T., Zeng, L., Cohen, J., and Kassel, D.B. (1999). A multiple electrospray interface for parallel mass spectrometric analyses of compound libraries. *Comb. Chem. High-Throughput Screening* 2: 327–334.

95 De Biasi, V., Haskins, N., Organ, A. et al. (1999). High-throughput liquid chromatography/mass spectrometric analyses using a novel multiplexed electrospray interface. *Rapid Commun. Mass Spectrom.* 13: 1165–1168.

96 Covey, T.R. (1995). Analytical characteristics of the electrospray ionization process. In: *Biochemical and Biotechnological Applications of Electrospray Ionization Mass Spectrometry*, Chapter 2, ACS Symposium Series 619 (ed. A.P. Snyder), 21–59. Washington, DC: American Chemical Society.

97 Cooks, R.G., Ouyang, Z., Takats, Z., and Wiseman, J.M. (2006). Ambient mass spectrometry. *Science* 311: 1566–1570.

98 Feider, C.L., Krieger, A., DeHoog, R.J., and Eberlin, L.S. (2019). Ambient ionization mass spectrometry: recent developments and applications. *Anal. Chem.* 91 (7): 4266–4290. https://doi.org/10.1021/acs.analchem.9b00807.

99 Harris, G.A., Galhena, A.S., and Fernández, F.M. (2011). Ambient sampling/ionization mass spectrometry: applications and current trends. *Anal. Chem.* 83 (12): 4508–4538. https://doi.org/10.1021/ac200918u.

100 Alberici, R.M., Simas, R.C., Sanvido, G.B. et al. (2010). Ambient mass spectrometry: bringing MS into the real world. *Anal. Bioanal. Chem.* 398: 265–294. https://doi.org/10.1007/s00216-010-3808-3.

101 Weston, D.J. (2010). Ambient ionization mass spectrometry: current understanding of mechanistic theory; analytical performance and application areas. *Analyst* 4: https://doi.org/10.1039/b925579f.

102 Ifa, D.R., Wu, C., Ouyang, Z., and Cooks, R.G. (2010). Desorption electrospray ionization and other ambient ionization methods: current progress and preview. *Analyst* 4: https://doi.org/10.1039/b925257f. Critical Review.

103 Van Berkel, G.J., Pasilis, S.P., and Ovchinnikova, O. (2008). Established and emerging atmospheric pressure surface sampling/ionization techniques for mass spectrometry. *J. Mass Spectrom.* 43: 1161–1180. https://doi.org/10.1002/jms.1440.

104 Zimmerman, T.A., Monroe, E.B., Tucker, K.R. et al. (2008). Imaging of cells and tissues with mass spectrometry: adding chemical information to imaging. *Methods Cell Biol.* 89: 361–390. https://doi.org/10.1016/S0091-679X(08)00613-4.

105 Van Berkel, G.J. and Kertesz, V. (2015). An open port sampling interface for liquid introduction atmospheric pressure ionization mass spectrometry. *Rapid Commun. Mass Spectrom.* 29 (19): 1749–1756.

106 Van Berkel, G.J. and Kertesz, V. (2017). Rapid sample classification using open port sampling interface with liquid introduction atmospheric pressure ionization mass spectrometry. *Rapid Commun. Mass Spectrom.* 31 (3): 281–291.

107 Cody, R.B. and Laramee, J.A. (2005). Versatile new ion source for the analysis of materials in open air under ambient conditions. *Anal. Chem.* 77: 2297–2302.

108 Yu, S., Crawford, E., Tice, J. et al. (2009). Bioanalysis without sample cleanup or chromatography: the evaluation and initial implementation of direct analysis in real time ionization mass spectrometry for the quantification of drugs in biological matrixes. *Anal. Chem.* 81: 193–202.

109 McEwen, C.N., McKay, R.G., and Larsen, B.S. (2005). The analysis of solids, liquids, and biological tissues using solids probe introduction at atmospheric pressure on commercial LC/MS instrumentation. *Anal. Chem.* 77: 7826–7831.

110 Robb, D.B., Covey, T.R., and Bruins, A.P. (2000). Atmospheric pressure photoionization: an ionization method for liquid chromatography/mass spectrometry. *Anal. Chem.* 72: 3653–3659.

111 Picard, P., Carrier, V., Lessard, D., and Paquim, R. (2005). Atmospheric pressure laser diode thermal desorption mass spectrometry: a new high-performance ionization source. In *53th ASMS Conference on Mass Spectrometry and Allied Topics*, ASMS. San Antonio, TX: American Society of Mass Spectrometry (June 2005). https://www.asms.org/publications/abstracts-and-proceedings.

112 Haarhoff, Z., Wagner, A., Picard, P. et al. (2016). Coupling laser diode thermal desorption with acoustic sample deposition to improve throughput of mass spectrometry-based screening. *J. Biomol. Screening* 21: 165–175.

113 Picard, P. (2022). Laser diode thermal desorption-mass spectrometry (LDTD-MS): fundamentals and applications of sub-second analysis in drug discovery environment, Chapter 12. In: *High-Throughput Mass Spectrometry in Drug Discovery* (ed. C. Liu and H. Zhang), 393–422. Wiley.

114 Van Berkel, G.J., Kertesz, V., Koeplinger, K.A. et al. (2008). Liquid microjunction surface sampling probe electrospray mass spectrometry for detection of drugs and metabolites in thin tissue sections. *J. Mass Spectrom.* 43 (4): 500–508.

115 Van Berkel, G.J., Sanchez, A.D., and Quirke, J.M.E. (2002). Thin-layer chromatography and electrospray mass spectrometry coupled using a surface sampling probe. *Anal. Chem.* 74 (24): 6216–6223.

116 Van Berkel, G.J., Kertesz, V., and King, R.C. (2009). High-throughput mode liquid microjunction surface sampling probe. *Anal. Chem.* 81 (16): 7096–8101. https://doi.org/10.1021/ac901098d.

117 Kertesz, V. and Van Berkel, G.J. (2010). Fully automated liquid extraction-based surface sampling and ionization using a chip-based robotic nanoelectrospray platform. *J. Mass Spectrom.* 45 (3): 252–260. https://doi.org/10.1002/jms.1709.

118 Oberlies, N.H., Knowles, S.L., Amrine, C.S.M. et al. (2019). Droplet probe: coupling chromatography to the in situ evaluation of the chemistry of nature. *Nat. Prod. Rep.* 36: 944–959.

119 Hsu, C.C., ElNaggar, M.S., Peng, Y. et al. (2013). Real-time metabolomics on living microorganisms using ambient electrospray ionization flow-probe. *Anal. Chem.* 85: 7014–7018.

120 Morato, N.M. and Cooks, R.G. (2021). Inter-platform assessment of performance of high-throughput desorption electrospray ionization mass spectrometry. *Talanta Open* https://doi.org/10.1016/j.talo.2021.100046.

121 Wleklinski, M., Loren, B.P., Ferreira, C.R. et al. (2018). High-throughput reaction screening using desorption electrospray ionization mass spectrometry. *Chem. Sci.* 9: 1647–1653.

122 Zhang, W. (2022). Developments and applications of DESI-MS in drug discovery, Chapter 13. In: *High-Throughput Mass Spectrometry in Drug Discovery* (ed. C. Liu and H. Zhang), 423–452. Wiley.

123 Liu, J., Hua, W., Liu, C. et al. (2018). Acoustic-open port-MS: a new anlaytical system capable of in-situ measurement of substrate-to-product conversion in protein matrices with high speed, sensitivity, and reproducibility. *Presented at the SLAS Conference on Laboratory Automation*, San Diego (2 February 2018). https://www.slas.org/events/past-conferences/.

124 Liu, C., Zhang, H., Ghislain, L. et al. (2018). A high-throughput mass spectrometry plate reader: acoustic droplet ejection to an open-port probe sampling interface. *Presented at the SLAS Conference on Laboratory Automation*, San Diego (2 February 2018). https://www.slas.org/events/past-conferences/.

125 Liu, C., Van Berkel, G.J., Kovarik, P. et al. (2021). Fluid dynamics of the open port interface for high speed nanoliter volume sampling mass spectrometry. *Anal. Chem.* 93 (24): 8559–8567. https://doi.org/10.1021/acs.analchem.1c01312.

126 Liu, C., Van Berkel, G.J., Cox, D.M., and Covey, T.R. (2020). Operational modes and speed considerations of an acoustic droplet dispenser for mass spectrometry. *Anal. Chem.* 92 (24): 15818–15826.

127 ElNaggar, M.S. and Van Berkel, G.J. (2011). Liquid microjunction surface sampling probe fluid dynamics: characterization of an analyte plug formation operational mode. *J. Am. Soc. Mass Spectrom.* 22 (10).

128 ElNaggar, M.S., Barbier, C., and Van Berkel, G.J. (2011). Liquid microjunction surface sampling probe fluid dynamics: computational analysis of coaxial intercapillary positioning effects. *J. Am. Soc. Mass Spectrom.* 22 (7): 1157–1166.

129 van Breemen, R.B., Huang, C.-R., Nikolic, D. et al. (1997). Pulsed ultrafiltration mass spectrometry: a new method for screening combinatorial libraries. *Anal. Chem.* 69: 2159–2164.

130 Johnson, B.M., Nikolic, D., and van Breemen, R.B. (2002). Applications of pulsed ultrafiltration-mass spectrometry. *Mass Spectrom. Rev.* 21: 76–86.

131 Rush, M.D., Walker, E.M., Burton, T., and van Breemen, R.B. (2016). Magnetic microbead affinity selection screening (MagMASS) of botanical extracts for inhibitors of 15-lipoxygenase. *J. Nat. Prod.* 79: 2898–2902. https://doi.org/10.1021/acs.jnatprod.6b00693.

132 Meng, J. (2022). Off-line affinity selection mass spectrometry and its application in lead discovery, Chapter 8. In: *High-Throughput Mass Spectrometry in Drug Discovery* (ed. C. Liu and H. Zhang), 269–296. Wiley.

133 Zheng, H. (2022). In-line affinity selection mass spectrometry. In: *High-Throughput Mass Spectrometry in Drug Discovery*, Chapter 9 (ed. C. Liu and H. Zhang), 297–316. Wiley.

134 Covey, T. and Douglas, D.J. (1993). Collision cross sections for protein ions. *J. Am. Soc. Mass Spectrom.* 4: 616–623.

135 Ganguly, A.K., Pramanik, B.N., Tsarboloulos, A. et al. (1992). Mass spectrometric detection of the noncovalent GDP-bound conformational state of the human H-ras protein. *J. Am. Chem. Soc.* 114: 6559–6560.

136 Huang, E.C., Pramanik, B.N., Tsarbopoulos, A. et al. (1993). Application of electrospray mass spectrometry in probing protein-protein and protein ligand noncovalent interactions. *J. Am. Soc. Mass Spectrom.* 4: 624–630.

137 Ganem, B., Li, Y.T., and Henion, J.D. (1991). Detection of noncovalent receptor-ligand complexes by mass spectrometry. *J. Am. Chem. Soc.* 113: 6294–6296.

138 Ganem, B., Li, Y.T., and Henion, J.D. (1991). Observation of noncovalent enzyme-substrate and enzyme-product complexes by ion spray mass spectrometry. *J. Am. Chem. Soc.* 113: 7818–7819.

139 Mengxuan, J. (2022). Native mass spectrometry in drug discovery and development. In: *High-Throughput Mass Spectrometry in Drug Discovery*, Chapter 10 (ed. C. Liu and H. Zhang), 317–346. Wiley.

140 Covey, T.R. and Blough, M. (1999). Organizers of "ASMS workshop on LC/MS technologies; strategies for achieving ultrahigh-throughput LC/MS. Fast serial vs parallel chromatography". *47th American Society for Mass Spectrometry and Allied Topics*. Chicago, IL: American Society of Mass Spectrometry (June 1999). https://www.asms.org/publications/abstracts-and-proceedings.

141 Fouda, H., Nocerini, M., Schneider, R., and Gedutis, C. (1991). Quantitative analysis by high-performance liquid chromatography atmospheric pressure chemical ionization mass spectrometry: the determination of the renin inhibitor C&80,794 in human serum. *J. Am. Soc. Mass Spectrom.* 2: 165–167.

142 Gilbert, J.D., Hand, E.L., Yuan, A.S. et al. (1992). Determination of L-365,260, a new cholecystokinin receptor (C CK-B) antagonist in plasma by liquid chromatography/atmospheric pressure chemical ionization mass spectrometry. *Biol. Mass Spectrom.* 21: 63–68.

143 Bernert, J.T. Jr., Turner, W.E., Pirkle, J.L. et al. (1997). Development and validation of sensitive method for determination of serum cotinine in smokers and nonsmokers by liquid chromatography/atmospheric pressure ionization tandem mass spectrometry. *Clin. Chem.* 43 (12): 2281–2291.

144 Covey, T.R., Schneider, B.B., Javaheri, H. et al. (2010). ESI, APCI, and MALDI a comparison of the central analytical figures of merit: sensitivity, reproducibility, and speed. In: *Electrospray and MALDI Mass Spectrometry: Fundamentals, Instrumentation, Practicalities, and Biological Applications*, Chapter 13, 2e (ed. R.B. Cole), 443–490. Wiley.

145 French, J.B., Etkin, B., and Jong, R. (1994). Monodisperse dried microparticulate injector for analytical instrumentation. *Anal. Chem.* 66: 685–691.

146 Cole, M.J., Janiszewski, J.S., and Fouda, H.G. (2002). Electrospray mass spectrometry in drug metabolism and pharmacokinetics. In: *Applied Electrospray Mass Spectrometry* (ed. B.N. Pramanik, A.K. Ganguly, and M.L. Gross), 211–249. New York: Marcel Dekker, Inc.

147 Janiszewski, J.S., Liston, T.E., and Cole, M.J. (2008). Perspectives on bioanalytical mass spectrometry and automation in drug discovery. *Curr. Drug Metab.* 9 (9): 986–994.

148 Janiszewski, J., Schneider, R., Kapinos, B. et al. (2012). Development of a high-speed, multiplexed sample-delivery instrument for LC-MS/MS bioanalysis. *Bioanalysis* 4 (9): 1039–1056.

149 Whalen, K., Courtney, J.C., Smith, D. et al. (2001). A specialized high-throughput sample delivery system utilizing a dual-spray LC/MS interface for the bioanalysis of 500 samples/hour. *49th ASMS Conference on Mass Spectrometry and Allied Topics*, ThPl 232, Chicago, IL (27–31 May 2001).

150 Hiller, D.L., Brockman, A.H., Goulet, L. et al. (2000). Application of a non-indexed dual sprayer pneumatically assisted electrospray source to the high-throughput quantitation of drugs in biological fluids. *Rapid Commun. Mass Spectrom.* 14: 2034–2038.

151 Whalen, K.M., Rogers, K.J., Cole, M.J., and Janiszewski, J.S. (2000). AutoScan: an automated workstation for rapid determination of mass and tandem mass spectrometry conditions for quantitative bioanalytical mass spectrometry. *Rapid Commun. Mass Spectrom.* 14: 2074–2079.

152 Van Pelt, C.K., Corso, T.N., Schultz, G.A. et al. (2001). A four-column chromatography system for isocratic or gradient LC/MS analysis. *Anal. Chem.* 73: 582–588.

153 Onorato, J.M., Henion, J., Lefebvre, P.M., and Kiplinger, J.P. (2001). SRM-LC-MS determination of idoxifene and its pyrrolidone metabolite in human plasma using robotic high-throughput sequential injection. *Anal. Chem.* 73: 119–125.

154 Kapinos, B., Liu, J., Piotrowski, M. et al. (2017). Development of a high-performance, enterprise-level, multimode LC-MS/MS autosampler for drug discovery. *Bioanalysis* 9 (21): 1643–1654.

155 Bretschneider, T., Ozbal, C., Holstein, M. et al. (2019). RapidFire BLAZE-mode is boosting ESI-MS toward high-throughput-screening. *SLAS Technol.* 24 (4): 386–393.

156 Chassaing, C., Stafford, H., Luckwell, J. et al. (2005). A parallel turbulent flow chromatography-tandem mass spectrometry method for the analysis of a pharmaceutical compound in plasma. *Chromatographia* 62 (1, 2): 17–24.

157 Huang, E., Henion, J., and Covey, T.R. (1990). Packed-column supercritical fluid chromatography-mass spectrometry with ionization at atmospheric pressure. *J. Chromatogr.* 511: 257–279.

158 Baker, T.R. and Pinkston, J.D. (1998). Development and application of packed-column supercritical fluid chromatography/pneumatically assisted electrospray mass spectrometry. *J. Am. Soc. Mass Spectrom.* 9: 498–509.

159 Schneider, B.B. and Covey, T.R. (2007). Multiple sprayers in electrospray mass spectrometry. In: *The Encyclopedia of Mass Spectrometry*, vol. 6 (ed. M. Gross and R. Caproli), 456–467. Amsterdam: Elsevier.

160 Hindsgaul, O. and Schriemer, D.C. (2003). Device for delivery of multiple liquid sample streams to a mass spectrometer. US Patent 6,621,075.

161 Bayliss, M.K., Little, D., Mallett, D.N., and Plumb, R.S. (2000). Parallel ultrahigh flow rate liquid chromatography with mass spectrometric detection using a multiplexed electrospray source for direct, sensitive determination of pharmaceutical in plasma at extremely high-throughput. *Rapid Commun. Mass Spectrom.* 14: 2039–2045.

162 Wang, L., Mann, T.D., Little, D. et al. (2001). Evaluation of a four channel multiplexed electrospray triple quadrupole mass spectrometer for the simultaneous validation of methods in four preclinical matrixes. *Anal. Chem.* 73 (8): 1740–1747.

163 Covey, T.R., Thomson, B., and Jolliffe, C.L. (2004). Parallel sample introduction electrospray mass spectrometer with electronic indexing through multiple ion entrance orifices. US Patent 6,784,422.

164 Mylchreest, I.C., Hemenway, T.A.E., and Hartford, R.E. (2002). Mass spectrometer with a plurality of ionization probes. US Patent 6,501,073.

165 Covey, T.R., Jong, R. and Weiss, A. (2001). Fast fluidic switching system for multiple sample introduction mass spectrometry. *49th ASMS Conference on Mass Spectrometry and Allied Topics*. Chicago, IL: American Society of Mass Spectrometry (June 2001). https://www.asms.org/publications/abstracts-and-proceedings.

166 Londo, T., Grippo, P., Bantz, D. et al. (2006). Novel sample handling and introduction technology decouples throughput and carryover by utilizing autonomous syringes. *54th ASMS Conference on Mass Spectrometry and Allied Topics*, ThPl 232, Seattle, Washington (28–32 May 2006).

167 Haslam, C., Hellicar, J., and Dunn, A. (2016). The evolution of MALDI-TOF mass spectrometry toward ultrahigh-throughput screening: 1536-well format and beyond. *J. Biomol. Screening* 21: 176–186.

168 Simon, R.P., Winter, M., Kleiner, C. et al. (2020). MALDI-TOF mass spectrometry-based high-throughput screening for inhibitors of the cytosolic DNA sensor cGAS. *SLAS Discovery* 25: 372–383.

169 Schneider, B.B., Nazarov, E.G., Londry, F. et al. (2016). Differential mobility spectrometry/mass spectrometry history, theory, design optimization, simulations, and applications. *Mass Spectrom. Rev.* 35: 687–737. https://doi.org/10.1002/mas.21453.

170 Guevremont, R. and Purves, R.W. (1999). Atmospheric pressure ion focusing in a high-field asymmetric waveform ion mobility spectrometer. *Rev. Sci. Instrum.* 70: 1370–1383.

171 Hoaglund, C.S., Valentine, S.J., Sporleder, C.R. et al. (1998). Three-dimensional ion mobility/TOFMS analysis of electrosprayed biomolecules. *Anal. Chem.* 70 (11): 2236–2242.

172 Wyttenbach, T., Kemper, P.R., and Bowers, M.T. (2001). Design of a new electrospray ion mobility mass spectrometer. *Int. J. Mass Spectrom.* 212 (1–3): 13–23.

173 Ibrahim, Y.M., Baker, E.S., Danielson, W.F. III et al. (2015). Development of a new ion mobility time-of-flight mass spectrometer. *Int. J. Mass Spectrom.* 377: 655–662.

174 Russ, J., Moro, D., Sillero, J.A. et al. (2010). IMS-MS studies based on coupling a differential mobility analyzer (DMA) to commercial API-MS systems. *Int. J. Mass Spectrom.* 298: 30–40.

175 Giles, K., Pringle, S.D., Worthington, K.R. et al. (2004). Applications of a travelling wave-based radiofrequency-only stacked ring ion guide. *Rapid Commun. Mass Spectrom.* 18: 240.

176 Pringle, S.D., Giles, K., Wildgoose, J.L. et al. (2007). An investigation of the mobility separation of some peptide and protein ions using a new hybrid quadrupole/travelling wave IMS/oa-ToF instrument. *Int. J. Mass Spectrom.* 261 (1): 1–12.

177 Michelmann, K., Silveira, J.A., Ridgeway, M.E., and Park, M.A. (2015). Fundamentals of trapped ion mobility spectrometry. *J. Am. Soc. Mass Spectrom.* 26 (1): 14–24.

178 Wojcik, R., Nagy, G., Attah, I.K. et al. (2019). SLIM ultrahigh resolution ion mobility spectrometry separations of isotopologues and isotopomers reveal mobility shifts due to mass distribution changes. *Anal. Chem.* 91 (18): 11952–11962. https://doi.org/10.1021/acs.analchem.9b02808.

179 Giles, K., Ujma, J., Wildgoose, J. et al. (2019). A cyclic ion mobility-mass spectrometry system. *Anal. Chem.* 91 (13): 8564–8573.

180 Krylov, E.V. (2003). Comparison of the planar and coaxial field asymmetrical waveform ion mobility spectrometry (FAIMS). *Int. J. Mass Spectrom.* 225: 39–51.

181 Krylov, E.V., Nazarov, E.G., and Miller, R.A. (2007). Differential mobility spectrometer: Model of operation. *Int. J. Mass Spectrom.* 266: 76–85.

182 Krylov, E.V. (2012). Differential mobility spectrometer: optimization of the analytical characteristics. *Int. J. Ion Mobil. Spectrom.* 15: 85–90.

183 Schneider, B.B., Covey, T.R., Coy, S.L. et al. (2010). Planar differential mobility spectrometer as a pre-filter for atmospheric pressure ionization mass spectrometry. *Int. J. Mass Spectrom.* 298: 45–54.

184 Schneider, B.B., Covey, T.R., Coy, S.L. et al. (2010). Control of chemical effects in the separation process of a differential mobility mass spectrometer system. *Eur. J. Mass Spectrom.* 16: 57–71.

185 Schneider, B.B., Covey, T.R., Coy, S.L. et al. (2010). Chemical effects in the separation process of a differential mobility/mass spectrometer system. *Anal. Chem.* 82: 1867–1880.

186 Schneider, B.B., Nazarov, E.G., and Covey, T.R. (2012). Peak capacity in differential mobility spectrometry: effects of transport gas and gas modifiers. *Int. J. Ion Mobil. Spectrom.* 15: 141–150.

187 Schneider, B.B., Covey, T.R., and Nazarov, E.G. (2013). DMS-MS separations with different transport gas modifiers. *Int. J. Ion Mobil. Spectrom.* 16: 207–216.

188 Ieritano, C., Rickert, D., Featherstone, J. et al. (2021). The charge-state and structural stability of peptides conferred by microsolvating environments in differential mobility spectrometry. *J. Am. Soc. Mass Spectrom.* 32 (4): 956–968. https://doi.org/10.1021/jasms.0c00469.

189 Parson, W.B., Schneider, B.B., Kertesz, V. et al. (2011). Rapid analysis of isomeric metabolites by differential mobility spectrometry-mass spectrometry. *Rapid Commun. Mass Spectrom.* 25: 3382–3386.

190 Covey, T.R. (2022). Where have all the ions gone, long time passing? Atmospheric pressure ionization and tandem quadrupole mass spectrometers. Sensitivity gains since the mid 1970s. A perspective. *Rapid Commun. Mass Spectrom.* https://doi.org/10.1002/rcm.9354.

191 Little, J., Wempe, M., and Buchanan, C.M. (2006). LC-MS for drug metabolism studies: examining lipid matrix ionization effects. *J. Chromatogr. B* 833: 219–230.

192 Beach, D.G. and Gabryelski, W. (2013). Linear and nonlinear regimes of electrospray signal response in analysis of urine by electrospray ionization-high field asymmetric waveform ion mobility spectrometry-MS and implications for nontarget quantification. *Anal. Chem.* 85: 2127–2134.

193 Covey, T.R. (2019). System for high speed analysis with matrix suppression detection/correction. *67th ASMS Conference on Mass Spectrometry and Allied Topics*, Atlanta, GA (2–6 June 2019).

194 Polson, C., Sarkar, P., Incledon, B. et al. (2003). Optimization of protein precipitation based upon effectiveness of protein removal and ionization effect in liquid chromatography-tandem mass spectrometry. *J. Chromatogr. B* 785: 263–275.

195 King, R., Bonfiglio, R., Fernandex-Metzler, C. et al. (2000). Mechanistic investigation of ionization suppression in electrospray ionization. *J. Am. Soc. Mass Spectrom.* 11: 942–950.

196 Bonfiglio, R., King, R.C., Olah, T.V., and Merkle, M. (1999). The effects of sample preparation methods on the variability of the electrospray ionization response for model drug compounds. *Rapid Commun. Mass Spectrom.* 13: 1175–1185.

197 Abersold, R., Bures, E.J., Namchuk, M. et al. (1992). Design, synthesis, and characterization of a protein sequencing reagent yielding amino acid derivatives with enhanced detectability by mass spectrometry. *Protein Sci.* 1: 494–503.

198 Quirke, J.M.E., Van Berkel, G.J., and Adams, C.L. (1994). Derivatization for electrospray ionization mass spectrometry. 1. Alkyl halides, alcohols, phenols, thiols, and amines. *Anal. Chem.* 66: 1302–1315.

199 Van Berkel, G.J. and Asano, K.J. (1994). Derivatization for electrospray ionization mass spectrometry. 2. Aromatic and highly conjugated molecules. *Anal. Chem.* 66: 2096–2102.

200 Van Berkel, G.J., Quirke, J.M.E., Tigani, R.A. et al. (1998). Derivatization for electrospray ionization mass spectrometry. 3. Electrochemically ionizable derivatives. *Anal. Chem.* 70 (8): 1544–1554.

201 Van Berkel, G.J., Quirke, J.M.E., and Adams, C.L. (2000). Derivatization for electrospray ionization mass spectrometry. 4. Alkenes and alkynes. *Rapid Commun. Mass Spectrom.* 14: 849–858.

202 Star-Weinstock, M., Williamson, B.L., Dey, S. et al. (2012). LC-ESI-MS/MS analysis of testosterone at sub-picogram levels using a novel derivatization reagent. *Anal. Chem.* 84 (21): 9310–9317. https://doi.org/10.1021/ac302036r.

203 Bammesberger, S., Ernst, A., Losleben, N. et al. (2013). Quantitative characterization of noncontact microdispensing technologies for the sub-microliter range. *Drug Discovery Today* 18 (9/10): 435–446.

204 Hadimioglu, B., Stearns, R., and Ellison, R. (2015). Moving liquids with sound: the physics of acoustic droplet ejection for robust laboratory automation in life sciences. *J. Lab. Autom.* 1–15. https://doi.org/10.1177/2211068215615096.

205 Pu, F., Radosevich, A.J., Sawicki, J.W. et al. (2021). High-throughput label-free biochemical assays using infrared matrix-assisted desorption electrospray ionization mass spectrometry. *Anal. Chem.* 93: 6792–6800.

206 Radosevich, A.J., Pu, F., Cheng-Yen, D. et al. Ultrahigh-throughput ambient MS: direct analysis at 22 samples per second by infrared matrix-assisted desorption electrospray ionization mass spectrometry. *bioRxiv* https://doi.org/10.1101/2021.10.25.465730.

207 Ryumin, P. and Cramer, R. (2018). The composition of liquid atmospheric pressure matrix-assisted laser desorption/ionization matrices and its effect on ionization in mass spectrometry. *Anal. Chem. Acta* 1013: 1–11.

208 Krenkel, H., Brown, J., Richardson, K. et al. (2022). Ultrahigh-throughput sample analysis using liquid atmospheric pressure matrix-assisted laser desorption/ionization mass spectrometry. *Anal. Chem.* https://doi.org/10.1021/acs.analchem.1c05614.

209 Kempa, E.E., Smith, C.A., Li, X. et al. (2020). Coupling droplet microfluidics with mass spectrometry for ultrahigh-throughput analysis of complex mixtures up to and above 30 Hz. *Anal. Chem.* 92 (18): 12605–12612.

210 Sinclair, I., Stearns, R., Pringle, S. et al. (2016). Novel acoustic loading of a mass spectrometer: toward next-generation high-throughput MS screening. *J. Lab. Autom.* 21: 19–26.

211 Sinclair, I., Bachman, M., Addison, D. et al. (2019). Acoustic mist ionization platform for direct and contactless ultrahigh-throughput MS analysis of liquid samples. *Anal. Chem.* 91: 3790–3794.

212 Häbe, T.T., Liu, C., Covey, T.R. et al. (2020). Ultrahigh-throughput ESI-MS: sampling pushed to six samples per second by acoustic ejection mass spectrometry. *Anal. Chem.* 92 (18): 12242–12249.

213 DiRico, K.J., Hua, W., Liu, C. et al. (2020). Ultrahigh-throughput acoustic droplet ejection-open port interface-mass spectrometry for parallel medicinal chemistry. *ACS Med. Chem. Lett.* 11 (6): 1101–1110. https://doi.org/10.1021/acsmedchemlett.0c00066.

214 Zhang, H., Liu, C., Hua, W. et al. (2021). Acoustic ejection mass spectrometry for high-throughput analysis. *Anal. Chem.* 93: 10850–10861. https://doi.org/10.1101/2020.01.28.923938.

215 Wagner, A., Zhang, J., Liu, C. et al. (2020). Ultrahigh-throughput and chromatography-free bioanalysis of polar analytes with acoustic ejection mass spectrometry. *Anal. Chem.* 92 (19): 13525–13531.

216 Simon, R.P., Häbe, T., Ries, R. et al. (2021). Acoustic ejection mass spectrometry – a fully automatable technology for high-throughput screening in drug discovery. *SLAS Discovery* 26 (8): 961–973.

217 Wen, X., Liu, C., Ghislain, L. et al. (2021). Direct analysis from phase-separated liquid samples using ADEOPI-MS: applicability to high-throughput screening for inhibitors of diacylglycerol acyltransferase 2. *Anal. Chem.* 93: 6071–6079. https://doi.org/10.1021/acs.analchem.0c04312.

218 Zhang, J., Zhang, Y., Liu, C. et al. (2021). Acoustic ejection/full-scan mass spectrometry analysis for high-throughput compound quality control. *SLAS Technol.* 26 (2): 178–188.

219 Guo, Y., Forbush, M., Covey, T.R. et al. (2021). High-throughput analysis from complex matrices: acoustic ejection mass spectrometry from phase-separated fluid samples. *Metabolites* 11: 789. https://doi.org/10.3390/metabo11110789.

220 Chu, W. (2022). Acoustic ejection mass spectrometry. development, applications, and future perspective. *Biomed. Chromatogr.* 36 (2): e5278.

221 Zhang, H., Hua, W., Liu, J. et al. (2018). Acoustic-OPP-MS – the next generation high-throughput bioanalytical platform for drug discovery. *Presented at the SLAS Conference on Laboratory Automation*, Washington, DC (2 February 2019). https://www.slas.org/events/past-conferences/.

222 Zhang, H. (2018). *Acoustic-OPP-MS – The Next Generation High-Throughput Bioanalytical Platform for Drug Discovery*. Orlando, FL: Pittcon.

223 Hua, W. (2018). Acoustic-open port-mass spectrometry enabled HTS: assay development for choline transporter (CHT) uptake function assessment. *66th ASMS Conference on Mass Spectrometry and Allied Topics*, San Diego, CA (3–7 June 2018).

224 Zhang, H., Chang, L., Jianhua, L. et al. (2018). Acoustic-OPP-MS: the next generation bioanalytical platform for drug discovery with ultrahigh throughput. *66th ASMS Conference on Mass Spectrometry and Allied Topics*, San Diego, CA (3–7 June 2018).

225 Chang, L., Cox, D., Covey, T.R. et al. (2019). High-throughput ESI-MS enabled by the acoustic droplet ejection to the open port probe sampling interface. *Presented at the HT-ADME Conference*, Cambridge, MA (20 June 2019).

226 Santanilla, A.B., Egalado, E.L.R., Pereira, T. et al. (2015). Nano-scale high-throughput chemistry for the synthesis of complex molecules. *Science* 347 (6217): 49–53. https://doi.org/10.1126/science.1259203.

227 Rienzoa, M., Jacksona, S.J., Chaoa, L.K. et al. (2021). High-throughput screening for high-efficiency small-molecule biosynthesis. *Metab. Eng.* 63: 102–215. https://doi.org/10.1016/j.ymben.2020.09.004.

228 Si, T., Lihao, F., and Jianzhi, Z. (2020). Recent advances in high-throughput mass spectrometry that accelerate enzyme engineering for biofuel research. *BMC Energy* 2: 1. https://doi.org/10.1186/s42500-020-0011-8.

229 Drouin, N., Liu, C., Cox, D. et al. (2021). High-throughput mass spectrometry assay for early chick gender determination. https://vimeo.com/529385576 (accessed 29 March 2023).

230 Lee, E.D., Muck, W., Henion, J.D., and Covey, T.R. (1989). Real-time reaction monitoring by continuous-introduction ion-spray tandem mass spectrometry. *J. Am. Chem. Soc.* 111: 4600–4604.

231 Knight, W.B., Swiderek, K.M., Sakuma, T. et al. (1993). Electrospray ionization mass spectrometry as a mechanistic tool in enzymology: mass of human leucocyte elastase and its β-lactam-derived inhibition complex. *Biochemist* 32: 2031–2035.

232 Underwood, D.J., Green, B.G., Chabin, R. et al. (1995). Mechanism of inhibition of human leucocyte elastase by β-lactams. 3. Use of electrospray ionization mass spectrometry and two-dimensional NMR techniques to identify β-lactam-derived inhibition complexes. *Biochemist* 34: 14344–14355.

233 Bravo-Veyrat, S. and Hopfgartner, G. (2021). Mass spectrometry based high-throughput bioanalysis of low molecular weight compounds: are we ready to support personalized medicine? *Anal. Bioanal. Chem.* 414 (1): 181–192. https://doi.org/10.1007/s00216-021-03583-2.

234 Thomson, B.A. The heated nebulizer LC/MS interface. In: *Encyclopedia of Mass Spectrometry*, vol. 6 (ed. M.L. Gross and R.M. Caproli). Elsevier.

235 Caproli, R.M., Fan, T., and Cottrell, J.S. (1986). Continuous-flow sample probe for fast atom bombardment mass spectrometry. *Anal. Chem.* 58: 2949–2954.

236 Ito, Y., Takeuchi, T., Ishii, D., and Goto, M. (1985). Direct coupling of micro-high-performance liquid chromatography with fast atom bombardment mass spectrometry. *J. Chromatogr.* 346: 161–166.

Section 2

LC-MS

2

The LeadSampler (LS-1) Sample Delivery System: Integrated Design and Features for High-Efficiency Bioanalysis

Brendon Kapinos[1] *and John Janiszewski*[2]

[1] *Pfizer Worldwide Research and Development, Groton, CT, USA*
[2] *National Center for Advancing Translational Sciences (NCATS), Rockville, MD, USA*

2.1 Introduction

Biopharmaceutical drug discovery has been revolutionized by recent advancements in foundational disciplines that have shortened cycle times and expanded the scope of actionable chemical space [1]. Integration of novel hit-identification methods, such as affinity selection-mass spectrometry (ASMS) and DNA-encoded libraries (DEL), can rapidly identify pharmacologically interesting compounds by assessing target engagement [2]. Along with fragment-, phenotypic-, and virtual-screening methods, these technologies can lower the cost and reduce the lead generation (LG) timeframe in comparison with traditional HTS [3–5]. Project teams often deploy multiple strategies in parallel as they seek to identify active small molecule entities (SME). The emerging paradigm, depicted in Figure 2.1, significantly expands the types of samples and screening applications that use liquid chromatography–mass spectrometry (LC-MS) as their endpoint detector [6]. Over the past decade, High-Throughput liquid chromatography and tandem mass spectrometry (HT-LC/MS/MS) has emerged as the analytical technique of choice for screening applications in drug discovery. Advancements in mass spectrometry technology are largely responsible for the rapid expansion of drug discovery screening applications that rely on "medium"-throughput LC/MS bioanalysis to deliver high-quality quantitative data [7, 8]. Chemogenomic and lead optimization studies of all types fall into this category of 96- and 384-well assays requiring analysis of ~100–10,000 samples per study [9, 10]. Coordination and communication between client and support groups is important in building robust LC-MS data pipelines given time and resource constraints encountered in pharma R&D. A key challenge

High-Throughput Mass Spectrometry in Drug Discovery, First Edition. Edited by Chang Liu and Hui Zhang.
© 2023 John Wiley & Sons, Inc. Published 2023 by John Wiley & Sons, Inc.

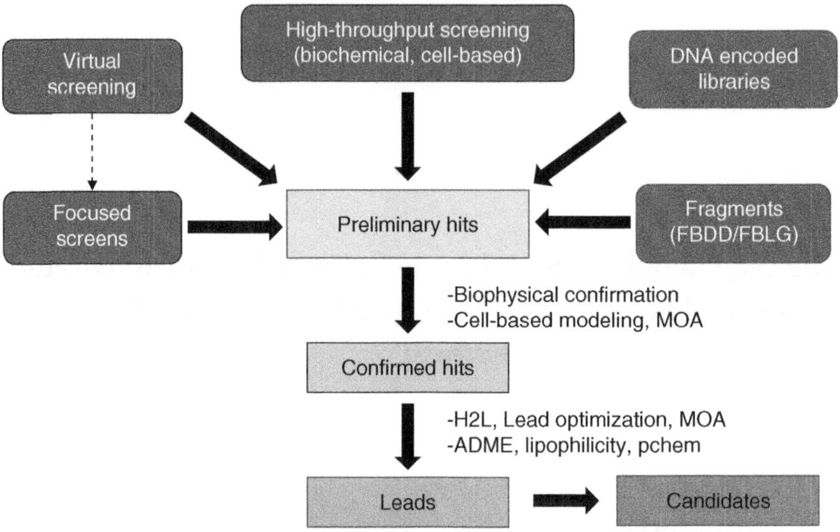

Figure 2.1 Drug discovery LC-MS-based screening pathways. Although project teams share objectives, each team's target biology and therapeutic hypothesis are unique. LC-MS is a versatile endpoint readout that can deliver high-quality data across a wide range of screening models throughout project lifecycles.

is to individually tailor the right mix of screening methods for each project to facilitate data integration and provide a multi-objective quality assessment of active chemistry [11–13]. Current practice tends to stitch together LC/MS/MS-derived chemical and biological assay results to build a decision-making narrative around target biology and active chemistry to establish confidence in forward rationale from project inception [14]. A broad range of plate-based ADME (absorption, distribution, metabolism, and excretion), safety, and physicochemical property screens are used routinely to evaluate the fitness of SMEs. The widespread adoption of automated liquid-handling platforms, expertise in cell culture, functional genomics, and automation allow rapid development of custom assay pipelines to address chemistry, biology, and translation issues as they arise during a project lifecycle [15]. Project teams have the freedom to creatively construct a panel of assays best suited to assign value and/or discharge risk as the chemistry under study is understood and refined [16]. The primary bioanalytical requirements in this scenario are throughput and flexibility to match LC-MS methods with incoming sample sets in a seamless fashion.

To manage multiple projects, assay types, and formats slated for HT-LC/MS/MS analysis, bioanalytical managers often need to choose between LC/MS/MS systems designed for high-throughput (~10–15 seconds per sample) or those capable of high-performance separations [17]. The latter may feature operation at ultrahigh pressure (UHPLC-MS), lower flow rates (1–20 μL/min, sub μL/min range), or in combination. Often, attributes that increase chromatographic resolution or enhance detection sensitivity can limit throughput [18–21]. When comparing systems, there are numerous tradeoffs and instruments are typically purchased for specific tasks. While a mix of mass spectrometers with different vendor software and functionality within a bioanalytical screening team may prove advantageous in broadening the types of studies undertaken, this arrangement can make automation and integration of the respective screening workflows more difficult and reduce throughput.

Despite significant advances in mass spectrometry instrumentation, there are very few commercially available options for sample delivery that have sufficient flexibility to address the multitude of challenges encountered in discovery bioanalysis [22]. In context of Figure 2.1, computationally derived "virtual" screening sets having a smaller number of compounds, may be a cost-effective way to explore chemical space. Screens testing 1000 total SMEs can often be completed in one day, while a 10,000 compound screen may take one to two weeks depending on the specific LC-MS methodology used [74]. Virtual screening collections are built using a computational framework that typically has a focused structural and/or property-based rationale [23]. Once active chemistry is identified, the team may already have ideas on what to make next, or follow-up Lead Optimization (LO) screening may be requested to confirm hypotheses. The "traditional" LG to LO project stages are run concurrently. To pursue LO ideas, well-constructed bioassays can be developed and brought online rapidly to address specific translationally important issues that arise around chemical biology, MOA, and biomarker selection pertaining to the model under study [24–27]. Rather than a fixed panel of established screens, project teams can develop or test several cell models "on-the-fly" to discern an endpoint that can be used to infer clinical efficacy. These applications typically require high-performance gradient LC methods that can separate co-dosed inhibitors and/or quantify multiple ordinates per sample [28–30]. The requirements are for rapid methods development and high-throughput analysis to turnaround data quickly.

In this chapter, we describe development of a Sample Delivery system, the "Leadsampler-1" (The LS-I is sold commercially by Sound Analytics, Niantic, CT), that was designed to meet the challenges associated with bioanalytical support of modern drug discovery. The system supports high-performance of UHPLC, microflow, gradient, and trap and elute LC methods while enabling integration into an industrialized pipeline that includes compound management, liquid handling, and cell culture automation. The system's control software "LeadScape" was designed to manage multiple study designs in parallel and associate samples, compounds, methods, and users automatically during batch submission. The associated meta data is merged with raw data during chromatographic review and exported to downstream clients in user-configurable formats. Development of the system, design criteria, key features and applications will be discussed in context of biopharmaceutical drug discovery and describe LS-1/LeadScape as a platform upon which emerging biotechnology can be developed and deployed.

2.2 Hardware and System Design

The LS-1 was designed to achieve high-performance LC gradient separations at sufficient throughput to enable the broadest range of plate-based screening studies across drug discovery project lifecycles. Earlier versions of the system were focused on achieving robust performance along with high per sample throughput (~10–15 seconds per sample), using a "trap and elute" LC approach [31, 32]. While these systems, the DAAS (Dual-Arm Autosampler) and ADDA (Apricot Designs Dual-Arm), could run gradient methods at relatively high-throughput (~1.0–3.0 minute per sample), the overall system volume was limiting for both LC performance and throughput. Fluid lines between the injection valves, located on the side of the instrument, scheduling valves, and the ion source added volume and negatively impacted performance for gradient LC methods that were ordinarily run at lower flow rates, in the range 0.5–0.8 mL/min. System volume was much less of an issue for trap and elute methods that are run at higher flow rates, 0.8–1.5 mL/min. Additional volume equates to wasted time, especially for lower flow gradient methods. Reducing system volume was a major design objective. In addition, the LS-1 was designed for ease-of-use, straightforward diagnosis, and maintenance by laboratory staff. To industrialize the system, LC plumbing and valves were made easily accessible, with the expectation that bioanalytical staff can troubleshoot and correct any problems that arise and get the system back online quickly without needing a service call. The first prototype LeadSampler autosampler is shown in Figure 2.2. The instrument features an Omni liquid handling robot integrated with two XLP 6000 syringe pumps from Cavro/Tecan for aspiration and dispense sampling. Rheodyne UHPLC valves handle sample and fluid routing. Network and instrument communications are connected via an "Electronics box" (the ebox). The ebox is tethered to the LS-1 via a 2-m cable; it can be placed in a location that is convenient for peripheral device communication. Decoupling these functional units (i.e. autosampler and electronics) helps free up space and access to the liquid handling deck. The system can accommodate up to four injection ports (6-port, 2-position valves), an additional four valves, three 2-position and one multi-position valve, can be installed on the system and scheduled in code. The syringe pumps and valves can be replaced as integrated assemblies thereby minimizing downtime in the event service is required. The standard onboard valves have 1/16 in. outer diameter (OD) ports and accept standard LC tubing and fittings.

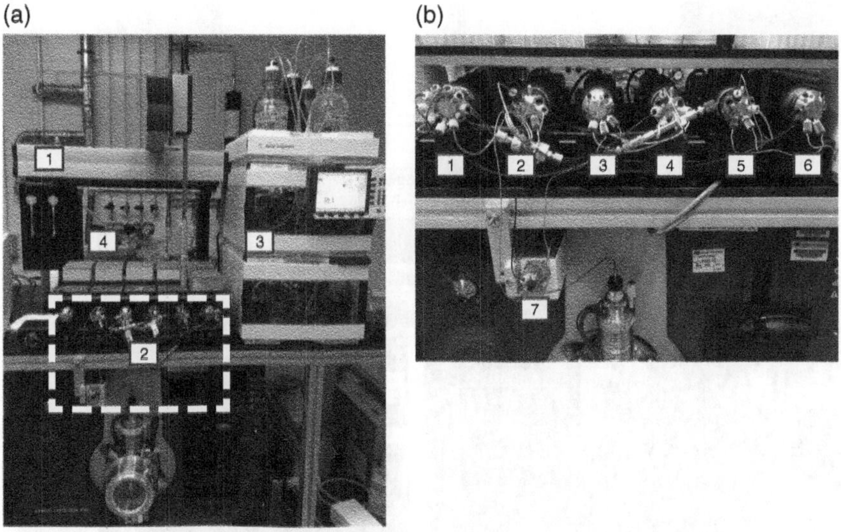

Figure 2.2 Prototype LeadSampler platform. (a) System overview. 1. Cavro® Omni Robotic Arm, 2. Valve Manifold and ion source inlet, 3. Stacked 1290 Binary UHPLC pumps (Agilent), 4. Network communication cables and CompactRIO controllers are housed in a tethered "Electronics-box." (b) Multimode LC plumbing: 1. 10-port "trap-and-elute" diverter valve, 2. 6-port MS/MS tuning injection valve, 3. 10-port gradient diverter valve, 4. 6-port "trap-and-elute" injection valve, 5. 6-port UHPLC gradient injection valve, 6. 10-port pump flow selector, 7. 4–1 multi-position stream selector, ESI grounding lead to metal LC fitting at ion source inlet, 0.004″ ID tubing.

The LS1's compact design allows placement of the unit within a small work envelope that includes HPLC pumps and mass spectrometer (Figure 2.3). This configuration minimizes system volume such that throughput and performance are enhanced across LC methods, including microflow LC in the range 5–20 μL/min. A reduction in band broadening, a result of lowered system dispersion, provides "UHPLC-Like" [33–35] chromatographic performance and high-throughput. To demonstrate the impact of these design features, identical analyses were performed on the LS-1 using "standard" versus optimized LC plumbing configurations. A set of physiochemically diverse small molecule substrates were analyzed using a fast LC gradient method (1-minute cycle time). The test mix was analyzed iteratively following adjustments to pre- and post-column system volume, including in-line mixer volume and transfer tubing internal diameter (ID), column outlet tubing, and ESI electrode ID. System configurations were tested by carefully measuring pre- and post-column volumes associated with each variation to determine the optimal plumbing configuration (Table 2.1). Chromatographic performance was assessed using peak capacity, retention factor, and peak widths at half height for benchmarking and comparison [36]. The largest improvement in peak capacity

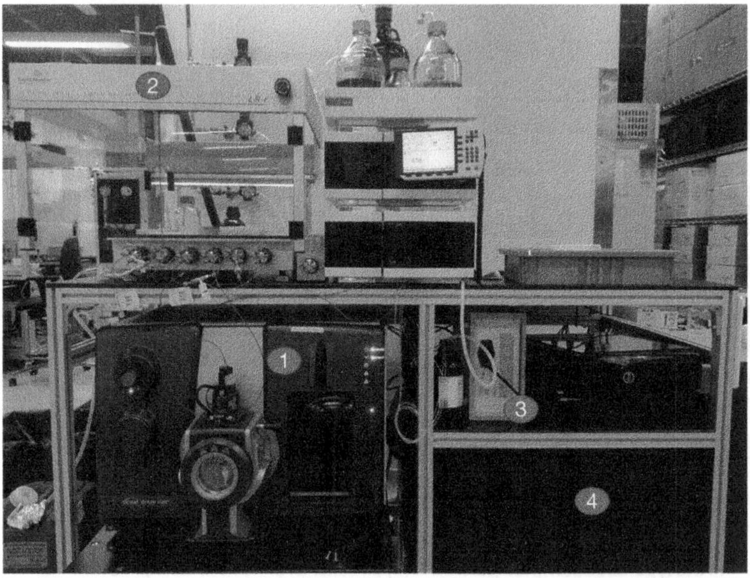

Figure 2.3 The current LeadSampler platform. The Sciex 6500 mass spectrometer (1) is nested underneath the LS-1 (2). The overall dimensions of the LS-1 cart are 32 × 72 × 42 in (D × W × H). The ebox (3) rests on a shelf above the rough pump enclosure (4). Just out of the frame is a swing arm keyboard and monitor tray at about counter height.

was achieved by optimizing post-column volume. This was accomplished by installing short, narrow-bore transfer tubing (10–20 cm × 100 μ ID) to interface column outlet and downstream diverter valve. Additional improvement was realized when dimensions of transfer tubing leading to the ion source were decreased, and further when ESI electrode ID was reduced. Reduction of gradient mixing volume and associated transfer tubing lengths in turn affected delay time. In comparing chromatographic performance, it was found that the LS-1 in its fully optimized configuration delivered superior analyte separation and peak shape. Peak width was reduced by roughly 35% on average compared to the standard system (Figure 2.4a–b). These studies highlighted system volume as the primary determinant of chromatographic performance and throughput. Significant gains in LC performance in terms of peak capacity and chromatographic resolution were realized without modifying LC flow rate or gradient program. The LS-1's design and optimized fluidics were the primary drivers for achieving superior LC performance and high-throughput. Enhanced LC separations can increase the data content per sample when quantifying multiple ordinates, and can be particularly useful for applications where analytes must be separated from closely eluting matrix components to avoid ion suppression [37]. These features broaden the types of assays and applications amenable to LC-MS-based screening in drug discovery.

Table 2.1 Volumes associated with system components and plumbing.

		Standard	65 μ electrode	Bypass tee	Standard, optimized tubing 1	Standard, optimized tubing 2	External mixer (10 μL)	External mixer (2 μL)	No mixer	Optimized
Pump/mixer	Pump	10	10	10	10	10	10	10	10	10
	Transfer tube[a]	7.6	7.6	7.6	7.6	0	0	0	0	0
	Mixer	35	35	35	35	0	10	2	0	0
Pre-column	Pump . . . pump valve[b]	20.4	20.4	20.4	20.4	2.8	6.3	6.3	2.8	2.8
	Pump valve . . . inj port	2.5	2.5	2.5	1.6	1.6	2.5	2.5	2.5	1.6
	Inj port . . . column[c]	9.5	9.5	9.5	2.4	2.4	9.5	9.5	9.5	2.4
Post-column	DIVERTER . . . stream sel.	2.5	2.5	2.5	1.6	1.6	2.5	2.5	2.5	1.6
	Stream sel . . . tee	7								
	Tee . . . electrode	5.7								
	125 μ electrode	2.5		2.5	2.5	2.5				
	Stream sel . . . electrode[d]		1.8	3.2	3.2	3.2	1.8	1.8	1.8	1.8
	65 μ electrode		0.66				0.66	0.66	0.66	0.66
Total volume (μL)		102.7	90.0	93.2	84.3	24.1	43.3	35.3	29.8	20.9

The "optimized" setup reduced system volume fivefold compared with standard configuration.
[a] Green spirally tube from pump outlet to Jet Weaver.
[b] 20.4 μL = Agilent tubing. 6.3/2.8 μL tubing vol. Either side of mixer.
[c] 9.5 μL = typical sampler tubing vol. (0.005″), 2.4 μL = LS-1 (0.004″).
[d] 3.2 = 125 μ electrode w/0.005″ versus 0.66 μ electrode and 75 μm ID, 1/32 fitting.

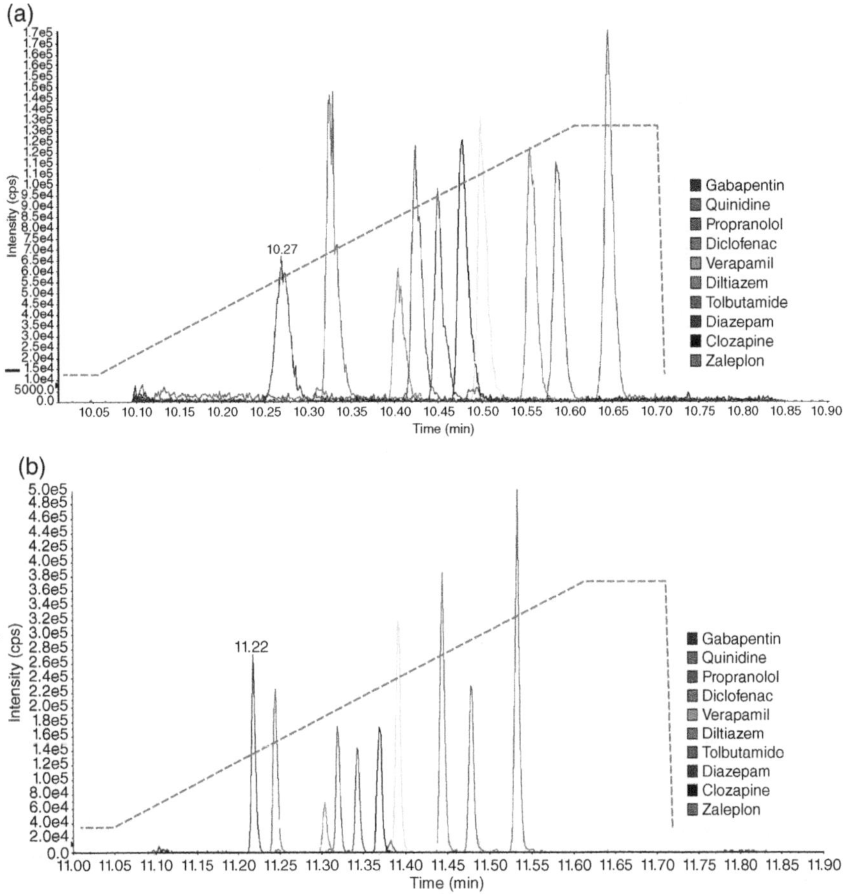

Figure 2.4 (a) LC trace with "standard" configuration: HALO C18 direct-connect (20 × 2.1 mm), 55C temp, 0.005″ID tubing pump valve to injector (20 cm), injector to column, div to stream select (20 cm), stream selector to source tee. Agilent JetWeaver mixer (35uL) + 0.17 × 900 mm tubing to pump valve, standard TIS electrode, 0.005″ID source tubing, grounding tee included. (b) "Optimized" configuration: HALO C18 direct-connect (20 × 2.1 mm), 55C temp, Optimized plumbing throughout (0.004″ID), bypasses Jetweaver on 1290 (MarvelX 100u × 350 mm tube to injector), 65 μ electrode.

2.3 Software Integration

LeadScape is the primary interface for the LS-1 sample delivery system. The earliest prototype versions of LeadScape, nicknamed "Evalution" and "AutoScan," were developed in-house at Pfizer [38–41]. AutoScan was developed to automate compound tuning and MS/MS methods development, its functionality is similar to Sciex's "Automaton" software [42]. AutoScan stored acquired MS/MS settings in an MS-Access database. Evalution was used to automate batch submission. It connected directly to AutoScan's database of stored methods for automatic download and creation of injection sequences. Evalution was also used for chromatographic peak review and integration of "multiply-injected" data acquisition files. Injecting multiple samples into an open acquisition file significantly reduces the number of files to manage and review. Depending on a user's preference, a screening run comprising ten 384-well plates (3840 samples total) can be split into 10 files, each with 384 injected samples. Alternatively, each file may contain 24 injections, 16-files per plate, 160 files total. Nearly any multiply injected sampling pattern that makes sense to the user could be setup in the software. Partitioning the data in this way can expedite chromatographic review. By arranging peak response in a format that provides experimental context (e.g. IC50), the data can be viewed in relation to the endpoint readout. The graphical depiction of raw MS signal (intensity, peak shape, %CV of Internal Standard, S/N, fidelity of integration, etc.) provides figures of merit in qualifying each discrete multiply injected data set (Figure 2.5). The overall workflow, from receipt of samples, through layout of an injection sequence, retrieval of MS/MS methods, batch submission and scheduling followed by post acquisition data review and export, comprise the linked sequence of events that were automated using AutoScan and Evalution. These tasks are time-consuming and prone to error, they become nearly

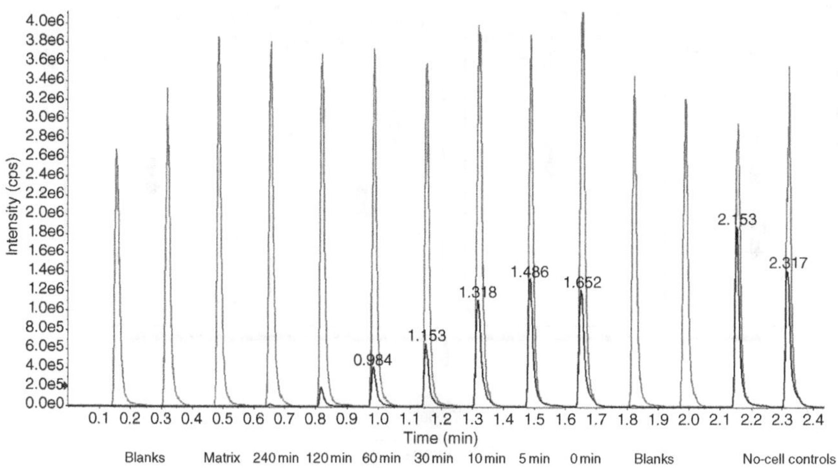

Figure 2.5 Bioanalytical data from high-throughput in vitro pharmacokinetic screen collected with LS1 and LeadScape. Multiply injected, extracted ion chromatogram provides simultaneous measurement of internal standard (gray trace) and test compound (black trace) per timepoint from primary human hepatocyte clearance study.

impossible to manage manually or in semiautomated fashion once scaled to 1000s of samples. The original versions of AutoScan and Evalution were written using a combination of AppleScript and FaceSpan coding software for use with Sciex's "MassChrom" operating system on the Apple MacIntosh computer platform. The AutoScan and Evalution modules were migrated to a Visual Basic 6.0 code base around 2003 [43]. The functional units were released commercially as the "Optimize" and "Analyze" modules within DiscoveryQuant (DQ) software. DQ was released in 2007 by Sciex and Sound Analytics. LeadScape and DQ share nearly identical functionality, user interfaces, and workflow automation. LeadScape adds control of the LS-1 autosampler and LC system to integrate LC-MS analysis into the drug discovery bioanalytical screening process. LeadScape and DQ were designed as enterprise products, users are linked together by a central SQL server database that stores MS/MS conditions (Figure 2.6). As of 2018, the DQ-database had over 600,000 compound entries and connects to approximately 50 Sciex mass spectrometers at Pfizer's R&D site in Groton, CT.

The design of LeadScape was strongly influenced by the emergence of Combinatorial Chemistry and High-Throughput screening at Pfizer circa 1997–2000 [44, 45]. At that time, multiple teams, Medicinal chemistry, Drug Metabolism, Compound Management and Small molecule purification, were engaged in building the infrastructure to maximally utilize these technologies in

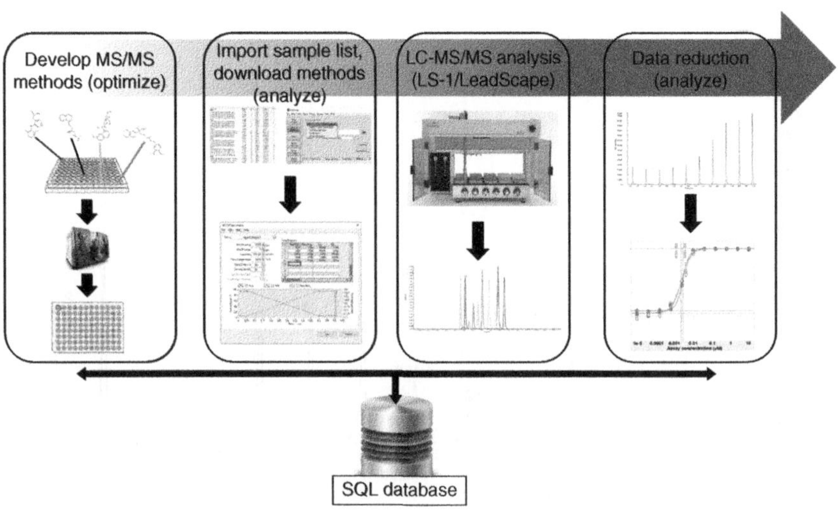

Figure 2.6 Flow of information through key touchpoints in the LeadScape bioanalytical workstream.

LG and optimization screening assays [46–52]. High-throughput ADME screening was developed to guide the design of SMEs that have the desired activity, but also suitable potency and duration of action. Plate-based in vitro ADME models are surrogate measures of in vivo properties. They are valuable in differentiating lead structural chemistries and to confirm hypotheses [53–61]. LC-MS provides a high-quality quantitative readout but generally at much lower throughput when compared with HTS. The challenge was to make LC-MS fast enough to cover an adequate subset of lead chemistry to guide drug design, selection, and enable decision-making therein.

From the outset of this effort, text files indexing the contents of each screening well were used across respective teams to transfer information. Spreadsheet-based plate files were the easiest way to move ADME screening forward within the broader organizational effort to build a drug discovery screening platform [62, 63]. To that end, LeadScape was designed to import lists of compounds in text file format, 96- and 384-well, including complex sample layouts, automatically [63, 41, 43]. Typical sample information captured includes compound, project, date and time, and submitter identities.

LeadScape's top-level menu (Figure 2.7) requests a minimum amount of information from the user prior to commencing high-throughput bioanalysis [64]. The design minimizes user interaction needed to acquire LC-MS runs of 1–3840 samples. The Batch queue remains active during runtime. Additional samples sets can be submitted or the queue can be paused to remove or add plates to the deck. LeadScape schedules analysis across multiple interconnected experimental modules, including *Optimize* and *Analyze*, as well as respective *Review* modules, on-the-fly. LeadScape automation reflects the development of the *Optimize-Analyze-Review* workflow at Pfizer.

The batch building process starts with the "Select Compounds" user interface as depicted in Figure 2.8. Automated options for building an injection sequence and

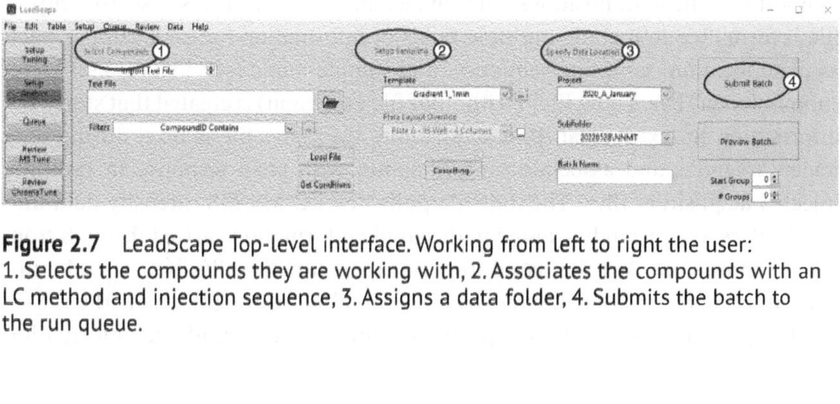

Figure 2.7 LeadScape Top-level interface. Working from left to right the user: 1. Selects the compounds they are working with, 2. Associates the compounds with an LC method and injection sequence, 3. Assigns a data folder, 4. Submits the batch to the run queue.

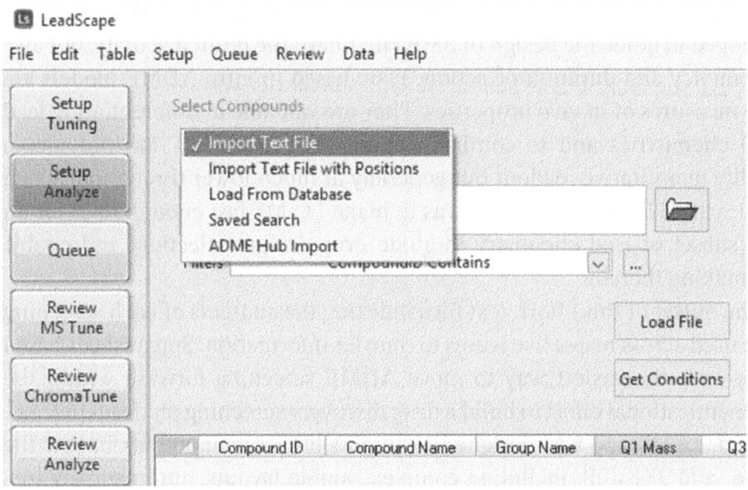

Figure 2.8 The user can import a text file list of compounds and query the data base to associate each compound with an MS/MS method. The "import text file with positions" option requires a formatted text file that includes sample locations and specifies the injection sequence. Additional options allow user to pull compounds or batch files directly from the central database, or the user can search for a specific compound. A list of compounds or metabolites (such as Krebs cycle intermediates) can be flagged within the database and pulled into the batch using a flag (suffix or prefix attached to each metabolite).

submitting it to the batch queue require MS methods for each compound to be present in a LeadScape-accessible database. Methods consist of precursor ion, fragment ions, DP and CE values, for each discrete entry. Storage of LC conditions is optional. Optimal MS/MS settings can be derived experimentally using the LeadScape Optimize Module or manually entered. During quantitative bioanalysis, additional sample information (metadata) can be recorded using the "Import text file with positions" option (see Figure 2.9). In this case, the import file is formatted to include multiple fields of information for each sample, such as plate location, sampling sequence, and sample name (see Figure 2.9.) Upon submission, a "batch data file" (with *.bdf file extension) is created that stores the information in text format. Post-acquisition, the "bdf" file is used during data review to index and associate sample names with peak response in multiply injected acquisition files. The .bdf file pulls raw LC/MS/MS intensity data into the *Analyze Review* module along with sample information carried through from batch submission. Text file-based import–export functionality is a critical automation component that drives overall sampling throughput of the LS-1/LeadScape workstation. This type of low-level human-readable information

(a) Import file

Sample name	Sample ID	Compound ID	Stream	Group name	Plate location	Vial position	Plate code
BLK1		13C5-Glu2	Stream 1	set1	Plate 2	1	384 -std
PLT_8_A01	A1	13C5-Glu2	Stream 1	set1	Plate 1	1	384 -std
PLT_8_A02	A2	13C5-Glu2	Stream 1	set1	Plate 1	2	384 -std
PLT_8_A03	A3	13C5-Glu2	Stream 1	set1	Plate 1	3	384 -std
PLT_8_A04	A4	13C5-Glu2	Stream 1	set1	Plate 1	4	384 -std
PLT_8_A05	A5	13C5-Glu2	Stream 1	set1	Plate 1	5	384 -std

(b) Batch data file (*.bdf)

Sample name	Sample ID	Wiff filename	Analyte cmpd ID	Analyte mass	IS cmpd ID	IS mass	Analyte RT	IS RT	Analyte area	IS area
BLK1		20220525_YQ___Q0000529_P2.wiff	13C5-Glu2	153.10/88.10	D3Glu	151.00/87.00	0.75	0.75	0	802616.92
PLT_8_A01	A1	20220525_YQ___Q0000529_P2.wiff	13C5-Glu2	153.10/88.10	D3Glu	151.00/87.00	2.73	2.73	0	644754.44
PLT_8_A02	A2	20220525_YQ___Q0000529_P2.wiff	13C5-Glu2	153.10/88.10	D3Glu	151.00/87.00	4.73	4.73	515308.89	665450.84
PLT_8_A03	A3	20220525_YQ___Q0000529_P2.wiff	13C5-Glu2	153.10/88.10	D3Glu	151.00/87.00	6.73	6.73	1040266.36	669225.71
PLT_8_A04	A4	20220525_YQ___Q0000529_P2.wiff	13C5-Glu2	153.10/88.10	D3Glu	151.00/87.00	8.73	8.73	1022363.27	656617.81
PLT_8_A05	A5	20220525_YQ___Q0000529_P2.wiff	13C5-Glu2	153.10/88.10	D3Glu	151.00/87.00	10.73	10.73	741884.11	654460.4

(c) Export file format

Sample ID	Analyte area	IS area
	0	802616.92
A1	0	644754.44
A2	515308.89	665450.84
A3	1040266.36	669225.71
A4	1022363.27	656617.81
A5	741884.11	654460.4

Figure 2.9 Import text file with positions. Users can customize import and export formats for batch submission and data processing. The import file (a) requires Plate Location and Vial Position to guide sampling sequence, MRM methods are pulled from the SQL database, additional fields specify sample names and project specific information. The Batch Data File (b) records imported information along with timestamps and raw MS/MS signal counts post-processing. The Export File (c) can be formatted for direct import into an endpoint calculation software (e.g. GraphPad Prism).

transfer was continually improved over software development cycles. Figure 2.10 depicts import–export formatting utilities used in LeadScape. These features allow users to kick-off a 3840-sample run (10× 384-well plates) in about 10 minutes. All LeadScape batch submissions, from 10 to 3840 samples, take about the same amount of time to setup and submit. The LeadScape batch queue can be added to and rearranged during acquisition. Figure 2.11 depicts the LeadScape queue during runtime.

In retrospect, using simple broadly accessible spreadsheet-based lists of information (for example, well ID, sample name, and MW) to automate the LC-MS screening process has worked out well. Dictating import formats to client groups is difficult to manage due to constantly evolving science, assays, and projects. LeadScape's formatting tools allow teams to quickly set up customized data import and export formats to fit the given project's data pipeline. Moving information accurately and quickly are the primary goals.

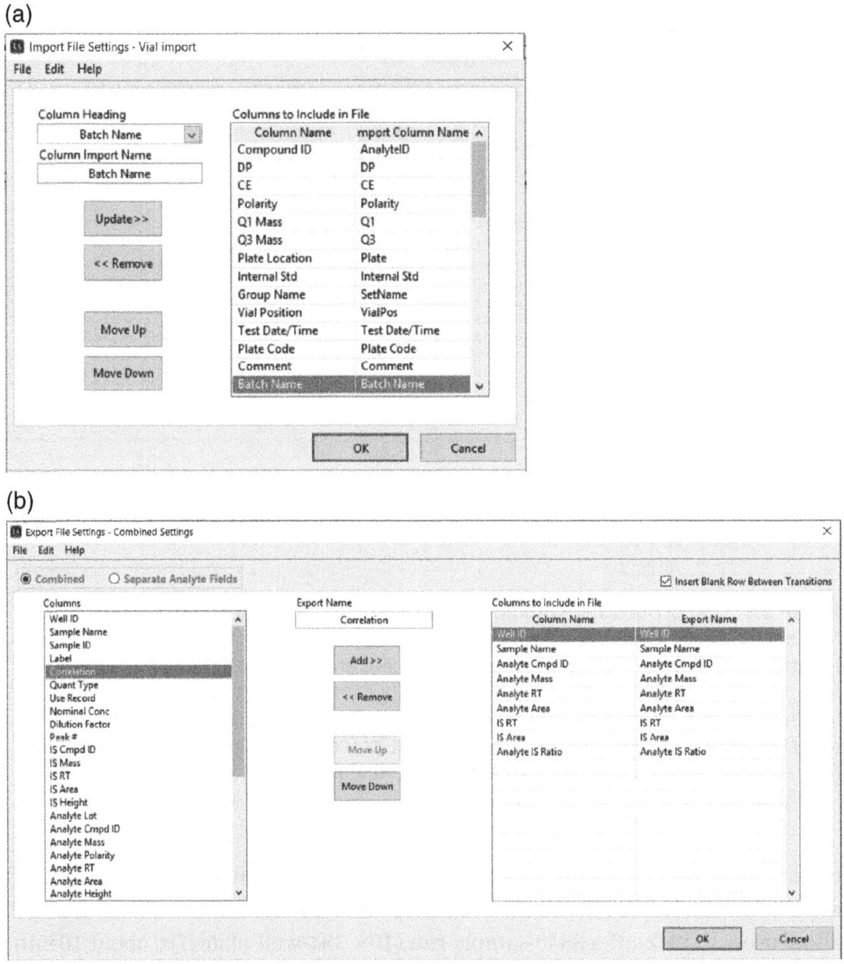

Figure 2.10 LeadScape user interfaces for configuring input and data formatting. (a) Import file settings can be configured and accessed for flexibility across workflows. (b) Highly flexible data file formatting enables seamless integration with LIMS following review in LeadScape.

2.4 Enabling Emerging Techniques

High-throughput, in vitro ADME, and biopharmaceutical property assessment assays require costly consumables for assay execution, consume liters of solvent, and generate enormous quantities of solid and liquid hazardous waste. A potential avenue for reducing overhead costs can include assay miniaturization. Shifting

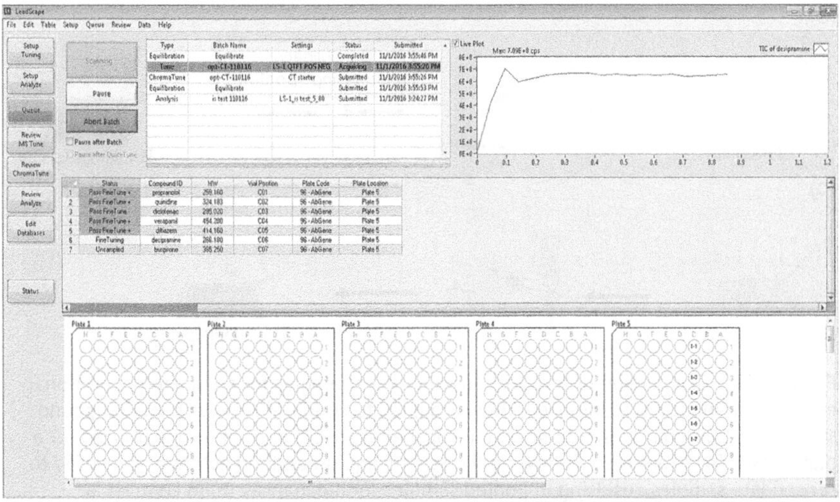

Figure 2.11 LeadScape queue displays status of past, current, and future submitted batches, and corresponding sample metadata.

from 96 to 384, or even to 1536-well, assay formats can greatly reduce the amount of reagent consumed per experiment. This can translate to significant cost savings, especially in the setting of ADME screening wherein expensive cellular fractions are utilized for purposes of in vitro–in vivo correlation. However, assay miniaturization can be hindered by time-consuming development and validation, as well as technical limitations presented by liquid-handling automation, consumables, or evaporation when considering an assay with very low incubation volumes. A miniaturized assay can generate less sample volume but dealing with very low sample volumes is not always practical or cost effective.

An alternate and/or complimentary strategy for increasing workflow efficiency is to intentionally combine endpoints, achieving higher throughput by virtue of sample multiplexing [65–68]. In context of high-throughput ADME screens, the combined endpoints can be small molecule lead chemistries that differ in m/z but have similar chromatographic retention properties. This approach can be implemented in several ways, including at the outset of an in vitro assay – which can be referred to as "assay-cassetting" – as well as immediately following an experiment, termed "sample-pooling." While both approaches yield samples containing multiple analytes, the former describes a co-dosed experiment design, such that a single sample contains several test substrates. In contrast to singleton assays that generate samples each with a single unique compound (Figure 2.12a), assay cassetting utilizes compound compression at the point of stock plate preparation (Figure 2.12b). The cassetted configuration would therefore have

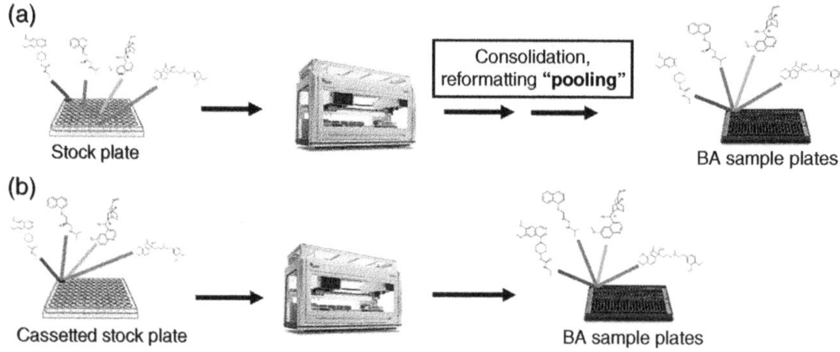

Figure 2.12 Comparison of sample-consolidation strategies for enhancing operational efficiency of in vitro screening workflows coupled to LC-MS/MS analysis. (a) Compound stock plate containing a single novel chemical entity (NCE) in each well is utilized for a plate-based in vitro screen. Following the screen, liquid-handling automation is used to consolidate samples, resulting in pooled samples for bioanalysis. (b) In contrast, a cassetted stock plate, containing $n = 4$ NCEs per well, can be prepared prior to an in vitro assay, resulting in sample plates for bioanalysis that each contain $n = 4$ NCEs per well.

n-fold greater capacity from a single assay. This approach yields significant savings in various dimensions. For example, given a fixed number of compounds for screening each week, when compared to a singleton assay, a cassetted version would require fewer runs and resources, such as assay reagents, plates, tips, FTE, and instrument time. While application of this strategy to some in vitro discovery assays could be contraindicated (e.g. for metabolism assays wherein co-dosing could introduce drug–drug interactions), optimization at this level can reduce R&D resource costs without requiring a change in liquid-handling instrumentation or assay methodology to implement.

Historically, efforts to improve LC-MS/MS efficiency have focused on slashing injection-to-injection cycle time. Most often, very short columns or trap cartridges are used to minimize re-equilibration time required by chromatographic media following each injection, reducing overall cycle time at the expense of LC separation [69, 70]. These methods typically use high LC flow rates (≥1.0 mL/min); a single run can consume considerable solvent and generate liters of hazardous liquid waste. This "traditional" approach to achieving high-throughput comes at considerable cost.

Micro-flow LC-MS/MS (μFLC-MS/MS) touts numerous benefits compared to conventional LC-MS/MS analysis, including reduced solvent consumption and waste generation, and enhanced sensitivity in many instances. In contrast to conventional LC, μFLC typically utilizes flows below 50 μL/min. The shift in LC flow rate renders extra-column volume a critical parameter, and underscores the need for LC pumping systems specifically designed for precise low-flow solvent delivery. These technical considerations have slowed μFLC adoption for high-throughput applications [71]. Commercially available microflow systems offer poor sample throughput and are dedicated to specific use cases limiting their applicability. The LS-1 was designed to address the need for a high-throughput microflow LC-MS system that can be used in an industrial setting. Critical system parameters were identified, optimized, and subsequently incorporated into the sample delivery platform.

Preliminary development and testing focused on assessing μFLC performance with existing instrumentation and equipment. Agilent 1290 Infinity pumps were situated as close as possible to the sample delivery system to minimize tubing lengths running from LC pump outlet to injector valve. Tubing ID was reduced from 170 to 100 μ, and the standard 35 μL JetWeaver mixer was replaced with an external 1 μL in-line mixer to reduce delay volume. Post-column dead volume was identified as a major contributor to band-broadening and reduced chromatographic performance. A source of post-column dead volume being the MS/MS electrode itself. To address this, a 65 μ electrode replaced the standard 150 μ ID version. Following system configuration changes, performance was assessed via analysis of a cocktail of small-molecule substrates at 50 μL/min on a 30×0.5 mm HALO C18 column using a linear gradient. The total cycle time was 90 seconds. Alternative solvent-delivery instruments, designed expressly for lower flow rates, were also tested, including Eksigent ExpressHT pumps. Under identical analytical conditions, the Eksigent pumps delivered superior analyte separation, peak shape, and narrower peak widths compared with the Agilent 1290. We attribute enhanced performance to microfluidic flow control, which provides highly precise LC gradients at capillary flow rates [72]. The "Zirconium Ultra" pump (Prolab Instruments, CH) has microflow performance specifications that are similar to the Eksigent system. The Zirconium pump has about twofold greater solvent capacity (~2700 μL) compared with the Express HT (1300 μL). Of particular importance is the Zirconium's form factor, which allowed installation underneath the LS-1. This geometry places the pump's outlet very close to a μFLC injector valve, minimizing delay volume (see Figure 2.13a, b). Optimizing pre-column tubing ID significantly impacted delay time and hence, peak capacity – this effect was magnified greatly at flows of 1–5 μL/min, where eliminating 1.15 μL of pre-column volume resulted in peak elution nearly 20s earlier [73].

Incorporating the SCIEX "OptiFlow" ion source further improved μFLC performance and ease-of-use compared with the standard IonDrive/TurboV Sciex source design. Manual adjustment of the ion source spray tip was eliminated on the OptiFlow design. Instead, discrete probes and electrodes are used depending on flow rate range [70]. Optimization of ion source parameters for microflow application was perceived to be a limitation of the IonDrive/TurboV source design. Table 2.2 Lists dimensions and volumes of optimized system components.

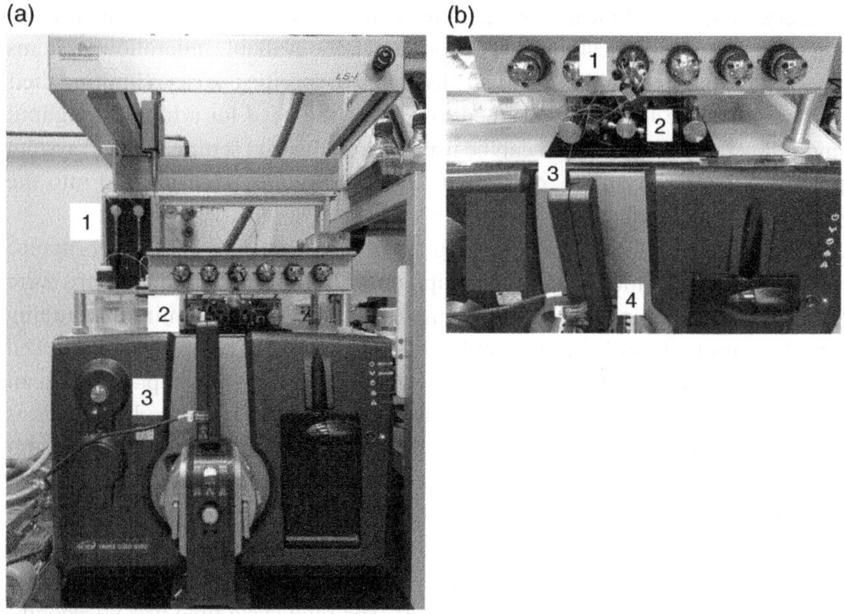

Figure 2.13 LeadSampler system optimized specifically for micro-flow LC-MS/MS. (a): 1. Current LeadSampler platform, 2. Prolab microflow LC pump installed under LeadSampler, 3. SCIEX 6500$^+$ TripleQuad equipped with OptiFlow ion source. (b): Optimized microflow LC flow-path. 1. 6-port injection valve equipped with low volume sample loop, 2. Microflow pump mixing tee outlet placed as close as possible to injection valve, 3. Transfer tubing comprised of finger-tight, UHPLC NanoViper tubing, 4. Microflow column installed directly on ESI electrode.

Table 2.2 Volumes of flow path components identified in Panel B.

Component	Description	Dimensions	Volume (μL)
1	Sample loop	$50\,\mu \times 150\,\text{mm}$	0.29
2	Mixing tee outlet	$50\,\mu \times 70\,\text{mm}$	0.14
3	Transfer tubing	$50\,\mu \times 350\,\text{mm}$	0.69
4	ESI electrode	$25\,\mu \times 124\,\text{mm}$	0.06

In addition, Interfacing the μFLC column directly with the source virtually eliminated post-column volume as the short ESI electrode itself was the only added volume [72]. This configuration resulted in profound enhancements to μFLC capabilities on the LS-1 platform (Figure 2.14a, b). Successive testing of flow path, flow rate, and column ID parameters resulted in a robust, high-throughput μFLC

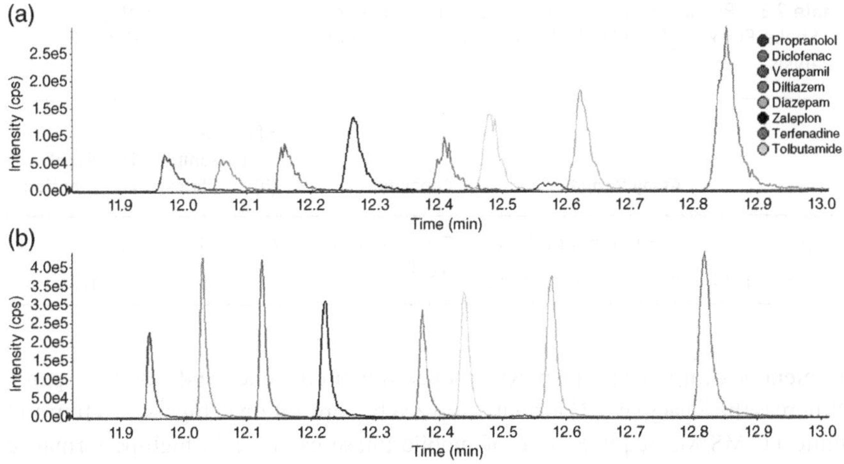

Figure 2.14 Impact of post-column system volume on microflow LC-MS/MS chromatographic performance. (a) Analysis of a small-molecule standard cocktail with LS-1 system equipped with standard IonDrive MS/MS source. (b) Sample analyzed with same LS-1 system equipped with OptiFlow MS/MS source. Mobile phase A: 0.1% formic acid in water, mobile phase B: 0.1% formic acid in acetonitrile. Flow rate: 4 μL/min, linear gradient from 4 to 85%B over 3, 5-minutes cycle time per injection.

method wherein throughput and performance were identical to that achieved at conventional flow rates.

Bioanalytical throughput was then revisited with the goal of optimizing endpoint collection – data per injected sample – rather than sampling rate. Through cassetting or pooling, the number of discrete endpoints contained in a sample can be increased. Effective throughput, a function of both LC cycle time and endpoints per sample, argues that sample collection rate is not the sole determinant of throughput. Therefore, LC methods with vastly different cycle times can capture an identical number of endpoints per unit time. This exercise clearly demonstrates value added by utilizing sample consolidation strategies for achieving the desired throughput.

This approach was applied to a 96-well plate-based shake-flask LogD (SFLogD) assay wherein μFLC-MS/MS was used for separation and analysis of cassetted samples. The SFLogD assay provides distribution coefficient determination for $n = 720$ SMEs each week. The legacy assay, conducted in singleton format ($n = 1$ test compound per well), must therefore be run across eight plates per week to meet portfolio needs. However, when reconfigured for cassetting, $n = 4$ SMEs are co-dosed in each well, an equivalent number of endpoints are delivered from two-assay plates per week, leading to significant savings in consumables (plates, tips, and reagents) and FTE. This cassetted-μFLC-MS/MS workflow also has

Table 2.3 Resource savings and operational efficiency resulting from reconfiguration of Shake-Flask LogD (SFLogD) for assay-cassetting coupled to microflow LC-MS/MS analysis.

	Assay format	Capacity	FTE time (h)	No. sample plates	LC solvent consumption
Legacy	$n = 1$ compound/well	720 compounds/week	6	24	3.7 L
Cassette-μfLC	$n = 4$ compounds/well		1	6	33.8 mL

tremendous impact on bioanalytical operational efficiency and resource spend. Whereas the legacy ($n = 1$) format, analyzed by conventional, high-speed trap and elute LC-MS/MS requires 3.7 L of mobile phase each week, high-performance μFLC-MS/MS developed on the LS-1 needs only 34 mL – over 100× less – to deliver identical effective throughput (Table 2.3). Traditionally, throughput and operating cost in biopharmaceutical research and development were viewed as inexorably linked; conventional approaches sought to increase cadence, volume of work, or sampling speed to increase throughput, contributing operational expense including additional instrumentation, consumables, and FTE. However, coupling sample-consolidation with high-performance μFLC-MS/MS results in highly efficient, resource-sparing, and cost-effective high-throughput analysis.

2.5 Concluding Remarks

At first glance, the LS1 sample-delivery system could be considered a bioanalytical "Swiss-Army knife," an integrated platform capable of on-the-fly support of an array of functions and LC modes necessary in discovery bioanalysis. The system was purposely designed to be open and uncluttered. Its simple design provides bioanalytical chemists a framework to configure multiple LC-MS methods that are best suited for managing a dynamic screening queue. Likewise, LeadScape software can be used creatively to develop customized screening workflows that connect samples, users, and data. The flexibility of the platform can be demonstrated by its ability to drive industrial bioanalysis across business critical ADME screening assays. Table 2.4 lists a range of assays used in Lead Optimization screening; note that microflow (10–50 μL/min) and standard flow rate (50–1000 μL/min) bioanalysis can be mixed in the runtime batch queue. The ability of a single platform to support a broad scope of bioanalytical methodologies provides researchers remarkable agility and opportunities to disrupt the bioanalytical status quo.

Table 2.4 Summary of in vitro pharmacokinetic screens, associated methodologies, and formats supported by a bioanalytical laboratory positioned in early drug discovery.

Assay	Sample burden	Endpoints per sample	Study runtime (h)	LC method, cycle time	Runs per week
Human hepatocyte CL	11 × 384-well plates	1	15	Trap-and-elute, 10 s/inj	4
Human liver microsome CL	12 × 384-well plates	1	17	Trap-and-elute, 10 s/inj	1
RRCK (passive permeability)	5 × 96-well plates	4	13	Micro-flow LC, 60 s/inj	2–4
PgP/BCRP (active transport)	7 × 384-well plates	2	19	Micro-flow LC, 60 s/inj	1
Apparent solubility	5 × 96-well plates	1	3.5	Trap-and-elute, 15 s/inj	4
Chemical stability	4 × 384-well plates	1	6	Trap-and-elute, 15 s/inj	2
Shake-flask LogD	3 × 96-well plates	4	13	Micro-flow LC, 60 s/inj	2
Multi-species microsome CL	12 × 384-well plates	1	17	Trap-and-elute, 15 s/inj	2

Multi-ordinate assays are configured for sample consolidation and coupled to microflow-LC-MS/MS analysis.

References

1 Fontana, S. (2014). Approaches for high-throughput pharmacokinetic screening of low-molecular-weight drug candidates. *Expert Opin. Drug Metab. Toxicol.* 10 (2): 139–142.

2 Foley, T.L., Burchett, W., Chen, Q. et al. (2021). Selecting approaches for hit identification and increasing options by building the efficient discovery of actionable chemical matter from DNA-encoded libraries. *SLAS Discovery* 26 (2): 263–280.

3 Vincent, F., Loria, P., Pregel, M. et al. (2015). Developing predictive assays: the phenotypic screening "rule of 3". *Sci. Transl. Med.* 7: 293ps15.

4 Jones, L.H. and Bunnage, M.E. (2017). Applications of chemogenomic library screening in drug discovery. *Nat. Rev. Drug Discovery* 16: 285–296.

5 Parker, C.G., Galmozzi, A., Wang, Y. et al. (2017). Ligand and target discovery by fragment-based screening in human cells. *Cell* 168: 527–41. e29.

6 Holenz, J. and Stoy, P. (2019). Advances in lead generation. *Bioorg. Med. Chem. Lett.* 29: 517–524.

7 Eliuk, S. and Makarov, A. (2015). Evolution of orbitrap mass spectrometry instrumentation. *Annu. Rev. Anal. Chem.* 8 (1): 61–80.

8 Covey, T. (2022). Where have all the ions gone, long time passing? Tandem quadrupole mass spectrometers with atmospheric pressure ionization. Sensitivity gains since the mid 1970s. A perspective. *Rapid Commun. Mass Spectrom.* e9354.

9 Comess, K.M., McLoughlin, S.M., Oyer, J.A. et al. (2018). Emerging approaches for the identification of protein targets of small molecules – a practitioners' perspective. *J. Med. Chem.* 61: 8504–8535.

10 Gerry, C.J. and Schreiber, S.L. (2018). Chemical probes and drug leads from advances in synthetic planning and methodology. *Nat. Rev. Drug Discovery* 17: 333–352.

11 Schneider, P., Walters, W.P., Plowright, A.T. et al. (2020). Rethinking drug design in the artificial intelligence era. *Nat. Rev. Drug Discovery* 19: 353–364.

12 Wager, T.T., Hou, X., Verhoest, P.R., and Villalobos, A. (2016). Central nervous system multiparameter optimization desirability: application in drug discovery. *ACS Chem. Neurosci.* 2016 (7): 767–775.

13 Ekins, S., Honeycutt, J.D., and Metz, J.T. (2010). Evolving molecules using multi-objective optimization: applying to ADME/Tox. *Drug Discovery Today* 15 (11, 12): 451–460.

14 James, T., Sardar, A., and Anighoro, A. (2020). Enhancing chemogenomics with predictive pharmacology. *J. Med. Chem.* 63: 12243–12255.

15 Schneider, G. (2018). Automating drug discovery. *Nat. Rev. Drug Discovery* 17: 97–113.

16 Peakman, M.-C., Troutman, M., Gonzales, R. et al. (2014). Experimental screening strategies to reduce attrition risk. In: *Attrition in the Pharmaceutical Industry: Reasons, Implications and Pathways Forward* (ed. A. Alex, C.J. Harris, and D.A. Smith), 180–214. Hoboken, NJ: Wiley.

17 Shou, W. and Zhang, J. (2010). Recent development in high-throughput bioanalytical support for in vitro ADMET profiling. *Expert Opin. Drug Metab. Toxicol.* 6 (3): 321–336.

18 Mazzeo, J.R., Neue, U.D., Kele, M., and Plumb, R.S. (2005). A new separation technique takes advantage of sub-2-μm porous particles. *Anal. Chem.* 77 (23): 460–467.

19 Plumb, R.S., McDonald, T., Rainville, P.D. et al. (2021). High-throughput UHPLC/MS/MS-based metabolic profiling using a vacuum jacketed column. *Anal. Chem.* 93: 10644–10652.

20 Kapinos, B., Janiszewski, J., Zhang, H. et al. (2017). Development of high-performance micro-flow LC-MS/MS methodology and application to a high-throughput screening workflow. *Proceedings of the 65th ASMS Conference on Mass Spectrometry and Allied Topics (TP 153)*, Indianapolis, IN (4–8 June 2017).

21 Christianson, C.C., Johnson, C.J.L., and Needham, S.R. (2013). The advantages of microflow LC-MS/MS compared with conventional HPLC-MS/MS for the analysis of methotrexate from human plasma. *Bioanalysis* 5 (11): 1387–1396.

22 Covey, T.R. (2022). Forty year evolution of high-throughput mass spectrometry a perspective. In: *High-Throughput Mass Spectrometry in Drug Discovery* (ed. C. Liu and H. Zhang), 1–40. Wiley.

23 Neves, B.J., Braga, R.C., Melo-Filho, C.C. et al. (2018). QSAR-based virtual screening: advances and applications in drug discovery. *Front. Pharmacol.* 9: 1275.

24 Darvas, F., Dorman, G., Urge, L. et al. (2001). Combinatorial chemistry. Facing the challenge of chemical genomics. *Pure Appl. Chem.* 73: 1487–1498.

25 Rodrigues, A.D. (1997). Preclinical drug metabolism in the age of high-throughput screening: an industrial perspective. *Pharm. Res.* 14: 1504–1151.

26 Bleicher, K.H., Böhm, H.-J., Müller, K., and Alanine, A.I. (2003). Hit and lead generation: beyond high-throughput screening. *Nat. Rev. Drug Discovery* 2: 369–378.

27 Maurer, T.S., Edwards, M., Hepworth, D. et al. (2022). Designing small molecules for therapeutic success: a contemporary perspective. *Drug Discovery Today* 27: 538–546.

28 Plise, E., Gaffney, K., Jorski, J. et al. (2019). Dual-stream LC coupled with 'plug and play' automation for routine bioanalysis in drug discovery. *Proceedings of the 67th ASMS Conference on Mass Spectrometry and Allied T2opics (WP 246)*, Atlanta, GA (2–6 June 2019).

29 Piotrowski, M.A., Dumlao, D., Keefer, J. et al. (2017). Simultaneous determination of key glycolysis and tricarboxylic acid cycle intermediates in fluxomics studies by DMS-LC/MS/MS. *Proceedings of the 65th ASMS Conference on Mass Spectrometry and Allied Topics (WP483)*, Indianapolis, IN (4–8 June 2017).

30 Dumlao, D.S., Piotrowski, M., Shields, J. et al. (2015). Understanding the contribution of glutamine in fatty acid biosynthesis using 13C-stable labelled metabolites in conjunction with an integrated high-throughput methodology. *Proceedings of the 63rd ASMS Conference on Mass Spectrometry and Allied Topics (MP17 352)*, St. Louis, MO (2–6 June 2015).

31 Janiszewski, J.S., Rogers, K.J., Whalen, K.M. et al. (2001). A high-capacity LC/MS system for the bioanalysis of samples generated from plate-based metabolic screening. *Anal. Chem.* 73: 1495–1501.

32 Janiszewski, J., Schneider, R., Kapinos, B. et al. (2012). Development of a high-speed, multiplexed sample-delivery instrument for LC-MS/MS bioanalysis. *Bioanalysis* 4 (9): 1039–1056.

33 Neue, U.D., Marchand, D.H., and Snyder, L.R. (2006). Peak compression in reversed-phase gradient elution. *J. Chromatogr. A* 1111 (1): 32–39.

34 Neue, U.D., Kele, M., Bunner, B. et al. (2009). Ultra-performance liquid chromatography technology and applications. *Adv. Chromatogr.* 23 (48): 99–104.

35 Vanderlinden, K., Broeckhoven, K., Vanderheyden, Y., and Desmet, G. (2016 Apr). Effect of pre-and post-column band broadening on the performance of high-speed chromatography columns under isocratic and gradient conditions. *J. Chromatogr. A* 15 (1442): 73–82.

36 Kapinos, B., Liu, J., Piotrowski, M. et al. (2017). Development of a high-performance, enterprise-level, multimode LC-MS/MS autosampler for drug discovery. *Bioanalysis* 9 (21): 1643–1654.

37 Furey, A., Moriarty, M., Bane, V. et al. (2013). Ion suppression; a critical review on causes, evaluation, prevention and applications. *Talanta* 115: 104–122.

38 Whalen, K.M., Rogers, K.J., Cole, M.J., and Janiszewski, J.S. (2000). AutoScan: an automated workstation for rapid determination of mass and tandem mass spectrometry conditions for quantitative bioanalytical mass spectrometry. *Rapid Commun. Mass Spectrom.* 14: 2074–2079.

39 Whalen, K., Gobey, J., and Janiszewski, J. (2006). A centralized approach to tandem mass spectrometry method development for high-throughput ADME screening. *Rapid Commun. Mass Spectrom.* 20: 1497–1503.

40 Whalen, K.M., Rogers, K.J., Janiszewski, J.S. et al. (2000). Eva_Lution: a software tool for high-speed quantitative processing and review of LC/MS data collected from high-throughput ADME screens. *Proceedings of the 48th ASMS Conference on Mass Spectrometry and Allied Topics (MPJ 303)*, Long Beach, CA (11–15 June 2000).

41 Whalen, K.M., Janiszewski, J.S., Ainley, S.A. et al. (2007). Automatic MS/MS methods development using an information dependent scanning protocol to enhance sensitivity for high-throughput ADME screening and drug discovery. *Proceedings of the 55th ASMS Conference on Mass Spectrometry and Allied Topics (ThPK 171)*, Indianapolis, IN (3–7 June 2007).

42 Mahan, E.A., Dube, N., Geddes, K. et al. (2007). Optimized DMPK sample throughput and quality: NanoMate and automaton to group discovery compounds based on experimentally-determined analytical characteristics. *Proceedings of the 55th ASMS Conference on Mass Spectrometry and Allied Topics (WPP 254)*, Indianapolis, IN (2–7 June 2007).

43 Whalen, K.M., Janiszewski, J.S., and Cole, M.J. (2004). Evalution II: software automation for industrialized high-throughput bioanalytical mass spectrometry in support of ADME-HTS. *Proceedings of the 52nd ASMS Conference on Mass Spectrometry and Allied Topics (WPN 257)*, Nashville, TN (23–27 May 2004).

44 Spencer, R.W. (1998). High-throughput screening of historic collections: observations on file size, biological targets, and file diversity. *Biotechnol. Bioeng.* 61 (1): 61–67.

45 Clark, R.D., Patterson, D.E., Soltanshahi, F. et al. (2000). Visualizing substructural fingerprints. *J. Mol. Graph. Model.* 18: 404–411.

46 Hiller, D.L., Zuzel, T.J., Williams, J.A., and Cole, R.O. (1997). Rapid scanning technique for the determination of optimal tandem mass spectrometric conditions for quantitative analysis. *Rapid Commun. Mass Spectrom.* 11: 593–597.

47 Cole, M.J., Janiszewski, J.S., and Fouda, H.G. (2002). Electrospray mass spectrometry in drug metabolism and pharmacokinetics. In: *Applied Electrospray Mass Spectrometry* (ed. B.N. Pramanik, A.K. Ganguly, and M.L. Gross), 211–249. New York: Marcel Dekker, Inc.

48 Lombardino, J.G. and Lowe, J.A. (2004). The role of the medicinal chemist in drug discovery – then and now. *Nat. Rev. Drug Discovery* 3: 853–862.

49 Chandler, C.E., Wilder, D.E., Pettini, J.L. et al. (2003). CP-346086: an MTP inhibitor that lowers plasma cholesterol and triglycerides in experimental animals and in humans. *J. Lipid Res.* 44 (10): 1887–1901.

50 Benson, N., Boyd, H.F., Everett, J.R. et al. (2005). Nanostore: a concept for logistical improvements of compound handling in high-throughput screening. *SLAS Discovery* 10 (6): 573–580.

51 Hay, B.A., Abrams, B., Zumbrunn, A.Y. et al. (2007). Aminopyrrolidineamide inhibitors of site-1 protease. *Bioorg. Med. Chem. Lett.* 17: 4411–4414.

52 Hawkins, J.L., Robbins, M.D., Warren, L.C. et al. (2008). Pharmacologic inhibition of site 1 protease activity inhibits sterol regulatory element-binding protein processing and reduces lipogenic enzyme gene expression and lipid synthesis in cultured cells and experimental animals. *J. Pharmacol. Exp. Ther.* 326 (3): 801–808.

53 Thompson, N.T. (2000). Early ADME in support of drug discovery: the role of metabolic stability studies. *Curr. Drug Metab.* 1: 215–241.

54 Lin, J., Sahakian, D.C., de Morais, S.M. et al. (2003). The role of absorption, distribution, metabolism, excretion and toxicity in drug discovery. *Curr. Top. Med. Chem.* 3 (10): 1125–1154.

55 Gao, F., Johnson, D.L., Ekins, S. et al. (2002). Optimizing higher throughput methods to assess drug-drug interactions for CYP1A2, CYP2C9, CYP2C19, CYP2D6, rCYP2D6, and CYP3A4 in vitro using a single point IC50. *J. Biomol. Screening* 7 (4): 373–382.

56 Banker, M.J., Clark, T.H., and Williams, J.A. (2003). Development and validation of a 96-well equilibrium dialysis apparatus for measuring plasma protein binding. *J. Pharm. Sci.* 92: 967–974.

57 Kassel, D.B. (2004). Applications of high-throughput ADME in drug discovery. *Curr. Opin. Chem. Biol.* 8 (3): 339–345.

58 Jenkins, K.M., Angeles, R., Quintos, M.T. et al. (2004). Automated high-throughput ADME assays for metabolic stability and cytochrome P450 inhibition profiling of combinatorial libraries. *J. Pharm. Biomed. Anal.* 34 (5): 989–1004.

59 Wright, P., Chassaing, C., Cussans, N. et al. (2006). Evolution of an open-access quantitative bioanalytical mass spectrometry service in a drug discovery environment. *Biomed. Chromatogr.* 20: 585–596.

60 Hop, C.E., Cole, M.J., Davidson, R.E. et al. (2008). High-throughput ADME screening: practical considerations, impact on the portfolio and enabler of in silico ADME models. *Curr. Drug Metab.* 9 (9): 847–853.

61 Stepan, A.F., Mascitti, V., Beaumont, K., and Kalgutkar, A.S. (2013). Metabolism-guided drug design. *MedChemComm* 2013 (4): 631–652.

62 Pereira, D.A. and Williams, J.A. (2007). Origin and evolution of high-throughput screening. *Br. J. Pharmacol.* 152 (1): 53–61.

63 Janiszewski, J.S., Liston, T.E., and Cole, M.J. (2008). Perspectives on bioanalytical mass spectrometry and automation in drug discovery. *Curr. Drug Metab.* 9 (9): 986–994.

64 Zhang, H., Liu, J., Kapinos, B. et al. (2017). LeadScape: an automated platform for LC/MS/MS methods development, storage and distribution. *Proceedings of the 65th ASMS Conference on Mass Spectrometry and Allied Topics (MP 146)*, Indianapolis, IN (4–8 June 2017).

65 Fandino, A., Naegele, E., Buckenmaier, S., and Glatz, B. (2009). Metabolic stability study using cassette analysis and polarity switching in an ultra high performance liquid chromatography (UHPLC)-triple quadrupole system. *Proceedings of the 57th ASMS Conference on Mass Spectrometry and Allied Topics (MPR 406)*, Philadelphia, PA (4–8 June 2009).

66 Zhang, H., Zelesky, V., Schneider, R., and Janiszewski, J. (2010). Improving the throughput of in vitro ADME screens using cassette analysis and polarity switching. *Proceedings of the 58th ASMS Conference on Mass Spectrometry and Allied Topics (TP21 497)*, Salt Lake City, UT (23–27 May 2010).

67 Di, L., Whitney-Pickett, C., Umland, J.P. et al. (2011). Development of a new permeability assay using low-efflux MDCKII cells. *J. Pharm. Sci.* 100 (11): 4974–4985.

68 Zhang, H., Funk, C., Federico, J. et al. (2012). Fully integrated high-throughput system for ADME screens using cassette analysis. *Proceedings of the 60th ASMS Conference on Mass Spectrometry and Allied Topics (ThP16 349)*, Vancouver, CN (20–24 May 2012).

69 Luippold, A., Arnhold, T., Jorg, W. et al. (2011). Application of a rapid and integrated analysis system (RIAS) as a high-throughput processing tool for in vitro ADME samples by liquid chromatography/tandem mass spectrometry. *J. Biomol. Screen.* 16 (3): 370–377.

70 Zhang, J., Vath, M., Ferraro, C. et al. (2013). A high-speed liquid-chromatography/tandem mass spectrometry platform using multiplexed multiple-injection chromatography controlled by single software and its application in discovery ADME screening. *Rapid Commun. Mass Spectrom.* 27: 731–737.

71 Barricklow, J., Tweed, J., Holliman, C.L., and Ramanathan, R. (2020). Evaluation of OptiFlow-MS/MS for bioanalysis of pharmaceutical drugs and metabolites. *Bioanalysis* 12 (1): 23–34.

72 Kapinos, B., Janiszewski, J., Lootsma, W. et al. (2015). High-throughput, dual-stream micro-flow LC/MS/MS bioanalysis. *Proceedings of the 63rd ASMS Conference on Mass Spectrometry and Allied Topics (ThP11 246)*, St. Louis, MO (2–6 June 2015).

73 Kapinos, B., Janiszewski, J., Nemec, B. et al. (2018). Design and integration of a high-performance micro-flow LC-MS/MS system. *Proceedings of the 66th ASMS Conference on Mass Spectrometry and Allied Topics (MP 191)*, San Diego, CA (3–7 June 2018).

74 Hoxie, N., Qiu, X., Kales, S. et al. (2023). Development of a High-Throughput Dual Stream LC/MS/MS method to screen for inhibitors of Glutamate Carboxypeptidase II (GCPII). *Rapid Commun. Mass Spectrom.* (Submitted)

3

Evolution of Multiplexing Technology for High-Throughput LC/MS Analyses

Adam Latawiec

SCIEX, Concord, ON, Canada

3.1 Introduction and Historical Developments

The growth of analytical chemistry technology over the last 30–40 years has been truly staggering. Improvements in electronics, optics, and mechanical technologies have enhanced virtually all of the modern high-performance liquid chromatography (HPLC), gas chromatography (GC), and mass spectrometry (MS) technologies currently utilized for routine quantitative and qualitative analyses in a diversity of scientific and industrial fields. Even more impressive is that the scientific foundations of technologies such as MS go back over 120 years with early work on the deflection of charged particles by magnetic fields [1]. Liquid chromatography can similarly trace its roots in the separation of plant pigments over a century ago [2]. Elementary school experiments with inks and paper chromatography provide a striking visual example of the capacity of chromatography to reveal hidden components. From early LC instruments in the late 1960s such as the Varian LCS-100 and Waters ALC-100, high-pressure pumps and injection ports enabled routine HPLC analyses at 6000 psi (400 Bar) within a few short years [3, 4]. Current HPLC technologies can provide stable pump pressures almost up to 22,000 psi (1500 Bar) and reliable systems with 14,000 psi (1000 Bar) are relatively commonplace. The rapid conversion of traditional HPLC to ultrahigh performance liquid chromatography (UHPLC) regimes allowed for optimizing linear velocities, and taking advantage of developments in reverse-phase particle technology. First in ever-decreasing particle sizes – first 5–10 μm irregular particles,

High-Throughput Mass Spectrometry in Drug Discovery, First Edition. Edited by Chang Liu and Hui Zhang.
© 2023 John Wiley & Sons, Inc. Published 2023 by John Wiley & Sons, Inc.

then spherical particles of 5, 3, 2, and eventually sub-2 μm particles [5]. In conjunction, the emergence of nonporous shell particles further improved the capacity of reverse-phase columns so that short columns of 20–50 mm could easily provide and indeed improve on, the separations traditionally obtained on 250 mm columns while shortening chromatographic run times, allowing for significant improvements in sample throughput. Whereas early reverse-phase HPLC was limited to a small number of phases such as the C18, C8, and CN types, new robust silica phases including cyclic alkyl, aryl, mixed aryl–alkyl, fluorinated, polar, polar imbedded, and ionic phases allowed for the separation of more polar and ionic compounds and the effective separations of more complex mixtures of analytes. By one account, over 170 different reverse-phase chemistries were introduced between 1970 and 2010 [6]. To the outside observer, the field of HPLC developed rapidly in a few decades but on closer inspection the progress was mostly by evolution along many different lines rather than by simply a single revolutionary HPLC concept or technology.

Mass spectroscopic technology continued apace with magnetic sector instruments such as the LBK 9000 being sold commercially in 1965 [7]. The technology reached a level of robustness such that small scanning magnetic sector mass spectrometer with a mass range of 12–100 and 45–200 m/z was even incorporated into the 1975 Viking mission to Mars (albeit at a cost of around nine million dollars) [8, 9]. Eventually, ionization techniques such as electrospray ionization (ESI) and atmospheric pressure chemical ionization (APCI) were improved to the point of becoming the standard sample introduction techniques for the vast majority of LC-MS workflows and supplanting an array of novel but ultimately limited sample introduction techniques such as Finnigan MAT Moving Belt LC-MS interface [10] and Waters Fast Atom Bombardment [11]. Research-oriented instruments such as the SCIEX TAGA 6000E/3000 triple quadrupole API in the late 1980s developed eventually into more mainstream instruments such as the API III, API 300, and joined rivals such as the VG B10-Q by Waters. Robust LC-MS/MS (tandem mass spectrometers or triple quads) instruments such as the API III were introduced by SCIEX in 1989 [12]. The fundamental science of the early 1990s established the basics of the mechanical, electronic, ion optics, detectors, and vacuum systems that are easily recognizable in the present day. By 1995, such systems could be described as reliable and routine analytical tools [13]. The ability of LC/MS and LC/MS/MS to analyze polar and thermally labile species with relative ease, allowed LC/MS/MS to dominate areas such as proteomics, metabolomics, clinical, and drug discovery laboratories. The advantage of direct analysis also positioned LC/MS to displace traditional GC/MS techniques that required specialty columns or derivatization techniques and worked best on compounds resistant to thermal degradation [14]. In a similar vein, faster triple quad Multiple Reaction Monitoring (MRM) acquisition systems which required dwell times of only a few milliseconds combined with retention time-based MRM scheduling experiments have expanded analytical runs to hundreds and even thousands of compounds. Naturally all of these impressive technical advances have given the analytical chemist an ever-increasing variety of chromatographic methods and LC-MS techniques which could be optimized for increases in sample throughput.

Ultimately, irrespective of the speed and efficiencies of the sample injection or analysis portions of the chromatographic process, the time required for elution and subsequent re-equilibration of the HPLC system results in only a small portion of the chromatogram contains the analytes of interest. This "analytes of interest" chromatographic window is variable but typically accounts for less than half the entire chromatographic run. The idea of alternating between the active elution window of one system while the other HPLC system is in the process of re-equilibration and collecting the data sequentially on a single mass spectrometer naturally gained much interest [15]. Systems of this nature comprising of two or more HPLC systems connected by way of a selector valve with appropriate hardware and software would form the basis for what is commonly referred to as multiplexing. Such multiplexed workflows were originally developed for LC coupled with ultra-violet (UV), fluorescence (FLD), and single quadrupole systems usually developed as "in-house" systems targeting research-based workflows. Other alternative early multiplexing techniques incorporating multiple source inlets to a single mass spectrometer were also developed. Although multiplexing was an attractive concept for academic papers and publications, commercialization of high-throughput multiplexing was primarily driven by developments in online sample processing rather than simple multiplexing.

3.2 Developments Toward Fully Integrated Multiplexing Systems

Previous online sample preparation techniques were often based on solid-phase extraction (SPE) which employed similar particles and bonded phase chemistries. The general principles focused on duplicating the adsorption and laminar flows of conventional column chromatography except using smaller length SPE columns. An alternative for online sample processing was developed in 1995 by Cohesive Technologies exploiting their proprietary TurboFlow™ chromatography (a type of perfusion chromatography) whereby smaller analytes of interest would remain in the large channels of silica-based chromatographic particles. Such TurboFlow columns were characterized by large particle sizes (typically 50–150 μm versus less than 10 μm for conventional laminar flow columns). The use of the larger diameter particles allowed for significantly higher flow rates (cf. 1.5–5.0 mL/min) with minimal backpressure whereby the larger molecular weight species of the sample mixture are flushed rapidly from the column to waste. In practice, the TurboFlow technology exhibits a molecular weight exclusion in the region of 10,000 Da. This proved ideal for biological matrices such as urine, serum, or plasma which were often the target of drug discovery programs screening large numbers of samples. Because of the larger particle size and relatively smaller pores, the residence time for higher molecular weight products is typically in the order of 300 ms allowing for their rapid removal. The MW exclusion profile was considered superior to competing techniques such as restricted access media (RAM) and SPE [16]. Once the higher molecular weight compounds are removed from the sample stream with a typical 100% aqueous mobile phase, elution of lower molecular weight compounds could then be controlled by changing both flow and mobile phase composition to selectively concentrate analytes of interest on to the analytical column. The use of TurboFlow columns and associated LC hardware gained considerable market penetration as a sample preparation technique by virtue of providing high-speed sample extraction with better removal of matrix interferences compared to either protein precipitation technique (PPT) or SPE for plasma, serum, and whole blood matrices. The use of TurboFlow technology for the analysis of isoquinoline-based pharmaceuticals in urine samples [17] provided one of the earliest references. Test experiments showed that a single TurboFlow method could be used with a variety of matrices such as phosphate-buffered saline (PBS), bovine serum albumin (BSA), and rat plasma which minimized the requirement for different matrix specific extraction techniques and the advantages, uses, and applications of TurboFlow and related technologies were the subject of several comprehensive review articles [18, 19].

In 2000, Cohesive applied for a patent for a complete system whereby "a detector is included for the detecting of compounds in the samples from each of the chromatography systems along with a valve positioned between the detector and the plurality of the chromatographic systems. . ." and introduced the term "multiplexing" to describe multiple LC systems feeding into a single mass spectrometric system [20] (Figure 3.1). This combination of several liquid chromatography systems with a single mass spectrometer became the Aria™ system which combined both specialized software and instrumentation designed specifically for multiplexing. Although "multiplexing" was a common term to describe such analytical systems whereby a selected portion of the LC run was diverted to a mass spectrometer, this approach was also referred to in the scientific literature as "staggered" or "MISER" (Multiple Injections in a Single Experimental Run) chromatography. Originally commercially released with two LC streams (2-plex), the Cohesive system was later elaborated into a 4-plex system (four LCs feeding into a single mass spectrometer). The patent description provided the basis for the typical multiplexed systems commercialized by in succession by companies such as Cohesive (later Thermo Scientific™), SCIEX, Agilent, and Shimadzu among

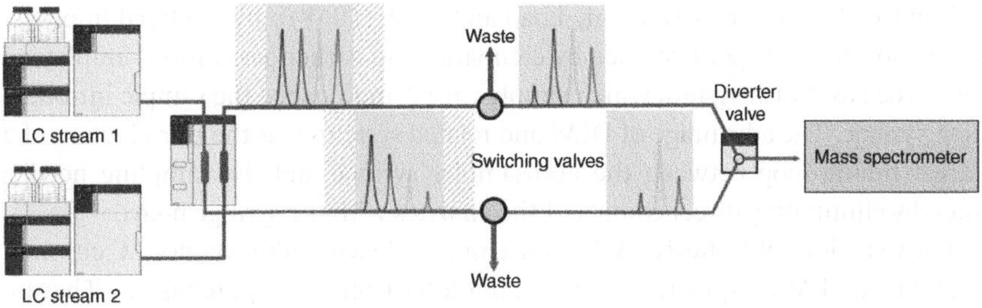

Figure 3.1 Schematic representation of the general multiplexing process for a two stream (duplex) LC system without SPE. Injections from each stream are staggered and only selected portions of the LC chromatograms are collected by the mass spectrometer data system. Highlighted areas of LC chromatograms are diverted to the mass spectrometer; dashed areas are diverted to waste by way of column switching valves in each stream. The mass spectrometer throughput is greatly improved by collecting only relevant portions of the chromatogram.

others. Waters Corporation introduced a rival "multiplexing" technique primarily for sample purification employing a multiplexed (MUX) electrospray ion source using independent probes which failed to gain widespread adoption compared to traditional selector valve and single ion spray probe designs [21]. Initial workflows for high-throughput LC-MS/MS assays focused on in-line processing techniques whereby samples could be pre-concentrated and purified on suitable TurboFlow or SPE cartridges then back-flushed onto conventional reverse-phase analytical columns.

The Thermo Scientific Aria™ MX was a stand-alone multiplexing unit available as LX (multiplexing technology only) and TLX (incorporating both multiplexing and TurboFlow technologies) with a 400 bar LC system. Subsequent development of the Transcend™ and Transcend™ II in 2006 expanded these capabilities largely by introducing High Pressure LC systems operating in the 600 bar range. Nonetheless, a Transcend TLX-4 system reaches a level of complexity necessitating four sets of binary pumps and column heaters as well as 10 switching valves (nine 2-position 6-port valves and one 8-position selector valve). The resulting complexity of the LC components for a 4-plex system probably reaches the limits of sample throughput versus complexity and is a factor in the continued prevalence of 2-plex systems in commercial laboratories requiring both high-throughput and high reliability. Thermo Transcend systems continued to employ either the PAL HTC or HTC autosamplers (CTC Analytics AG) or alternatively variants produced by LEAP Technologies. Such autosampler systems gradually evolved to meet the increasing throughput requirements of multiplexing to incorporate two or more injector ports and eventually dual injection heads to decrease cycle time delays. Carryover performance, initially an issue with CTC Fast Wash systems, was minimized by the use of Dynamic Load and Wash (DLW) which offered improvements to carryover performance by comparison to earlier generations employing the basic Fast Wash system which simply rinsed and flushed the sample introduction syringe. The advantage of DLW and related systems was the introduction of a larger tubing loop between the autosampler syringe and the sampling needle, thereby eliminating direct contact of the analytes with the syringe body.

Early versions of Cohesive Aria were compatible with other vendor LC chromatography and MS acquisition systems, but after Cohesive's purchase by Thermo Scientific, later versions supported only HPLC pumps from either Thermo or Agilent. Mass Spectrometer and HPLC compatibility would be tied thereafter to only the specific Multiplexing Vendor. Over time, as a result of both mergers and competition, the integration of multiplex functionality with MS acquisition and data analysis software became limited with respect to the compatibility with other rival vendors of LC and MS software. The current approach is largely to provide "turnkey" systems with multiplexing often being simply an additional feature of the MS vendors acquisition software.

Both the Aria and the later Transcend systems largely targeted the ADME (Administration, Distribution, Metabolism, Excretion) drug discovery and drug development markets. Utilizing a Transcend TLX2 system, a Testosterone and Androstenedione assay from a 50 μL Plasma injection could be performed in half the time including sample extraction. Optimized on a 4-plex system, the assay would only require 45 seconds. Similarly, a longer assay including Testosterone, Androstenedione, and eight metabolites could be performed in 5.5 minutes versus the traditional 18 minutes analytical run [22]. An example of Fentanyl and Norfentanyl from urine using an Aria TLX-4 system provided a throughput rate of one sample every 86 seconds using a method runtime of 5.5 minutes and a resulting LOQ of 0.5 ng/mL in human urine [23]. At Pittcon 2014, Thermo Fisher Scientific showed numerous examples of multiplexing assays of drug discovery compounds on triple quadrupole instruments such as the TSQ Vantage™ and lipidomic, antiretroviral assays on accurate mass instruments such as the Q Exactive™ [24].

3.3 Broadening Customer Options

In 2008, SCIEX (then operating as Applied Biosystems) introduced their MPX™ multiplexing system in 2-plex only. Similar in function to the Aria and Transcend systems, the MPX differed slightly by the inclusion of an additional "loading" pump which operated in addition to the normal autosampler to facilitate sample introduction. Although available in an online SPE configuration to allow for sample enrichment and purification, the primary focus of the MPX system was to showcase enhanced software features. The multiplexing capabilities could be accessed through a specialized MPX interface or by way of SCIEX Cliquid®. The Cliquid® user interface was developed to allow for a simple four-step process to set-up, acquire, and process samples. The inclusion of full audit trail and traceability within the MPX software encouraged early adoption by existing users of SCIEX Analyst® software. The addition of a "Walk Up" mode allowed for easy addition of samples to an existing sample queue and the order of the samples in the existing queue could be easily modified by the user to accommodate changes in sample priority. Further flexibility of the MPX system allowed for the operation of different analytical methods on either stream – for example, reverse phase and HILIC (Hydrophilic interaction liquid chromatography) methods [25] and the ability to multiplex automatically longer and shorter run times. For example, one stream could be set up for a 10 minute run and the other stream could alternate with two 5 minute runs ensuring that maximum throughput could be realized in manner comparable to the competing Aria and Transcend systems from Thermo Fisher Scientific. The MPX duplex system would also divert samples to an available LC stream in the event of LC system errors or malfunctions. Additional end-user flexibility supported MPX systems configured with either stand-alone Shimadzu SIL-20AC or SIL-20ACXR autosamplers coupled with rack changers offering up to 12 well plates (96 or 384 format) per stream at either 440 or 660 bar maximum pressure. Users could also configure MPX systems to operate with several variants of the CTC HTC and HTS PAL autosamplers. The flexibility of the SCIEX MPX system was largely targeted toward the emerging drug testing laboratories with customers developing assays for OH-Vitamin D, food residues, or drug screening assays. For example, an MPX-based 25-OH-Vitamin assay using serum from bar coded tubes and internal standards were processed using an 8-channel pipettor. The samples were then subjected to a two-step protein precipitation sequence, followed by the transferring of the supernatant to a 96-well SPE plate. The loaded plates could then be placed directly into an autosampler. The use of three dual-channel MPX systems and associated AB SCIEX API-5000 MS allowed for the identification and quantitation of some 44 sampler per hour per stream, or a combined throughput of some 2500 per day [26]. In cases where a simple protein precipitation and spin down were sufficient, the use of MPX and a SCIEX Q-TRAP 5500 could result in sample cycle times of 1.5 minutes per sample with accurate quantitation of 25-OH Vitamin D3 and 25-OH-Vitamin D2 in the range of 0.5–100 ng/mL in serum [27]. Although MPX was primarily geared towards support of SCIEX triple quadrupole instrumentation such as the API 3200, 4000, 5000, and 5500, MPX systems were initially also made available for HRMS instruments such as the 4600 Triple TOF™ using MPX TF 1.0. The use of accurate mass coupled with existing SCIEX software packages such as Peak View® and Master View® provided initial HRMS screening information. The addition of accurate mass TOF-MS/MS product ion scan libraries, mass error, isotope matching, MS/MS library matching, and retention time allowed for greater certainty in identifying closely related species. As ordinary isotope ratio patterns can be altered for many analytes in actual urine samples, this underscored the utility of additional MS/MS information and was effectively used in a multiplexed screening assay of some 40 synthetic cannabinoids and their metabolites. An effective MPX screening assay using a SCIEX Triple TOF 4600 resulted in a significant sample throughput with a 50 seconds injection to injection cycle time [28].

Although the two major players in the multiplexing market (Thermo and SCIEX) developed multiplexing systems for slightly different markets, with Thermo leveraging the online sample preparation advantages of its TurboFlow technology and SCIEX focusing on robustness and software enhancements, other major MS manufacturers such as Agilent and Shimadzu also provided entries into the multiplexing market. In 2014, Agilent introduced its own variant of multiplexing, the StreamSelect in 2014 focusing and highlighting the 4-plex capability. Using existing Agilent Infinity 1260/1290 series pumps, column ovens, an HTC/HTS derived autosampler equipped with either 2, 3, or 4 injection ports, and a 6400 Series triple quadrupole mass spectrometer, the entire system was largely configured as a turnkey system using exclusively Agilent components and software. The 4-plex system was even available with a specialized laboratory StreamSelect bench to provide a less imposing footprint. The StreamSelect, as a result of its sole vendor configuration and software, could be easily adapted from 2-plex to 3-plex, and to 4-plex by simply adding additional pumps and column thermostats. Installing the Agilent Mass Hunter acquisition and processing software, the user was prompted to install the MassHunter StreamSelect software and the multiplexing was immediately configured and available for use through the StreamSelect Console. As the common configuration of the Agilent TCC column thermostat included a switching port, the inclusion of online SPE sample cleanup capabilities to the StreamSelect system could be accomplished in a relatively straightforward fashion. Two of the key features advertised for the StreamSelect system were equal quantitative performance across all streams and intelligent error handling. The intelligent error handling feature largely monitored the status of all streams and automatically diverted samples to the next available LC stream in the event of solvent leaks or pump overpressure. One of the earliest throughput examples for the Agilent StreamSelect was provided in a poster session at the 2014 MSACL conference and outlined the analysis of Δ^9-tetrahydrocannabinol in oral fluids [29]. Using a 2-plex system equipped with a 6460 QQQ mass spectrometer, excellent linearity and less than 2% area difference was observed between the two streams. The acquisition window of 0.9 minutes within a 5 minute LC limited the maximum sample throughput in this 2-plex example but nonetheless was used to showcase stream reproducibility and accuracy of quality control samples.

A 4-plex system would later in 2019 demonstrate a sample throughput of 97 seconds per analysis for a panel of six cannabinoids from a 6 minute LC runtime or a sample throughput improvement of approximately 3.7 times. Robustness was demonstrated by acquiring 1000 samples over the course of 27 hours over the four LC streams. Retention time reproducibility was between 0.43% and 1.54% depending on the analyte [30]. A multiplexed analysis of several fentanyl analogs and metabolites from whole blood would use a 2-plex StreamSelect system to extract a 2.5 minute window from a 13 minute LC run [31]. Other examples, included the multiplexing of a previously developed 25-Hydroxyvitamin D3 method, shortening the analytical run time from 5 to 2.5 minutes [32].

As early as 2001, Shimadzu offered versions of multiplexing with parallel gradient analyses controlled with a single SCL-10AVP unit and up to eight pumps (four binary gradients) using the SIL-5000 autosampler (an HTC PAL-type autosampler) equipped with a 4-channel injection valve. These early offerings were primarily designed for operation with optical detectors or other vendor MS/MS units. An example of this is a high-throughput ADME microsomal stability screening of compound libraries using an eight-channel autosampler on to eight microbore columns diverted to a Applied Biosystems MDS-SCIEX API 165 mass spectrometer. Some 100,000 samples were analyzed using this parallel configuration with Retention Time coefficient of variations (CV) averaging about 0.9% with CVs of mass spectrometric peaks averaging around 5.7% [33]. With the introduction of improved triple quadrupole mass spectrometers such as the Shimadzu 8030 Triple Quadrupole Mass Spectrometer in 2010, Shimadzu began to focus on developing an improved multiplexing solution for customers based on the newer Nexera branded HPLC components.

In late 2015, Shimadzu launched a multiplexing (2-plex system) package highlighting an extended steroid panel. The Shimadzu Nexera MX multiplex LC-MS system with Dual Stream Technology (MX-DST) allows two separate analysis systems to perform overlapping sample injections. The design supports either an injection cycle of two batches of samples at the same time or to inject one batch of samples using the two separate streams of the Nexera MX system. In an example steroid panel application, serum samples were prepared using supported liquid extraction (SLE) followed by elution with dichloromethane. The evaporated extracts were then reconstituted with methanol:water and analyzed using the Dual Stream system using Restek Raptor Biphenyl (2.7 μm, 50 × 3 mm) columns coupled to a Shimadzu LC-MS-8050 mass spectrometer. Limit of Quantitation (LOQ) ranged from 0.5 pg/mL for Progesterone and Estrone to 100 pg/mL for DHEA sulfate [34]. Rather than using a dual injector autosampler or two separate autosamplers, the Nexera MX multiplex system used a SIL-30ACMP autosampler which synchronized the injections to the two analytical streams. The study highlighted the very good overlap between the relative response of both streams as well as the retention time stability of the chromatographic peaks. The study also considered the influence on the results when using the two separate streams for one batch compared to the use of only one stream for one batch. A further example showing the throughput gains for Levosalbutamol, a widely used bronchodilator used in asthma treatment, was also shown. Plasma samples (300 μL) were extracted with methyl tertiary butyl ether (MTBE) and evaporated to dryness. The reconstituted samples were then analyzed using the Nexera MX multiplex system equipped with Grace Altima HILIC columns coupled to a Shimadzu LC-MS-8060 mass spectrometer. An LOQ of 0.1 pg/mL was achieved with a two-fold throughput improvement compared to the original 6 minutes chromatographic run [35]. Although taking advantage of the speed improvements of the single injector, the MX Dual Stream multiplexing design does not allow use of an alternate stream in the event of autosampler failure. Furthermore, the sample capacity of the SIL-30ACMP is limited to 6 MTP well plates with no options for expansion of plate capacity.

With the availability of newer Nexera XR 40 series UHPLC components, Shimadzu developed its Nexera QX multiplexing solution by combining the Nexera 40 series pumps and a single SIL-40 dual port autosampler introduced in 2020 highlighting an injection time of <7 seconds. Although still utilizing a single autosampler, the availability of two stream-dedicated injection ports allowed the QX to divert samples to any available stream. Additionally, the installation of up to three separate plate changers allowed for improved sample capacity (now a maximum of 44 microtiter plates) over the previous MX multiplexing configuration. Examples with a panel of antiarrhythmic agents on Shim-pack XR-ODS III reverse-phase columns using an LC-MS-8060NX showed excellent linearity, analytical performance between the two streams, and extremely low carryover (<0.0005% and <0.0006% for both streams) [36]. The Nexera QX system also highlighted Shimadzu's LabSolution Insight software for automated data review with over 75 QA/QC flagging controls and configurable color coding available to the user.

For the customer interested in high-throughput sample analysis, all major triple quad mass spectrometer manufacturers offer robust and effective solutions. Decisions over which system is "best" are often constrained by which systems offer the greatest Return On Investment (ROI). Many vendors supply ROI online calculators. These ROI calculators typically ask users to input values such as the numbers of hours per day that the mass spectrometer is used, length of LC method (with the assumption that only 50% of the LC run contains usable information), how many samples are run per day, and the revenue per sample. However, in addition to the basic multiplexing ROI parameters, there are numerous additional constraints when addressing very short-duty cycles – less than 30 seconds. In such rapid multiplexing scenarios, the impact of instrument and column parameters such as gradient delay, flow rate, rinse cycle time, and autosampler speed on high-throughput liquid chromatography performance need to also be considered both in their impacts to sample throughput but also in the ROI calculations. A recent review of the contributions of cycle time, void volumes, and software-related delays for ballistic gradients as low as 12 seconds examined the contributions to overall throughput [37]. Also, in the cases of very short-duty cycles, the type of autosampler, injector design, sample compartment style, carryover, and robustness will also impact multiplexing systems where several hundred samples are the required daily throughput from each stream [38]. Users are welcome to examine the ROI calculators from multiplexing vendors on their own and create hypothetical usage and revenue maximization scenarios.

3.4 Workflow and End-User Considerations

Although many multiplexing options are available, multiplexing typically only make financial sense if there are a large number of unanalyzed samples at the end of a workday. If the number of samples is well matched to the capacity of the available mass spectrometers, then multiplexing is probably unnecessary. Multiplexing will obviously positively impact sample throughput when expressed in number of samples per system daily. Increases of two, three, and fourfold are possible although the end user needs to consider the additional maintenance and costs of three additional LC systems in a 4-plex system.

However, many users find that although multiplexing with even a 2-plex system provides them with the required improvement in analytical data acquisition throughput, the typical multiplexing system still does very little to improve data throughput. More data invariably results in simply more data to be processed, manually checked, and reported at the beginning of the next workday. Although the graphical user interfaces of all current multiplexing software packages can quickly flag and color code results to facilitate the evaluation of acceptable and failed samples, all of these processes only occur after the samples have been run and processed. Depending of the workflows, the evaluation, review, and reprocessing of samples becomes a major bottle neck to the release of approved and validated data. This is particularly the case for the analysis of drugs of abuse in so-called "Pain Panels" which are usually analyses of urine samples for common opiate and opiate analogs. Drug testing programs in the United States are either federally mandated or nonmandated. In the first group (e.g. Department of Transportation), the employer is required by federal regulation to drug test its employees. In the second category, employers choose to test for reasons other than federal requirements. Some private employers that are not mandated to test under federal authority have instituted employee drug testing programs to create drug-free workplaces. Both programs also formalized the role of the medical review officer (MRO), a specialist who is an integral part of a drug testing program, determines the cause of positive drug test results (interference, other prescription drugs, etc.) and often counsels the employee. The participating laboratories can begin to screen the large number of urine specimens by way of an automated enzyme immunoassay (EIA) procedure. Putatively positive samples are then further processed using an approved "urine confirmation methodology," typically LC-MS/MS. Overall, studies indicate that confirmation is required in approximately 33% of samples [39].

In such laboratories, the presence of extremely high concentration samples, either as a result of narcotic abuse or by deliberate tampering, can often create direct contamination or cross contamination. In the case of multiplexing systems where the same sample injector feeds into two or more injection valves, this carryover contamination affects the ability of the analyst to determine whether subsequent low-level samples in an analytical result are due to endogenous analytes or simple as a result of carryover [40].

In 2017, SCIEX introduced the updated MPX 2.0 with an additional software feature called Automated Carryover Monitoring. This software improvement addressed three important issues encountered when multiplexing large numbers of highly variable samples. First, the flagging of high concentration samples; second, dealing with carryover contamination of the multiplexed streams; and last, samples showing potential carryover contamination. Although all existing multiplexing software had processing capabilities to view such information, SCIEX MPX 2.0 allowed changes to the sample queue based on data being obtained in "real time." In general, an MPX 2.0 sample batch with Automated Carryover Detection enabled begins with a series of Blank and calibration samples. Prior to the run, the user defines in the software User Interface a Low threshold where the Blank sample is deemed "clean" and a High threshold that either exceeds the calibration limit or a concentration limit that may result in carryover – these can be defined as either concentration units or region height. The region height option allows flagging of samples when a sample analyte has drifted out of the retention time window used for peak determination and quantitation but is still likely to indicate a large concentration of analyte. This feature is particularly useful for cases where an extremely high analyte peak is "split" as a result of MS detector saturation. This results in either the wrong retention time assignment or with the correct retention time assignment but only to a partial section of the entire peak. During the analysis, the concentration values of each analyte are determined at the end of each sample run. The resulting concentration values are determined and compared to the user inputted Low and High thresholds. If the Blank samples are below the Low threshold, the analysis continues. For samples where the analytes are above the Low threshold but below the High threshold, the analysis on each stream continues. When a sample on one of the multiplexed streams exceeds the High threshold for any of the targeted analytes, two things occurs. First, the sample is flagged to indicate that the result exceeds the calibration or carryover limits; second, the analysis stream pauses and the system injects a Blank sample to verify that the stream remains uncontaminated. If this Blank result is below the Low threshold, the analysis continues. Otherwise up to two additional Blank samples are injected in an effort to reduce the observed carryover. If after three Blank injections, the system continues to show contamination, the stream is aborted and the remainder of the samples continues on the other stream. In the case of samples that immediately follow a high sample, the user cannot immediately be sure whether the resulting value is due to carryover or endogenous analyte. The Automated Carryover Detection logic flags these samples and adds these samples to the end of the analytical queue for reanalysis at the end of the batch after determining that the stream is uncontaminated. The ability to flag high sample concentration, monitor carryover, and reinject suspected carryover samples greatly improves sample reporting and reduces the number of samples requiring reanalysis. Rather than performing these functions after reviewing the analytical data, these functions are performed automatically during the run. The MPX 2.0 streams only operate as long as the Carryover logic assesses them as "clean" and flagged samples are easily identified either for dilution or re-extraction [41, 42]. In many respects, features such as Automated Carryover Detection represents a future path for improvements to multiplexing throughput, whereby the focus moves away from simply analyzing larger and larger numbers of samples to evaluating and processing the data of large sample sets and modifying the sample queues to minimize sample reanalysis and rework. Eventually, systems incorporating sample dilution and internal standard addition will enable a much more significant sample throughput than simply adding another liquid chromatographic stream to a single mass spectrometer.

3.5 Conclusion

In summary, although multiplexing with either 2-plex or 4-plex systems will often provide large gains in sample throughput, many other considerations factor into whether overall data throughput is improved simultaneously. Mechanically it is feasible to imagine compact and robust 6-plex or 8-plex systems allowing for even greater utilization of mass spectrometer capabilities; however, the increased use of software enabling data-dependent automation for real-time sample dilution, reanalysis, and possibly even re-extraction will likely offer the next major incremental step in improving sample throughput in laboratory environments.

References

1 Thomson, J.J. (1913). Rays of positive electricity. *Proc. R. Soc. London A* A89: 1–20. https://doi.org/10.1098/rspa.1913.0057.

2 AAPS (2019). A history of how HPLC became common in pharmaceutical testing. *AAPS* (1 October).

3 McDonald, P. (2016). Waters corporation: fifty years of innovation in analysis and purification. *Science History Institute* (2 June).

4 Pomeranz, Y. and Meloan, C.E. (1994). High-performance liquid chromatography and ion chromatography. In: *Food Analysis*, 324–351. Boston, MA: Springer https://doi.org/10.1007/978-1-4615-6998-5_21.

5 Majors, R. and Baltrus, J. (2020). The top 10 game changers in HPLC history. *The Analytical Scientist* (9 October).

6 Majors, R.E. (2018). HPLC column technology: the first 50 years. *Labcompare* (28 September). https://www.labcompare.com/10-Featured-Articles/353029-HPLC-Column-Technology-The-First-50-Years

7 Majors, R.E. (2015). Historical developments in HPLC and UHPLC column technology: the past 25 years. *LCGC N. Am.* 33 (11): 818–840.

8 Biemann, K. (2007). On the ability of the Viking gas chromatograph-mass spectrometer to detect organic matter. *Proc. Natl. Acad. Sci. U.S.A.* 104 (25): 10310–10313.

9 Mukhopadhyay, R. The viking GC/MS and the search for organics on Mars. *ACS Anal. Chem.* 79 (19): 7249–7256. https://doi.org/10.1021/ac071972t.

10 Pullen, F. (2010). The Fascinating history of the development of LC-MS; a personal perspective. *Chromatography Today*, 4–6 February/March 2010. https://www.chromatographytoday.com/article/hplc-uhplc/31/unassigned-independent-article/the-fascinating-history-of-the-development-of-lc-ms-a-personal-perspective/601 (accessed 23 February 2023).

11 Claeys, M. and Claereboudt, J. (2017). Fast atom bombardment in mass spectrometry. In: *Encyclopedia of Spectroscopy and Spectrometry*, 3e (ed. J.C. Lindon, G.E. Tranter, and D.W. Koppenaal), 581–587. https://doi.org/10.1016/B978-0-12-803224-4.00320-4. Elsevier.

12 Thomson, B.A. (1998). Atmospheric pressure ionization and liquid chromatography/mass spectrometry – together at last. *J. Am. Soc. Mass Spectrom.* 9: 187–193.

13 Niessen, W.M.A. and Tinke, A.P. (1995). Liquid chromatography-mass spectrometry general principles and instrumentation. *J. Chromatogr. A* 703 (1, 2): 37–57. https://doi.org/10.1016/0021-9673(94)01198-N.

14 Perez, E.R., Knapp, J.A., Horn, C.K. et al. (2016). Comparison of LC-MS-MS and GC-MS analysis of benzodiazepine compounds included in the drug demand reduction urinalysis program. *J. Anal. Toxicol.* 40 (3): 201–207.

15 Greb, S.K. and Singh, R.J. (2011). LC-MS/MS in the clinical laboratory – where to from here? *Clin. Biochem. Rev.* 32 (1): 5–31.

16 Couchman, L. (2012). Turbulent flow chromatography in bioanalysis: a review. *Biomed. Chromatogr.* 26 (8): 892–905 https://doi.org/10.1002/bmc.2769.

17 Ayrton, J., Dear, G., Leavens, W.J., and Mallett, D.N. (1997). The use of turbulent flow chromatography/mass spectrometry for the rapid, direct analysis of a novel pharmaceutical compound in plasma. *Rapid Commun. Mass Spectrom.* 11 (18): 1953–1958.

18 Chassaing, C. and Robinson, S. (2009). Turbulent flow chromatography: an evolving solution for bioanalysis. *Chromatogr. Today* 2: 20–24.

19 Wang, P. (2009). *High-Throughput Analysis in the Pharmaceutical Industry.* CRC Press.

20 Brann, J. (2006). Multi column chromatography system. US Patent 6, 635,173, filed 23 March and issued 15 May 2007. Assigned to Cohesive Technologies.

21 Sage, A.B., Giles, K., Bateman, R., and Preece, S. (1999). High-throughput parallel LC/MS analysis of multiple liquid streams for drug discovery applications using a multiplexed electrospray interface. *Sage* 7: 9. https://www.gimitec.com/file/mmp56la3.pdf (accessed 23 February 2023).

22 Colletti, J.D., Redor-Goldman, M.M., Pomperada, A.E. et al. (2020). Sample multiplexing: increased throughput for quantitation of total testosterone in serum by liquid chromatography-tandem mass spectrometry. *Clin. Chem.* 66 (9): 1181–1189.

23 Espourteille, F.A. (2009). Quantitation of fentanyl and norfentanyl from urine using on-line high-throughput system. *Thermo Fisher Scientific.* Application Note: 457. https://assets.fishersci.com/TFS-Assets/CMD/Application-Notes/AN457(1).pdf (accessed 23 February 2023).

24 Fink, J. (2014). So how fast is your LC/MS analysis. *Pittcon-2014 Seminar*, Chicago, USA (2–6 March). http://apps.thermoscientific.com/media/cmd/hypersite-events/Pittcon-2014/seminars/Sem%2014-How-Fast-is-Your-LC-MS-Analysis-Fink.pdf (accessed 23 February 2023).

25 Taylor, A. and Schreiber, A. (2019). Multiplexing two different food residue methods using HILIC and reverse phase chromatography in the same LC-MS/MS run. https://sciex.com/content/dam/SCIEX/pdf/brochures/Multiplexing_1550210.pdf (accessed 23 February 2023).

26 Stone, J. (2016). High-throughput serum 25-hydroxy vitamin D testing with automated sample preparation. *Methods Mol. Biol.* 1378: 301–320. https://doi.org/10.1007/978-1-4939-3182-8_32.

27 Seegmiller, J.C. and Goodman, K.J. Fast LC-MS-MS analysis of 25-hydroxyvitamin D2 and 25-hydroxyvitamin D3. https://www.sigmaaldrich.com/CA/en/technical-documents/protocol/analytical-chemistry/small-molecule-hplc/analysis-of-25-hydroxyvitamin-d2 (accessed 23 February 2023).

28 Taylor, A.M., He, X., Moshin, J.E., and Liu, H-F. Investigating the high-throughput capabilities of a high resolution accurate mass method using the AB SCIEX triple TOF 4600 LC/MS/MS system for toxicology urine screening. https://public.jenck.com/notijenck/uploads/monitoreo-toxicologico-de-orina-con-un-sistema-lc-ms-ms-tripletof-4600.pdf (accessed 23 February 2023).

29 Szczesniewski, A., Stone, P. and McCann, K. Increased throughput for the analysis of Δ^9-tetrahydrocannabinol in oral fluids using triple quadrupole mass spectrometry coupled with an automated dual-channel HPLC system. www.agilent.com/cs/library/posters/Public/MSACL_Poster_Oral_Fluids_QQQ_Mass_Spec.pdf (accessed 23 February 2023.

30 Szczesniewski, A. and McCann, K. Fast multiplexed analysis of cannabinoids and their metabolites in urine using MassHunter StreamSelect LC-MS system, ASMS 2019, WP158, Atlanta, GA, USA (2–6 June 2019). https://www.agilent.com/cs/library/posters/public/Agilent_ASMS_2019_WP158_Poster.pdf (accessed 23 February 2023)

31 Strayer, K.E., Antonides, H.M., Juhascik, M.P. et al. (2018). LC-MS/MS-based method for the multiplex detection of 24 fentanyl analogues and metabolites in whole blood at sub ng mL^{-1} concentrations. *ACS Omega* 3 (1): 514–523. https://doi.org/10.1021/acsomega.7b01536.

32 McCann, K., Nene, S., McIntyre, D., et al. Maximizing triple quadrupole mass spectrometry productivity with the Agilent StreamSelect LC/MS system. https://www.agilent.com/cs/library/applications/5991-2900EN.pdf (accessed 23 February 2023)

33 Xu, R., Nemes, C., Jenkins, K.M. et al. (2002). Application of parallel liquid chromatography/mass spectrometry for high-throughput microsomal stability screening of compound libraries. *J. Am. Soc. Mass Spectrom.* 13 (2): 155–165. https://doi.org/10.1016/S1044-0305(01)00342-7.

34 Loftus, N., Moreau, S., Levi, M., and Gruning, A. (2016). High speed UHPLC determination of multiple steroids in human plasma using the Nexera MX system for multiplex analysis. *MSCAL 2016 EU Abstract.* https://www.shimadzu.eu/sites/shimadzu.seg/files/poster_steroids_human_serum_lcmsms_msacl2016.pdf (accessed 23 February 2023)

35 Shelar, A., Lohar, A., Sutar, P. et al. Highly sensitive multiplexed analysis of Levosalbutamol from plasma using LC/MS/MS, ASMS 2016 WP 716. https://www.ssi.shimadzu.com/sites/ssi.shimadzu.com/files/pim/pim_document_file/ssi/applications/posters/15045/LC-MS-8060%20Application%20-%20wP-716_Levosalbutamol%20in%20plasma_P.pdf (accessed 23 February 2023).

36 Davide, V., Satoru, W., Shinji, T. et al. Nexera™ QX: improving the efficiency of routine LC-MS/MS assay with multiplex UHPLC platform. *Shimadzu Technical Report C190-E265.* https://www.ssi.shimadzu.com/sites/ssi.shimadzu.com/files/

pim/pim_document_file/ssi/others/14304/C190-E265_Technical_Report_Nexera_QX.pdf (accessed 23 February 2023).

37 Gilar, M., McDonald, T.S., and Gritti, F. (2017). Impact of instrument and column parameters on high-throughput liquid chromatography performance. *J. Chromatogr. A* 1523: 215–223.

38 Steiner, F., Dong, M.W., and Paul, C. (2019). HPLC autosamplers: perspectives, principles, and practices. *LCGC N. Am.* 37 (8): 514–529.

39 Langman, L.J. and Jannetto, P.J. ed. (2017). Using clinical laboratory tests to monitor drug therapy in pain management patients. *AACC Academy Laboratory Medicine Practice Guidelines.* https://www.aacc.org/-/media/Files/Science-and-Practice/Practice-Guidelines/Pain-Management/LMPGPain-Management20171220.pdf?%20la=en&hash=19670524407619F78999AB60731A24CB4901939D (accessed 23 February 2023).

40 Pesce, A., West, C., West, R. et al. (2012). Analytical considerations when monitoring pain medications by LC-MS/MS. *J. Anal. Bioanal. Tech.* S5: 003. https://doi.org/10.4172/2155-9872.S5-003.

41 RUO-MKT-03-6974-B ©2019 DH Tech. Dev. Pte. Ltd. https://sciex.com/content/dam/SCIEX/pdf/brochures/MPX-2-oView.pdf (accessed 23 February 2023).

42 Kafle, A., Latawiec, A., Burrows, C. et al. (2019). Society of forensic toxicologists. *Meeting Abstracts, P23: Improving Analytical Throughput Using Automated Carryover Monitoring*. https://www.soft-tox.org/assets/docs/SOFT_2019_meeting_abstracts.pdf (accessed 23 February 2023)

Section 3

ESI-MS Without Chromatographic Separation

4

Direct Online SPE-MS for High-Throughput Analysis in Drug Discovery

Andrew D. Wagner and Wilson Z. Shou

Lead Discovery and Optimization, Bristol-Myers Squibb Company, Princeton, NJ, USA

4.1 Introduction

Mass spectrometry (MS) is a desirable readout method for many in vitro assays in drug discovery [1–5]. These assays include target activity, ADME (absorption, distribution, metabolism, excretion) profiling, and safety pharmacology. From an analytical detection perspective, they can be usually categorized into two types: substrate-specific and compound-specific assays [6, 7]. In a substrate-specific assay, a single substrate analyte is monitored for all samples. Whereas in a compound-specific assay, samples from many compounds of different physicochemical properties need to be analyzed. Mass spectrometry with atmospheric pressure ionization (API/MS), including the techniques of electrospray ionization (ESI) and atmospheric pressure chemical ionization (APCI), is generically applicable to compounds and substrates of diverse physicochemical properties. Therefore, it is suitable for detection of samples from both assay types. MS is label-free, which eliminates the needs and cost associated with obtaining (and in the case of radiolabels, disposing of) labeled reagents. It enables the use of native substrates, which have more physiological relevance to downstream assays and in vivo studies. MS-based detection also reduces false positives due to interferences from fluorescent compounds in the library [8]. The main drawback of MS-based detection is that MS is a serial technique requiring sample clean-up (typically in the form of liquid chromatography, LC), therefore the analysis speed is slower than optical readout such as fluorescence and luminescence, when assays are performed in high-density (384- or 1536-well) microtiter plates in a parallel fashion with automation. High-throughput MS analysis of discovery samples was a big challenge to the bioanalytical community in the

High-Throughput Mass Spectrometry in Drug Discovery, First Edition. Edited by Chang Liu and Hui Zhang.
© 2023 John Wiley & Sons, Inc. Published 2023 by John Wiley & Sons, Inc.

early 2000s. At the time, conventional, single-stream LC-MS at several minutes per sample was too slow for many early discovery applications, due to its autosampler overhead, initial wash, gradient time, and also column re-equilibration time. Several multiplexed LC approaches were developed at the time to improve duty cycle. Multiplexed LC-MS approaches operated in parallel [9] or staggered parallel [10] modes reached a speed of ~20–30 seconds per sample, but they added more complexity to the system, and had channel-to-channel variability. Direct ionization techniques, including MALDI, LDTD, DESI, DART, etc., have also since been attempted with very high-throughput, but they had typically lower data quality due to ionization suppression effect [11], and less compound coverage in the cases of MALDI and LDTD [12]. In addition, the need to deposit liquid samples onto a solid support prior to MS analysis added an extra liquid handling step (and the associated cost) to the overall assay process. Recent advancements in these direct ionization approaches have made them more amenable to screening applications in drug discovery, but they are not the focus here and instead are covered in Chapters 11, 12, and 13 of this book.

4.2 History of the Development of Direct Online SPE-MS

Solid-phase extraction (SPE) is an effective sample preparation approach for bioanalysis [13], but it has been mostly performed offline with cartridges or plates prior to analysis with LC-MS. Onlinc SPE or onlinc extraction, inspired by multidimensional LC, has been established with both homemade and commercial systems [14]. The SPE sorbents used in online extraction include restricted access medium (RAM), monolithic, and the typical LC-like sorbents such as C8, C18, and mixed mode. Cohesive and Spark Holland [15–17] represent the two commercial efforts in automating SPE and putting it directly online with LC-MS/MS analysis. The Cohesive Aria systems utilize multiple-use extraction columns (operated under so-called Turboflow mode), whereas the Spark Prospect and Symbiosis systems use disposable cartridges that could be swapped out during the run. In both approaches, the SPE eluent is then directed into an LC column for LC-MS/MS analysis, which usually requires the dilution with a more aqueous solvent to reduce solvent strength and focus the SPE eluent on the LC column head. Multiplexing is also available for Cohesive systems (up to 4×) to enhance throughput [18, 19]. These systems eliminate the offline SPE and enable the direct analysis of biological samples (urine, plasma, and blood), and they remain commercially available to this day. However, the speed of such analysis, typically at several minutes per sample, is still too slow for many drug discovery studies.

Since many drug discovery samples are reasonably clean, it has been long recognized that direct SPE alone, instead of SPE and LC, can be used as an effective separation step before MS analysis. Early attempts involve either offline SPE in a 96-well plate format and then direct flow injection analysis (FIA) into MS [20], or offline SPE using a special SPE card followed by direct elution into MS from the card [21] for high-speed analysis. However, the multistep nature of this bioanalytical process has made it somewhat difficult to be implemented widely in discovery laboratories.

Two groups independently reported the combination of the two aforementioned concepts, namely direct SPE without LC and online extraction, for high-throughput MS bioanalysis. Bu et al. [22–24] have reported the direct injection/on-line guard cartridge extraction/tandem mass spectrometry (DI-GCE/MS/MS) for Caco-2 and CYP inhibition sample analysis. Their set-up consisted of a guard cartridge and two switching valves configured in the backflush mode. The in vitro samples were directly injected onto a guard cartridge, with salts and other un-retained materials washed away to waste. The analytes were then eluted directly into the MS operated under ESI mode. Gradient elution was still used in this case, resulting in an injection-to-injection time of 2.5 minutes. Therefore, sample pooling was implemented to enhance throughput [22–24].

Janiszewski et al. [25] first reported their direct online SPE-MS approach, later referred to as "trap and elute," for analysis of metabolic stability samples [25]. The SPE cartridge had large particles (40 μm) and therefore could be operated under high flow rates. The pumps were operated at 1.5 mL/min in isocratic mode. Configured in a multiplexed (2×) mode, this homemade direct online SPE system (Figure 4.1) achieved an injection-to-injection speed of 50 seconds. All samples from the same compound were acquired into the same data file, eliminating the need (and the time it would take) to open and close data files between sample injections. This innovative online SPE approach has subsequently been emulated by several other groups [26, 27], and has since become a widely adopted platform for high-throughput in vitro ADME assay support. This multiplexed platform has also recently been further optimized, and has become available commercially from Apricot Designs and Sound Analytics [28, 29]. It should be noted that no elaborate chromatographic separation is performed with the direct online SPE approach; therefore, isomeric interference could cause potential problems. However, in early drug discovery, the risk of isomeric interference causing false positives is low, since there are usually follow-up experiments with more definitive (LC-MS) detection.

Another hugely successful direct online SPE-MS system is the Agilent RapidFire platform. Developed by BioTrove Inc with initial funding by PFE and AMG using MIT technology [30], it has a single-stream, direct online SPE configuration equipped with a fast sample stage, specialized valves, and injectors (described in detail in the next section) that have enabled a cycle time of as fast as five seconds per sample. BioTrove initially operated as a contract research organization (CRO)

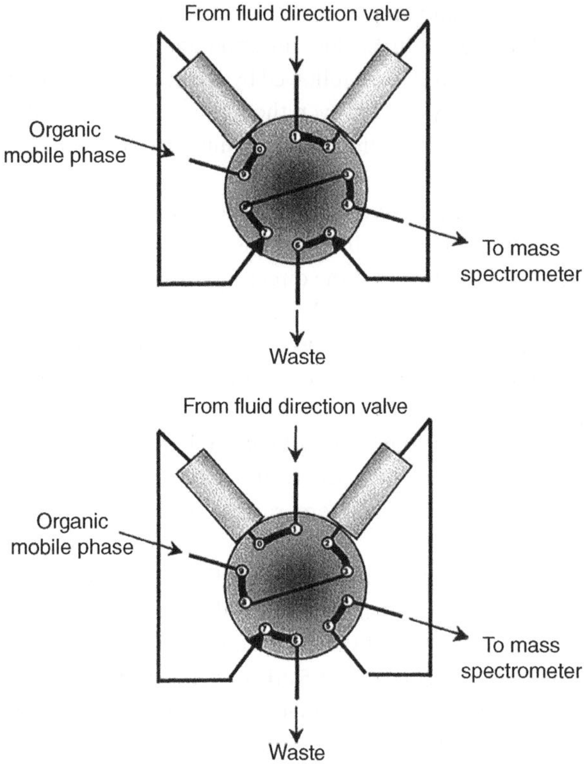

Figure 4.1 Direct online SPE with a dual-cartridge configuration, as first reported by Janiszewski et al. *Source:* Reproduced with permission from Janiszewski et al. Janiszewski et al. [25]/American Chemical Society.

with sample analysis service using the RapidFire instrument (c. 2004). Subsequently in 2006, the first-generation RapidFire instrument was commercialized. In 2009, the RapidFire product line was spun off from Biotrove into a new company called Biocius Life Sciences. Then, in 2011, Biocius was acquired by Agilent, which has maintained and continued to develop the RapidFire product line ever since. In 2015, Pure Honey Technologies was established as a separate business to again offer the RapidFire-based sample analysis service.

4.3 Hardware Details and Data Processing

The RapidFire system combines advanced microfluidic hardware components with high-speed robotics to achieve efficient online SPE of liquid samples prior to mass spectrometric analysis. By implementing robust system hardware components and a highly optimized sample flow path, the RapidFire platform has proven to be an effective and reliable sample delivery tool for MS-based analysis within drug discovery. Throughout multiple generations of RapidFire platform upgrades, the central hardware components have remained relatively unchanged, with evolutionary enhancements to capacity, flexibility, and software capabilities launched with each new iteration.

At the core of the RapidFire system are three high-speed, 6-port valves controlling the flow path of sample and solvents throughout the system, as well as a robotic stage for maneuvering the aspiration position. A 4-port valve is located at the back of the platform for directing the flow of vacuum pressure (Figure 4.2). The three main valves and aspiration needle (or sipper) are attached to a fixed manifold that moves vertically along the z axis (up and down) in order to move the sipper in and out of the aspiration source. A robotic stage with space for a single sample plate, a wash station with two separate wash ports (organic and aqueous), and a waste reservoir sits below this manifold and moves horizontally along the x and y axes in order to position individual sample wells (or wash/waste location) directly below the aspirator. These main hardware components have remained relatively unchanged over multiple generations of RapidFire platforms.

There are three main valve states that control the path of sample and solvent flow: (a) aspirate/re-equilibrate, (b) wash/load, and (c) elute (Figure 4.3). The system cycles through these three states utilizing user-defined timings for each step. During the aspiration step (Figure 4.3a), the sipper is lowered into the source well and overfills the sample loop on Valve 1 with the liquid sample. An optical sensor ensures that the sample loop has been overfilled and triggers the system to stop further aspiration. Next, Valve 1 is switched to the wash/load state (Figure 4.3b) to transfer the contents of the sample loop to the SPE

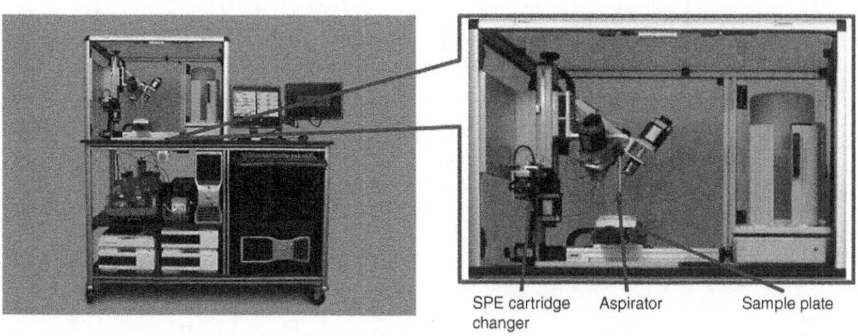

Figure 4.2 RapidFire instrument (RapidFire 365) picture. *Source:* BioTrove, Inc. https://www.selectscience.net/product-news/biotrove-launches-rapidfire-300-for-in-vitro-adme-discovery/?artID=18262, last accessed 31 October 2022.

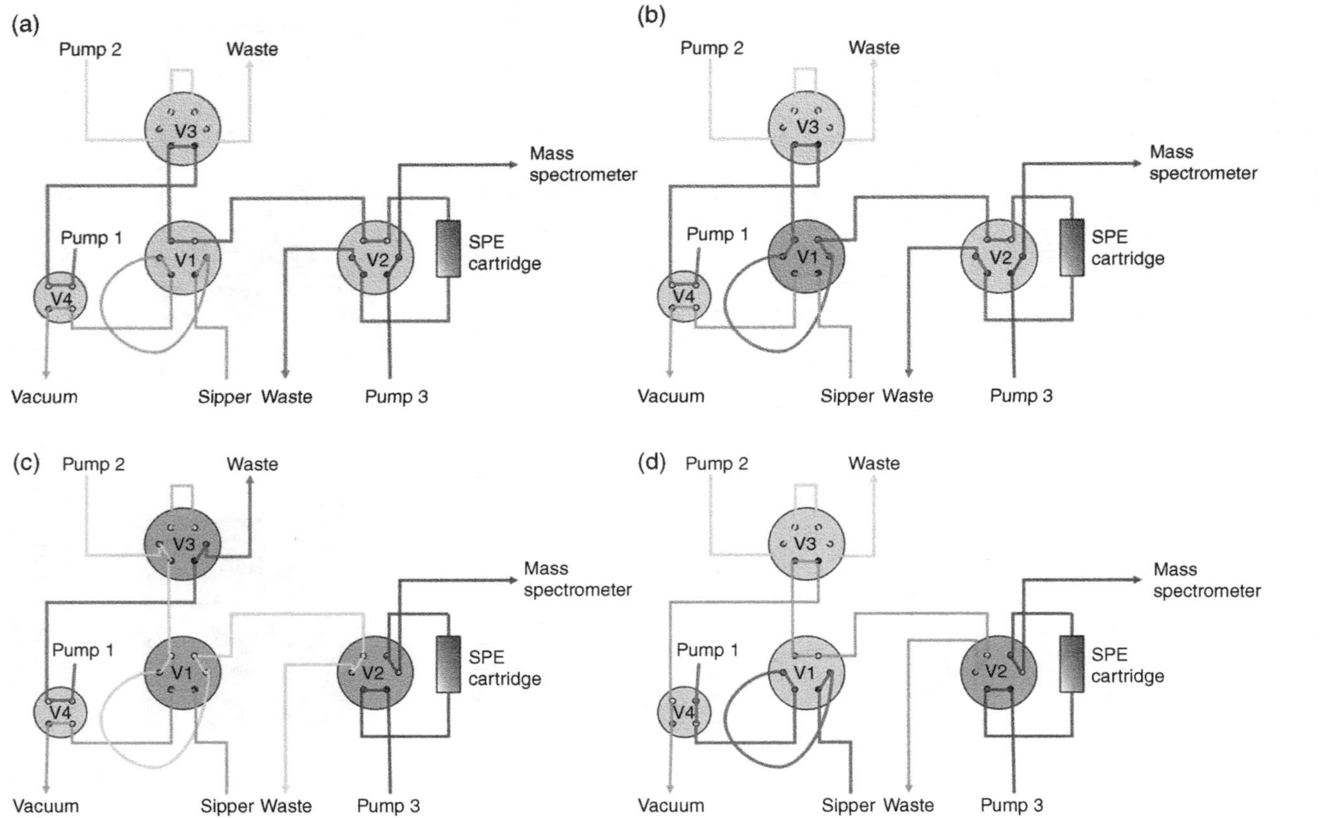

Figure 4.3 Valve positions and flow paths during RapidFire operation: (a) aspirate/re-equilibrate, (b) wash/load, (c) elute, and (d) flush sipper.

cartridge. With an appropriate method, the analytes are retained on the SPE cartridge whereas un-retained sample components pass through directly to waste. The wash/load state is followed by the switching of both Valves 2 and 3 for the elute step (Figure 4.3c). During this step, retained components on the SPE cartridge are back-flushed into the mass spectrometer for detection. Also during the elute step, the solvent from Pump 2 switches in line with the sample loop to clean it before the next aspiration. The final step is the re-equilibration state (Figure 4.3a) to ready the valves and sipper for the next sample. Concurrently, while the system is cycling through these three states, the stage positions the wash stations below the sipper so that an aspiration can occur at both the organic and aqueous locations. This happens during each sample extraction cycle in order to rinse the sipper and minimize carryover between wells. A separate, more thorough sipper flush cycle can also be performed following the completion of each plate (or at any time determined by user), in which Valve 4 switches Pump 1 directly in line with the sipper to perform a high-flow backflush (Figure 4.3d) directly into the waste reservoir. One aspect of the RapidFire system that has evolved over the years is the capacity of plates/samples that can be loaded on the platform during uninterrupted batch analysis. Sample plates are transferred in and out of the sample stage via a robotic arm. Early generations of the RapidFire system (RapidFire 200/300/360) use a Twister plate handling robot with two plate stacks (1 input, 1 output) and an 18 plate capacity. The RapidFire 365 uses a BenchBot sample handler and four plate stacks which increases the system capacity to 63 plates. The RapidFire 400 system incorporates a new plate handling robot with temperature-controlled plate storage and capacity for 64 standard height sample plates or up to 90 1536-well plates. The RapidFire 400 is the first system to offer compatibility with 1536-well plates – all previous generations of the system only support 96- and 384-well plate formats.

Flexibility of operation has also improved over multiple generations of the platform. The RapidFire system comes equipped with three high-pressure LC pumps for controlling load and elute solvent flow and a single peristaltic pump for controlling the flow of wash solvents. The first three generations of the RapidFire system (200/300/360) all use Agilent 1200 series isocratic pumps for controlling mobile-phase solvent flow. Newer generations of the platform (365 and 400) use Agilent 1260 Infinity quaternary pumps allowing the user to switch between different mobile phases and increase assay diversity during batch analysis. Adding to flexibility of the platform has been an increase in the number of SPE cartridges available on the system. The SPE cartridges used on the RapidFire system are available in multiple chemistries with bed volumes ranging from 4 to 20 μL. The original RapidFire 200 system accommodates only a single SPE cartridge which needs to be swapped manually. The RapidFire 300/365 systems are equipped with a 6-cartridge changer, allowing the user the ability to switch SPE cartridges during batches. Having this capability enables access to diverse SPE chemistries during a batch and/or the ability to have auxiliary cartridges of the same chemistry readily available. The RapidFire software enables user-defined SPE cartridge switching during batch analysis, as well as triggering of cartridge changes when backpressure exceeds a set threshold or using a user-defined sample count/cartridge has been reached. The 365 and 400 systems increased cartridge capacity further to a maximum of 12.

RapidFire instruments launched prior to the 365 model are compatible with triple-quadrupole mass spectrometers from Sciex, Thermo, and Agilent. These systems all have a two-computer configuration, with one computer controlling the mass spectrometer data acquisition and the other controlling the RapidFire online SPE system. The most recent RapidFire 400 systems are only available with Agilent mass spectrometers (both triple-quadrupole and time-of-flight systems), along with a single-computer option to control both the SPE components and the MS data acquisition.

Data generated during RapidFire analysis is always acquired into a single file containing all sample information from an entire 96-, 384-, or 1536-well plate (Figure 4.4). In order to process these data files, it is necessary to have software capable of identifying and integrating each individual peak within the acquisition raw data. RapidFire Integrator software is often used for the data processing step, and is capable of processing data from multiple mass spectrometer instruments (SCIEX Analyst, Thermo Xcalibur, and Agilent MassHunter). The RapidFire Integrator software uses a sequence file generated during RapidFire analysis which includes aspiration timing information and is necessary to perform peak splitting prior to integration. Multiple users have also reported using third party software applications to perform the data processing [32, 33].

Most RapidFire instruments are operated as stand-alone, sample analysis-only systems. Luippold et al. designed and developed a more integrated system, referred to as "Rapid and Integrated Analysis System" (RIAS), by integrating the core RapidFire components with additional hardware features such barcode scanning and automated plate loading, as well as software tools to enable automated MS/MS optimization, data processing, and overall workflow management. With these impressive features, the RIAS system has become the universal analytical system for in vitro ADME samples at Boehringer Ingelheim [32, 34]. More recently, the same group has further integrated RIAS into an even more comprehensive automation platform to include additional liquid handlers, a plate sealer, a centrifuge, and a freezer to enable fully automated standard/QC preparation, protein precipitation, and sample analysis for discovery PK studies [35].

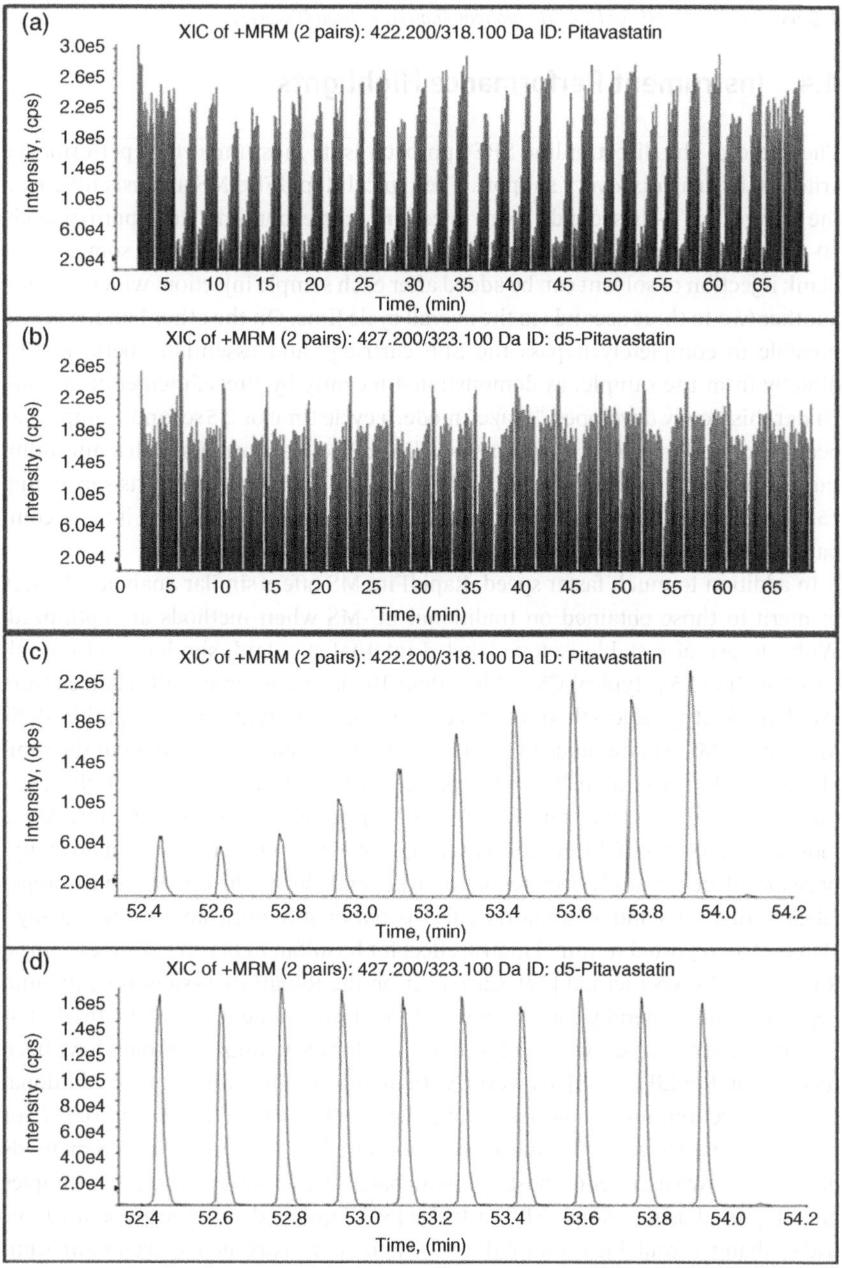

Figure 4.4 Typical RapidFire-MS data from a transporter inhibition assay, with all samples from a 384-well plate acquired to the same data file: (a) pitavastatin from full 384-well plate, (b) d5-pitavastatin from 384-well plate, (c) a zoomed-in 10-point IC_{50} curve for pitavastatin, and (d) corresponding d5-pitavastatin peaks from a 10-point IC_{50} curve. *Source:* Reproduced with permission from Wagner et al. [31]/John Wiley & Sons.

4.4 Instrument Performance Highlights

The speed of the direct online SPE approach is its most important performance criterion in drug discovery support. The typical RapidFire MS analysis time is in the range of 5–10 seconds per injection, corresponding to approximately 30–60 minutes per 384-well plate. If excessive carryover is observed, sometimes a blank injection of solvent can be added after each sample injection, which can add another two to three seconds to the overall cycle time. On the other hand, it is also possible to completely bypass the SPE cartridge and essentially perform FIA directly from the sample, as demonstrated recently by Bretschneider et al. [36]. Under this newly developed "Blaze" mode, a cycle time of 2.5 seconds sample has been reported. RapidFire supports 96-well and 384-well formats with a minimum sample volume of ~40 μL. The recently launched RapidFire 400 starts supporting 1536-well plates with a minimum sample volume of ~10 μL, making it more compatible with typical high-throughput screening workflows.

In addition to much faster speed, RapidFire MS offers similar analytical figures of merit to those obtained on traditional LC-MS when methods are optimized. With the use of a stable isotope-labeled internal standard, similar precision (all CVs less than 15%, typical CVs of less than 10%), and accuracy (all relative errors less than 15%, typical less than 10%) for digoxin were reported by both RapidFire-MS/MS and LC-MS/MS methods [33]. In terms of signal stability throughout the sample run, a CV of less than 10% was reported for the internal standard of pitavastatin, a substrate used in a transporter inhibition assay [31]. Since RapidFire MS is mostly used in a non-GLP, discovery setting, formal characterization of matrix suppression effect has rarely been reported. In general, due to the use of stable isotope-labeled internal standards, matrix effect is reasonably mitigated in these assays. Danso et al. reported minimal matrix effect for busulfan in plasma samples using a RapidFire-MS/MS method [37]. Carryover on the RapidFire system is compound dependent and generally consistent with those from gradient LC-MS methods. For CYP inhibition assays, carryover levels of 1–4% for the various substrates have been reported on RapidFire [38]. However, with method optimization and/or additional solvent injection, much lower carryover levels (0.03% for digoxin and 0.07% for cyclosporine A) have been achieved on RapidFire [33, 39]. RapidFire-MS methods have also shown excellent robustness, with batch sizes of several thousand samples often reported using a single SPE cartridge [39]. Equipped with the automated cartridge changer available on RapidFire 300 and later instruments, even more samples can be analyzed in a single batch with prescheduled cartridge changes. Since RapidFire uses the highly selective and specific MS as its detection method, the assay performance is usually excellent. The reported Z′ values from HTS assays are almost always more than 0.7 [40, 41] (a high-quality assay generally has a Z′ value of >0.5), and the hit confirmation rates are more than 90% from initial assay to retest. Despite its speed advantage over LC-MS, the typical RapidFire cycle time of ~10 seconds is still considered relatively slow for primary screening of the entire compound library (>1 million compounds). Therefore, RapidFire-MS methods are often deployed as a follow-up screen after the optical plate reader-based primary screen, in order to maximize its high quality while at the same time limit the number of samples requiring analysis [8, 42–44].

4.5 Applications

HTS is the first major area in drug discovery where direct online SPE-MS has been used extensively as a readout method. In 2004, Özbal et al. [45] published the first paper describing a RapidFire-MS method to screen for inhibitors of acetylcholinesterase. A speed of four to five seconds per sample was reported, and several inhibitors were identified from a small library. Since the publication of this proof-of-concept paper, a number of groups from many biopharmaceutical companies have reported the use of RapidFire-MS as a label-free readout for enzyme-based HTS assays for targets including AKT1/PKBα, phosphatidylserine decarboxylase, stearoyl-CoA desaturase, Acetyl-Coenzyme A Carboxylase, Histone Acetyltransferases, among many others [40, 41, 46–54]. In addition to simple enzyme-based biochemical assays, RapidFire-MS methods have also been reported for HTS assays in heavier matrices such as microsomes, serum, whole blood, and tissue extracts [42, 55–57]. Most of these screens were conducted in 96- or 384-well plates, and the typical library size was around 20,000–50,000 compounds (so-called "focused library"), although larger library sizes up to 310,000 compounds have also been reported [49, 54]. Screening of larger compound libraries of more than one million compounds with RapidFire-MS has not been reported to date. This is due to [1] the speed limitation of RapidFire (it would take four months to run 1,000,000 samples on one instrument at 10 seconds per sample), and [2] the somewhat large sample volume (~40 μL) requirement of previous generations of RapidFire systems, which is not compatible with the 1536-well format typically used in full-deck HTS. Multiplexing by combining the screening collection into eight-compound pools before the assay [51], or post-assay sample pooling from up to four HTS assays [58, 59] prior to RapidFire-MS analysis has been reported, which has improved the throughput by up to 8× without negatively impacting assay quality. With these multiplexed approaches, libraries sized at about 700,000 compounds can be screened with RapidFire-MS [51, 59].

HT-ADME is another area where the direct online SPE-MS approach has found widespread utilities [6, 60]. As mentioned previously, HT-ADME assays are either probe-specific or compound-specific, and direct online SPE-MS has demonstrated suitability for both assay types.

Drug–drug interaction (DDI) assays are commonly used in drug discovery to assess the potential of discovery compounds to change the pharmacokinetics (PK) of coadministered medications, and are typical examples of probe-specific assays. Due to the need to generate IC_{50} curves with multiple dosed concentrations, and the large number of enzymes that require assessment, these DDI assays tend to generate a large number of samples (~1000) per assay. Conceptually, these DDI assays are very similar to HTS assays; therefore, the direct online SPE approach is an excellent HT-MS approach for sample analysis. For example, RapidFire has become a standard platform for sample analysis of Cytochrome P450 (CYP) inhibition assays, where probe metabolites are analyzed along with their stable isotope-labeled internal standards from microsomal samples. A number of groups from several companies have independently reported RF-MS methods to support CYP inhibition assays, with a typical speed of 5–10 seconds per sample [32, 38, 61]. Similarly, Wagner et al. [31, 33, 62] have reported the use of RapidFire for the analysis several transporter inhibition assays, where transporter-specific probe substrates are monitored to assess the inhibitory potential of discovery compounds. The RapidFire-MS methods have been demonstrated to generate equivalent data to those from LC-MS methods, but with a more than 15× speed advantage.

Many of the HT-ADME assays require the analyses of samples from many different discovery compounds within a batch. These compound-specific assays are more challenging analytically, since compounds have different physicochemical properties and could require different separation and MS conditions on triple-quadrupole instruments to analyze [63, 64]. While LC-MS with a generic gradient is the standard method for these assays, direct online SPE-MS has also been attempted by several groups. Compound-specific HT-ADME assays including metabolic stability [32, 65], permeability [34], cell uptake [66], and protein binding [67] have all been reported using RapidFire methods. Since RapidFire became an Agilent product, it has been available with a TOF-MS configuration, which could further enhance the speed of compound-specific analysis due to the elimination of method development of selected reaction monitoring (SRM) conditions on triple-quadrupole instruments [60]. In addition to in vitro ADME samples, RapidFire MS methods have also been demonstrated for the analysis of in vivo samples [68].

4.6 Others

Clinical diagnostics is another area with a large sample load as well as a very fast turn-around-time requirement. Therefore, it shares many commonalities with various high-throughput analyses in drug discovery to which direct online

SPE-MS methods are applicable. Emerging RapidFire's applications in clinical diagnostics have been recently reviewed [69], and additional RapidFire methods for immunosuppressants [39], drugs of abuse [70], alkylating agents [37], anticonvulsants [71], and sulfonamides [72] have been reported. In addition, biomarker analyses from cells [73, 74] and plasma/blood [75, 76] using the RapidFire platform have also been reported.

4.7 Future Perspectives

Direct online SPE MS approaches using either commercial (RapidFire and others) or homemade system have been, and will continue to be a mainstay for MS-based screening in drug discovery. Due to its online sample cleanup capability, this approach still offers the best bioanalytical quality in terms of accuracy, reproducibility, and resistance to ionization suppression effect among all non-LC-based high-throughput mass spectrometric platforms. The main limitations of the platform are the relatively slow speed (at 5~10 seconds per sample) and the need for relatively large sample volume (>10 μL), which have made conducting full-deck primary high-throughput screens difficult on this platform. However, with the recent emergence of direct acoustic mass spectrometric approaches capable of direct analysis from 1536-well plates using only nanoliter (nL) sample volumes at sub one second per sample speed [77–81], performing direct full-deck primary screens with MS have now become practical. Therefore, one can envision an HTS screening strategy where primary screening of the entire library is conducted with acoustic MS, and direct online SPE-MS is then applied for follow-up/hit assessment as well as focused deck screens. In addition to HTS, we also fully expect online SPE-MS to remain as the "gold standard," with its combination of fast speed and high data quality, for HT-ADME and biomarker analysis in drug discovery and adjacent fields for years to come.

References

1 Roddy, T.P., Horvath, C.R., Stout, S.J. et al. (2007). Mass spectrometric techniques for label-free high-throughput screening in drug discovery. *Anal. Chem.* 79 (21): 8207–8213.

2 Kempa, E.E., Hollywood, K.A., Smith, C.A., and Barran, P.E. (2019). High-throughput screening of complex biological samples with mass spectrometry – from bulk measurements to single cell analysis. *Analyst* 144 (3): 872–891.

3 Wingfield, J. and Wilson, I.D. (2016). Advances in mass spectrometry within drug discovery. *J. Biomol. Screening* 21 (2): 109–110.

4 Pu, F., Elsen, N.L., and Williams, J.D. (2020). Emerging chromatography-free high-throughput mass spectrometry technologies for generating hits and leads. *ACS Med. Chem. Lett.* 11 (11): 2108–2113.

5 McLaren, D.G., Shah, V., Wisniewski, T. et al. (2021). High-throughput mass spectrometry for hit identification: current landscape and future perspectives. *SLAS Discovery* 26 (2): 168–191.

6 Shou, W.Z. and Zhang, J. (2010). Recent development in high-throughput bioanalytical support for in vitro ADMET profiling. *Expert Opin. Drug Metab. Toxicol.* 6 (3): 321–336.

7 Zhang, J. and Shou, W.Z. (2013). Mass spectrometry for quantitative in vitro ADME assays. In: *Mass Spectrometry for Drug Discovery and Drug Development* (ed. W.A. Korfmacher), 97–113. Wiley.

8 Adam, G.C., Meng, J., Rizzo, J.M. et al. (2014). Use of high-throughput mass spectrometry to reduce false positives in protease uHTS screens. *J. Biomol. Screening* 20 (2): 212–222.

9 Bayliss, M.K., Little, D., Mallett, D.N., and Plumb, R.S. (2000). Parallel ultrahigh flow rate liquid chromatography with mass spectrometric detection using a multiplex electrospray source for direct, sensitive determination of pharmaceuticals in plasma at extremely high-throughput. *Rapid Commun. Mass Spectrom.* 14 (21): 2039–2045.

10 Murphy, K., Bennett, P.K., and Duczak, N. (2012). High-throughput quantitation of large molecules using multiplexed chromatography and high-resolution/accurate mass LC-MS. *Bioanalysis* 4 (9): 1013–1024.

11 Chandler, J., Haslam, C., Hardy, N. et al. (2016). A systematic investigation of the best buffers for use in screening by MALDI-mass spectrometry. *SLAS Discovery* 22 (10): 1262–1269.

12 Haarhoff, Z., Wagner, A., Picard, P. et al. (2015). Coupling laser diode thermal desorption with acoustic sample deposition to improve throughput of mass spectrometry-based screening. *J. Biomol. Screening* 21 (2): 165–175.

13 Kole, P.L., Venkatesh, G., Kotecha, J., and Sheshala, R. (2011). Recent advances in sample preparation techniques for effective bioanalytical methods. *Biomed. Chromatogr.* 25 (1,2): 199–217.

14 Xu, R.N., Fan, L., Rieser, M.J., and El-Shourbagy, T.A. (2007). Recent advances in high-throughput quantitative bioanalysis by LC-MS/MS. *J. Pharm. Biomed. Anal.* 44 (2): 342–355.

15 Alnouti, Y., Srinivasan, K., Waddell, D. et al. (2005). Development and application of a new on-line SPE system combined with LC-MS/MS detection for high-throughput direct analysis of pharmaceutical compounds in plasma. *J. Chromatogr. A* 1080 (2): 99–106.

16 Jones, R.L., Owen, L.J., Adaway, J.E., and Keevil, B.G. (2012). Simultaneous analysis of cortisol and cortisone in saliva using XLC-MS/MS for fully automated online solid phase extraction. *J. Chromatogr. B* 881, 882: 42–48.

17 Koal, T., Sibum, M., Koster, E. et al. (2006). Direct and fast determination of antiretroviral drugs by automated online solid-phase extraction-liquid chromatography-tandem mass spectrometry in human plasma. *Clin. Chem. Lab. Med.* 44 (3): 299–305.

18 Ceglarek, U., Lembcke, J., Fiedler, G.M. et al. (2004). Rapid simultaneous quantification of immunosuppressants in transplant patients by turbulent flow chromatography combined with tandem mass spectrometry. *Clin. Chim. Acta* 346 (2): 181–190.

19 Chassaing, C., Stafford, H., Luckwell, J. et al. (2005). A parallel micro turbulent flow chromatography-tandem mass spectrometry method for the analysis of a pharmaceutical compound in plasma. *Chromatographia* 62 (1,2): 17–24.

20 Zheng, J.J., Lynch, E.D., and Unger, S.E. (2002). Comparison of SPE and fast LC to eliminate mass spectrometric matrix effects from microsomal incubation products. *J. Pharm. Biomed. Anal.* 28 (2): 279–285.

21 Inman, B.L., Chovan, L.E., Dandliker, P.J., and Lau, Y.Y. (2007). Solid phase extraction as a faster alternative to HPLC: application to MS analysis of metabolic stability samples. *J. Pharm. Sci.* 96 (6): 1619–1624.

22 Bu, H.Z., Magis, L., Knuth, K., and Teitelbaum, P. (2000). High-throughput cytochrome P450 (CYP) inhibition screening via cassette probe-dosing strategy. I. Development of direct injection/on-line guard cartridge extraction/tandem mass spectrometry for the simultaneous detection of CYP probe substrates and their metabolites. *Rapid Commun. Mass Spectrom.* 14 (17): 1619–1624.

23 Bu, H.Z., Magis, L., Knuth, K., and Teitelbaum, P. (2001). High-throughput cytochrome P450 (CYP) inhibition screening via a cassette probe-dosing strategy. VI. Simultaneous evaluation of inhibition potential of drugs on human hepatic isozymes CYP2A6, 3A4, 2C9, 2D6 and 2E1. *Rapid Commun. Mass Spectrom.* 15 (10): 741–748.

24 Bu, H.Z., Poglod, M., Micetich, R.G., and Khan, J.K. (2000). High-throughput Caco-2 cell permeability screening by cassette dosing and sample pooling approaches using direct injection/on-line guard cartridge extraction/tandem mass spectrometry. *Rapid Commun. Mass Spectrom.* 14 (6): 523–528.

25 Janiszewski, J.S., Rogers, K.J., Whalen, K.M. et al. (2001). A high capacity LC/MS system for the bioanalysis of samples generated from plate-based metabolic screening. *Anal. Chem.* 73 (7): 1495–1501.

26 Yan, Z., Lu, C., Wu, J.-T. et al. (2005). Validation of a high-throughput absorption, distribution, metabolism, and excretion (ADME) system and results for 60 literature compounds. *Rapid Commun. Mass Spectrom.* 19 (9): 1191–1199.

27 Kerns, E.H., Kleintop, T., Little, D. et al. (2004). Integrated high capacity solid phase extraction-MS/MS system for pharmaceutical profiling in drug discovery. *J. Pharm. Biomed. Anal.* 34 (1): 1–9.

28 Janiszewski, J., Schneider, R., Kapinos, B. et al. (2012). Development of a high-speed, multiplexed sample-delivery instrument for LC-MS/MS bioanalysis. *Bioanalysis* 4 (9): 1039–1056.

29 Kapinos, B., Liu, J., Piotrowski, M. et al. (2017). Development of a high-performance, enterprise-level, multimode LC-MS/MS autosampler for drug discovery. *Bioanalysis* 9 (21): 1643–1654.

30 https://www.agilent.com/cs/library/flyers/public/Agilent%20Rapidfire%20WAD%20Intro.pdf (accessed 19 January 2021).

31 Wagner, A.D., Elkin, L., Mosure, K. et al. (2016). Development of a high-throughput mass spectrometry based analytical method to support an in vitro OATP1B1 inhibition screening assay. *Rapid Commun. Mass Spectrom.* 30 (15): 1787–1796.

32 Luippold, A.H., Arnhold, T., Jörg, W., and Süssmuth, R.D. (2010). An integrated platform for fully automated high-throughput LC-MS/MS analysis of in vitro metabolic stability assay samples. *Int. J. Mass Spectrom.* 296 (1): 1–9.

33 Wagner, A.D., Kolb, J.M., Özbal, C.C. et al. (2011). Ultrafast mass spectrometry based bioanalytical method for digoxin supporting an in vitro P-glycoprotein (P-gp) inhibition screen. *Rapid Commun. Mass Spectrom.* 25 (9): 1231–1240.

34 Luippold, A.H., Arnhold, T., Jörg, W. et al. (2011). Application of a rapid and integrated analysis system (RIAS) as a high-throughput processing tool for in vitro ADME samples by liquid chromatography/tandem mass spectrometry. *J. Biomol. Screening* 16 (3): 370–377.

35 Runge, F., Mayer-Wrangowski, S., Bretschneider, T., et al. (2020). In iBA – intelligent bioanalysis, striving towards the lab of the future. *68th ASMS Conference on Mass Spectrometry and Allied Topics* (1–12 June 2020), Virtual.

36 Bretschneider, T., Ozbal, C., Holstein, M. et al. (2019). RapidFire BLAZE-mode is boosting ESI-MS toward high-throughput-screening. *SLAS Technol.* 24 (4): 386–393.

37 Danso, D., Jannetto, P.J., Enger, R., and Langman, L.J. (2015). High-throughput validated method for the quantitation of busulfan in plasma using ultrafast SPE-MS/MS. *Ther. Drug Monit.* 37 (3): 319–324.

38 Lim, K.B., Özbal, C.C., and Kassel, D.B. (2010). Development of a high-throughput online solid-phase extraction/tandem mass spectrometry method for cytochrome P450 inhibition screening. *J. Biomol. Screening* 15 (4): 447–452.

39 Grote-Koska, D., Czajkowski, S., and Brand, K. (2015). Performance of the new RapidFire system for therapeutic monitoring of immunosuppressants. *Ther. Drug Monit.* 37 (3): 400–404.

40 Forbes, C.D., Toth, J.G., Özbal, C.C. et al. (2007). High-throughput mass spectrometry screening for inhibitors of phosphatidylserine decarboxylase. *J. Biomol. Screening* 12 (5): 628–634.

41 Hutchinson, S.E., Leveridge, M.V., Heathcote, M.L. et al. (2011). Enabling lead discovery for histone lysine demethylases by high-throughput RapidFire mass spectrometry. *J. Biomol. Screening* 17 (1): 39–48.

42 Leveridge, M.V., Bardera, A.I., LaMarr, W. et al. (2012). Lead discovery for microsomal prostaglandin E synthase using a combination of high-throughput fluorescent-based assays and RapidFire mass spectrometry. *J. Biomol. Screening* 17 (5): 641–650.

43 Takamiya, M., Sakurai, M., Teranishi, F. et al. (2016). Lead discovery for mammalian elongation of long chain fatty acids family 6 using a combination of high-throughput fluorescent-based assay and RapidFire mass spectrometry assay. *Biochem. Biophys. Res. Commun.* 480 (4): 721–726.

44 Meng, J., Lai, M.-T., Munshi, V. et al. (2015). Screening of HIV-1 protease using a combination of an ultrahigh-throughput fluorescent-based assay and RapidFire mass spectrometry. *J. Biomol. Screening* 20 (5): 606–615.

45 Özbal, C.C., LaMarr, W.A., Linton, J.R. et al. (2004). High-throughput screening via mass spectrometry: a case study using acetylcholinesterase. *ASSAY Drug Dev. Technol.* 2 (4): 373–382.

46 Quercia, A.K., Lamarr, W.A., Myung, J. et al. (2007). High-throughput screening by mass spectrometry: comparison with the scintillation proximity assay with a focused-file screen of AKT1/PKBα. *J. Biomol. Screening* 12 (4): 473–480.

47 Soulard, P., McLaughlin, M., Stevens, J. et al. (2008). Development of a high-throughput screening assay for stearoyl-CoA desaturase using rat liver microsomes, deuterium labeled stearoyl-CoA and mass spectrometry. *Anal. Chim. Acta* 627 (1): 105–111.

48 Rye, P.T., Frick, L.E., Ozbal, C.C., and Lamarr, W.A. (2011). Advances in label-free screening approaches for studying histone acetyltransferases. *J. Biomol. Screening* 16 (10): 1186–1195.

49 Leveridge, M., Collier, L., Edge, C. et al. (2015). A high-throughput screen to identify LRRK2 kinase inhibitors for the treatment of Parkinson's disease using RapidFire mass spectrometry. *J. Biomol. Screening* 21 (2): 145–155.

50 Izquierdo, M., Lin, D., O'Neill, S. et al. (2020). Development of a high-throughput screening assay to identify inhibitors of the major M17-Leucyl aminopeptidase from *Trypanosoma cruzi* using RapidFire mass spectrometry. *SLAS Discovery* 25 (9): 1064–1071.

51 Langsdorf, E.F., Malikzay, A., Lamarr, W.A. et al. (2009). Screening for antibacterial inhibitors of the UDP-3-*O*-(R-3-hydroxymyristoyl)- *N*-acetylglucosamine deacetylase (LpxC) using a high-throughput mass spectrometry assay. *J. Biomol. Screening* 15 (1): 52–61.

52 Uchiyama, N., Dougan, D.R., Lawson, J.D. et al. (2017). Identification of AHCY inhibitors using novel high-throughput mass spectrometry. *Biochem. Biophys. Res. Commun.* 491 (1): 1–7.

53 Yoneyama-Hirozane, M., Matsumoto, S.-I., Toyoda, Y. et al. (2017). Identification of PARP14 inhibitors using novel methods for detecting auto-ribosylation. *Biochem. Biophys. Res. Commun.* 486 (3): 626–631.

54 Qi, J., Masucci, J.A., Lang, W. et al. (2017). Novel LC/MS/MS and high-throughput mass spectrometric assays for monoacylglycerol acyltransferase inhibitors. *SLAS Discovery* 22 (4): 433–439.

55 Highkin, M.K., Yates, M.P., Nemirovskiy, O.V. et al. (2011). High-throughput screening assay for sphingosine kinase inhibitors in whole blood using RapidFire® mass spectrometry. *J. Biomol. Screening* 16 (2): 272–277.

56 Holt, T.G., Choi, B.K., Geoghagen, N.S. et al. (2009). Label-free high-throughput screening via mass spectrometry: a single cystathionine quantitative method for multiple applications. *ASSAY Drug Dev. Technol.* 7 (5): 495–506.

57 Bretschneider, T., Luippold, A.H., Romig, H. et al. (2017). Ultrafast and predictive mass spectrometry-based autotaxin assays for label-free potency screening. *SLAS Discovery* 22 (4): 425–432.

58 Leveridge, M., Buxton, R., Argyrou, A. et al. (2013). Demonstrating enhanced throughput of RapidFire mass spectrometry through multiplexing using the JmjD2d demethylase as a model system. *J. Biomol. Screening* 19 (2): 278–286.

59 McManus, J., He, T., Gavigan, J.-A. et al. (2018). A robust multiplex mass spectrometric assay for screening small-molecule inhibitors of CD73 with diverse inhibition modalities. *SLAS Discovery* 23 (3): 264–273.

60 Miller, V.P. (2012). SPE–MS analysis of absorption, distribution, metabolism and excretion assays: a tool to increase throughput and steamline workflow. *Bioanalysis* 4 (9): 1111–1121.

61 Brown, A., Bickford, S., Hatsis, P. et al. (2010). High-throughput analysis of in vitro cytochrome P450 inhibition samples using mass spectrometry coupled with an integrated liquid chromatography/autosampler system. *Rapid Commun. Mass Spectrom.* 24 (8): 1207–1210.

62 Wagner, A., Elkin, L., Mosure, K., et al. (2013). In development of high-throughput, label-free analytical methods to determine transporter inhibition using RapidFire™-MS/MS. *2nd Annual Conference of Society for Laboratory Automation and Screening (SLAS)*, Orlando, FL (January 2013).

63 Zhang, J., Shou, W.Z., Vath, M. et al. (2010). An integrated bioanalytical platform for supporting high-throughput serum protein binding screening. *Rapid Commun. Mass Spectrom.* 24 (24): 3593–3601.

64 Zhang, J., Vath, M., Ferraro, C. et al. (2013). A high-speed liquid chromatography/tandem mass spectrometry platform using multiplexed multiple-injection chromatography controlled by single software and its application in discovery ADME screening. *Rapid Commun. Mass Spectrom.* 27 (7): 731–737.

65 Wu, X., Wang, J., Tan, L. et al. (2012). in vitro ADME profiling using high-throughput RapidFire mass spectrometry: cytochrome P450 inhibition and metabolic stability assays. *J. Biomol. Screening* 17 (6): 761–772.

66 Gordon, L.J., Allen, M., Artursson, P. et al. (2016). Direct measurement of intracellular compound concentration by RapidFire mass spectrometry offers insights into cell permeability. *J. Biomol. Screening* 21 (2): 156–164.

67 Amaral, A., Saran, C., Amin, J., and Hatsis, P. (2016). A comparison of LC-MS/MS and a fully integrated autosampler/solid-phase extraction system for the analysis of protein binding samples. *J. Biomol. Screening* 21 (6): 620–625.

68 Jian, W.Y., Romm, M.V., Edom, R.W. et al. (2011). Evaluation of a high-throughput online solid phase extraction-tandem mass spectrometry system for in vivo bioanalytical studies. *Anal. Chem.* 83 (21): 8259–8266.

69 Jannetto, P.J. and Langman, L.J. (2016). High-throughput online solid-phase extraction tandem mass spectrometry: is it right for your clinical laboratory? *Clin. Biochem.* 49 (13): 1032–1034.

70 Neifeld, J.R., Regester, L.E., Holler, J.M. et al. (2016). Ultrafast screening of synthetic cannabinoids and synthetic cathinones in urine by rapidfire-tandem mass spectrometry. *J. Anal. Toxicol.* 40 (5): 379–387.

71 Korman, E., Langman, L.J., and Jannetto, P.J. (2015). High-throughput method for the quantification of lacosamide in serum using ultrafast SPE-MS/MS. *Ther. Drug Monit.* 37 (1): 126–131.

72 Veach, B.T., Mudalige, T.K., and Rye, P. (2017). RapidFire mass spectrometry with enhanced throughput as an alternative to liquid–liquid salt assisted extraction and LC/MS analysis for sulfonamides in honey. *Anal. Chem.* 89 (6): 3256–3260.

73 Dittakavi, S., Mahadevan, L., Chandrashekar, D.V. et al. (2020). High-throughput screening assay for the quantification of Cer d18:1/16:0, d18:1/24:0, d18:1/24:1, d18:1/18:0, d18:1/14:0, d18:1/20:0, and d18:1/22:0 in HepG2 cells using RapidFire mass spectrometry. *Biomed. Chromatogr.* 34 (5): e4790.

74 Rye, P.T. and LaMarr, W.A. (2015). Measurement of glycolysis reactants by high-throughput solid phase extraction with tandem mass spectrometry: characterization of pyrophosphate-dependent phosphofructokinase as a case study. *Anal. Biochem.* 482: 40–47.

75 Razavi, M., Frick, L.E., LaMarr, W.A. et al. (2012). High-throughput SISCAPA quantitation of peptides from human plasma digests by ultrafast, liquid chromatography-free mass spectrometry. *J. Proteome Res.* 11 (12): 5642–5649.

76 Shimada, T., Kelly, J., LaMarr, W.A. et al. (2014). Novel heparan sulfate assay by using automated high-throughput mass spectrometry: application to monitoring and screening for mucopolysaccharidoses. *Mol. Genet. Metab.* 113 (1): 92–99.

77 DiRico, K.J., Hua, W., Liu, C. et al. (2020). Ultrahigh-throughput acoustic droplet ejection-open port interface-mass spectrometry for parallel medicinal chemistry. *ACS Med. Chem. Lett.* 11 (6): 1101–1110.

78 Sinclair, I., Stearns, R., Pringle, S. et al. (2015). Novel acoustic loading of a mass spectrometer: toward next-generation high-throughput MS screening. *J. Lab. Autom.* 21 (1): 19–26.

79 Zhang, H., Liu, C., Hua, W. et al. (2021). Acoustic ejection mass spectrometry for high-throughput analysis. *Anal. Chem.* 93 (31): 10850–10861.

80 Häbe, T.T., Liu, C., Covey, T.R. et al. (2020). Ultrahigh-throughput ESI-MS: sampling pushed to six samples per second by acoustic ejection mass spectrometry. *Anal. Chem.* 92 (18): 12242–12249.

81 Wagner, A., Zhang, J., Liu, C. et al. (2020). Ultrahigh-throughput and chromatography-free bioanalysis of polar analytes with acoustic ejection mass spectrometry. *Anal. Chem.* 92 (19): 13525–13531.

5

Acoustic Sampling for Mass Spectrometry: Fundamentals and Applications in High-Throughput Drug Discovery

Chang Liu[1], *Lucien Ghislain*[2], *Jonathan Wingfield*[3], *Sammy Datwani*[2], *and Hui Zhang*[4]

[1] *SCIEX, Concord, ON, Canada*
[2] *Beckman Coulter Life Sciences, San Jose, CA, USA*
[3] *Mechanistic and Structural Biophysics, Discovery Sciences, R&D, AstraZeneca, Cambridge, UK*
[4] *Entos Inc., Department of Analytical Technologies, Entos, San Diego, CA, USA*

Chang Liu, Lucien Ghislain, and Jonathan Wingfield were contributed equally.

5.1 Introduction

The use of mass spectrometry (MS) as a tool for biomedical research has grown significantly over recent decades. The development of mass spectrometers fitted with electrospray sources has enabled analyses of a wide range of pharmaceutically relevant biomolecular species, from druglike small molecules to large proteins and cellular metabolites [1, 2]. Typically, the mass spectrometer is coupled to high-performance liquid chromatography (HPLC), to separate analytes of interest from complex mixtures and characterize them in one process [3]. The electrospray mechanism generates either positively or negatively charged ions, depending on the ionization mode, generating molecular ions with very little fragmentation, thus simplifying the resulting datasets [4].

The technology has been applied across a wide range of disciplines including drug metabolism [5], proteomics [6], metabolomics [7, 8], lipidomics [9], and imaging [10]. The increasing ease of use and robustness of MS systems, together with improved software tools to support data interpretation, have delivered this instrumentation into the hands of the bench scientist. While the breadth of application areas for MS has grown, throughput has not evolved at the same pace.

High-Throughput Mass Spectrometry in Drug Discovery, First Edition. Edited by Chang Liu and Hui Zhang.
© 2023 John Wiley & Sons, Inc. Published 2023 by John Wiley & Sons, Inc.

The requirement for lengthy sample separation techniques to reduce the signal suppression from biological matrices before injection into the mass spectrometer has been the rate-limiting step. There have been improvements in HPLC system design, but few have resulted in faster run times and separation times in less than 10 seconds per sample are not possible [11]. Even if chromatographic separation is not required, the mechanics of getting samples into the mass spectrometer still limit sample loading time.

There are MS techniques that do not require sample separation and typically rely on a laser to deliver ionization energy and volatilize the analytes from the sample matrix. Without a separation requirement, these systems can take advantage of the speed of the laser. Matrix-assisted laser desorption ionization (MALDI) relies on secondary ionization of the sample [12]. The laser energy is absorbed by either nanostructures in the plate surface topography or a matrix molecule (usually substituted cinnaminic or picolinic acids) [13]. These excited matrix molecules in turn ionize the target molecule through a charge transfer mechanism. Peptides, small proteins, lipids, and oligonucleotides can readily be ionized via MALDI and analysis speeds of one seconds per sample can be achieved [14, 15]. However, MALDI cannot match the performance of electrospray ionization (ESI), which can be used with a wider range of small molecules. Laser diode thermal desorption (LDTD) desorbs samples directly into the gas phase via a thermal pathway [16]. However, LDTD in the literature has been mainly limited to small molecules; the thermal nature of this technique and the use of an ambient pressure chemical ionization (APCI) system make it unsuitable for both modified peptides and cellular metabolites in biochemical screening.

The team from AstraZeneca/Waters/Labcyte introduced a breakthrough approach for high sampling speed directly from standard microplates using the acoustic ejection technology [17–19]. To improve the ionization efficiency, the acoustic energy was adjusted to create a plume of mist with picoliter-sized droplets from the sample well. The mist was then aspirated into an MS sampling orifice through a heated transfer tube for desolvation. High voltage applied at the tube entrance induces charge on the mist droplets. This innovative acoustic mist ionization (AMI) technology offers contactless sampling at a speed faster than one sample per second directly from 384-well microplates and has been utilized successfully in multiple high-throughput drug discovery workflows.

In 2018, the Pfizer/SCIEX/Labcyte team reported another type of acoustic MS, acoustic droplet ejection open-port interface mass spectrometry (ADE-OPI-MS) [20], which is also called acoustic ejection mass spectrometry (AEMS) [21], utilizing acoustic ejection of nanoliter-sized sample droplets from high-density plates with both accurate and precise droplet volume control. This is the same technology used in commercial liquid handling automation systems for compound management [22]. A continuous liquid stream in an open-port interface (OPI) [23] captures the ejected droplets, dilutes the sample matrix in-line, and delivers sample to a standard ESI-MS for high-throughput analysis. Acoustic droplet ejection (ADE) provides reproducibility, and in-line matrix dilution reduces ionization suppression in many drug discovery assays. Since its introduction, ADE-OPI-MS has been adapted in many high-throughput analysis workflows. This chapter summarizes the working principle, system tuning and method development approaches, as well as representative applications of both AMI-MS and ADE-OPI-MS acoustic sampling approaches.

5.2 Technology Overview

Here, we address the working principle and the system configuration of both AMI-MS and ADE-OPI-MS technologies.

5.2.1 AMI-MS

The earliest proof-of-concept AMI-MS instrument used a Waters SQDII as the detector, a length of capillary wrapped with a trace heating wire and fitted to the inlet on the sample block (Figure 5.1a). The other end of the capillary was positioned next to a single well cut from a 384-well plate that was seated above the Echo® transducer [24]. At 90°, a secondary electrospray ionization (SESI) source was used to ionize the samples generated by the acoustic spray event. A sample of sulfadimethoxine was placed in the sample well and the transducer was tuned to spray the sample with a range of power offsets. Using the basic Echo setting, the 2.5 nL droplets were sprayed and no ion signal was seen, as the power offset was increased the droplet size decreased and ions were detected. The signal at the detector increased with power indicating that smaller droplets were formed as power increased (Figure 5.1b).

While this system demonstrated the ability to use acoustics to generate a spray of droplets that could be ionized, the use of a secondary spray was considered suboptimal. The SESI source would require considerable tuning and the data quality from this early test was suboptimal.

The second iteration of prototype had many similarities to the proof-of-concept system, the capillary was shortened to improve sensitivity, positioned above the sample well and the cross spray removed. The transducer was connected to a standard Echo 555 system which provided the electronics and fluidics control via

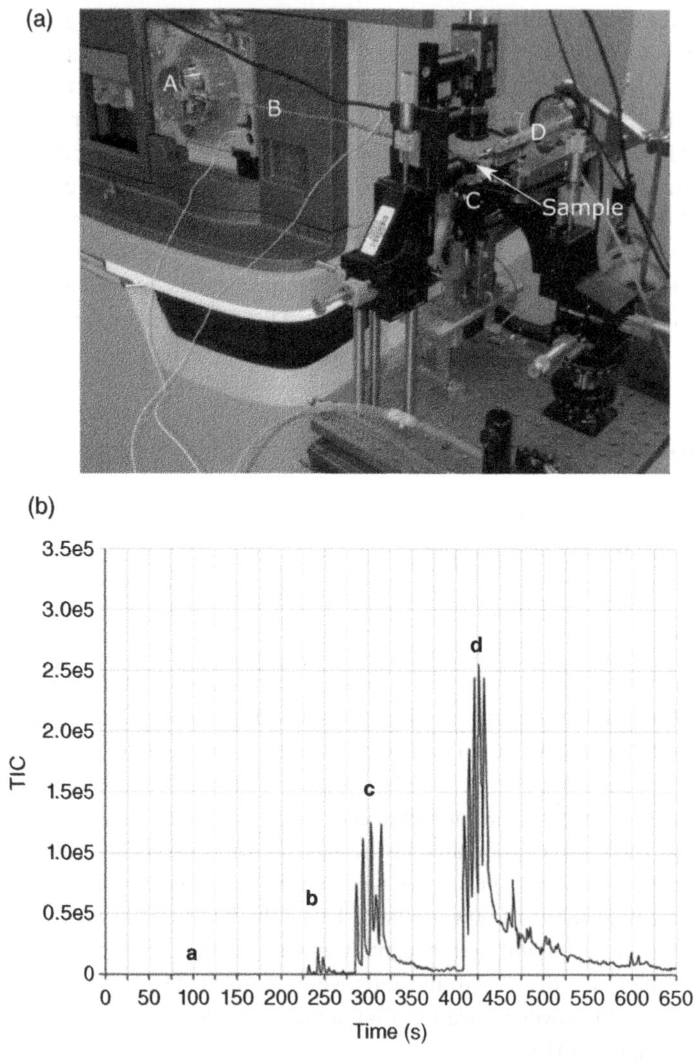

Figure 5.1 (a) A – Source block of Waters SQDII, B – capillary transfer line with trace heating, C – transducer from an Echo 555 system, D – SESI solvent spray. Sample was placed in a single well plate positioned over the transducer. (b) A mass chromatogram, at *m/z* 311, of various acoustic ejections of sulfadimethoxine. Initially, no signal was generated using the standard Echo 555 2.5 nL droplet (a). In altering the operating parameters of the acoustic pulse to produce successively smaller droplet sizes, an increasing MS signal is seen at 230 (b), 300 (c), and 425 (d) seconds, indicating a link between droplet size and ionization efficiency. The multiple peaks and high background signal produced during this experiment were attributed to suboptimal conditions for both the SESI mixing stage and the MS transfer tube. The *Y* axis, total ion current (TIC), is in arbitrary units.

an umbilical cable. Charge was provided to the sample directly via an electrode in the well. The configuration of this prototype is shown in Figure 5.2a.

The data generated by this prototype unit using a caffeine sample is shown in Figure 5.2b. The signal rapidly increases from baseline when the acoustic transducer is switched on. As soon as the transducer is switched off, the signal immediately falls back to baseline and remains stable between power cycling the transducer. Stability of the baseline between ejections suggests that there is no carryover or leakage of ions between firing events.

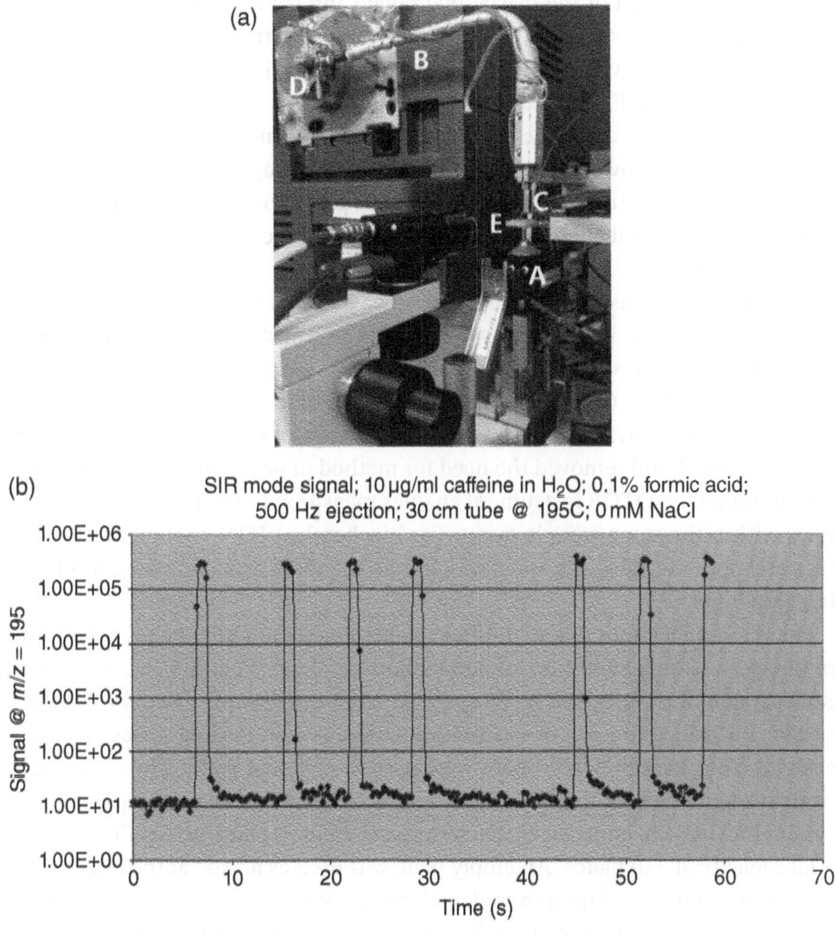

Figure 5.2 (a) A – Echo 555 transducer, B – heated transfer capillary, C – collection nozzle, D – mass spectrometry source block, E – cut down 384-well plate. (b) Data generated with a caffeine sample. The rapid onset and fall of signal coincide with the firing of the acoustic transducer.

Clearly the requirement for an electrode in each well is not practical for high-throughput screening (HTS). The question was how to get the charge into the well while isolating the charge from the mass detector. Initially, a resistive glass ion guide capillary was fitted at the end of the transfer capillary. This was used to supply a high voltage to push charge into the well and set up a field gradient in the liquid sample as shown in the schematic in Figure 5.3a. When charge is pushed into the well, opposite charge builds up on the surface of the fluid sample such that the droplets generated carry that charge into the capillary where they are dried and generate an ion beam. The issue with this approach is that charge builds up in the source well of the plastic microplate and is not dissipated. Eventually, droplets do not carry sufficient charge and at this point, the ion beam collapses. To avoid this issue, the voltage supplied to the glass capillary was rapidly switched between polarities (Figure 5.3b). While initially successful, there were obvious issues with the glass capillary, the main issue being condensation of samples on the glass capillary over time. This resulted in a steady rise in baseline signal between transducer firing events. To overcome this issue, the glass capillary was replaced with a metal charging cone insulated from the transfer capillary with a ceramic connector.

To enable plate-based screening, a simple X-Y stage was added to the prototype to position the appropriate well under the transfer line. The first usable "screening" version of the system is shown in Figure 5.4 and incorporated a Waters Xevo G2-XS QToF mass detector rather than the original SQDII. The value of the QTOF was that it enables a wide range of masses to be analyzed in each tone burst and removed the need for method of development for each analyte of interest. A QTOF instrument enables measurement of two (or more) analytes within the same sample event. For biochemical HTS, this means we can measure both the substrate and product from the same acoustic event, minimizing variability.

For HTS applications where multiple plates were required to be processed, the prototype was integrated into an automation envelope (Figure 5.4b). An Access Robot system was controlled by Tempo™ scheduling software to feed the AMI-MS system. To reduce the risk of automation failures when feeding plates, a simple array of plate stackers and a robot were used to pick and place. The robot picks from the top of each of the three input stacks and places plates coming out of the system into the 4th empty stack. Each stack can hold 50 plates giving the system a total capacity of 150 plates. An empty plate was used as a "lid" at the top of each stack to prevent evaporation from the test plates. While this approach introduced a time penalty as each "lid" plate has to go through the AMI-MS system, it did prove to be more robust than using either a lid on every plate or plate seals. With no electronic handshake between the robot and the storage locations, the system proved to be fast. A small number of temporary put down locations ensured that

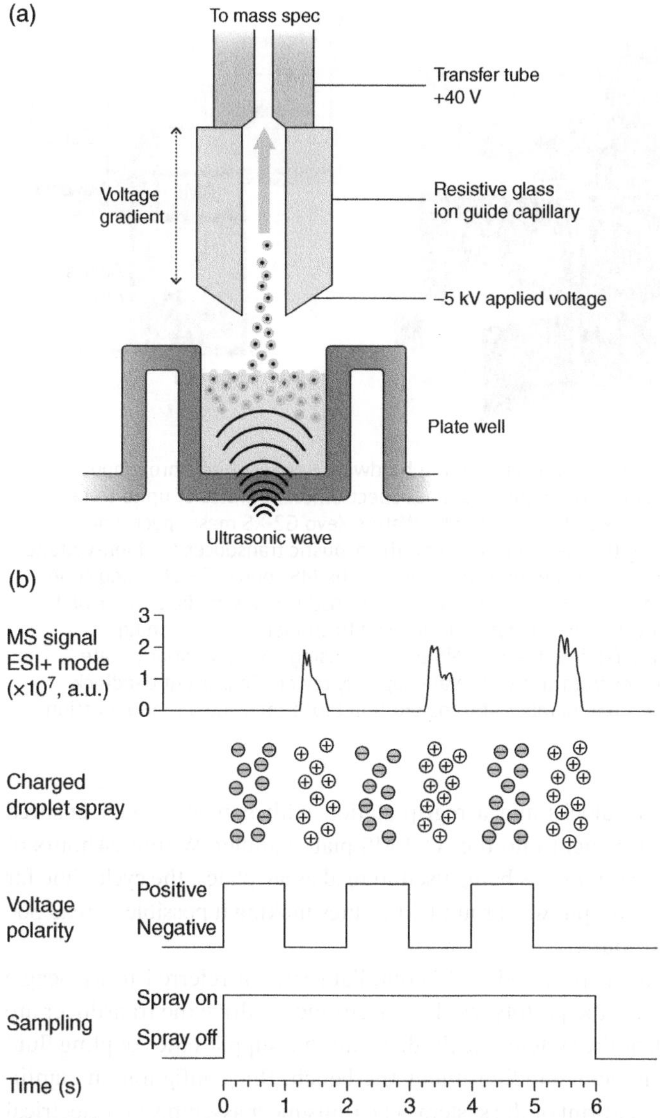

Figure 5.3 (a) Schematic of early AMI-MS charging system. The resistive glass ion guide enables a charge to be pushed into the well to set up the field gradient. The high voltage applied to the glass capillary is isolated from the transfer capillary which is carrying the voltage applied at the ion block. (b) A graphic to demonstrate the requirement for polarity switching with AMI-MS. The top panel shows the signal in the detector when the spray events happen. The signal peaks quickly but as charge builds in the well over time, the signal decays. When polarity is switched, there are no ions of the correct polarity generated to be seen at the detector (negative ions in this example) even though the transducer is constantly firing.

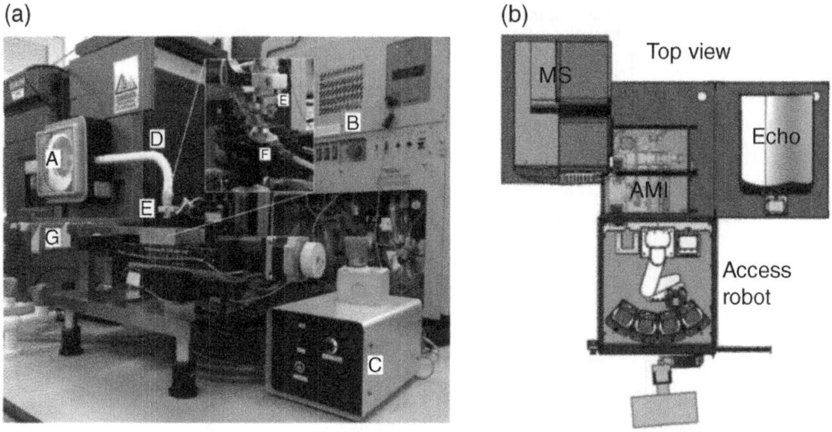

Figure 5.4 (a) Acoustic mist ionization (AMI) hardware used for high-throughput contactless delivery of liquid sample to a mass spectrometer at rates of up to three samples per second. A – Modified source of a Waters Xevo G2-XS mass spectrometer, B – Echo 555 providing the electronics driving the acoustic transducer, C – high-voltage power supply, D – heated transfer line connected to the MS source, E – charging cone providing high voltage to source well, F – acoustic transducer assembly, G – X–Y plate carrier. (b) A schematic view of the automation used to support HTS screening applications. Plates are fed into the AMI-MS platform using a simple robotic plate handler that picks plates from static 50-plate capacity towers. To prevent deadlock situations, there are a small number of transient hotel locations and a regrip location with a barcode reader.

the system never deadlocks and a regrip station with barcode reader ensured reliable pick and placement into the AMI-MS plate handler. Within 24 hours of the installation, the system was being used to read assay plates, the cycle time for a plate from input to output was around 4 minutes making it possible to read 300 plates in around 20 hours.

Ultimately, Beckman provided a CE-compliant system referred to as Beagle (Figure 5.5) to replace the prototypes. The electronics to drive the transducer are housed in the back of the system, the fluidics unit that supplies the coupling fluid to the transducer is floor-standing under the bench; this configuration significantly reduces the footprint of the system. The transducer assembly and electrical components are based on the next generation of Echo acoustic dispensers, providing improved droplet generation and higher transfer rates than the early generation of prototypes. The control software is a modified version of Echo Cherry Pick software making it easy to operate for anyone with experience of running a standard Echo dispenser. The control software manages not only the acoustic tone burst but also the configuration of the mass detector setup, so the user does not have to interact directly with the MassLynx software.

Figure 5.5 CE-compliant version of AMI-MS system.

5.2.2 ADE-OPI-MS

5.2.2.1 System Description

In addition to AMI, another process utilizing acoustic energy for noncontact transfer of a liquid sample out of a microplate is ADE. ADE was developed for high-throughput plate reformatting and is used widely for compound logistics in the pharmaceutical industry [25]. In an ADE process, discrete, nanoliter-sized droplets are transferred with high precision and accuracy from source to an inverted destination. The droplet size is precisely controlled at the desired volume from 1 to 25 nL [26]. The most commonly used volume, especially on the Echo system, is 2.5 nL. In the ADE-OPI-MS system, the target destination plate and its moving stage is replaced by a fixed-position OPI well-aligned with the acoustic transducer.

OPI was introduced by Van Berkel et al. [27]. It has a simple coaxial dual-tube structure, and the inner tube is directly connected to the nebulizer-gas-assisted atmospheric pressure ion source of MS (e.g. ESI or APCI). A low-pressure solvent pump delivers a continuous stream through the tubing annulus to the sampling region (i.e. open-port) where fluid is directly aspirated into the ion source driven by the Venturi aspiration force for direct MS analysis. The OPI interface captures, mixes, and dilutes the sample and delivers it to the ionization source. The benefits of both ambient ionization technologies, high analytical throughput and simplified sample preparation, and the advantages of the classic ion sources of MS (e.g. ESI) including high sensitivity, reproducibility, and wide compound coverage can be maintained simultaneously. Since its introduction, OPI has been used as a universal interface between raw samples and the classic ESI or APCI ion source of MS, including the direct sampling of tissues [28–32], laser ablation particles [33–37], solid-phase micro-extraction (SPME) devices [38–40], magnetic particles [41], and aerosols [42]. In addition, the OPI has been reported as a convenient MS sampling interface for discrete liquid droplets, from nanoliters [43] to the microliter range [44–46].

A typical ADE-OPI-MS system is illustrated in Figure 5.6. The stainless steel, electrically grounded outer tube of the OPI defines the droplet capture region with the inner diameter (ID) of 950 μm. The inner tube has the outer diameter (OD) of 300 μm and ID of 250 μm, the open-end of the inner tube is 300 μm recessed from the outer tube. The inner tube is directly connected with the electrode in the nebulizer gas-assisted ion source, where the gas pressure is regulated at 90 psi (10–12 L/min). The OPI with the sampling port facing downward is placed above the sample well-plate, with the *x*–*y* position well aligned with the acoustic transducer.

5.2.2.2 System Tuning and Assay Development

The analytical performance of the ADE-OPI-MS system is dependent on the system tuning condition. In such a system, ESI nebulizer gas serves two purposes. First, the conventional use of the nebulizer gas assists with creating a mist at the

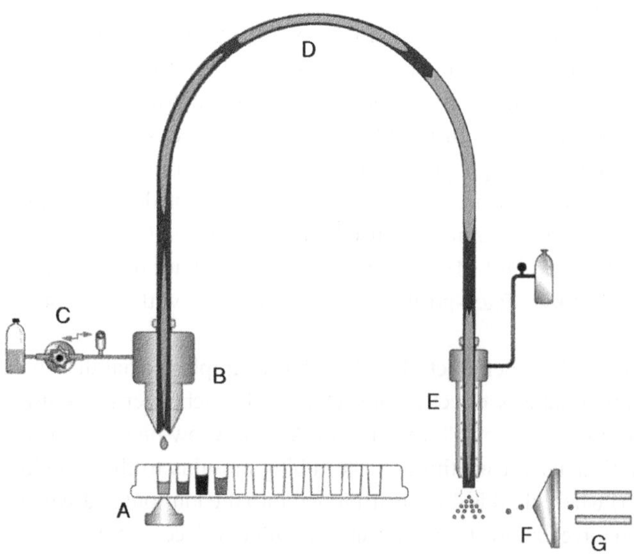

Figure 5.6 Diagram of an ADE-OPI-MS system. (A) Acoustic transducer. (B) Droplet capture region of OPI. (C) Pump delivering fluid to the OPI. (D) An OPI fluid transport tube. (E) Nebulizer gas-assisted Venturi pump drawing the transport fluid from (B). (F) Atmosphere to vacuum aperture. (G) Mass spectrometer. *Source:* Reproduced with permission from Liu et al. [47]/American Chemical Society.

outlet of the ESI for efficient, high flow rate ion production. Second, the ESI nebulizer gas is used as a Venturi pump to transport the capture fluid from the capture region of the OPI. This gas accelerates and expands from the exit of the pressurized nebulizer nozzle and reaches sonic velocities within approximately 1 mm. The observation of multiple Mach discs creating multiple low-pressure regions was reported [48]. The position of the electrode tip relative to the Mach disc is a key parameter for determining the strength of the Venturi aspiration force. As part of the optimization process, the Venturi pump is tuned to deliver the maximum pressure drop by adjusting the position of the ESI electrode in the nebulizer gas expansion. For gas pressures available in a typical ADE-OPI-MS at 90 psi (10–12 L/min), the optimized protrusion was about 300 μm. The detailed procedure for the protrusion optimization workflow was described by Liu et al. [48].

To quantitively capture the acoustically ejected droplets within the capture region of the OPI, the acoustic transducer, sample well, and OPI capture region need to be aligned. The alignment between the acoustic transducer central axis and the sample well center is automatically optimized as part of an ADE ejection preparation step. This step may include evaluation of the reflection of an acoustic signal from either the lowest point of the sample meniscus or from the sample well walls. The OPI was then aligned with the acoustic transducer in the *X*- and *Y*-axis. Several methods have been introduced for this purpose. One way is to monitor the generated MS signal from dispensed sample droplets while moving the OPI to determine the optimized position [48]. Another alignment method is to monitor the acoustic signal intensity reflected from an alignment pin within the OPI, while moving the OPI to locate the optimized position in alignment with the acoustic transducer axis. The off-alignment tolerance is based on the ability to eject a droplet from the sample surface into the capture region of the OPI and, accordingly, is related to the separation distance between the liquid sample meniscus and the OPI. For a 1 mm ID port and a separation distance of 12 mm between the sample surface and the OPI, a ±20 μm position accuracy is recommended to ensure the reliable capture of the droplets in the OPI.

Increasing the carrier solvent flow rate with a fixed Venturi aspiration force results in four OPI fluid dynamics conditions at the capture region as the flow rate is ramped [48]. When the inlet flow rate is low, the fluid surface is deeply funneled with an air core in the OPI transfer tubing. In this supercritical condition, the air would disintegrate into segmenting bubbles, resulting in unstable chronograms and widened peaks. On the other extreme, when the inlet flow rate is higher than the capacity of the aspiration rate, the capture region is in an unstable transition (a narrow range) mode, or dome condition. This is also not desired for the AEMS system setup. The fluid surface in the critical condition (flow rate higher than the supercritical condition but lower than the transition/dome condition) is the region where the ion current is stable with the narrowest peak shape. In the critical condition, the inlet flow rate is higher than the supercritical condition but lower than the transition/dome condition. A flow rate mapping process is recommended as part of the system optimization workflow.

In addition to the system optimization discussed above, several assay development steps have been reported to improve the analytical performance. The optimization of the carrier flow through additives (e.g. salts and acid) or solvent choice (e.g. methanol and acetonitrile) has been demonstrated with enhanced sensitivity [49–51] and/or peak shape improvement [49, 52]. In addition, the determination of the sample ejection volume for assay execution is another assay development step to balance assay sensitivity and ionization suppression [50]. The ejection delay time adjustment according to the assay concentration range is another factor to be considered in the assay development process [47].

5.2.2.3 ADE-OPI-MS Automated Data Processing and Automation Integration

In the commercial ADE-OPI-MS system, the data processing step is automatically triggered by the completion of the data acquisition, and the data acquisition of the following run is initiated simultaneously. Therefore, data processing does not require additional user operation or increased assay turnaround time. The results table containing the intensity information of each sample is generated automatically for review and the laboratory information management system (LIMS) to access. The ADE-OPI-MS system has been integrated into various automation environments. The first report was with the Beckman Biomek i-Series liquid handling station, where a freshly prepared plate could be loaded onto a prototype AEMS system for direct analysis [52]. A commercial system was controlled by the Cellario™ (HighRes Biosolutions, Beverly, MA) [53] or Momentum™ (Thermo Scientific, San Jose, CA) [51] workflow scheduling software for HTS assays.

5.3 System Performance

5.3.1 AMI-MS Performance

A typical Echo 555 transducer is set to fire with a frequency of 500 Hz. For AMI-MS applications, this is raised to 1400 Hz for the original prototypes. This allows a higher flow rate of sample into the MS improving sensitivity while maintaining high-throughput. For most HTS applications, only 5 nL of sample is fired from each well and the higher frequency of firing enables three wells to be sampled every second. This exceeds the initial requirement of one sample per second sampling rate. Using only a 5 nL dispense with this higher frequency removes the need to switch polarity of the electrodes as the sampling event is so fast that the well does not have time to fully charge.

The most recent instruments supplied by Beckman can operate at frequencies up to 3000 Hz, which in theory provides even higher throughputs to be achieved (>5 samples per second). However, working at such a high frequency presents a challenge for the physics of the fluidics. As the transducer reaches the highest frequency, we noted a drop in signal intensity. We speculate that this is due to pushing too much power into the sample and collapsing the mound of liquid from which the droplets are generated. Lowering the power offset applied to the fluid restores the signal and improves the sensitivity of the system allowing 3000 Hz to be fully utilized.

Connecting the acoustics to the MS source block is a heated capillary. It is typically heated to 300 °C with a cone gas flow to 50 L/h. The high temperature and the counter flow of gas aid the drying of the droplets to generate the ion beam. Reducing the temperature of the transfer line helps prevent thermal instability of analytes. In particular, tertiary phosphate groups tend to be unstable at high temperatures, so dropping to 275 °C may help though this has to be balanced with a loss of sensitivity. Reducing or increasing the cone gas flow rate will affect the transit time of droplets in the transfer line; slowing down the droplets will increase the drying time for the sample and this may help de-solvate larger molecules. For most samples, a flow rate of 50 L/h at 300 °C works well.

To understand the variability of the signal generated with each tone burst from the transducer, a small peptide in TRIS-based buffer was prepared and dispensed into all wells of a 384-well Echo compatible plate. The samples were loaded into the mass detector and the signal from each well compared. Using only the raw intensity of the ^{13}C peak for the peptide, the coefficient of variations (CV) was 23%. However, using a ratio of the ^{12}C:^{13}C peaks from each well, the CV reduced to 3.5% (Figure 5.7).

Initially, the early prototype systems were validated with small molecule analytes such as warfarin, adenosine triphosphate (ATP), and *S*-adenosyl-L-methionine (SAM) (Figure 5.8a). Using the ratio of analyte to a labeled internal standard showed that there was good linearity of signal over four log units of concentration with typical limits of detection around nanomolar levels. For HTS applications, assays are typically run with analyte concentrations near the K_m for substrate and enzyme turnover is usually set to be around 20% substrate. Under these conditions, the system should work, provided the substrate K_m is in the μM range. To reduce the variability in signal, a ratio of two analytes is used, this can be either substrate: product or product: internal standard. Since the substrate is typically in excess, it can be challenging to follow loss of substrate, so following the product formation is a better option. The use of a time-of-flight mass detector as the analyzer helps since it is possible to measure both analytes in the same acoustic tone burst, further reducing signal variability. Typical extracted spectra from example test wells are shown in Figure 5.8b. The mass shift between

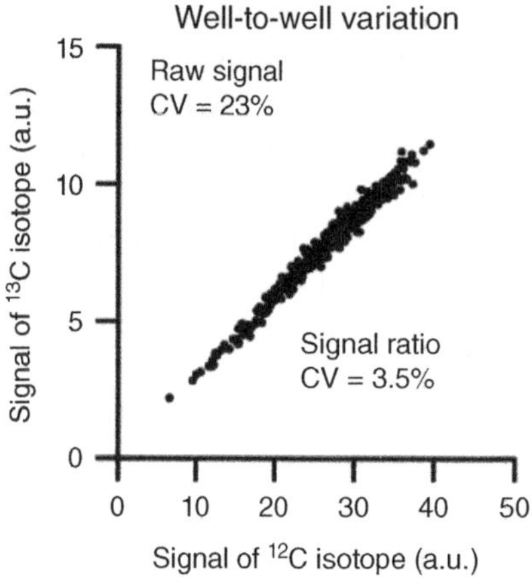

Figure 5.7 Well-to-well variability assessment for a 384-well plate containing a peptide. Samples were fired at the maximum throughput ~2 minutes per plate. The ^{12}C and ^{13}C peak intensities for the extracted ion spectra from each well were used to calculate the CV for both the individual ^{12}C peak or for the ratio of the ^{12}C:^{13}C peaks.

substrate and product can be significant; in the case of a peptide phosphorylated by a kinase, the mass shift would be 80 Da. For test conditions where an internal standard is used for the ratio calculations, the mass shift between peaks of interest might be only a few Da.

Having demonstrated that the prototype AMI-MS unit had sufficient sensitivity to support HTS applications, the system was tested with full 384-well plates containing "mock" assay reagents. These tests enabled a more accurate assessment of throughput. A plate was analyzed multiple times by the mass detector; a single data file was opened on MassLynx for the whole experiment rather than a new file for each plate read. Using a single data file reduced the time lost by the system opening and closing folders within the Windows software. The typical read time for a 384-well plate firing 5 nL from each well was about two minutes. The total ion current (TIC) for an example run is shown in Figure 5.9a. Each cluster of peaks is a single 384-well plate. Increasing the time resolution, it is possible to see the chromatograms from individual wells. The sampling rate of three samples per second is easily achievable. The gaps between packages of chromatograms are the loading and unloading times required to feed plates into the system.

Taking a single plate (Figure 5.9b,c) and extracting the spectra for the substrate peak shows there is a result for each well on the plate with one exception. This demonstrates how easy it is to identify an inhibitor on a plate. Note that in this example there was no attempt to reduce the variability in signal by using a ratio of two assay components. Even in this worst-case scenario, it was easy to identify the "active" well.

In addition to small molecule metabolites, the AMI-MS platform can ionize a range of peptides from polymers made of three amino acids through to larger peptides in the 2000 Da size range. Interestingly, the system is not very useful for

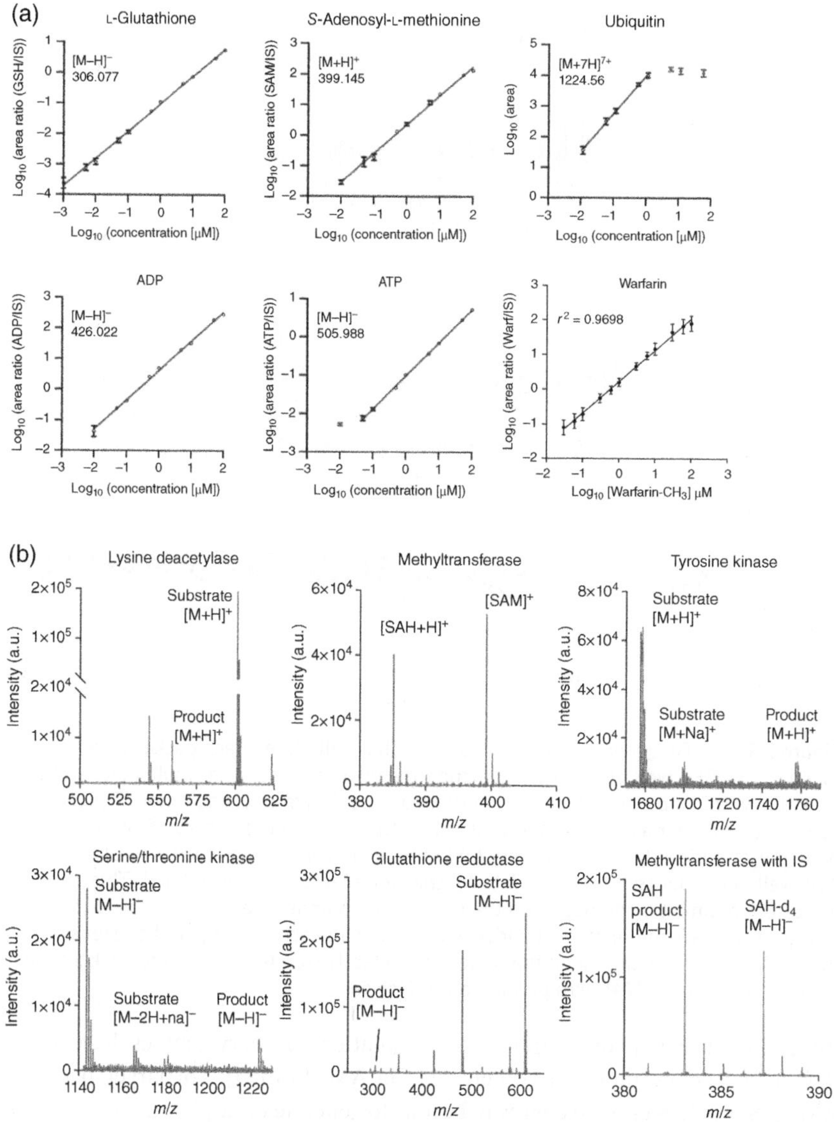

Figure 5.8 (a) Linearity of signal from example biomolecular analytes of interest. Where possible, labeled internal standards were used to generate ratio data, results shown are from $n = 6$ replicates acquired at a rate of 0.4 second per sample. (b) Typical extracted mass spectra from example biochemical assays acquired directly from a 384-well plate after the addition of a stop buffer. Substrate product pairs are marked except for the final methyltransferase example where an internal standard was used for ratio calculations. *Source:* Adapted from Sinclair et al. [19].

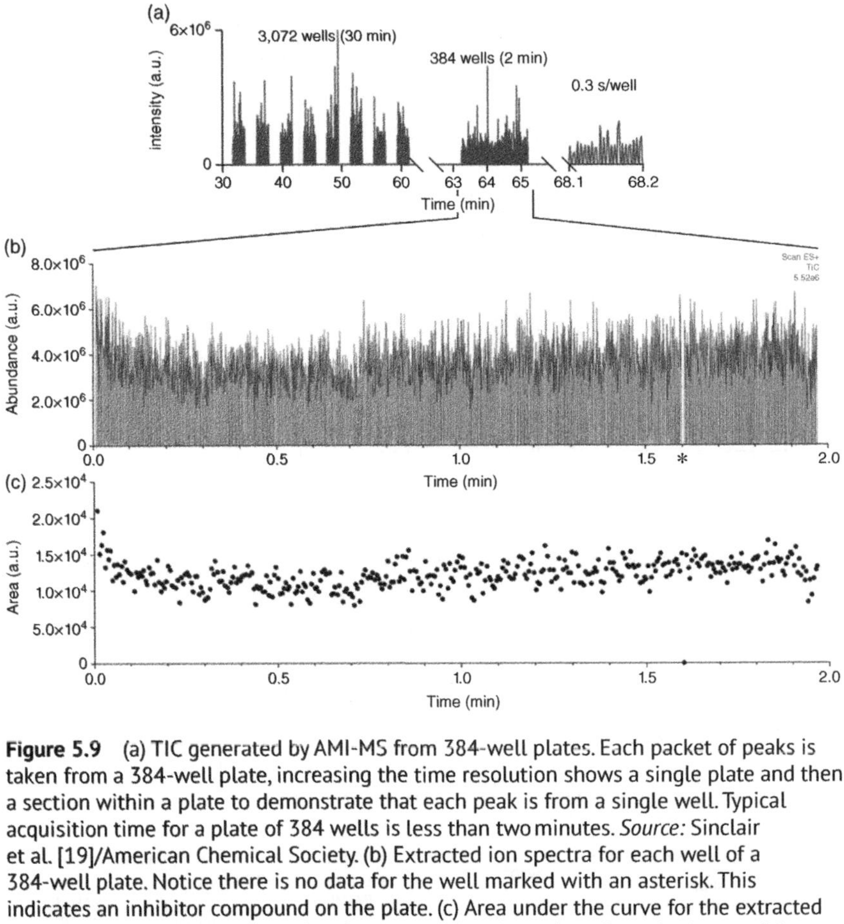

Figure 5.9 (a) TIC generated by AMI-MS from 384-well plates. Each packet of peaks is taken from a 384-well plate, increasing the time resolution shows a single plate and then a section within a plate to demonstrate that each peak is from a single well. Typical acquisition time for a plate of 384 wells is less than two minutes. *Source:* Sinclair et al. [19]/American Chemical Society. (b) Extracted ion spectra for each well of a 384-well plate. Notice there is no data for the well marked with an asterisk. This indicates an inhibitor compound on the plate. (c) Area under the curve for the extracted ion spectra shows the relative abundance of analyte in each well, again the asterisk marks the "active" compound in the plate which clearly stands out from the background. *Source:* Sinclair et al. [24]/with permission of Elsevier.

direct measurement of amino acids in solution; primary amines have to be derivatized before they can be detected via AMI-MS [54]. For larger peptides, AMI-MS provides enough energy to be able to generate multiple charge species of peptides present in solution. When working with aqueous peptide solutions in the 10–50 μM concentration range, it is not uncommon to see double or triple charge states. Using a commercially available LRRKtide peptide, the +4 charge state of the peptide is the most abundant species identified by AMI-MS, significantly more is present than either the +3 or +2 charge states (Figure 5.10a).

These types of multiple-charged ionization patterns are consistent with an electrospray-like ionization modality. Larger proteins with a mass up to 24 kDa can also be detected by AMI-MS; a classical multi-charge spectrum for ubiquitin

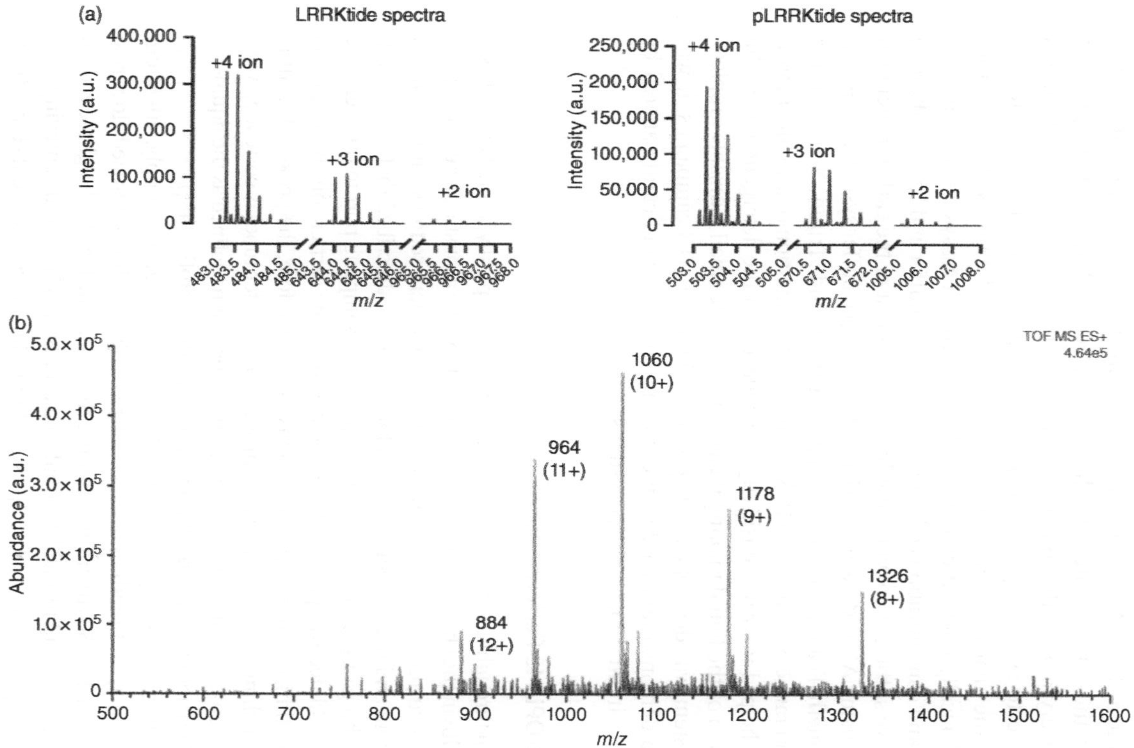

Figure 5.10 Peptide and protein spectra from AMI-MS. (a) Multiple-charged species of a commercially available peptide and its phosphorylated counterpart as shown. *Source:* Adapted from Sinclair et al. [24]. The LRRKtide peptide was dissolved in water at a concentration of 10 μM and fired from a 384-well plate. The 3+ and 4+ ions are the most abundant species. (b) Extracted ion spectrum obtained for ubiquitin clearly showing the 8, 9, 10, 11, and 12 charge states. This spectrum is typical of that produced by commercially available electrospray sources. *Source:* Sinclair et al. [24]/with permission of Elsevier.

protein is shown in Figure 5.10b. However, for proteins greater than 24 kDa, AMI-MS is not able to generate clean spectra since it is challenging for this system to de-solvate the hydration shell of these larger proteins. While the ionization modality may be similar to ESI, the fact that it cannot generate spectra from the larger proteins demonstrates that there are also differences. It is probable that the droplet size generated by AMI-MS, while small enough to enable the generation of an ion beam may be larger than the spray generated by ESI.

The advantage of this AMI system is speed, using the acoustics to load samples directly from the source plate into the mass detector. However, this approach means that everything in the source well is fired into the detector and this limits sensitivity due to signal suppression from the biological matrix. To minimize suppression, the analytes of interest must be in minimal buffer solutions rather than physiologically relevant conditions. In biochemical assays, we can manage the number and concentrations of components in the buffer and balance these to ensure we have sufficient enzyme activity for HTS. Most of the early biochemical HTS screens using AMI-MS were carried out in a simple buffer system containing tris(hydroxymethyl)-amino-methane (TRIS) pH 7.4 and Triton X-100 in the concentration range of 1–10 μM [55]. Surprisingly, the addition of a small amount of detergent helped improve sensitivity; we believe that the detergent lowers the surface tension of the liquid making it easier to generate small ionizable droplets. Since the system was developed to support biochemical HTS campaigns, it was important to demonstrate that enzyme kinetics are not significantly altered by running assays in minimized buffer systems.

5.3.2 ADE-OPI-MS Performance

The analytical throughput of the ADE-OPI-MS system is typically limited by the baseline peak-width in the chronogram, before approaching the acoustic dispensing speed limit, which is currently about 3–6 Hz depending on system settings (time spent on stage movement and the acoustic ejection) [56]. Liu et al. described that most of the dilution stretching of the sample plug happens at the sampling port to transfer tube interface. This diluted sample volume is insensitive to the carrier solvent type or its flow rate at optimized operational conditions [48]. Therefore, the peak-width (time required for flushing the sampling plug through the transfer tube) is correlated with the carrier solvent flow rate at the optimized flow condition (defined as the liquid–air interface in the capture region presenting a critical-vortex profile [48]).

For a typical system setup with ~10–12 L/min nebulizer gas supply, 40–60 cm long transfer tube, and methanol as the carrier solvent, this optimized flow rate would be around 400–500 μL/min, enabling the peak-width at 1% height less than 1 second (or <0.4 second full width at half maximum, FWHM) and achieving 1 Hz analytical throughput [47, 48, 57]. Switching the carrier solvent type from methanol to a lower viscosity solvent such as acetonitrile allows a higher optimized flow rate (600–800 μL/min), and a sharper peak shape (FWHM ~0.2 second), so that a faster analytical throughput at 2 Hz could be demonstrated [48, 57]. The relationship between the carrier flow viscosity and the analytical throughput was discussed in detail by Liu et al. [48].

Increasing the nebulizer gas flow rate and/or shortening the transfer tube are other options to increase the operational flow rate and improve the analytical throughput. Häbe et al. reported a prototype system using 19 L/min nebulizer gas, acetonitrile/methanol mixture (78/22, v/v) as the carrier flow, and a 25 cm transfer line, which could increase the ADE-OPI-MS operational flow rate to 1 mL/min and reduce the FWHM to 110 ms (FW at 5% was 239 ms) and achieve about 3 Hz analytical throughput while maintaining baseline separation [56]. In another study, a flow rate of 2.5 mL/min and a FWHM of 17 ms (FW at 1% was 68 ms) were reported with a 4.8 cm long OPI transfer tube, demonstrating the possibility of 15 Hz analytical throughput for OPI [48]. However, this speed was validated with a pneumatic nanoliter-drop dispenser, since the inter-well sampling speed from an acoustic dispenser is currently limited to <6 Hz [56].

The MS chronogram peak-width discussed above was based on low sample loading volume (<20 drops) per sampling event. When the sampling events consist of multiple droplets (i.e. multiple high-speed droplet ejections), the FWHM would increase while the analytical throughput necessarily reduces [47]. Another factor impacting the analytical throughput with a given peak-width is the expected concentration dynamic range for adjacent ejections. The integration of a low intensity signal peak is poor if it is immediately after a 1000 times higher intensity ejection. A longer ejection delay is required for the accurate quantification of all samples when a wide concentration range is expected, and this sampling speed optimization could be a part of the assay development process.

The AEMS system eliminates the variations associated with injection valves and chromatographic separations. The high precision of ADE and ESI demonstrated reproducibility in different studies to be better than 8% for both inter- and intra-sample analyses [47, 57, 58]. Classic ESI also enables more than 3 orders of magnitude linear dynamic range without an internal standard, as well covering a wide segment of chemical space for both small and large molecules [57]. The non-contact nature of the ADE process and the continuous self-washing of the OPI device eliminate any carryover across samples [47, 57].

With nanoliter-volume sampling, the lower limit of quantitation (LLOQ) of the ADE-OPI-MS analysis was from sub-nanomolar to tens of nanomolar [57, 58]. It is not comparable with the LC-MS approach that utilizes the microliter-volume sample loading. However, this sensitivity that AEMS achieves is sufficient for many high-throughput drug discovery applications.

Ionization suppression is a key factor that may affect assay sensitivity for ADE-OPI-MS. The nanoliter-sized raw sample is diluted approximately 1000-fold within the OPI, which reduces the ionization suppression and allows the significantly simplified sample preparation for many sample matrices, including in vitro drug discovery samples containing high concentrations of salts and proteins, and biological samples such as plasma and urine [52, 57]. Liu et al. reported a method for measuring ionization suppression through monitoring the signal of a standard spiked into the continuous carrier flow. This was observed to be less than 50% for rat plasma. This method was also used to accurately determine the droplet size acoustically dispensed, by measuring the single-droplet level [47]. The direct loading of a complex matrix containing high concentration salts and proteins can create a high risk of the OPI clogging. Including a washing cycle with aqueous solvents could significantly enhance the system robustness when processing these types of samples [51, 56].

5.4 Applications

5.4.1 High-Throughput Screening

5.4.1.1 AMI-MS for HTS

The first full HTS screening, consisting of around two million small molecules, was run with the automation system against a histone deacetylase (HDAC) target. A commercial HDAC substrate, acetylated RGK tripeptide with an AMC (7-amino-4-methylcoumarin) fluorescent tag yielding a total mass of 601.31 Da, was purchased [19]. HDAC cleaves the acetyl group from the substrate, generating a product mass of 559.30 Da (Figure 5.11a and b). Since the enzyme was sensitive to detergent, the assay was run in 100 mM Tris, pH 7.4 with 15 mg/mL bovine serum albumen to stabilize the enzyme rather than Triton X100. Initially, 10 μL of a solution containing 2.4 nM enzyme in buffer was added to each well with test compound (or controls). The enzyme/compound mixture was allowed to mix for 30 minutes before the reaction was initiated with a 10 μL addition of 40 μM substrate in the buffer. The reaction progressed for 90 minutes before the addition of 30 μL of stop solution (0.2% v/v acetic acid). The completed assay plates were collected and stacked into the automation system. When 150 plates were ready, the AMI-MS system was started. Standard data acquisition settings were used, the transfer line temperature was set to 300 °C, and scan range set between 500 and 650 Da with a scan time of 60 ms in positive ion mode. For this assay, an enhancement window was set at 550 Da. All 150 plates were read into a single data file and post processed to generate the spectra for each individual well. With the acoustics set to fire 5 nL at 1400 Hz from each well and a charging cone voltage of 3 kV, it took around three minutes to load each plate of samples into the detector. The total running time for a typical 150 plate batch was around nine hours. The full collection screen was completed in around 20 working days.

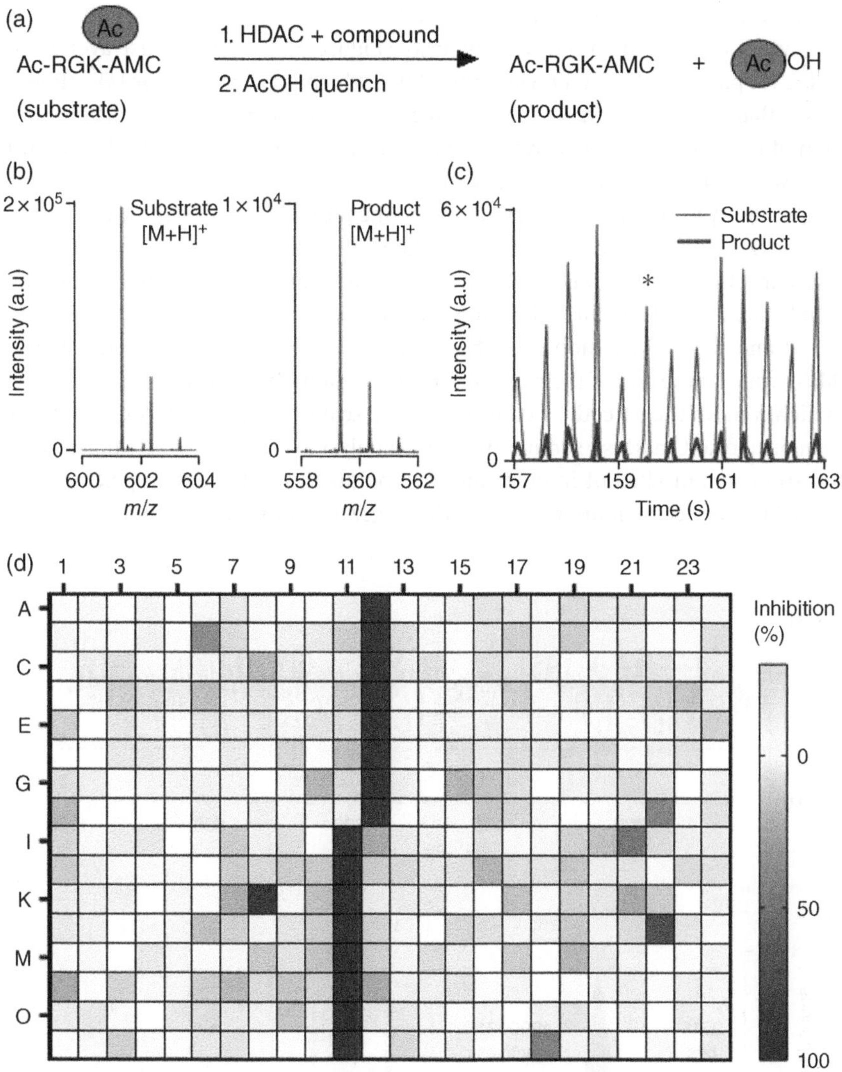

Figure 5.11 HDAC inhibition screen. (a) Overview of the HDAC assay. (b) Example raw spectra extracted from a single well showing the mass of the substrate and product pair. (c) Overlay of the substrate and product present in representative wells of a 384-well test plate, the asterisk highlights a well containing a compound which inhibits the enzyme activity as there is no product present. (d) Heat map of a test plate, control wells were in columns 11 and 12 containing either DMSO or inhibitor compound. Full inhibition is seen in well K8 as the test compound completely inhibits the HDAC enzyme activity.

The hit rate for the screen was around 0.3% with the majority of the active inhibitors confirmed in subsequent tests (Figure 5.12). The robust average Z′ value for plates in the screen was around 0.6, which is acceptable for HTS though lower than you might expect for some technology formats. However, the elimination of technology artifacts and false positives together with the label-free nature of AMI-MS made this a successful campaign.

The ability to slice the chromatograms to the individual plates and wells is critical for HTS applications. As previously stated, we open a single acquisition file at the start of the run and all the data from the 150-plate batch is stored within this single file. The Echo records the plate barcode and the well position for each firing event and this information must be parsed into the raw data file generated by MassLynx. At the end of the run, the single raw data file is sliced into smaller files which contain the barcode, well position, and scans for each well. A web application on a local GlassFish service allows the analyst to enter the mass ranges of the substrate and product of interest and then points this to the data repository. The GlassFish service then filters through the target files and sums the signal intensity

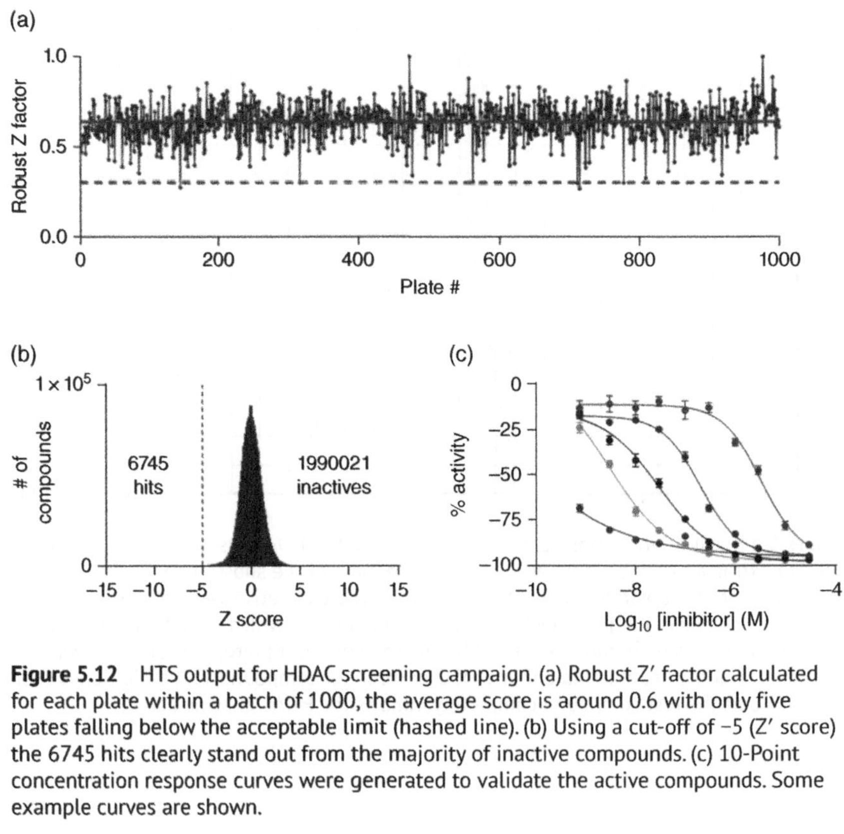

Figure 5.12 HTS output for HDAC screening campaign. (a) Robust Z′ factor calculated for each plate within a batch of 1000, the average score is around 0.6 with only five plates falling below the acceptable limit (hashed line). (b) Using a cut-off of −5 (Z′ score) the 6745 hits clearly stand out from the majority of inactive compounds. (c) 10-Point concentration response curves were generated to validate the active compounds. Some example curves are shown.

for each scan within the mass ranges that have been specified for each well on each plate. Finally, the sum intensity for each target mass is output into a csv file, one per plate. These processed files are used as the input source files for corporate data analysis tools within HTS. A schematic representation of the data analysis process is shown in Figure 5.13 [19].

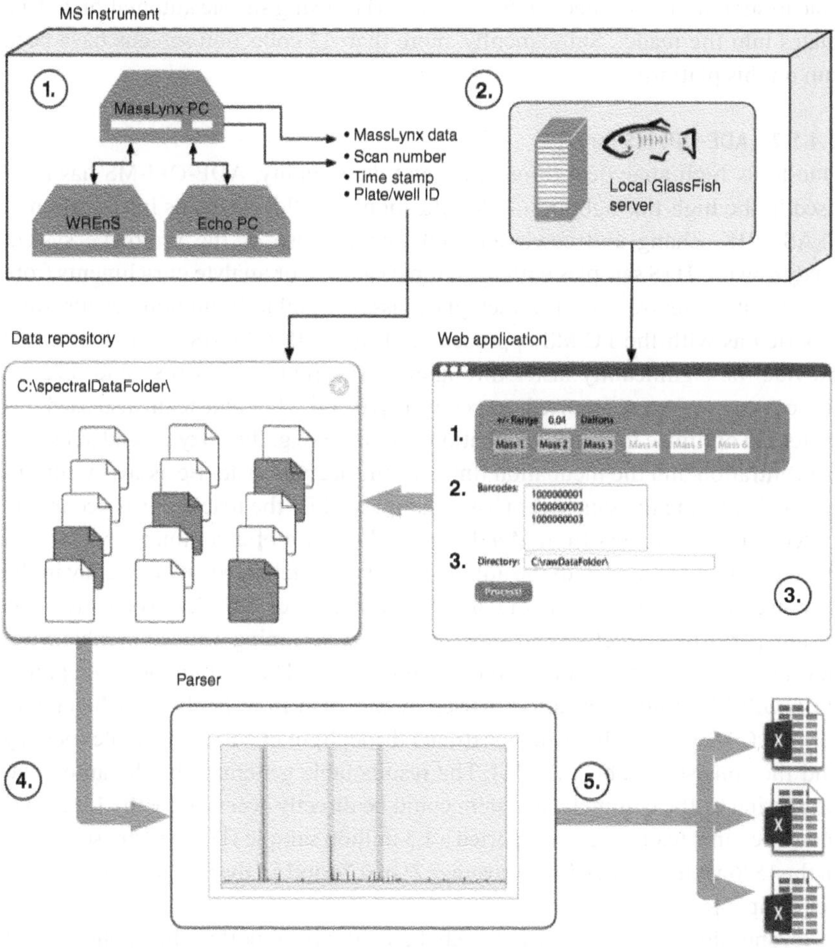

Figure 5.13 A representation of the original data analysis workflow for AMI-MS. 1 – The echo PC passes the plate barcode, time stamp, and well position data into the MassLynx raw data file as chromatograms are being acquired. Post run, the single raw data file is sliced into smaller files containing the scans for each well on each plate. 2 – The GlassFish service runs locally to reduce time lost moving files and 3 – allows the analyst to specify the mass range for the substrate and product pair of interest and points this to the data file repository. 4 – Sum intensities from the scans within the target mass range specified are calculated and output to csv files which are saved with the plate barcode as the name 5.

Using this data analysis system, it was possible to generate the csv files for the 150-plate batch in around three hours. Subsequently, we moved to a commercial software to post process the single raw data file generated by MassLynx. This proved much faster and allowed the output csv files to be generated in around 8–15 minutes.

The HDAC screening campaign demonstrated that the AMI-MS platform was able to support a full collection biochemical HTS using simple automation to feed plates into the reader. Subsequently, more than 12 collection screens have been run on this platform.

5.4.1.2 ADE-OPI-MS for HTS

Due to its high analytical throughput and data quality, ADE-OPI-MS has been used in the high-throughput hit-identification workflow since its introduction at SLAS 2018. Zhang and coworkers reported the use of the prototype system analyzing the HTS samples without sample cleanup or analyte enrichments from an acetyltransferase and a demethylase assay. Similar inhibition results were reported as with the LC-MS approach, and the ADE-OPI-MS technology could provide the significantly faster throughput. The robust Z′ of 0.84 and 0.72 for these two assays were reported, respectively [20, 59]. The same team also used this system to optimize the assay incubation condition (e.g. the enzyme and substrate concentration and the incubation time) for the acetyltransferase assay, by continuously analyzing the same set of assay samples while the reaction was occurring.

Recently, researchers from Merck reported the use of a commercial Echo MS system fully integrated with the automation system delivered an HTS assay in the 1536-well format. A robust Z′ of 0.62 was reported over 200 1536-well plates. The system performance before and after the primary screening was comparable without any maintenance during the analysis [53]. The automation integrated ADE-OPI-MS system was also reported by the team from Boehringer Ingelheim for the cGAS assay with the demonstrated throughput of >65,000 samples per day and the consistent robust Z′ [51]. The results table generated by the auto-data-processing of the commercial system could be directly accessed by the LIMS system. The same team recently reported a 1.5 million sample HTS primary screening in the 1536-well format, with the average Z′ of 0.74 and analysis speed of 0.9 second per sample [60].

Beyond biochemical HTS assays, Zhang et al. also reported the application of the AEMS technology to the cell transporter assay for hit identification [57]. Good quantitation performance was demonstrated across a wide dynamic range of concentrations without using an internal standard for signal correction. No significant ionization suppression was observed for the direct analysis of the cellular matrix without sample cleanup. The identified hit ID was similar to the >10× slower LC-MS approach.

In another study, the team from Merck introduced the use of the AEMS system to identify inhibitors for diacylglycerol O-acyltransferase-2 (DGAT2), with the assistance of the liquid–liquid extraction [50]. Significant signal instability was observed for the target analytes when aqueous assay samples were analyzed directly, due to the low solubility of the analytes and their interactions with the well surface. To overcome this challenge, an organic phase (pentanol) was added to the sample plates for analyte extraction. The ADE module of the prototype system selectively generated droplets from the pentanol phase for the OPI-MS analysis without requiring any additional liquid handling. Stable Z′ and the consistent results across replicates have been reported, as demonstrated in Figure 5.14.

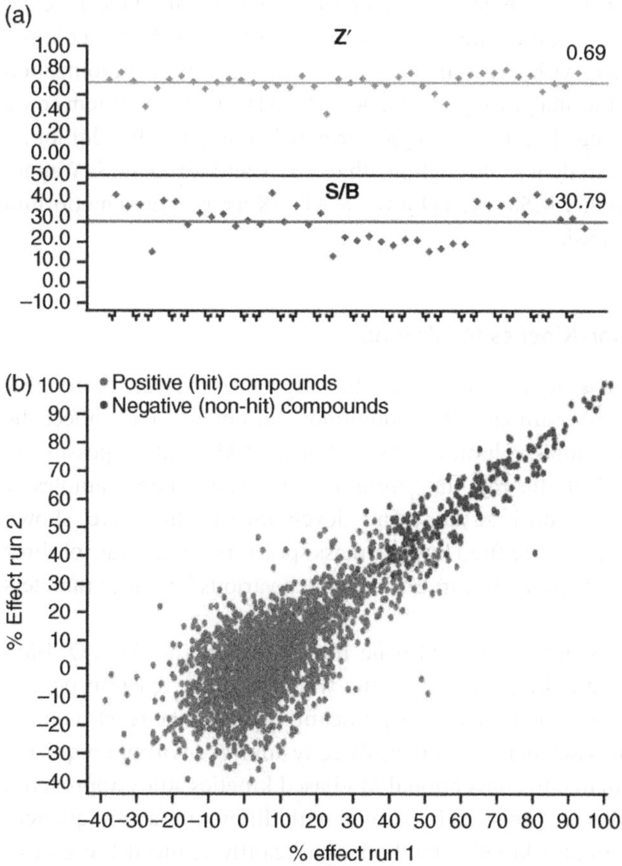

Figure 5.14 The application of the ADE-OPI-MS system for the DGAT2 screening assay. (a) Plots of Z′ and signal-to-background ratio of different screening plates. (b) Comparison of the inhibition effects determined from two replicate runs with AEMS. *Source:* Reproduced with permission from Wen et al. [50]/American Chemical Society.

5.4.2 High-Throughput ADME

The first utilization of the ADE-OPI-MS technology in ADME was for the cytochrome P450 drug–drug interaction assay. The samples were directly analyzed without any sample preparation. The IC50 data for CYP3A4, CYP2D6, and CYP1A2 showed good consistency with the analysis results from LC-MS [57]. Other reported ADME assays analyzed with the ADE-OPI-MS platform include a liver microsome stability assay [57, 61, 62] and a cell permeability assay [63].

ADE-OPI-MS is not only used as a faster alternative to LC-MS, but also brings additional advantages for some ADME assays. Small polar compounds are used as substrates in some transporter assays. However, these analytes are typically challenging to the LC-MS workflow due to their weak retention on the LC column. The co-elution of matrix with target analytes results in the significant ionization suppression reducing the analytical performance. The ADE-OPI-MS system could well-handle this challenge because of its greater matrix tolerance. In addition, it provides much faster analytical throughput than the standard LC-MS. Wagner et al. reported high-quality IC50 results for the MATE2K assay, where metformin was the target analyte [58].

5.4.3 In Situ Reaction Kinetics Monitoring

Prior to any assay development, the kinetics of the target enzyme were thoroughly tested to ensure that the minimal buffer conditions did not adversely affect the test system. The small sample volumes and speed of AMI-MS make it possible to run experiments kinetically in real time from a single well. Some examples of experiments carried out during the assay development phase are shown (Figure 5.15), 5 nL samples were fired into the mass spectrometer at a rate of three per second to generate the progress curves from which various kinetic parameters can be assessed.

Enzymatic reaction kinetics could also be monitored on the ADE-OPI-MS system by repeatedly sampling from the same well while the reaction occurs. Each sampling event does not induce any significant impact to the reaction incubation system, and the assay matrix could be directly analyzed without requiring sample cleanup. Compared to conventional MS-based kinetics study approaches where multiple samples are prepared in parallel with different reaction quenching time points, this in situ kinetics method significantly reduced the cost of expensive reagents and provides higher data quality by eliminating multiple liquid handling steps. In comparison with the fluorescence measurement-based plate-reader approach, the label design and coupling process could be eliminated

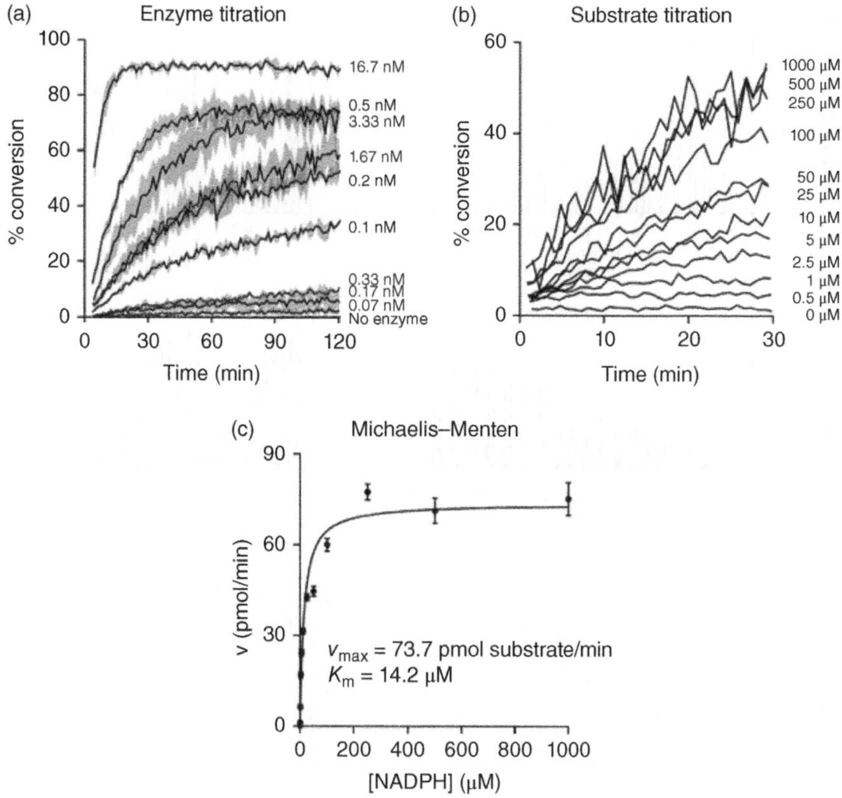

Figure 5.15 Examples of data generated to support assay development using AMI-MS. (a) A graph showing the time-dependent generation of product by differing amounts of enzyme. The black line is the mean value, and the gray area shows the standard deviation. Results were from three independent replications of the experiment. *Source:* Adapted from Sinclair et al. [55]. (b) Progress curves for an assay where a standard amount of enzyme has been added to each test well while the substrate concentration is varied. (c) Michaelis–Menten plot showing the kinetic parameters calculated from data generated by AMI-MS. *Source:* Sinclair et al. [55] / Taylor & Francis.

and a wider dynamic range could be achieved with ADE-OPI-MS [64]. Figure 5.16 is an example from the dextromethorphan metabolic stability assessment with human liver microsomes [57]. This complex assay matrix containing high salt and protein concentrations could be directly analyzed with the ADE-OPI-MS system with good quantitation performance. The derived K_m value agreed well with the literature.

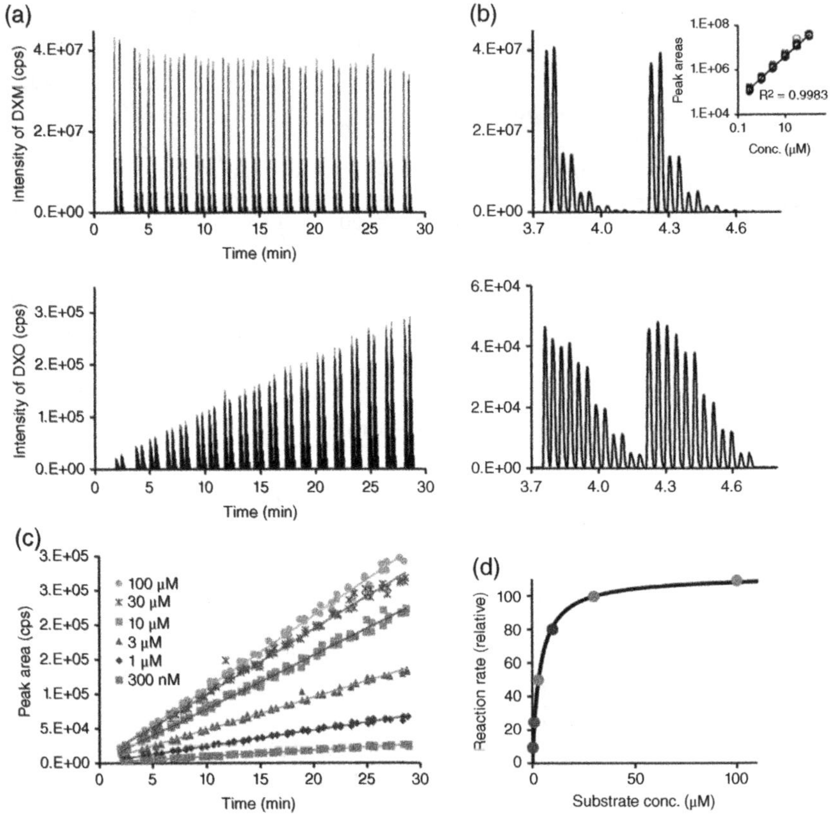

Figure 5.16 The application of the ADE-OPI-MS system for in situ kinetics study. (a) Chronograms of both the substrate (dextromethorphan) and product (dextrorphan) for the metabolic stability assessment reaction. (b) Expansion view of the reaction time between 3.7 and 4.8 minutes. The insert figure shows the peak area of the substrate versus the concentration. (c) The plots of the peak area of the product vs. the reaction time at different substrate concentration levels. (d) Michaelis–Menten fit of the reaction kinetics results. *Source:* Reproduced with permission from Zhang et al. [57] / American Chemical Society.

5.4.4 Bioanalysis

Because complex biological matrices induce significant ionization suppression in LC-MS, labor-intensive sample preparation is typically required before conventional bioanalysis workflow. The use of the ADE-OPI-MS system as a quantitative tool for bioanalysis was evaluated with the assessment of the ionization suppression for plasma samples [57]. Signal was reduced by only ~30% when straight plasma was injected without cleanup. Good quantitation performance was maintained over the concentration range from 4 nM to 20 μM of verapamil. This demonstrated the high tolerance of the ADE-OPI-MS system to a complex matrix like plasma. Ionization suppression was further reduced by diluting the plasma sample with an equal volume of the water. Sample preparation and method development were significantly reduced with typical procedures including protein precipitation, centrifugation, and multiple liquid handling steps eliminated. There have been several studies reporting the analysis of plasma-based pharmacokinetics study samples with the ADE-OPI-MS system [65–67]. Quantitation results were similar to LC-MS, with throughput significantly improved, assay development simplified, and sample preparation reduced or eliminated.

In a recent study, ADE-OPI-MS was reported to directly eject liquid samples from two-phase liquid systems in a standard microplate. This transfer of the upper fluid in a biphasic solution was demonstrated for high-throughput analysis of urine samples. The in-well sample cleanup through the liquid–liquid extraction further reduces the ionization suppression. High quantitation sensitivity, precision, and accuracy were demonstrated [68].

5.4.5 Compound QC

Many drug discovery assays require the addition and incubation of each compound in the biological reaction. The quality of the stock standard directly impacts the assay readout – any impurities and/or degradation of the standard may cause false positive or negative results. Therefore, high fidelity quality control (QC) of compound libraries is required. Due to the large number of samples and high chemical diversity, the analytical platform used in QC is required to provide both a wide coverage of the broad chemical space and high sample throughput. Thus, AEMS is a suitable technology fulfilling these requirements. The first report of using the AEMS system for a compound QC purpose was reported by Zhang et al. from Bristol-Myers Squibb with the Q1-scan data equalization mode [69]. The similar results they showed that the AEMS breadboard gave similar results to that acquired with LC-UV/MS and that it had approximately 30 times higher throughput and 1000-fold reduction in sample consumption.

In a recent study, teams from Pfizer and SCIEX reported the operation of MS in TOF-scan mode for compound QC based on a high-resolution mass spectra readout [70]. They described the use of automated data integration and heat-map-based visualization tools to determine the presence and to measure the intensity of the target compounds from each well. Other advanced data processing capabilities that have been described include a background mass spectra identification and subtraction algorithm developed to differentiate the ions of the sample from those of the background (e.g. from carrier solvent, ion source, etc.) as well as a spectra comparison tool to quantitively describe the spectra similarly across multiple

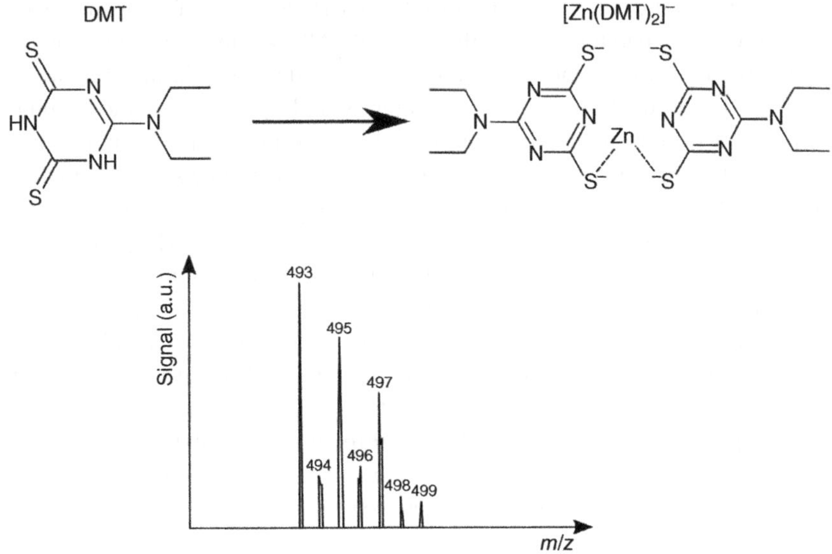

Figure 5.17 Use of metal ion chelator to identify zinc contamination in compound libraries. The example provided is using a small molecule chelator (DMT) which binds to zinc to generate a larger complex which then shows a characteristic spectral pattern using AMI-MS as a read out.

datasets. The background MS identification and subtraction algorithm provides better estimates of the purity of the target drug compounds. The spectra comparison tool enables the analyst to use this new parameter of the spectra similarity score versus the reference spectra for a description of the sample quality.

One common problem with compound libraries is that they can be contaminated with metal ions, which are often used as catalysts during the syntheses of compounds. Interestingly, many biological enzymes can be adversely affected by metal ions such as silver, platinum, and zinc. Therefore, there is a real need to quickly determine whether the activity seen in the assay was due to the compound or any potentially contaminating metals. Two metal chelators have been identified that can co-ordinate a range of metal ions. The resulting mass shift can be seen at the mass detector. The speed of AMI-MS ensures that the output from an HTS campaign can be assessed for metal ion contamination in less than a day (Figure 5.17).

5.4.6 Parallel Medicinal Chemistry

In addition to compound QC, full-scan mode ADE-OPI-MS has been utilized for the high-throughput readout of parallel synthesis samples including reaction optimization and nanoscale chemistry [57, 71]. Compared to the conventional

LC-UV/MS-based approach, the ADE-OPI-MS technology could provide similar information, but would consume much less sample (0.1%), while delivering significantly higher analytical throughput. This allows the results to be delivered to the chemist much faster and allows better direction of subsequent synthetic operations. In a follow-up study from the team at Pfizer [71], the ADE-OPI-MS system was also utilized for monitoring on-DNA nanomole-scale synthetic reactions. This monitoring is critical to confirm the results from DNA Encoded Library (DEL) workflows.

5.4.7 High-Content Screening

The advantage of using a TOF detector with the system is that it enables the user to scan across a large mass window. This can be particularly useful when analyzing complex mixtures. Direct infusion MS has been used in the past for high-throughput metabolomic analysis of cell lysates [72]. Having proven that AMI-MS was able to support biochemical HTS applications, it was only logical to investigate if the technology could be used with cell lysates. Initial tests were carried out using adherent cell lines. Around 7000 MCF7 (breast cancer) cells were seeded into each well of a 384-well plate and after 24 hours, the cell medium was removed, and the wells washed thoroughly with ammonium phosphate buffer. The washing was critical to reduce the effects of ion-suppressing components in the medium. After the final wash, the buffer was removed, and the cells immediately lysed using ice cold acidified water [73]. The lysate material was transferred to an Echo qualified 384-well plate and fired directly into the AMI-MS. Typically, for this application, the dispense volume was set to 100 nL and a mass scan range from 50 to 1000 *m/z* was used. The most abundant metabolic species present in the lysate are seen in Figure 5.18a. The spectra are particularly rich in the mass region from 600 to 900 and almost all of the common lipids are found in this region. Moving down the mass range, we can see signals from nucleotides and smaller metabolites including some amino acids. The smaller and more soluble metabolites are less well represented in these spectra. Using labeled internal standards, it was possible to identify some of these metabolites. Finally, we compared the metabolite profiles from untreated and compound-treated cells to see if it was possible to identify changes in metabolite levels.

Different cell lines were tested and we found that the composition of the lysis solution needs to be modified for the cell line being used. For example, HepG2 liver cells are more resistant to osmotic shock, so the lysis solution requires methanol (25 : 75 v/v methanol: water). The addition of a second solvent to the lysis buffer will likely change the extracted metabolites making it difficult to directly compare results from different cell lines. Suspension cells can also be used for metabolite analysis but more stringent washing steps may be required to remove all the cell medium.

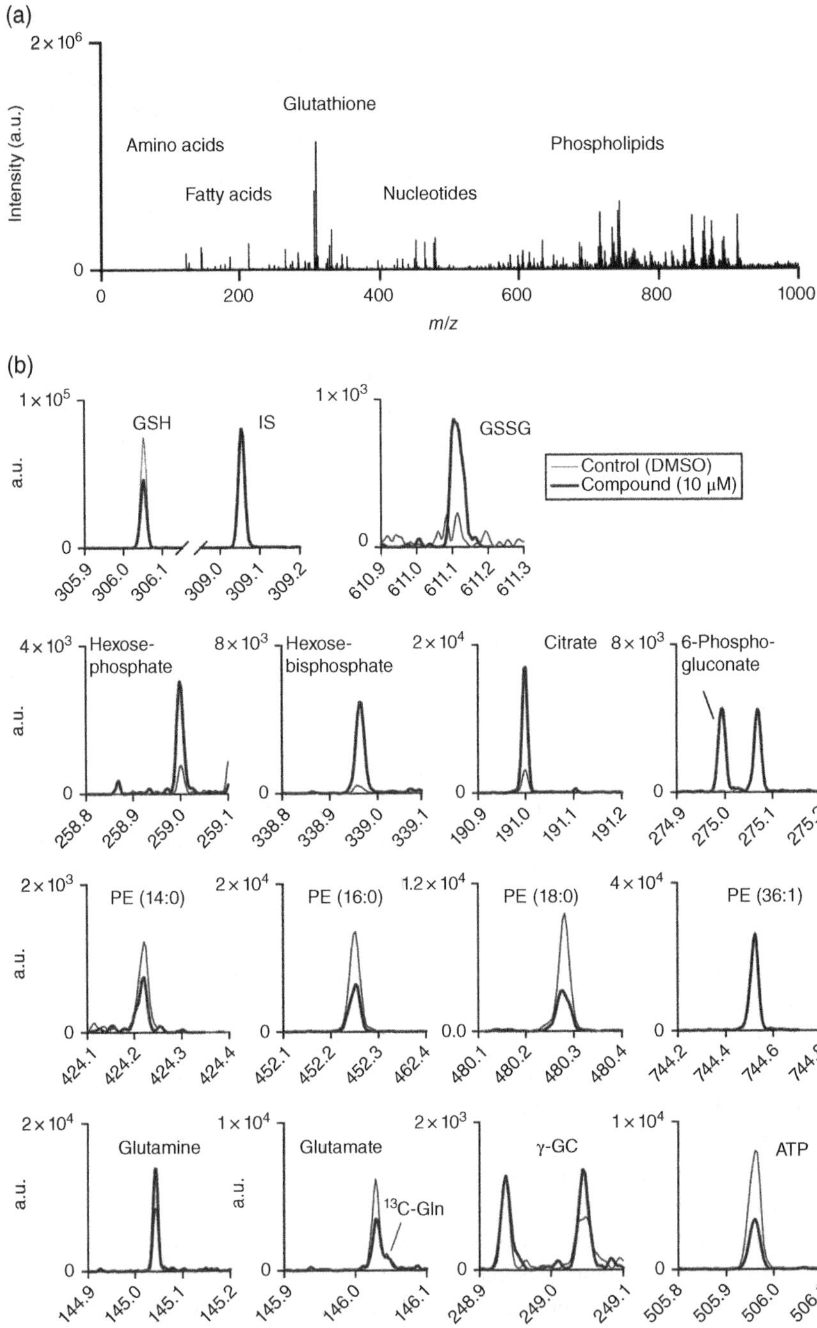

Figure 5.18 AMI-MS spectra from cell lysates. (a) Spectrum from MCF7 cell lysate. (b) Individual spectra for specific metabolites, untreated cells shown in gray while the same analyte from compound treated cells is overlayed in bold.

There are several advantages of using AMI-MS for this type of metabolite analysis. First, the requirement for only a few thousand cells makes it possible to test many conditions or set up multiple technical and biological replicates in a single plate. The sampling volume of only 100 nL ensures that multiple sampling events from the same well are possible; this brings an additional statistical power to analysis of AMI-MS data. Finally, the speed of the system enables time course data to be generated from cells treated with compounds in a way that has not been previously possible.

5.5 Challenges and Limitations

The utilization of acoustic sample loading methods to MS has demonstrated the advantageous features of high analytical throughput and the ease-of-integration with automation control. This innovative technology has been utilized in numerous high-throughput drug discovery workflows and demonstrates the trend to substitute AEMS for LC-MS platforms in many pharmaceutical companies. It also opens the door to utilizing MS in new application spaces, and enables novel workflows, for example, the QC of library-sized compound collections. We believe it is a transformational approach benefiting the drug discovery process.

However, these technologies have several limitations at the current development stage, causing challenges when utilized in certain drug discovery assays. These technologies illustrate a special type of flow-injection analysis (FIA)-MS where there is no additional separation beyond the mass-to-charge ratio for species discrimination. Therefore, an analyte of interest may not be distinguished from its isomers existing in the sample solution or generated from the in-source fragmentation process [74]. This may cause specificity and/or sensitivity issues in some assays. Due to the low sample loading volume, the LLOQ of the acoustic-MS system is less than the conventional LC-MS-based approach. It is adequate for many in vitro assays but might be challenging for the compounds with low ionization efficacy with complex matrices.

As with most other high-throughput MS techniques, the signal duration of the AEMS analysis is controlled to be short (one second or less) to separate the analyses of adjacent samples. To achieve the high precision and accuracy, the data acquisition cycle time needs to be kept below 100 ms to allow enough data points through the signal duration [75]. This limits the number of MS/MS transitions that can be monitored simultaneously. Another consequence of the fast data acquisition is that the MS data of a whole run (e.g. all sample wells from a microplate) has to be collected within the same data file. More complex algorithms are required for both data acquisition and data processing, especially when such methods are sample-well dependent.

Compared to ADE-OPI-MS, AMI-MS has several additional limitations beside the lack of commercial systems. Since AMI-MS is a direct infusion technique, a key limitation is that everything in the test well is sprayed into the detector. Therefore, like all direct infusion techniques, AMI-MS is susceptible to suppression from the matrix. For biochemical assays, this can be controlled by running assays in low salt buffers; however, it is necessary to check that the buffer used does not significantly change the characteristics of the enzyme. For enzymes that have a dependency on cationic cofactors such as kinases, magnesium chloride needs to be removed before the samples are sprayed into the detector. This can be done with the addition of ammonium phosphate which precipitates the magnesium ions. Since acoustic liquid handlers fire samples from the air–liquid interface at the top of the well, the precipitated magnesium is not present. While this approach works well for screening applications, the requirement for the ammonium phosphate does limit the use of AMI-MS for kinetic analysis of kinase activity. Another approach for minimizing the effect of the matrix is to have a dilution step at the end of the assay. Acoustic liquid handlers require a minimum 50 μL volume to fire samples. It is possible to run the assay in 10 μL and then dilute to 50 μL with water. This increase in volume is often enough to reduce suppression to an acceptable level.

Currently, the AMI-MS system is limited to 384-well rather than 1536-well formats. The theoretical limitation on moving to 1536-well format is likely to be around the charging of a 1536-well. As we found with the 384-well plate, if you push enough charge into the well, the surface charge accumulates and eventually reaches a level where it is no longer possible to generate the mound of liquid from which the spray event occurs. Since the volume of a 1536-well is smaller than a 384-well, the time to reach this charging point will be shorter. To overcome this limitation would require a faster voltage switching system or an alternative way to push charge into the well.

5.6 Conclusion

As novel high-throughput sampling methods to MS utilize the acoustic ejection technologies, both AMI-MS and ADE-OPI-MS have demonstrated significantly faster speed than the routine LC-MS and have demonstrated their wide applications in various high-throughput drug discovery workflows. The applicability of ADE-OPI-MS to a wide breadth of analyte classes and the ability to analyze unprocessed samples with minimal method development greatly enhance the speed of the entire experimental workflow while facilitating miniaturization and preservation of original samples. We believe that with further development of data processing competency and integration with the orthogonal dimension of selectivity for isomeric separations, these technologies will significantly advance drug-discovery workflows.

References

1 Fenn, J.B., Mann, M., Meng, C.K. et al. (1989). Electrospray ionization for mass spectrometry of large biomolecules. *Science* 246 (4926): 64–71.

2 Covey, T.R., Lee, E.D., Bruins, A.P. et al. (2008). Liquid chromatography/mass spectrometry. *Anal. Chem.* 58 (14): 1451A–1461A.

3 Bruins, A.P. (1991). Liquid chromatography-mass spectrometry with ionspray and electrospray interfaces in pharmaceutical and biomedical research. *J. Chromatogr. A* 554 (1, 2): 39–46.

4 Awad, H., Khamis, M.M., and El-Aneed, A. (2015). Mass spectrometry, review of the basics: ionization. *Appl. Spectrosc. Rev.* 50 (2): 158–175.

5 Wan, H. and Holmen, A.G. (2009). High-throughput screening of physicochemical properties and in vitro ADME profiling in drug discovery. *Comb. Chem. High-Throughput Screening* 12 (3): 315–329.

6 Bantscheff, M., Lemeer, S., Savitski, M.M. et al. (2012). Quantitative mass spectrometry in proteomics: critical review update from 2007 to the present. *Anal. Bioanal. Chem.* 404 (4): 939–965.

7 Mishur, R.J. and Rea, S.L. (2012). Applications of mass spectrometry to metabolomics and metabonomics: detection of biomarkers of aging and of age-related diseases. *Mass Spectrom. Rev.* 31 (1): 70–95.

8 Theodoridis, G.A., Gika, H.G., Want, E.J. et al. (2012). Liquid chromatography–mass spectrometry based global metabolite profiling: a review. *Anal. Chim. Acta* 20 (711): 7–16.

9 Han, X., Yang, K., and Gross, R.W. (2012). Multi-dimensional mass spectrometry-based shotgun lipidomics and novel strategies for lipidomic analyses. *Mass Spectrom. Rev.* 31 (1): 134–178.

10 Wu, C., Dill, A.L., Eberlin, L.S. et al. (2013). Mass spectrometry imaging under ambient conditions. *Mass Spectrom. Rev.* 32 (3): 218–243.

11 Quercia, A.K., LaMarr, W.A., Myung, J. et al. (2007). High-throughput screening by mass spectrometry: comparison with the scintillation proximity assay with a focused-file screen of AKT1/PKB alpha. *J. Biomol. Screening* 12 (4): 473–480.

12 Hillenkamp, F. and Peter-Katalinic, J. (2013). *MALDI MS: A Practical Guide to Instrumentation, Methods and Applications*. Wiley. (Book, Whole).

13 Gobom, J., Schuerenberg, M., Mueller, M. et al. (2001). Alpha-cyano-4-hydroxycinnamic acid affinity sample preparation. A protocol for MALDI-MS peptide analysis in proteomics. *Anal. Chem.* 73 (3): 434–438.

14 Tanaka, K., Waki, H., Ido, Y. et al. (1988). Protein and polymer analyses up to *m/z* 100,000 by laser ionization time-of-flight mass spectrometry. *Rapid Commun. Mass Spectrom.* 2 (8): 151–153.

15 van Kampen, J.J., Burgers, P.C., de Groot, R. et al. (2011). Biomedical application of MALDI mass spectrometry for small-molecule analysis. *Mass Spectrom. Rev.* 30 (1): 101–120.

16 Wu, J., Hughes, C.S., Picard, P. et al. (2007). High-throughput cytochrome P450 inhibition assays using laser diode thermal desorption-atmospheric pressure chemical ionization-tandem mass spectrometry. *Anal. Chem.* 79 (12): 4657–4665.

17 Belov, A.M., Kozole, J., Bean, M.F. et al. (2020). Acoustic mist ionization-mass spectrometry: a comparison to conventional high-throughput screening and compound profiling platforms. *Anal. Chem.* 92 (20): 13847–13854.

18 Kempa, E.E., Hollywood, K.A., Smith, C.A. et al. (2019). High-throughput screening of complex biological samples with mass spectrometry – from bulk measurements to single cell analysis. *Analyst* 144 (3): 872–891.

19 Sinclair, I., Bachman, M., Addison, D. et al. (2019). Acoustic mist ionization platform for direct and contactless ultrahigh-throughput mass spectrometry analysis of liquid samples. *Anal. Chem.* 91 (6): 3790–3794.

20 Zhang, H. (2018). Acoustic-OPP-MS - the next generation high-throughput bioanalytical platform for drug discovery. *SLAS 2018*, San Diego, CA (3–7 February 2018).

21 Shou, W.Z. Acoustic ejection mass spectrometry: development, applications and future perspective. *Biomed. Chromatogr.* 36 (2): e5278.

22 Hadimioglu, B., Stearns, R., and Ellson, R. (2016). Moving liquids with sound: the physics of acoustic droplet ejection for robust laboratory automation in life sciences. *J. Lab. Autom.* 21 (1): 4–18.

23 Van Berkel, G.J. and Kertesz, V. (2015). An open port sampling interface for liquid introduction atmospheric pressure ionization mass spectrometry. *Rapid Commun. Mass Spectrom.* 29 (19): 1749–1756.

24 Sinclair, I., Stearns, R., Pringle, S. et al. (2016). Novel acoustic loading of a mass spectrometer: toward next-generation high-throughput MS screening. *J. Lab. Autom.* 21 (1): 19–26.

25 Dawes, T.D., Turincio, R., Jones, S.W. et al. (2016). Compound transfer by acoustic droplet ejection promotes quality and efficiency in ultrahigh-throughput screening campaigns. *J. Lab. Autom.* 21 (1): 64–75.

26 Sackmann, E.K., Majlof, L., Hahn-Windgassen, A. et al. (2016). Technologies that enable accurate and precise nano-to milliliter-scale liquid dispensing of aqueous reagents using acoustic droplet ejection. *J. Lab. Autom.* 21 (1): 166–177.

27 Van Berkel, G.J., Sanchez, A.D., and Quirke, J.M.E. (2002). Thin-layer chromatography and electrospray mass spectrometry coupled using a surface sampling probe. *Anal. Chem.* 74 (24): 6216–6223.

28 Van Berkel, G.J., Kertesz, V., Koeplinger, K.A. et al. (2008). Liquid microjunction surface sampling probe electrospray mass spectrometry for detection of drugs and metabolites in thin tissue sections. *J. Mass Spectrom.* 43 (4): 500–508.

29 Prideaux, B., ElNaggar, M.S., Zimmerman, M. et al. (2015). Mass spectrometry imaging of levofloxacin distribution in TB-infected pulmonary lesions by MALDI-MSI and continuous liquid microjunction surface sampling. *Int. J. Mass Spectrom.* 377: 699–708.

30 Gill, E.L., Marks, M., Yost, R.A. et al. (2017). Monitoring dopamine ex vivo during electrical stimulation using liquid-microjunction surface sampling. *Anal. Chem.* 89 (24): 13658–13665.

31 Griffiths, R.L., Randall, E.C., Race, A.M. et al. (2017). Raster-mode continuous-flow liquid microjunction mass spectrometry imaging of proteins in thin tissue sections. *Anal. Chem.* 89 (11): 5683–5687.

32 Wu, Q., Huang, Z., Wang, Y. et al. (2020). Absolute quantitative imaging of sphingolipids in brain tissue by exhaustive liquid microjunction surface sampling–liquid chromatography–mass spectrometry. *J. Chromatogr. A* 1609: 460436.

33 Ovchinnikova, O.S., Bhandari, D., Lorenz, M. et al. (2014). Transmission geometry laser ablation into a non-contact liquid vortex capture probe for mass spectrometry imaging. *Rapid Commun. Mass Spectrom.* 28 (15): 1665–1673.

34 Cahill, J.F., Kertesz, V., and Van Berkel, G.J. (2015). Characterization and application of a hybrid optical microscopy/laser ablation liquid vortex capture/electrospray ionization system for mass spectrometry imaging with sub-micrometer spatial resolution. *Anal. Chem.* 87 (21): 11113–11121.

35 Cahill, J.F., Kertesz, V., and Van Berkel, G.J. (2016). Laser dissection sampling modes for direct mass spectral analysis. *Rapid Commun. Mass Spectrom.* 30 (5): 611–619.

36 Cahill, J.F., Kertesz, V., Weiskittel, T.M. et al. (2016). Online, absolute quantitation of propranolol from spatially distinct 20- and 40-μm dissections of brain, liver, and kidney thin tissue sections by laser microdissection–liquid vortex capture–mass spectrometry. *Anal. Chem.* 88: 6026–6034.

37 Cahill, J.F., Kertesz, V., Porta, T. et al. (2018). Solvent effects on differentiation of mouse brain tissue using laser microdissection "cut and drop" sampling with direct mass spectral analysis. *Rapid Commun. Mass Spectrom.* 32 (5): 414–422.

38 Gómez-Ríos, G.A., Liu, C., Tascon, M. et al. (2017). Open port probe sampling interface for the direct coupling of biocompatible solid-phase microextraction to atmospheric pressure ionization mass spectrometry. *Anal. Chem.* 89 (7): 3805–3809.

39 Liu, C., Gómez-Ríos, G.A., Schneider, B.B. et al. (2017). Fast quantitation of opioid isomers in human plasma by differential mobility spectrometry/mass spectrometry via SPME/open-port probe sampling interface. *Anal. Chim. Acta* 991: 89–94.

40 Tascon, M., Alam, M.N., Gómez-Ríos, G.A. et al. (2018). Development of a microfluidic open interface with flow isolated desorption volume for the direct coupling of SPME devices to mass spectrometry. *Anal. Chem.* 90 (4): 2631–2638.

41 Tascon, M., Singh, V., Huq, M. et al. (2019). Direct coupling of dispersive extractions with magnetic particles to mass spectrometry via microfluidic open interface. *Anal. Chem.* 91 (7): 4762–4770.

42 Swanson, K.D., Worth, A.L., and Glish, G.L. (2017). Use of an open port sampling interface coupled to electrospray ionization for the on-line analysis of organic aerosol particles. *J. Am. Soc. Mass Spectrom.* 29 (2): 297–303.

43 Van Berkel, G.J., Kertesz, V., and Boeltz, H. (2017). Immediate drop on demand technology (I-DOT) coupled with mass spectrometry via an open port sampling interface. *Bioanalysis* 9 (21): 1667–1679.

44 Van Berkel, G.J. and Kertesz, V. (2017). Rapid sample classification using an open port sampling interface coupled with liquid introduction atmospheric pressure ionization mass spectrometry. *Rapid Commun. Mass Spectrom.* 31 (3): 281–291.

45 Van Berkel, G.J., Kertesz, V., Orcutt, M. et al. (2017). Combined falling drop/open port sampling interface system for automated flow injection mass spectrometry. *Anal. Chem.* 89 (22): 12578–12586.

46 Sosnowski, P. and Hopfgartner, G. (2020). Application of 3D printed tools for customized open port probe-electrospray mass spectrometry. *Talanta* 215: 120894.

47 Liu, C., Van Berkel, G.J., Cox, D.M. et al. (2020). Operational modes and speed considerations of an acoustic droplet dispenser for mass spectrometry. *Anal. Chem.* 92 (24): 15818–15826.

48 Liu, C., Van Berkel, G.J., Kovarik, P. et al. (2021). Fluid dynamics of the open port interface for high-speed nanoliter volumc sampling mass spectrometry. *Anal. Chem.* 93 (24): 8559–8567.

49 Liddle, N., Baghla, R., and Liu, C. (2021). Novel methods for increasing throughput and detection of acoustic ejection mass spectrometry (AEMS). *ASMS 2021*, Philadelphia, PA (31 October–4 November 2021).

50 Wen, X., Liu, C., Ghislain, L. et al. (2021). Direct analysis from phase-separated liquid samples using ADE-OPI-MS: applicability to high-throughput screening for inhibitors of diacylglycerol acyltransferase 2. *Anal. Chem.* 93 (15): 6071–7079.

51 Simon, R.P., Häbe, T.T., Ries, R. et al. (2021). Acoustic ejection mass spectrometry: a fully automatable technology for high-throughput screening in drug discovery. *SLAS Discovery* 26 (8): 961–973.

52 Liu, C. (2019). High-throughput ESI-MS enabled by the acoustic droplet ejection to the open-port probe sampling interface. *SLAS 2019*, Washington, DC (2–6 February 2019).

53 Wen, X. (2021). Observations from our first lap around the block: successful execution of an HTS campaign on an automated Echo MS. *The 1st Chinese American Society for Mass Spectrometry Conference; Virtual Conference* (9–13 August 2021).

54 Rebane, R. and Herodes, K. (2012). Comparison of three buffer solutions for amino acid derivatization and following analysis by liquid chromatography electrospray mass spectrometry. *J. Chromatogr. A* (1245): 134–142.

55 Sinclair, I., Davies, G., and Semple, H. (2019). Acoustic mist ionization mass spectrometry (AMI-MS) as a drug discovery platform. *Expert Opin. Drug Discovery* 14 (7): 609–617.

56 Häbe, T.T., Liu, C., Covey, T.R. et al. (2020). Ultrahigh-throughput ESI-MS: sampling pushed to six samples per second by acoustic ejection mass spectrometry. *Anal. Chem.* 92 (18): 12242–12249.

57 Zhang, H., Liu, C., Hua, W. et al. (2021). Acoustic ejection mass spectrometry for high-throughput analysis. *Anal. Chem.* 93 (31): 10850–10861.

58 Wagner, A., Zhang, J., Liu, C. et al. (2020). Ultrahigh-throughput and chromatography-free bioanalysis of polar analytes with acoustic ejection mass spectrometry. *Anal. Chem.* 92 (19): 13525–13531.

59 Zhang, H. (2018). Acoustic-OPP-MS – the next generation high-throughput bioanalytical platform for drug discovery. *Pittcon 2018*, Orlando, FL (27 February–1 March 2018).

60 Simon, R.P. ed. (2022). Acoustic ejection mass spectrometry: a versatile technology for fast bioanalysis and fully automated high-throughput screening. *SLAS 2022*, Boston, MA (5–9 February 2022).

61 Hollenbeck, T., Thibodeaux, S., White, P. et al. (2020). Acoustic droplet ejection and open port interface for rapid analysis of metabolic stability assays. *J. Pharm. Sci.* 109 (11): 3285–3291.

62 Devenport, N., Walsh, N., and Wood, S. (2021). Accelerated pharmacokinetic profiling and metabolite monitoring using Echo® MS system. *SCIEX Technical Note.*

63 Kern, R. and Baghla, R. (2021). True high-throughput analysis of a transporter activity assay. *ASMS 2021*, Pennsylvania, PA (31 October–4 November 2021).

64 Gribbon, P. and Sewing, A. (2003). Fluorescence readouts in HTS: no gain without pain? *Drug Discovery Today* 8 (22): 1035–1043.

65 Baghla, R. and Kern, R. (2021). Ultrafast bioanalytical analysis using the echo ms system – two case studies. *ASMS 2021*, Philadelphia, PA (31 October–4 November 2021).

66 Zhang, H. (2020). Applications of ADE-OPI-MS, the ultrahigh-throughput platform for drug discovery. *SLAS 2020*, San Diego, CA (25–29 January 2020).

67 Bateman, K. (2020). Quantitative analysis using Echo-MS. *WRIB, Virtual Conference, WIRB 2020* (15 June–29 June 2020).

68 Guo, Y., Forbush, M., Covey, T.R. et al. (2021). High-throughput analysis from complex matrices: acoustic ejection mass spectrometry from phase-separated fluid samples. *Metabolites* 11 (11): 789–799.

69 Zhang, J., Zhang, Y., Liu, C. et al. (2021). Acoustic ejection/full-scan mass spectrometry analysis for high-throughput compound quality control. *SLAS Technol.* 26 (2): 178–188.

70 Liu, C. (2021). Automated mass spectra comparison algorithm for high-throughput compound QC applications. *ASMS 2021*, Philadelphia, PA (31 October–4 November 2021).

71 DiRico, K.J., Hua, W., Liu, C. et al. (2020). Ultrahigh-throughput acoustic droplet ejection-open port interface-mass spectrometry for parallel medicinal chemistry. *ACS Med. Chem. Lett.* 11 (6): 1101–1110.

72 Fuhrer, T. and Zamboni, N. (2015). High-throughput discovery metabolomics. *Curr. Opin. Biotechnol.* 31: 73–78.

73 Bachman, M., Sinclair, I., Ivanov, D. et al. (2021). Information-rich high-throughput cellular assays using acoustic mist ionisation mass spectrometry. *Analyst* 146 (1): 315–321.

74 Asakawa, D., Mizuno, H., and Toyóoka, T. (2017). Gas-phase stability of negatively charged organophosphate metabolites produced by electrospray ionization and matrix-assisted laser desorption/ionization. *J. Am. Soc. Mass Spectrom.* 28 (12): 2561–2568.

75 Bedford L, Schneider BB, Duchoslav E, et al. Acoustic ejection mass spectrometry (AEMS)/differential ion mobility for ultrahigh-throughput mass spectrometry. *ASMS 2021*, Pennsylvania, PA (31 October–4 November 2021).

6

Ion Mobility Spectrometry-Mass Spectrometry for High-Throughput Analysis

Dylan H. Ross, Aivett Bilbao, Richard D. Smith, and Xueyun Zheng

Earth and Biological Sciences Directorate, Pacific Northwest National Laboratory, Richland, WA, USA

6.1 Introduction of Ion Mobility Spectrometry

The field of ion mobility spectrometry (IMS) has skyrocketed in recent years in both fundamental advancement and applications. IMS separates ions of different size or shape based on the balance of two forces that impact the ion movement, namely, the electric field and the drag force from the collision with buffer gas molecules [1], and provides insights into the structure.

While IMS has experienced greatly increased interest over the past several decades, it is not a new analytical technique. In fact, its foundation and that of mass spectrometry (MS) can be traced back to the X-ray experiments of Thomson and Rutherford in 1896 and their study investigating the connection between charge carriers and atmospheric aerosol particles [2, 3]. Langevin began the experiments that first directly measured ion mobility and later his theoretical work has laid the foundation for current analysis of the diffusion of ion through gases [4] that we used in ion mobility theory today [1]. The first ion mobility instrument called "plasma chromatograph" was developed in 1970. Since then, IMS has evolved into a powerful analytical technique for detection of gas-phase samples in atmospheric pressure device. It has been significantly improved since. Modern IMS devices became portable and started to be used not only in laboratories but also in fields to rapidly screen chemical vapors for trace quantities of hazardous material, detection of explosives and chemical warfare agents in military operations and airports [5].

High-Throughput Mass Spectrometry in Drug Discovery, First Edition. Edited by Chang Liu and Hui Zhang.
© 2023 John Wiley & Sons, Inc. Published 2023 by John Wiley & Sons, Inc.

While IMS itself is very powerful, when coupled with MS, IMS-MS provides a powerful hybrid analytical tool. With the significant improvements in coupling IMS with MS after the introduction of electrospray ionization (ESI) and matrix-assisted laser desorption/ionization (MALDI) soft ionization techniques, more broad and complex sample systems can be studied by IMS-MS, but they were generally in home-built instruments in the academic field and not feasible for broad application. The first commercial IMS-MS instrument based on traveling wave was introduced by Waters in 2006 (Waters Synapt), and soon became a widely used platform in academic and industrial fields. Since then, many instrument vendors followed and developed their own IMS-MS platforms, notably trapped IMS (TIMS) coupled with MS by Bruker in 2009 and commercial drift tube IMS (DTIMS) coupled with MS by Agilent in 2014. The hyphenated IMS-MS provides high resolution, high sensitivity, and high-throughput sampling and has become a powerful and popular tool for applications ranging from national security analyses, patient screening, environmental monitoring, pharmaceutical, and numerous other areas [5–9].

Different variations of the electric field and stationary state of the buffer gas have given rise to multiple IMS-based platforms such as DTIMS [10–15], traveling wave IMS (TWIMS) [16], TIMS [17], overtone IMS (OIMS) [18, 19], differential IMS (DIMS) [20], field asymmetric IMS (FAIMS) [21–23], and transversal modulation IMS (TM-IMS) [24]. While each platform has been used in numerous and different applications, we will focus this chapter mainly on the classic platform DTIMS and provide representative examples where the other techniques have been applied. For further information on each technique, we direct the readers to more extensive review articles which describe each method in greater detail [4, 25–29].

6.2 IMS Fundamental and Experiment

6.2.1 Ion Mobility Theory

In classical DTIMS, the ions gently drift through the buffer gas under the influence of a weak electric field. The ions are accelerated along the field lines and scatter in random directions due to collision with the buffer gas and diffusion. Under the low field condition, the ions are given enough time to reach equilibrium with the gas molecules. As a result, the ions travel with constant velocity parallel to the field lines and their sizes grow due to diffusion. The constant drift velocity, v_d, is determined by the forward accelerating force of the field and the retarding decelerating force of collisions. It is proportional to the electric field E [1]:

$$v_d = K \cdot E \tag{6.1}$$

Using kinetic theory [1], the momentum transfer and energy gained by the ions from the electric field is balanced with the energy lost in buffer gas collisions. The collision cross section (CCS, σ) can be derived,

$$K = \frac{3q}{16N}\left(\frac{2\pi}{\mu k_B T}\right)^{1/2}\frac{1}{\sigma} \tag{6.2}$$

Here, q is the charge state, N is the buffer gas number density at STP, μ is the reduced mass of the ion–He collision, k_B is the Boltzmann constant, and T is the temperature. The mobility (K) and the CCS (σ) provide information about the three-dimensional configurations of the ions. This equation demonstrates that the ion mobility depends on the gas pressure, the temperature, the electric field, the size or shape of the analyte ions, and the interactions of the ion and the buffer gas. Under the same pressure and temperature, ions with a more compact structure which corresponds to a smaller cross section would have larger mobility and drift faster. In contrast, ions with an extended conformation which corresponds to a larger cross section would have smaller ion mobility and drift slower. Thus, mixtures of compact and extended structures can be separated (Figure 6.1).

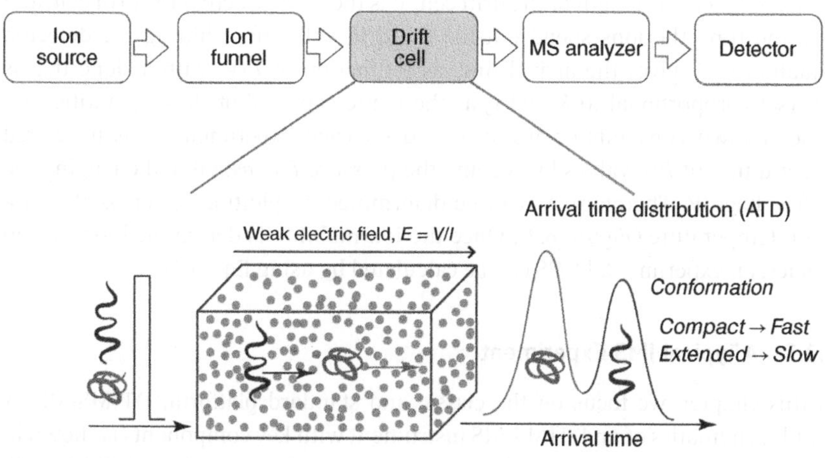

Figure 6.1 Schematics of a typical IMS-MS instrument and conformation separation. Ions with the same *m/z* but different sizes or shapes would travel differently during IMS where the ions with a compact structure travel faster and have a shorter arrival time, and ions with an extended structure travel slower and have a longer arrival time.

6.2.2 Collision Cross Section Measurement

During the experiment, the time the analyte ions spending inside and outside the drift cell with length l, before reaching the detector, is measured directly as arrival time, t_A. It is related to the drift velocity of the ion:

$$v_d = \frac{l}{t_d} = \frac{l}{(t_A - t_o)} \tag{6.3}$$

t_d is the time the analyte ions spending inside the drift cell with length l; t_o is the time the analyte ions spending after the drift cell before reaching the detector. The absolute ion mobility is dependent on the temperature (T) and the pressure (P) of the buffer gas, so it is typically converted to K_0, the standard or reduced mobility:

$$K_0 = K \cdot \frac{P}{760} \cdot \frac{273.16}{T} \tag{6.4}$$

The reduced mobility K_0 can be determined from the instrument parameters by using Eq. (6.2) and plotting t_A versus P/V [30]:

$$t_A = \frac{l^2}{K_0} \cdot \frac{273.16}{760T} \cdot \frac{P}{V} + t_0 \tag{6.5}$$

In Eq. (6.5), l is the length of the drift cell, V is the voltage across the drift cell, and t_0 is the time the ions spend outside the drift cell before hitting the detector. Equation (6.5) gives the arrival time as a function of P/V, with a slope that is inversely proportional to K and t_0 as the y-intercept. All of these quantities are either known constants or are measured for each experiment. t_A is measured under different P/V values by keeping the pressure P constant and changing the drift voltage V. Therefore, K_0 can be determined by plotting t_A versus P/V at a given temperature (Figure 6.2). Once the ion mobility is determined for a given species, an experimental CCS can be calculated by using Eq. (6.2).

6.2.3 A Typical IMS Experiment

In this chapter, we focus on the classic and standard platform, DTIMS-MS. A simple schematics of an DTIMS-MS instrument with key components is shown in Figure 6.1, which usually includes an ion source, an ion funnel, an IMS drift cell, a mass analyzer, and a detector. Briefly, for MS-only mode, ions are generated continuously by an ionization source, focused and guided through an ion funnel, pass through a temperature-controlled drift cell filled with buffer gas

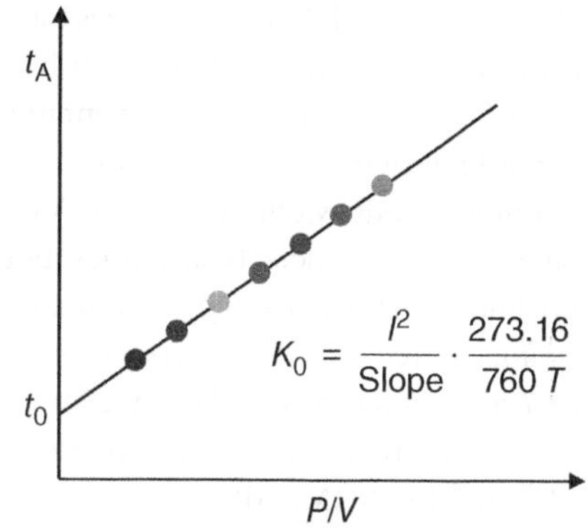

Figure 6.2 Example of an arrival time, t_A, versus *P*/*V* plot to determine the mobility of a given species. The slope of the curve is inversely proportional to the mobility and *y*-intercept is equal to t_0.

(e.g. nitrogen or helium), mass analyzed with a mass spectrometer, often a time-of-flight (TOF) or a quadrupole TOF (qTOF), allowing a mass spectrum to be obtained. For ion mobility measurements, the ions are stored in the ion funnel and pulsed into the drift cell. The ions exiting the drift cell are mass analyzed and detected by the detector to allow an arrival time distribution (ATD) to be recorded. The ATD can be related to the time the ions spend in the drift cell which is directly related to the ion mobility and CCS of the analyte ion [31]. The measured ion mobility and CCS provide information about the three-dimensional configurations of the ions.

6.3 IMS Analysis and Applications

IMS has several attractive advantages. First, IMS is able to resolve sequence and structural isomers that are indistinguishable in MS measurements. Second, mass spectral peak congestion is reduced by spreading ions out by their shapes and charge states. Third, an experimental CCS can be acquired for comparison with molecular modeling calculations, which may ultimately lead to a structural identification for a known or unknown molecule. Finally, the rapid IMS measurements are easily coupled with TOF-MS instruments for nested data acquisition of both shape and *m*/*z* information into a single IMS-MS experiment. The relatively short (millisecond scale) experimental timescale of IMS-MS measurements makes it particularly well-suited to high-throughput analysis.

6.3.1 Separation of Isomeric and Isobaric Species by IMS

One of the biggest and attractive features of IMS is its ability to distinguish isomeric species that cannot be resolved by MS alone. We have demonstrated the power of IMS-MS for isomer separation of small molecules including metabolites [32, 33], glycans [34], lipids [35], xenobiotics [36], and peptides [37]. In drug discovery studies, being able to separate and distinguish isomeric species is also critical as the molecular structure, e.g. *cis* or *trans* orientation could often lead to very different or even opposite bioactivities and therefore affect its potential use as a drug candidate. For example, dicaffeoylquinic acids (diCQAs) were recently found to have anti-HIV and other health beneficial bioactivities, but have been analytically challenging to identify and quantify since they commonly exist as a diverse mixture of positional and geometric (*cis/trans*) isomers. The structure and activity relationship for these *cis/trans* isomers has not been well understood, but it is known that the bioactivities of diCQAs are affected by the *cis/trans* isomerism. In one of our recent studies, we were able to use IMS-MS to separate the various *cis/trans* isomers that were induced by photochemistry (Figure 6.3a) [38]. Moreover, the experimental CCS for each isomer structure can be measured from DTIMS, and further compared with theoretical structures, allowing accurate differentiation and identification of those diCQA isomers. That study also revealed that the diCQAs naturally existed predominantly as *trans/trans* isomers, but after three hours of UV irradiation, *cis/cis*, *cis/trans*, *trans/cis*, and *trans/trans* isomers were all present in the mixture. These isomers were only partially separated by DTIMS, but were baseline separated by high resolution structures for lossless ion manipulation (SLIM) IMS (details about this technique will be covered in the latter Section 6.4 in this chapter). This was the first report of successful differentiation of *cis/trans* diCQA isomers individually and showed the great promise of IMS coupled with molecular modeling for determining the structure and activity relationships of different isomers in drug discovery studies [38].

Furthermore, IMS is also very powerful for separating molecules with nominal mass isobars. Many small molecules contain different isobar species from different classes that can be fragmented together when they co-elute or are directly injected in shotgun analyses. One of the analytical challenges for small molecule analyses is that many metabolites are similar in mass and hard to distinguish without high-resolution MS. For instance, L-tyrosine (exact mass: 181.0739), an amino acid, and glufosinate (exact mass: 181.0503), an herbicide, have the same nominal mass and very close residual masses. They will also be selected in the same MS/MS window (usually the narrowest window is 0.1 Da) so their fragmentation profiles will overlap. However, in DTIMS, these two molecules have completely different sizes and can be baseline separated in either positive or negative ion modes, with negative mode showing the best distinction (Figure 6.3b) [32].

6.3.2 High-Throughput IMS Measurements and Building a CCS Library

DTIMS is often described as a classic IMS due to its simple operation and its basics and theories that allow direct measurements of ion mobility and CCS. Therefore, it is often chosen as a primary method for CCS fingerprinting and applications.

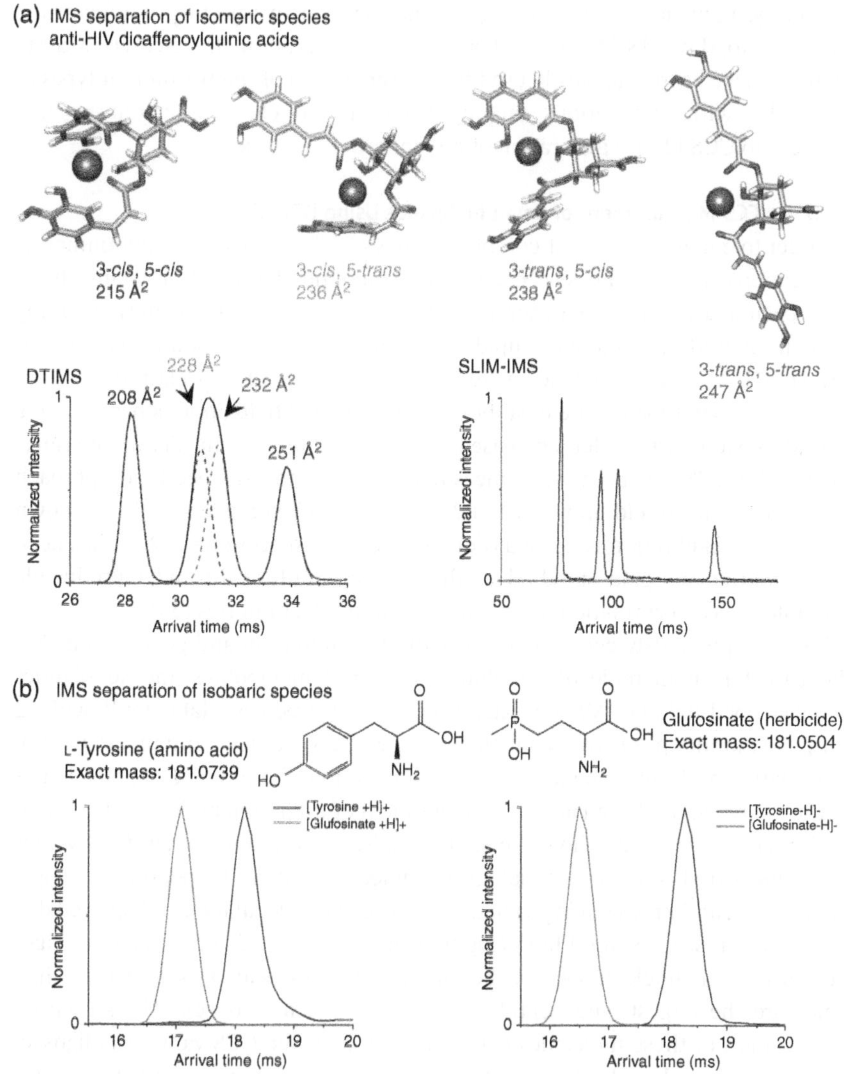

Figure 6.3 Examples of IMS separation of isomeric species separation (a) and isobaric species (b). *cis/trans* isomers of anti-HIV compound diCQAs were separated by IMS-MS; DTIMS can separate *cis/cis* and *trans/trans* isomers only while ultrahigh resolution SLIM IMS can resolve all isomers. By coupling with modeling structures and theoretical CCS values, the structure of each isomer can be assigned. Isobaric species l-tyrosine and Glufosinate which have the same nominal mass but very different molecular structures can be well separated by IMS. *Source:* (a) Reproduced with permission from Zheng et al. [38], Copyright 2017 American Chemical Society, (b) Reproduced from Zheng et al. [32] with permission from the Royal Society of Chemistry.

Currently, many measurements are performed by directly injecting or infusing the sample into the IMS-MS without a front-end separation. Injected-DTIMS-MS analyses have been extensively used for the separation of many different types of molecules with applications ranging from protein characterization to lipid extracts, to CCS library generation of metabolites [32, 39, 40].

6.3.2.1 CCS Measurement of Small Molecules Using DTIMS

In order to enable analysis of complex samples for biological and environmental applications, a CCS database is needed for compound identification. Many efforts have been dedicated to the generation of CCS databases for molecules including proteins, peptides, metabolites, lipids, glycans, and other compounds [15, 32, 33, 36, 41–52]. Most CCS libraries were generated on one of two platforms: either using DTIMS with single field calibration and stepped field approaches, or using TWIMS systems with reference-based calibration approach. In DTIMS measurements, the CCS can be directly measured by using the stepped-field approach (also called multi-field approach), independently of reference ions with known CCS and no calibration is needed (details described in Section 6.2.2). It has been considered as the gold standard method for straightforward and reproducible accurate CCS determination, even among different laboratories [53].

We have previously conducted a structural examination and generated a CCS database for small molecules including over 500 metabolites and xenobiotics using stepped-field DTIMS approach [32], which is freely available for download (http://panomics.pnnl.gov/metabolites). The CCS values contain structural information and can be beneficial for understanding the physical characteristics of the molecules and even allow the assignment of unknown molecules to certain molecular classes that follow the same trends. For example, when CCS values were measured for larger biological molecules such as peptides, sugars, nucleotides, and lipids having a single charge, each molecular class displayed distinct *m/z* versus CCS trend lines (Figure 6.4a). These trends illustrate that oligosaccharides and nucleotides have the smallest CCS values for a specific *m/z*, while lipids are the largest, and peptides fall somewhere in-between. The structural composition of these molecules plays a large role in the CCS values as oligosaccharides and nucleotides have cyclic structures and contain many heteroatoms, making them more compact while lipids contain more linear and aliphatic groups making them less structurally compact (manifesting as larger CCS values for a given *m/z*). The small molecule CCS values have a wide band that overlapped with the trend lines for the known larger biomolecules, illustrating structural compositions from ring to linear connectivity. The small molecules were cataloged into primary metabolites, secondary metabolites, and xenobiotics for further structural evaluations. The primary metabolites (e.g. amino acids, fatty acids, organic acids, carbohydrates, etc.) consisted of molecules directly involved in normal growth, development, and reproduction and have come from key metabolism

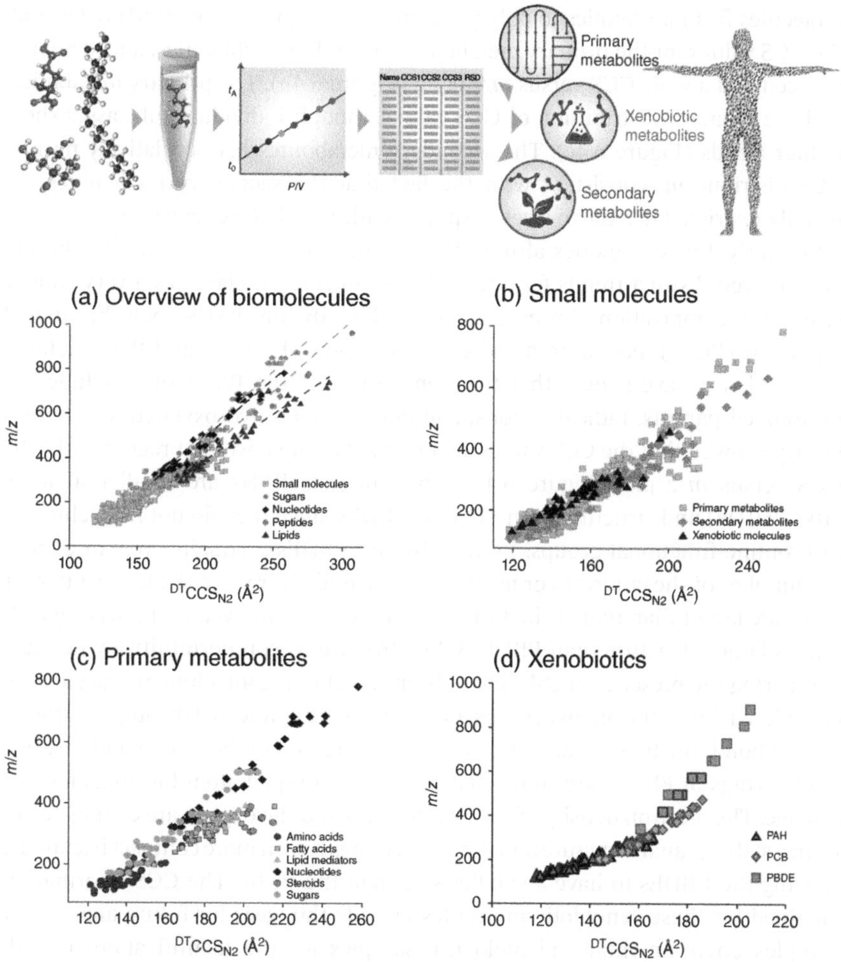

Figure 6.4 $^{DT}CCS_{N2}$ database for small molecules: (a) CCS values of small molecules were plotted with larger singly charged nucleotides, oligosaccharides, lipids, and peptides to observe structural trends; (b) CCS versus *m/z* plots for small molecules including primary metabolites, secondary metabolites, and xenobiotic molecules; (c) CCS versus *m/z* plots for primary metabolites with subclasses; (d) CCS versus *m/z* plots for xenobiotic molecules including PAHs, PCBs, and PBDEs. *Source:* (a–c) Reproduced from Zheng et al. [32] with permission from the Royal Society of Chemistry, and (d) Zheng et al. [36] with permission from Elsevier.

pathways such as glycolysis, the pentose phosphate pathway, and the tricarboxylic acid cycle. Secondary metabolites (alkaloids, terpenoids, flavonoids, etc.) were molecules not directly involved in these processes but affect the survivability of an organism [54]. The xenobiotics were foreign chemical substance not naturally produced by or expected within an organism and represented a range of molecules from antibiotics to polycyclic aromatic hydrocarbons (PAHs). Overall, the CCS values for the primary metabolites, secondary metabolites, and xenobiotics occupied a wide CCS versus *m/z* band (Figure 6.4b). The primary metabolites had the broadest distribution of CCS values, though individual subclasses show tighter trends (Figure 6.4c). The secondary metabolites had a relatively narrow CCS distribution, consistent with the fact that the secondary metabolites are usually restricted to a narrow set of species within a phylogenetic group.

Similarly, the xenobiotics also displayed a relatively narrow CCS distribution and showed distinct trends for each subclass since each class has a very similar structural composition. From an extended study on PAHs, polychlorinated biphenyls (PCBs), polybrominated diphenyl ethers (PBEDs), and their metabolites [36], we have shown that the trend lines for the PAHs that include the protonated parents, radical cations, and deprotonated hydroxylated metabolites clearly showed that the CCS values are crowded in the lower left region of the 2D CCS versus *m/z* plot (Figure 6.4d). This suggests PAHs are smaller in mass (100–350 Da) and structure than PCBs or PBDEs since they do not have chlorine or bromine functional groups. Both PCBs and PBDEs gradually grew in CCS as the number of the chlorines or bromines increased. Moreover, the CCS values for PCBs are larger than that of the PBDEs with the same *m/z* values, indicating PCB have a larger structure than PBDE. While this seems counterintuitive when just comparing the presence of chlorine or bromine, the ring attachment must also be considered. Since the phenyl groups for the PCBs are connected through a carbon–carbon bond, the two aromatic rings can rotate freely. On the other hand, the two phenyl rings in PBDEs are connected through an oxygen atom forming an ether linkage. The electron density of the phenyl rings and the oxygen are conjugated to form a stable planar structure and the phenyl rings have a more compact backbone, causing the PBDEs to have a smaller size than the PCBs. The CCS information obtained for these xenobiotic molecules provide the basis for future analyses of complex environmental and biological samples for the identification of both known and unknown molecules.

While the stepped-field DTIMS is the only approach that allows direct and accurate CCS determination, it is generally a more time-consuming and labor-intensive process and becomes impractical for generating CCS databases with thousands of millions of molecules. Thus, an alternative approach, single-field measurements with calibration using reference ions, is becoming the preferred and more practical procedure for estimating the CCS values of ions in complex samples. This is particularly the case when liquid chromatography (LC) and other front-end separation methods are coupled with DTIMS, as the time requirements of the stepped-field method are generally incompatible with such in-line separations. Briefly, the single-field DTIMS method is based on a linear regression using known CCS (from literature or stepped-field experiments) of calibrant species. For example, the Agilent ESI tune mix ions are often used as calibrant ions. The calibration allows CCS determination of other analytes from their arrival times measured under the same IMS conditions as the calibrants. An interlaboratory study of the Agilent DTIMS-MS platform demonstrated the consistency from stepped-field and single-field approaches, with 0.23% accuracy and 0.6% accuracy (both within 1% relative standard deviations) as long as the standard protocols were followed [53]. Therefore, it is possible that the single-filed approach would become the standard procedure for high-throughput CCS determination in DTIMS due to the possibility of automated data processing and compatibility with LC and other front-end separation techniques.

Besides using vendor software to manually extract ions and calculate the CCS values, software tools for automated CCS calculation from IMS measurements are being developed. For example, our group has previously developed software tools such as PIXiE [55] and AutoCCS [56] to allow robust and automated ion mobility arrival time extraction and CCS calculations from single field and stepped field approaches in DTIMS measurements, and from TWIMS measurements using reference-based nonlinear calibration approaches. These tools greatly facilitate the generation of large-scale experimental CCS databases, which allows us to apply IMS to examine the metabolism and pathways of these molecules in biological and environmental studies. Besides, experimental CCS databases can be used as training data set for machine learning-based CCS prediction.

6.3.2.2 CCS Measurements of Drug Compounds Using TWIMS

With other measurement approaches, calibration of drift times using standards of known CCS is a rapid method for obtaining CCS values. However, the choice of calibrants is always a challenge and could lead to bias in CCS values. For example, PolyAla, a series of singly charged polyalanines ($n = 2-21$), is often used to calibrate the CCS for peptides and later used in many other systems; however, people have noticed that it does not accurately represent small molecules like metabolites and drug compounds. For example, Hines et al. measured the CCS values of a large collection of drug and drug-like compounds (>1400 compounds) using TWIMS-MS and CCS calibration method [44]. It was found that calibration with PolyAla alone lead to higher CCS errors for the drug-like compounds ranging from m/z 100 to 600, while the combination of PolyAla and drug-like compounds for CCS calibration improved the CCS errors, but both numbers were noted within 1.5% of the DTIMS CCS values. The large collection of CCS values for these drug compounds revealed some interesting conformational landscapes. As shown in Figure 6.5, the drug and drug-like compounds distribute a large conformational space in the CCS versus mass plot when overlaid with that of lipids and peptides trends, indicating high structural diversity among these compounds (Figure 6.5a). More distinct trends were observed when the molecules were cataloged into different classes, suggesting

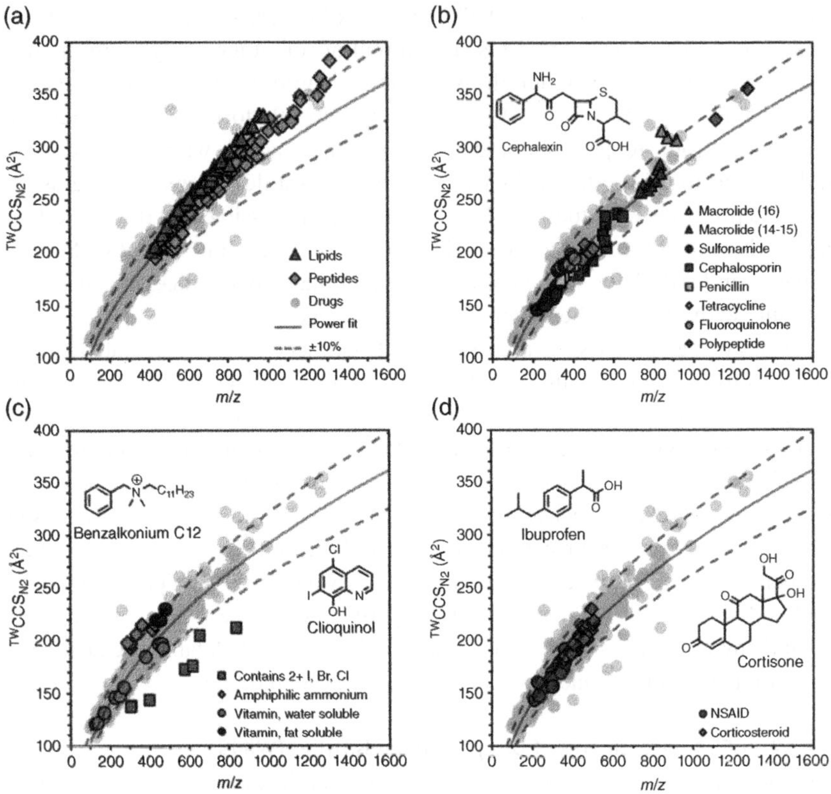

Figure 6.5 IM-MS conformational space plots showing the regions occupied by (a) lipids and peptides; (b) subclasses of antibiotics; (c) compounds of various densities; and (d) corticosteroid and nonsteroidal anti-inflammatory drugs (NSAIDs). Structures shown: cephalexin is a cephalosporin antibiotic, benzalkonium C12 is an amphiphilic ammonium, clioquinol is an antifungal drug, ibuprofen is a common NSAID, and cortisone is a common corticosteroid. *Source:* Reproduced from Hines et al. [44] with permission, Copyright 2017 American Chemical Society.

unique structure–bioactivity relationships for individual classes. For example, different subclasses of antimicrobials occupied in different regions of the CCS versus mass space (Figure 6.5b), which can be attributed to the common structural characteristics within each subclass. More specifically, some drug classes such as fluoroquinolones, penicillins, and cephalosporins which occupy a narrow mass range from 300 to 500 still display clearly separated trendlines in the CCS versus mass plot. Other drug classes also display characteristic tight groups in the CCS-mass plot. Additional common trends can be observed from this large collection. Lipid-like molecules tend to occupy large conformational space, such as fat-soluble vitamins (e.g. vitamins E and K), amphiphilic ammonium compounds (e.g. benzalkonium chloride), and lipophilic drugs (e.g. terfenadine). Heavy atom-containing molecules tend to occupy smaller space, such as clioquinol and erythrosine that contain heavy halogen atoms (Figure 6.5c). Overall, the CCS database of drug compounds allows the use of IMS-MS analysis for high-throughput drug metabolite identification and study of drug metabolism.

6.3.2.3 Large-Scale CCS Databases From Prediction Approaches

In recent years, machine learning-based CCS prediction methods have become popular for generating large-scale theoretical CCS databases for metabolites, lipids, and other compounds (e.g. MetCCS [43, 57], LipidCCS [58], LiPydomics [59], CCSbase [48], DeepCCS [60], DarkChem [61], etc.). These approaches generally involve the collection of large-scale experimental CCS values for groups of compounds, calculation of numerical representations of their chemical properties (molecular descriptors), and training regression models using various architectures (most popularly: support vector machines and artificial neural networks) that can predict CCS with high accuracy [62]. The most important distinguishing factors among these CCS prediction models are (i) the chemical space of compounds used for model training and (ii) the molecular descriptors/model architecture used to produce predictions. The chemical space of the training data determines range of compounds for which the model can be expected to make accurate predictions, and CCS prediction models range from specific to broad in terms of the chemical classes that are covered in their training data. Choice of molecular descriptors can have a large effect on the accuracy of CCS predictions, as they form the basis for mapping between CCS and underlying physiochemical properties. Machine learning architecture can also affect prediction accuracy, as different architectures may have different strengths and can be subject to different biases. Additionally, choice of molecular descriptors and model architecture can impose practical considerations, e.g. requiring proprietary software or minimum computing power. Some other recent examples of CCS prediction have made use of different approaches to machine learning-based CCS prediction by incorporation of quantum chemistry-based theoretical calculations (ISiCLE [63]) or prediction of CCS directly from peptide sequence as text (DeepCollisionalCross Section [64]). These technologies expand the possibilities for compound identification, which is a critical bottleneck in high-throughput and omics analyses.

6.3.3 LC-IMS-MS Analysis

Direct injection-IMS-MS works great for characterization of standard compounds or simple mixtures, but it often suppresses ionization when applied to complex sample analysis. Therefore, front-end separation approach such as LC is often required. To date, LC has been coupled with most IMS-MS including DTIMS, TWIMS, FAIMS, and TIMS. While LC separations require long minute to hour long timescales and low throughput, they also allow greater molecular separations and high peak capacity measurements. Furthermore, when combined, the three orthogonal separations (LC, IMS, and MS) provide three-dimensional spectra with LC elution times, IMS drift times (or CCS values), and m/z ratios for all detected ions in a sample [65, 66].

The advantages of LC-IMS-MS include the decreased chemical noise interferences in congested spectra due to the IMS separation allowing higher sensitivity measurements and the detection of low abundance ions even in the presence of species with much greater intensities [66–68]. Long LC separation times can be reduced due to the multiple orthogonal separations, while retaining the same or additional features as LC-MS [66], which is important for high-throughput studies. Consequently, LC-IMS-MS has been used to separate metabolites [42, 69], lipids [70, 71], glycans [72], proteins [73], peptides [66, 68, 74–78], and per- and polyfluoroalkyl substances (PFAS) [46] in complex biological and environmental samples and greatly improved the analytical sensitivity and specificity of LC-MS analyses, in addition to enhancing measurement dynamic range and providing reliable identification and quantitation of low abundance analyte species in complex biological matrices.

6.3.4 High-Throughput Analysis Using Rapidfire SPE-IMS-MS

While direct infusion often suppresses the ionization and therefore decreases the sensitivity due to the complexity of the samples, LC separation can be time-consuming and low throughput, and therefore not ideal for high-throughput screening. Recently, automated SPE techniques have been of interest for low peak capacity LC separations but with much higher throughput. SPE is one of the most frequently employed procedures to cleanup, extract, fractionate, and pre-concentrate biological and environmental samples [79]. SPE is also useful in desalting, derivatization, and buffer exchanging samples. Recently online and automated SPE techniques have become popular for high-throughput studies where the analytes of interest are retained on specific columns or cartridges, eluted with the appropriate solvents, and measured directly or determined in the eluate [80, 81]. The main attraction of online SPE is that it greatly reduces sample preparation time and enables automation of conditioning, washing, elution, and re-equilibration, thus increasing sample throughput. Analyte losses by evaporation are also often eliminated and the solvent consumption is much lower, reducing the risk of the exposure to infectious samples or toxic solvents. In addition, online SPE cartridges are often reusable, greatly decreasing material costs [80, 81].

One specific high-throughput automated SPE system is the Rapidfire SPE platform, which was originally introduced by BIOCIUS Life Sciences, Inc. and later acquired by Agilent Technologies (Santa Clara, CA). Briefly, the Rapidfire SPE system contains specially designed equipment, including an autosampler, LC pumps, SPE cartridges, and switching valves for ultrafast online sample preparation where 5–30 μL of sample is aspirated directly from 96- or 384-well assay plates and loaded onto the microscale SPE cartridge using specific buffers. The analytes of interest are retained in the cartridge while the salts and buffers are washed away. A valve is then switched to send the flow path to the mass spectrometer, and appropriate organic solvents are delivered to elute the compounds off the cartridges for MS analysis. Because the typical cycle time including loading, wash, elution, and re-equilibration is normally 10 seconds or less, these analyses are very attractive since they are 2–3 orders of magnitude faster than conventional GC or LC techniques. Another advantage of this approach is that the SPE cartridges can also be packed with different materials such as C4, C8, C18, graphitic carbon, cyano, phenyl, and HILIC, providing broad extraction for various analytes and enabling the capability to analyze complex environmental and biological samples such as urine and plasma. To date, the Rapidfire SPE system has been applied in several MS studies for ultrafast and high-throughput biological, biomedical, and drug discovery studies [82–87]. Recently, the SPE system was coupled with DTIMS (Instrument schematics shown in Figure 6.6 top panel), demonstrating the first high-throughput online SPE-DTIMS-MS platform [88]. The SPE-DTIMS-MS platform combined fast sampling and multidimensional separation, and allowed simultaneous targeted and global measurements for the detection of thousands of endogenous and exogenous metabolites in complex human biofluids in <1 minute total analysis time even when using several cartridge types. As shown in Figure 6.6, thousands of metabolite features were detected simultaneously in the discovery analyses. Isomeric metabolite features were well separated for $m/z = 327.197$ as shown in Figure 6.6b. Metabolite features with the same nominal mass were also distinguished by IMS ($m/z = 138.054$ and 138.129) as shown in Figure 6.6c. The 1-nM tiabendazole was detected in both human plasma and urine extracts with similar signal intensities indicating no matrix effects for the different biofluids. Overall the SPE-DTIMS-MS platform provided an appealing tool to quickly evaluate disease conditions and monitor environmental exposures for large sample numbers and will be extremely useful in future high-throughput studies [88].

With the availability of CCS databases built for metabolites and exobiotics, the high-throughput, high-sensitivity, and quantitative capabilities of the SPE-IMS-MS platform provide the ability to quickly perform endogenous metabolomic and exposomic pathway characterization, biomarker discovery, environmental exposure assessment, and inpatient screening following exposure prior to analyses with lower throughput methods if needed.

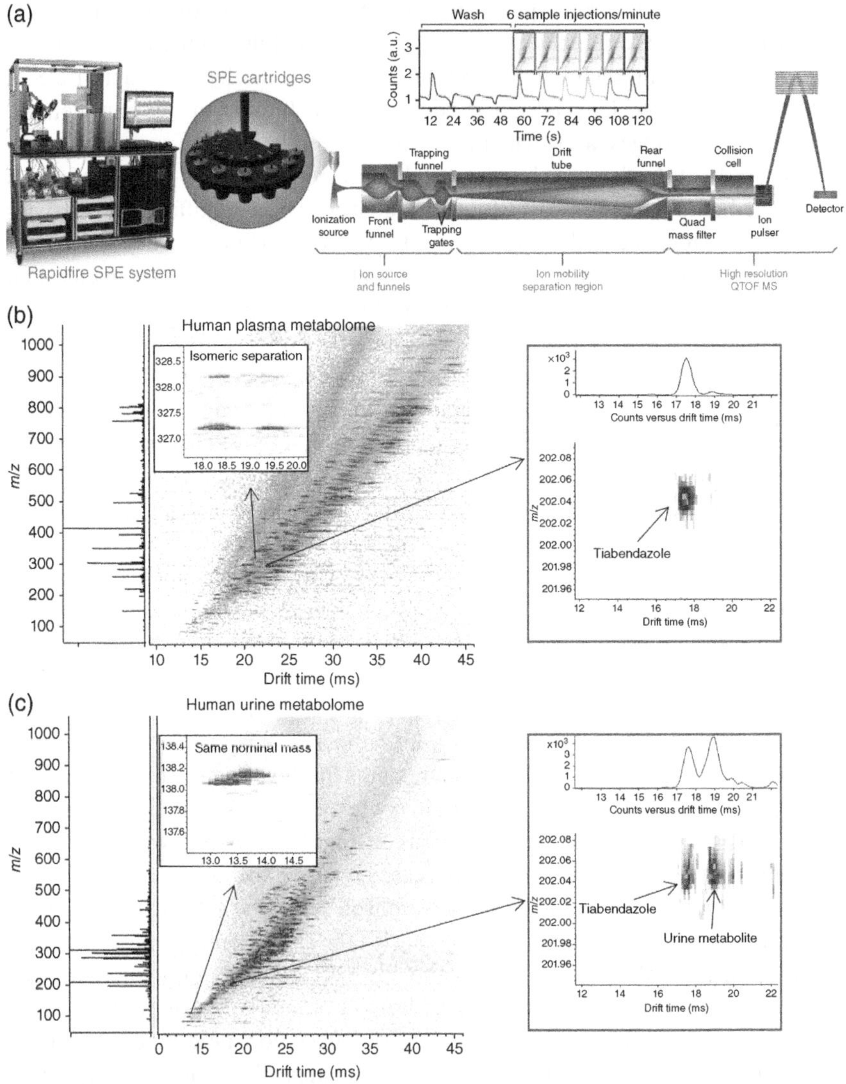

Figure 6.6 Coupling solid-phase extraction, ion mobility spectrometry, and mass spectrometry (SPE-IMS-MS) for measuring metabolites and xenobiotics: (a) the instrument schematics and 5 seconds analyses of (b) human plasma and (c) urine extracts with the spiked exogenous metabolites including 1 nM tiabendazole. *Source:* (a) [88] / Reproduced with permission from ELSEVIER / CC BY 4.0.

6.3.5 Software Tools for IMS Data Analysis

Analysis of the IMS data, particularly for complex samples and using multidimensional separations, has been challenging and limited the full utility of these measurements. Unlike LC-MS, where a plethora of tools with well-established algorithms exist, the incorporation of the additional IMS dimension requires upgrading existing computational pipelines and developing new algorithms to fully exploit the advantages of the technology. Fortunately, several of the popular and open-source bioinformatics tools originally designed for LC-MS have been extended to support the IMS information (e.g. drift time) such as Skyline [89–91], MS-DIAL [92], MZmine 3 [93], and OpenMS [94]. These tools are being used for interpreting the data obtained from complex multidimensional studies including IMS separation. New workflows supporting the IMS dimension are also available, such as AllCCS [51] using both experimental and theoretical CCS to annotate metabolites, and LipidIMMS Analyzer [95] for lipidomics analysis. With the added size and complexity of data resulting from integration of the IMS dimension, automated data processing is another area of increased effort in recent software development.

Our group has been developing new software programs to facilitate IMS data accessing and processing, including MZA data converter [96], PNNL PreProcessor [97], AutoCCS [56], DeIMoS [98], and others that are in development. We actively maintain and develop new functionality for the PNNL preprocessor [97], a tool specifically designed for IMS that houses various algorithmic implementations to generate new MS-files with enhanced signal quality and in the same instrument format. It supports different experimental approaches, including DTIMS and SLIM-IMS, and with LC or direct infusion analyses. The algorithms extend the dynamic range of the detection system, while reducing file sizes for faster and memory-efficient downstream processing. Specifically, multidimensional smoothing improves peak shapes of poorly defined low-abundance signals, and saturation repair reconstructs the intensity profile of high-abundance peaks from various analyte types. Other functionalities are data compression and interpolation, IMS demultiplexing, noise filtering by low intensity threshold and spike removal, and exporting of acquisition metadata. One of the novel functionalities being developed is the transformation of the raw IMS-MS data directly to the CCS/z space to facilitate data interpretation and standardization.

The large volume of multidimensional spectra collected by advanced and hybrid MS instrumentation (e.g. LC-IMS-MS analysis) is challenging current data storage and access technologies, therefore precluding informatics advancements. We have recently developed MZA [96], the mass-to-charge (m/z) generic data storage and access tool to allow simple programmatic access to raw MS-data from multiple instrument vendors and mass analyzers, and to facilitate software development and artificial intelligence research in multidimensional MS measurements.

6.4 High-Resolution SLIM-IMS Developments

To further benefit the multidimensional analyses and even increase the utility of DTIMS and IMS-MS studies, one challenge that must be addressed is the low separation power of most of current DTIMS and TWIMS measurements. Low separation power has been a key limitation which greatly restricts the range of molecular coverage from each measurement and makes the analysis of complex samples extremely difficult. In many cases, this is why other separation dimensions are added to an experiment. Higher IMS resolution promises to improve analysis coverage, separate similar structural isomers, and increase the overall dynamic range of IMS-MS measurements. In DTIMS, because the resolving power is proportional to the square root of the drift field and length [99], these parameters have been investigated extensively. Kemper et al. built a 2-m DTIMS that achieved a resolving power of >100; however, more length was still needed to separate many structurally similar isomers [100]. Increasing the drift tube length past 2 m unfortunately requires extremely high voltages making it impractical for many lab spaces. A cyclic multi-pass drift tube was constructed to address this challenge [101]. However, there was one constraint in their design that the measurable mobility range decreased with every cycle needed to achieve greater separation.

A new approach using SLIM was recently utilized to enable long-path TWIMS separations followed by MS analyses (Figure 6.7) [102, 103]. SLIM is created using electric field generated by arrays of electrodes fabricated using printed circuit board (PCB) technology on two planar surfaces. The two surfaces are aligned in parallel and the electric fields are generated by application of appropriate voltages to the electrodes to create the SLIM ion confinement regions or conduits through which ions can be moved, separated by IMS, or otherwise manipulated. This approach allows the construction of a 13-m SLIM IMS device by using traveling waves and a compact serpentine ion drift path, which provides more than an order of magnitude increase in ion travel length and therefore greatly improved separations. For example, as shown in Figure 6.7b, SLIM IMS-MS analysis of the peptide/carbohydrate/lipid mixture showed that different molecular classes were clearly separated by different trend lines with the doubly charged peptides arriving first, then singly charged carbohydrates, singly charged peptides, and finally the singly charged lipids arriving last due to the distinct backbone structures of each molecule type. Moreover, SLIM also allows multi-pass capabilities, Serpentine Ultralong Path with Extended Routing (SUPER) [104], providing previously unachieved resolution for molecule separation. Also, the sensitivity was greatly improved with more effective ion utilization during the ion accumulation process which enables better characterization of low abundance species and very small sample sizes, thus demonstrating SLIM IMS-MS's potential for high resolution and high-throughput measurements.

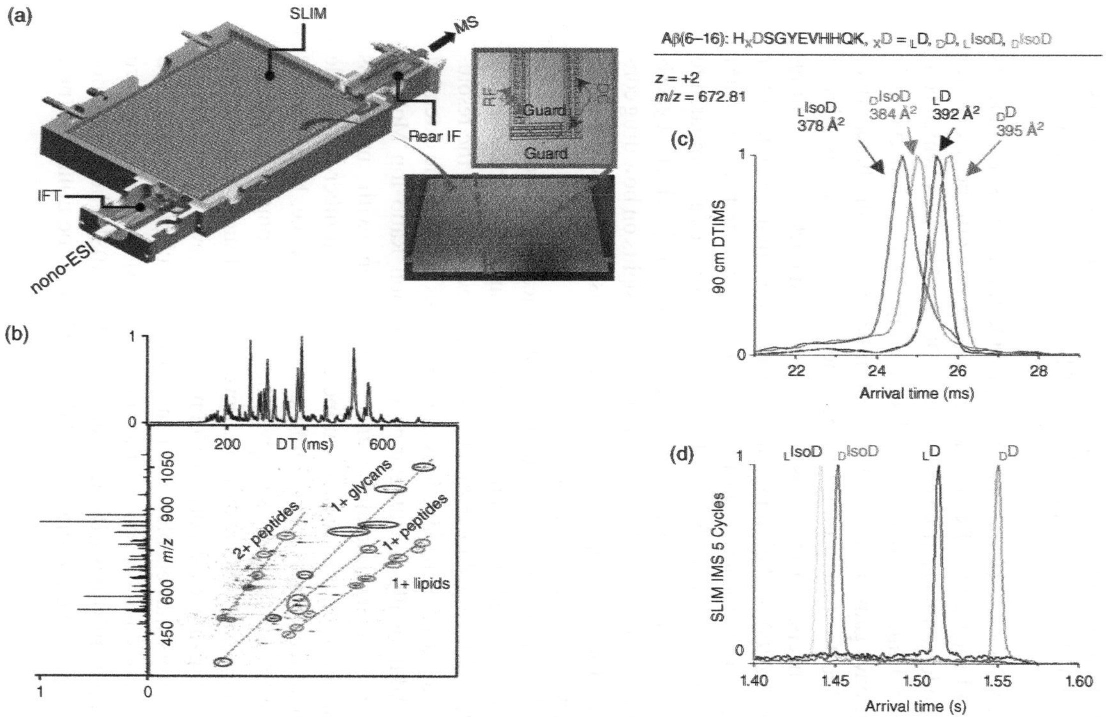

Figure 6.7 (a) A schematic and photograph of the serpentine 13-m SLIM IMS-MS platform. (b) The two-dimensional nested IMS-MS spectrum from the SLIM platform showing drift time (DT) versus *m/z* for the peptide/carbohydrate/lipid mixture analyzed. Different molecular classes typically separated by different trend lines due to the distinct backbone structures of each molecule type, with the 2+ peptides arriving first, followed by 1+ glycans, 1+ peptides, and finally the 1+ lipids. (c and d) IMS profiling of the Aβ [6–16] isomer variants, with the peptide sequence provided at the top of the panel. Individual Aβ [6–16] peptide isomer standards containing l-aspartic acid (l-Asp), d-aspartic acid (d-Asp), l-isoaspartic acid (l-isoAsp), and d-isoaspartic acid (d-isoAsp) were analyzed using a (c) 90 cm DTIMS-MS platform and (d) 5-cycle 67.5 m SLIM IMS-MS platform. The SLIM IMS-MS platform was able to baseline separate the isomers allowing accurate quantitation of each peptide. *Source:* (a and b) Reproduced with permission from Deng et al. [102], Copyright 2016 WILEY-VCH Verlag GmbH & Co. KGaA, Weinheim, and (c and d) from Zheng et al. [37] with permission from the Royal Society of Chemistry.

For example, D/L-form amino acids and isoAsp containing amyloid β peptide isomers, that were not separated by 90 cm DTIMS, were baseline separated in SLIM after five cycles (67.5 m) with a resolving power of ~350 [37]. The ability to fully separate these challenging Aβ peptide isomers is extremely beneficial for relative quantification of these specific peptides in a complex sample for Alzheimer's studies. The development of such rapid technique also supports the identification and quantification of D-form Asp and isoAsp residues in other proteins. Characterization of these isomers will enable a better way of determining molecular age and understanding the etiology of diseases. In addition to these advances, this technique will greatly facilitate studies of D-amino acid incorporation in peptides and allow a better appreciation of how often such modifications are occurring.

More recently, SLIM IMS-MS was shown to provide rapid and simultaneous characterization and separation of drug payload species on both the heavy and light chains of a model antibody–drug conjugate (ADC) [105]. ADCs, which consist of a monoclonal antibody that is covalently conjugated with a cytotoxic (anticancer) drug, have gained increased interest in biomedical and biopharmaceutical industry due to their promise to become alternative therapeutics to chemotherapy for cancers. However, it is very critical to characterize their drug-to-antibody ratios (DARs) or the average number of conjugated drugs to better evaluate the toxicity and efficacy of ADCs. This work demonstrated that SLIM IMS-based separations provide higher resolution of ADC species as compared to a commercial 1-m drift tube DTIMS-MS platform (Figure 6.8), which suggests SLIM IMS-MS-based measurements could provide new insights on how drug conjugation alters the structural properties of ADCs and thus help provide insights into their efficiency and selectivity as biotherapeutics.

The flexibility of SLIM opens the door to more sophisticated and extended ion manipulations. Multi-pass SLIM IMS modules provide capability for separating isomeric molecules with very similar structures. These devices allow the resolution to be further increased by passing ions multiple times through the extended serpentine path. Additionally, by integrating Compression Ratio Ion Mobility Programming (CRIMP) accumulation and compression with multi-pass serpentine path SLIM [106], it allows a great increase of the number of possible passes by occasionally compressing the diffused ion packets (i.e. broaden IMS peaks), thus not only improving the sensitivity but also providing much greater path lengths for even greater IMS resolution. Development of more sophisticated SLIM modules are ongoing, especially development of multilevel SLIM with ion elevators and escalators that enables wider mobility range ultrahigh-resolution ion mobility separations and expands on the ability of SLIM to obtain improved separations of complex mixtures with high sensitivity [107, 108]. The continuous development will further demonstrate the ability of SLIM to be utilized in new

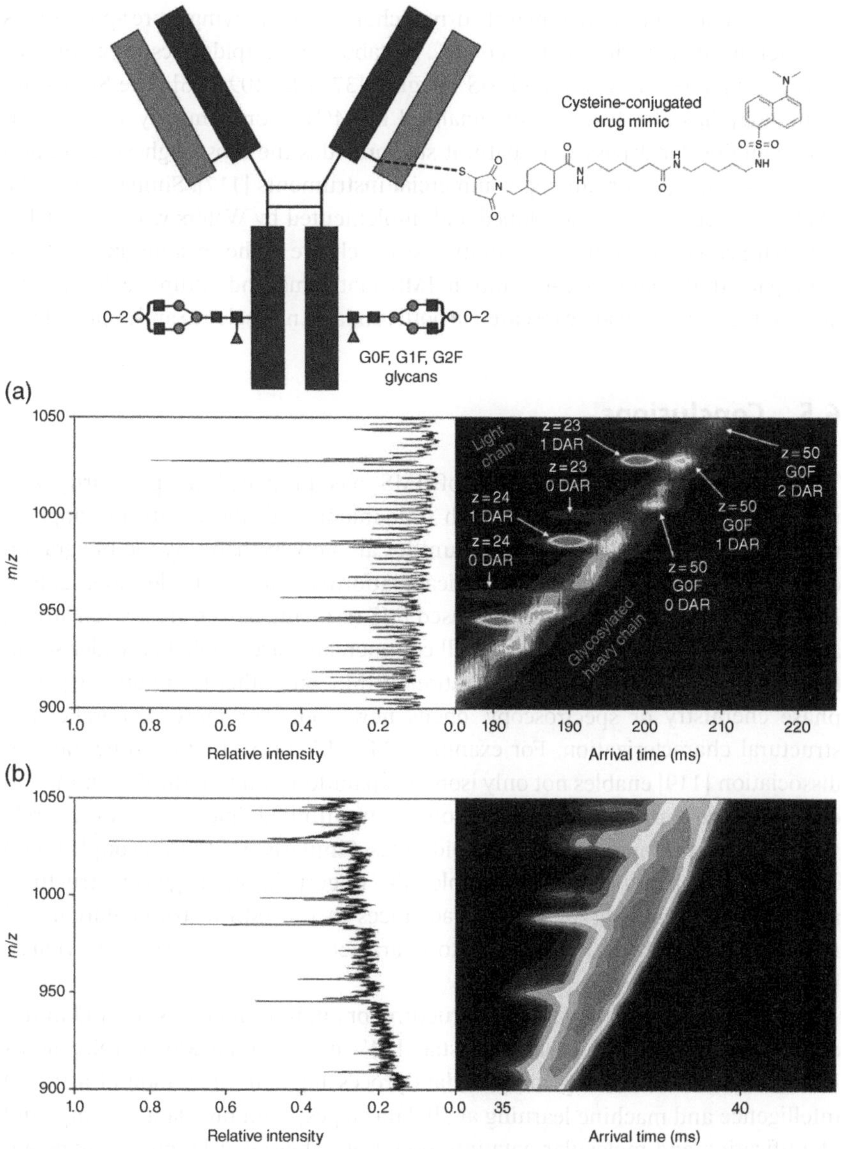

Figure 6.8 2D IMS-MS plot of the reduced antibody–drug conjugate after a 4.5 m SLIM IMS separation (a) and 1 m DTIMS separation (b). The light chain exhibits an overall lower charge state distribution compared to its heavy chain counterpart, as evidenced by less closely spaced IMS-MS peaks. *Source:* [105]/Reproduced with permission from American Chemical Society.

applications and address a range of current challenges, allowing more applications in different fields including proteomics, metabolomics, lipidomics, glycomics, as well as single cell research and MS imaging [37, 105, 109–116]. The SLIM-IMS technology has also been commercialized by MOBILion company. Although it only contains the 1-pass 13 m path, it still provides the most highest resolution (~300 resolving power) among commercial instruments [117]. Similarly, a cyclic TWIMS instrument was developed and implemented by Waters which has a 1 m path with capability to perform multi-pass to achieve higher resolution [118]. We anticipate that these high-resolution IMS platforms and future technological developments will advance research in both fundamental science and applications.

6.5 Conclusions

The versatility and orthogonality of IMS has propelled its popularity and utilization in the last decade, in both fundamental advancements and applications. Unique advantages and opportunities are now available by the integration of IMS in traditional MS-based analytical platforms. Continuous developments in IMS instrumentation including increased IMS resolving power, greater sensitivity, and alternative activation methods, all commercially accessible to a wider scientific community are expected. Integration of IMS with other techniques, e.g. gas-phase chemistry or spectroscopy, opens new doors for more comprehensive structural characterization. For example, IMS-MS coupled with ozone-induced dissociation [119] enables not only isomer separations but also the determination of carbon—carbon double bond positions in unsaturated lipids which were rarely reported by traditional LC-MS/MS lipidomics. Similarly, SLIM-IMS coupled with cryogenic infrared spectroscopy enables the fingerprinting of glycan structures and other molecules [120]. Further advances in the IMS instrumentation will reveal more structural information to characterize more molecules previously inaccessible.

The importance of software infrastructure for analysis of large scale and multi-dimensional IMS data cannot be overstated. We anticipate software developments in the areas of automating complex data processing and integration of artificial intelligence and machine learning at all data stages, from raw data to compound identification and molecular data interpretation. These developments are essential not only for streamlining existing high-throughput and omics data analysis workflows, but also to enable full utilization of the advancements in instrumentation and expansion of experimental applications. With the ever-increasing size and complexity of data that can be generated from these analyses, the availability of effective and user-friendly software tools that enable extraction of actionable biological insights is a critical bottleneck in need of continued and focused efforts.

High-Throughput IMS-MS analysis provides the ability to quickly perform molecule characterization and identification. While IMS-MS is not yet routinely utilized in drug discovery and pharmaceutical industry, there has been an increasing interest in high-throughput compound library screening and antibody characterization. With all the advancements and ongoing development in both IMS instrumentation technology and informatics, we foresee more and more exciting applications of high-throughput IMS analysis in different fields including omics studies, drug discovery, and clinical applications.

References

1 Mason, E. and McDaniel, E. (1988). *Transport Properites of Ions in Gases.* New York: Wiley.

2 Thomson, J.J. and Rutherford, E. (1896). XL. On the passage of electricity through gases exposed to Röntgen rays. *Lond. Edinb. Dublin Philos. Mag. J. Sci.* 42 (258): 392–407.

3 Rutherford, E. (1897). LIV. The velocity and rate of recombination of the ions of gases exposed to Röntgen radiation. *Lond. Edinb. Dublin Philos. Mag. J. Sci.* 44 (270): 422–440.

4 Cumeras, R., Figueras, E., Davis, C.E. et al. (2015). Review on ion mobility spectrometry. Part 1: current instrumentation. *Analyst* 140 (5): 1376–1390.

5 Mäkinen, M.A., Anttalainen, O.A., and Sillanpää, M.E.T. (2010). Ion mobility spectrometry and its applications in detection of chemical warfare agents. *Anal. Chem.* 82 (23): 9594–9600.

6 Armenta, S., Alcala, M., and Blanco, M. (2011). A review of recent, unconventional applications of ion mobility spectrometry (IMS). *Anal. Chim. Acta* 703 (2): 114–123.

7 Lapthorn, C., Pullen, F., and Chowdhry, B.Z. (2013). Ion mobility spectrometry-mass spectrometry (IMS-MS) of small molecules: separating and assigning structures to ions. *Mass Spectrom. Rev.* 32 (1): 43–71.

8 Lanucara, F., Holman, S.W., Gray, C.J., and Eyers, C.E. (2014). The power of ion mobility-mass spectrometry for structural characterization and the study of conformational dynamics. *Nat. Chem.* 6 (4): 281–294.

9 Chouinard, C.D., Wei, M.S., Beekman, C.R. et al. (2016). Ion mobility in clinical analysis: current progress and future perspectives. *Clin. Chem.* 62 (1): 124–133.

10 Cohen, M.J. and Karasek, F.W. (1970). Plasma Chromatography™ – a new dimension for gas chromatography and mass spectrometry. *J. Chromatogr. Sci.* 8 (6): 330–337.

11 Hoaglund, C.S., Valentine, S.J., Sporleder, C.R. et al. (1998). Three-dimensional ion mobility/TOFMS analysis of electrosprayed biomolecules. *Anal. Chem.* 70 (11): 2236–2242.

12 Wyttenbach, T., Kemper, P.R., and Bowers, M.T. (2001). Design of a new electrospray ion mobility mass spectrometer. *Int. J. Mass Spectrom.* 212 (1–3): 13–23.

13 Tang, K., Shvartsburg, A.A., Lee, H.N. et al. (2005). High-sensitivity ion mobility spectrometry/mass spectrometry using electrodynamic ion funnel interfaces. *Anal. Chem.* 77 (10): 3330–3339.

14 Ibrahim, Y.M., Baker, E.S., Danielson, W.F. et al. (2015). Development of a new ion mobility time-of-flight mass spectrometer. *Int. J. Mass Spectrom.* 377: 655–662.

15 May, J.C., Goodwin, C.R., Lareau, N.M. et al. (2014). Conformational ordering of biomolecules in the gas phase: nitrogen collision cross sections measured on a prototype high resolution drift tube ion mobility-mass spectrometer. *Anal. Chem.* 86 (4): 2107–2116.

16 Pringle, S.D., Giles, K., Wildgoose, J.L. et al. (2007). An investigation of the mobility separation of some peptide and protein ions using a new hybrid quadrupole/travelling wave IMS/oa-ToF instrument. *Int. J. Mass Spectrom.* 261 (1): 1–12.

17 Michelmann, K., Silveira, J.A., Ridgeway, M.E., and Park, M.A. (2015). Fundamentals of trapped ion mobility spectrometry. *J. Am. Soc. Mass Spectrom.* 26 (1): 14–24.

18 Ewing, M.A., CRP, C., Zucker, S.M. et al. (2015). Selected overtone mobility spectrometry. *Anal. Chem.* 87 (10): 5132–5138.

19 Zucker, S.M., Ewing, M.A., and Clemmer, D.E. (2013). Gridless overtone mobility spectrometry. *Anal. Chem.* 85 (21): 10174–10179.

20 Rus, J., Moro, D., Sillero, J.A. et al. (2010). IMS–MS studies based on coupling a differential mobility analyzer (DMA) to commercial API–MS systems. *Int. J. Mass Spectrom.* 298 (1–3): 30–40.

21 Brown, L.J. and Creaser, C.S. (2013). Field asymmetric waveform ion mobility spectrometry analysis of proteins and peptides: a review. *Curr. Anal. Chem.* 9 (2): 192–198.

22 Guevremont, R. (2004). High-field asymmetric waveform ion mobility spectrometry: a new tool for mass spectrometry. *J. Chromatogr. A* 1058 (1, 2): 3–19.

23 Kolakowski, B.M. and Mester, Z. (2007). Review of applications of high-field asymmetric waveform ion mobility spectrometry (FAIMS) and differential mobility spectrometry (DMS). *Analyst* 132 (9): 842–864.

24 Vidal-de-Miguel, G., Macía, M., and Cuevas, J. (2012). Transversal modulation ion mobility spectrometry (TM-IMS), a new mobility filter overcoming turbulence related limitations. *Anal. Chem.* 84 (18): 7831–7837.

25 Kanu, A.B., Dwivedi, P., Tam, M. et al. (2008). Ion mobility–mass spectrometry. *J. Mass Spectrom.* 43 (1): 1–22.

26 Cumeras, R., Figueras, E., Davis, C.E. et al. (2015). Review on ion mobility spectrometry. Part 2: hyphenated methods and effects of experimental parameters. *Analyst* 140 (5): 1391–1410.

27 May, J.C. and McLean, J.A. (2015). Ion mobility-mass spectrometry: time-dispersive instrumentation. *Anal. Chem.* 87 (3): 1422–1436.

28 Dodds, J.N. and Baker, E.S. (2019). Ion mobility spectrometry: fundamental concepts, instrumentation, applications, and the road ahead. *J. Am. Soc. Mass Spectrom.* 30 (11): 2185–2195.

29 Gabelica, V., Shvartsburg, A.A., Afonso, C. et al. (2019). Recommendations for reporting ion mobility mass spectrometry measurements. *Mass Spectrom. Rev.* 38 (3): 291–320.

30 Gidden, J., Baker, E.S., Ferzoco, A., and Bowers, M.T. (2005). Structural motifs of DNA complexes in the gas phase. *Int. J. Mass Spectrom.* 240 (3): 183–193.

31 Mason, E.A. and McDaniel, E.W. (2005). *Kinetic Theory of Mobility and Diffusion: Sections 5.1–5.2. Transport Properties of Ions in Gases*, 137–193. Weinheim: Wiley-VCH Verlag GmbH & Co. KGaA.

32 Zheng, X., Aly, N.A., Zhou, Y. et al. (2017). A structural examination and collision cross section database for over 500 metabolites and xenobiotics using drift tube ion mobility spectrometry. *Chem. Sci.* 8 (11): 7724–7736.

33 Zheng, X., Smith, F.B., Aly, N.A. et al. (2019). Evaluating the structural complexity of isomeric bile acids with ion mobility spectrometry. *Anal. Bioanal. Chem.* 411 (19): 4673–4682.

34 Zheng, X., Zhang, X., Schocker, N.S. et al. (2017). Enhancing glycan isomer separations with metal ions and positive and negative polarity ion mobility spectrometry-mass spectrometry analyses. *Anal. Bioanal. Chem.* 409 (2): 467–476.

35 Kyle, J.E., Aly, N., Zheng, X. et al. (2018). Evaluating lipid mediator structural complexity using ion mobility spectrometry combined with mass spectrometry. *Bioanalysis* 10 (5): 279–289.

36 Zheng, X., Dupuis, K.T., Aly, N.A. et al. (2018). Utilizing ion mobility spectrometry and mass spectrometry for the analysis of polycyclic aromatic hydrocarbons, polychlorinated biphenyls, polybrominated diphenyl ethers and their metabolites. *Anal. Chim. Acta* 1037: 265–273.

37 Zheng, X., Deng, L., Baker, E.S. et al. (2017). Distinguishing d- and l-aspartic and isoaspartic acids in amyloid β peptides with ultrahigh resolution ion mobility spectrometry. *Chem. Commun.* 53 (56): 7913–7916.

38 Zheng, X., Renslow, R.S., Makola, M.M. et al. (2017). Structural elucidation of *cis/trans* dicaffeoylquinic acid photoisomerization using ion mobility spectrometry-mass spectrometry. *J. Phys. Chem. Lett.* 8: 1381–1388.

39 Bernstein, S.L., Dupuis, N.F., Lazo, N.D. et al. (2009). Amyloid-β protein oligomerization and the importance of tetramers and dodecamers in the aetiology of Alzheimer's disease. *Nat. Chem.* 1 (4): 326–331.

40 Liu, Y.S., Valentine, S.J., Counterman, A.E. et al. (1997). Injected-ion mobility analysis of biomolecules. *Anal. Chem.* 69 (23): A728–A735.

41 Bush, M.F., Hall, Z., Giles, K. et al. (2010). Collision cross sections of proteins and their complexes: a calibration framework and database for gas-phase structural biology. *Anal. Chem.* 82 (22): 9557–9565.

42 Paglia, G., Williams, J.P., Menikarachchi, L. et al. (2014). Ion mobility derived collision cross sections to support metabolomics applications. *Anal. Chem.* 86 (8): 3985–3993.

43 Zhou, Z., Shen, X., Tu, J., and Zhu, Z.-J. (2016). Large-scale prediction of collision cross-section values for metabolites in ion mobility-mass spectrometry. *Anal. Chem.* 88 (22): 11084–11091.

44 Hines, K.M., Ross, D.H., Davidson, K.L. et al. (2017). Large-scale structural characterization of drug and drug-like compounds by high-throughput ion mobility-mass spectrometry. *Anal. Chem.* 89 (17): 9023–9030.

45 Hines, K.M. and Xu, L. (2019). Lipidomic consequences of phospholipid synthesis defects in *Escherichia coli* revealed by HILIC-ion mobility-mass spectrometry. *Chem. Phys. Lipids* 219: 15–22.

46 Dodds, J.N., Hopkins, Z.R., Knappe, D.R.U., and Baker, E.S. (2020). Rapid characterization of per- and polyfluoroalkyl substances (PFAS) by ion mobility spectrometry–mass spectrometry (IMS-MS). *Anal. Chem.* 92 (6): 4427–4435.

47 McCabe, J.W., Mallis, C.S., Kocurek, K.I. et al. (2020). First-principles collision cross section measurements of large proteins and protein complexes. *Anal. Chem.* 92 (16): 11155–11163.

48 Ross, D.H., Cho, J.H., and Xu, L. (2020). Breaking down structural diversity for comprehensive prediction of ion-neutral collision cross sections. *Anal. Chem.* 92 (6): 4548–4557.

49 Schroeder, M., Meyer, S.W., Heyman, H.M. et al. (2020). Generation of a collision cross section library for multi-dimensional plant metabolomics using UHPLC-trapped ion mobility-MS/MS. *Metabolites* 10 (1): 13.

50 Vasilopoulou, C.G., Sulek, K., Brunner, A.-D. et al. (2020). Trapped ion mobility spectrometry and PASEF enable in-depth lipidomics from minimal sample amounts. *Nat. Commun.* 11 (1): 331.

51 Zhou, Z., Luo, M., Chen, X. et al. (2020). Ion mobility collision cross-section atlas for known and unknown metabolite annotation in untargeted metabolomics. *Nat. Commun.* 11 (1): 4334.

52 Ross, D.H., Seguin, R.P., Krinsky, A.M., and Xu, L. (2022). High-throughput measurement and machine learning-based prediction of collision cross sections for drugs and drug metabolites. *J. Am. Soc. Mass Spectrom.* 33 (6): 1061–1072.

53 Stow, S.M., Causon, T.J., Zheng, X. et al. (2017). An interlaboratory evaluation of drift tube ion mobility-mass spectrometry collision cross section measurements. *Anal. Chem.* 89 (17): 9048–9055.

54 Wink, M. (2003). Evolution of secondary metabolites from an ecological and molecular phylogenetic perspective. *Phytochemistry* 64 (1): 3–19.

55 Ma, J., Casey, C.P., Zheng, X. et al. (2017). PIXiE: an algorithm for automated ion mobility arrival time extraction and collision cross section calculation using global data association. *Bioinformatics* 33: 1–8.

56 Lee, J.-Y., Bilbao, A., Conant, C.R. et al. (2021). AutoCCS: automated collision cross-section calculation software for ion mobility spectrometry–mass spectrometry. *Bioinformatics* 37 (22): 4193–4201.
57 Zhou, Z., Xiong, X., and Zhu, Z.-J. (2017). MetCCS predictor: a web server for predicting collision cross-section values of metabolites in ion mobility-mass spectrometry based metabolomics. *Bioinformatics* 33 (14): 2235–2237.
58 Zhou, Z., Tu, J., Xiong, X. et al. (2017). LipidCCS: prediction of collision cross-section values for lipids with high precision to support ion mobility–mass spectrometry-based lipidomics. *Anal. Chem.* 89 (17): 9559–9566.
59 Ross, D.H., Cho, J.H., Zhang, R. et al. (2020). LiPydomics: a python package for comprehensive prediction of lipid collision cross sections and retention times and analysis of ion mobility-mass spectrometry-based lipidomics data. *Anal. Chem.* 92 (22): 14967–14975.
60 Plante, P.-L., Francovic-Fontaine, É., May, J.C. et al. (2019). Predicting ion mobility collision cross-sections using a deep neural network: DeepCCS. *Anal. Chem.* 91 (8): 5191–5199.
61 Colby, S.M., Nuñez, J.R., Hodas, N.O. et al. (2020). Deep learning to generate in silico chemical property libraries and candidate molecules for small molecule identification in complex samples. *Anal. Chem.* 92 (2): 1720–1729.
62 Ross, D.H. and Xu, L. (2021). Determination of drugs and drug metabolites by ion mobility-mass spectrometry: a review. *Anal. Chim. Acta* 1154: 338270.
63 Colby, S.M., Thomas, D.G., Nuñez, J.R. et al. (2019). ISiCLE: a quantum chemistry pipeline for establishing in silico collision cross section libraries. *Anal. Chem.* 91 (7): 4346–4356.
64 Meier, F., Köhler, N.D., Brunner, A.-D. et al. (2021). Deep learning the collisional cross sections of the peptide universe from a million experimental values. *Nat. Commun.* 12 (1): 1185.
65 Sowell, R.A., Koeniger, S.L., Valentine, S.J. et al. (2004). Nanoflow LC/IMS-MS and LC/IMS-CID/MS of protein mixtures. *J. Am. Soc. Mass Spectrom.* 15 (9): 1341–1353.
66 Baker, E.S., Livesay, E.A., Orton, D.J. et al. (2010). An LC-IMS-MS platform providing increased dynamic range for high-throughput proteomic studies. *J. Proteome Res.* 9 (2): 997–1006.
67 Counterman, A.E., Hilderbrand, A.E., Barnes, C.A.S., and Clemmer, D.E. (2001). Formation of peptide aggregates during ESI: size, charge, composition, and contributions to noise. *J. Am. Soc. Mass Spectrom.* 12 (9): 1020–1035.
68 Valentine, S.J., Plasencia, M.D., Liu, X.Y. et al. (2006). Toward plasma proteome profiling with ion mobility-mass spectrometry. *J. Proteome Res.* 5 (11): 2977–2984.
69 Causon, T.J. and Hann, S. (2015). Theoretical evaluation of peak capacity improvements by use of liquid chromatography combined with drift tube ion mobility-mass spectrometry. *J. Chromatogr. A* 1416: 47–56.

70 Kyle, J.E., Zhang, X., Weitz, K.K. et al. (2016). Uncovering biologically significant lipid isomers with liquid chromatography, ion mobility spectrometry and mass spectrometry. *Analyst* 141 (5): 1649–1659.

71 Paglia, G., Angel, P., Williams, J.P. et al. (2015). Ion mobility-derived collision cross section as an additional measure for lipid fingerprinting and identification. *Anal. Chem.* 87 (2): 1137–1144.

72 Lareau, N.M., May, J.C., and McLean, J.A. (2015). Non-derivatized glycan analysis by reverse phase liquid chromatography and ion mobility-mass spectrometry. *Analyst* 140 (10): 3335–3338.

73 Sowell, R.A., Koeniger, S.L., Valentine, S.J. et al. (2004). Nanoflow LCAMS-MS and LCAMS-CID/MS of protein mixtures. *J. Am. Soc. Mass Spectrom.* 15 (9): 1341–1353.

74 Valentine, S.J., Kulchania, M., Barnes, C.A.S., and Clemmer, D.E. (2001). Multidimensional separations of complex peptide mixtures: a combined high-performance liquid chromatography/ion mobility/time-of-flight mass spectrometry approach. *Int. J. Mass Spectrom.* 212 (1-3): 97–109.

75 Liu, X.Y., Valentine, S.J., Plasencia, M.D. et al. (2007). Mapping the human plasma proteome by SCX-LC-IMS-MS. *J. Am. Soc. Mass Spectrom.* 18 (7): 1249–1264.

76 Moon, M.H., Myung, S., Plasencia, M. et al. (2003). Nanoflow LC/ion mobility/CID/TOF for proteomics: analysis of a human urinary proteome. *J. Proteome Res.* 2 (6): 589–597.

77 Baker, E.S., Burnum-Johnson, K.E., Ibrahim, Y.M. et al. (2015). Enhancing bottom-up and top-down proteomic measurements with ion mobility separations. *Proteomics* 15 (16): 2766–2776.

78 Baker, E.S., Burnum-Johnson, K.E., Jacobs, J.M. et al. (2014). Advancing the high-throughput identification of liver fibrosis protein signatures using multiplexed ion mobility spectrometry. *Mol. Cell. Proteomics* 13 (4): 1119–1127.

79 Sarafraz-Yazdi, A. and Amiri, A. (2010). Liquid-phase microextraction. *TrAC, Trends Anal. Chem.* 29 (1): 1–14.

80 Andrade-Eiroa, A., Canle, M., Leroy-Cancellieri, V., and Cerdà, V. (2016). Solid-phase extraction of organic compounds: a critical review (Part I). *TrAC, Trends Anal. Chem.* 80: 641–654.

81 Andrade-Eiroa, A., Canle, M., Leroy-Cancellieri, V., and Cerdà, V. (2016). Solid-phase extraction of organic compounds: a critical review. part II. *TrAC, Trends Anal. Chem.* 80: 655–667.

82 Lim, K.B., Özbal, C.C., and Kassel, D.B. (2010). Development of a high-throughput online solid-phase extraction/tandem mass spectrometry method for cytochrome P450 inhibition screening. *J. Biomol. Screening* 15 (4): 447–452.

83 Jian, W., Romm, M.V., Edom, R.W. et al. (2011). Evaluation of a high-throughput online solid phase extraction–tandem mass spectrometry system for in vivo bioanalytical studies. *Anal. Chem.* 83 (21): 8259–8266.

84 Hutchinson, S.E., Leveridge, M.V., Heathcote, M.L. et al. (2012). Enabling lead discovery for histone lysine demethylases by high-throughput RapidFire mass spectrometry. *J. Biomol. Screening* 17 (1): 39–48.

85 Lowe, D.M., Gee, M., Haslam, C. et al. (2014). Lead discovery for human kynurenine 3-monooxygenase by high-throughput RapidFire mass spectrometry. *J. Biomol. Screening* 19 (4): 508–515.

86 Rye, P.T. and LaMarr, W.A. (2015). Measurement of glycolysis reactants by high-throughput solid phase extraction with tandem mass spectrometry: characterization of pyrophosphate-dependent phosphofructokinase as a case study. *Anal. Biochem.* 482: 40–47.

87 Sawyer, W.S., Srikumar, N., Carver, J. et al. (2020). High-throughput antibody screening from complex matrices using intact protein electrospray mass spectrometry. *Proc. Natl. Acad. Sci. U.S.A.* 117 (18): 9851–9856.

88 Zhang, X., Romm, M., Zheng, X. et al. (2016). SPE-IMS-MS: an automated platform for sub-sixty second surveillance of endogenous metabolites and xenobiotics in biofluids. *Clin. Mass Spectrom.* 2: 1–10.

89 MacLean, B., Tomazela, D.M., Shulman, N. et al. (2010). Skyline: an open source document editor for creating and analyzing targeted proteomics experiments. *Bioinformatics* 26 (7): 966–968.

90 MacLean, B.X., Pratt, B.S., Egertson, J.D. et al. (2018). Using skyline to analyze data-containing liquid chromatography, ion mobility spectrometry, and mass spectrometry dimensions. *J. Am. Soc. Mass Spectrom.* 29 (11): 2182–2188.

91 Kirkwood, K.I., Pratt, B.S., Shulman, N. et al. (2022). Utilizing skyline to analyze lipidomics data containing liquid chromatography, ion mobility spectrometry and mass spectrometry dimensions. *Nat. Protoc.* **17**: 2415–2430.

92 Tsugawa, H., Ikeda, K., Takahashi, M. et al. (2020). A lipidome atlas in MS-DIAL 4. *Nat. Biotechnol.* 38 (10): 1159–1163.

93 Pluskal, T., Castillo, S., Villar-Briones, A., and Orešič, M. (2010). MZmine 2: modular framework for processing, visualizing, and analyzing mass spectrometry-based molecular profile data. *BMC Bioinform.* 11 (1): 395.

94 Meier, F., Brunner, A.-D., Frank, M. et al. (2020). diaPASEF: parallel accumulation–serial fragmentation combined with data-independent acquisition. *Nat. Methods* 17 (12): 1229–1236.

95 Zhou, Z., Shen, X., Chen, X. et al. (2019). LipidIMMS analyzer: integrating multi-dimensional information to support lipid identification in ion mobility – mass spectrometry based lipidomics. *Bioinformatics* 35 (4): 698–700.

96 Bilbao A, Ross DH, Lee J-Y, Donor M, Williams S, Zhu Y, et al. MZA: a data conversion tool to facilitate software development and artificial intelligence research in multidimensional mass spectrometry. J. Proteome Res. 2023; 22 (2), 508–513. doi: https://doi.org/10.1021/acs.jproteome.2c00313

97 Bilbao, A., Gibbons, B.C., Stow, S.M. et al. (2022). A preprocessing tool for enhanced ion mobility–mass spectrometry-based omics workflows. *J. Proteome Res.* 21 (3): 798–807.

98 Colby, S.M., Chang, C.H., Bade, J.L. et al. (2022). DEIMoS: an open-source tool for processing high-dimensional mass spectrometry data. *Anal. Chem.* 94 (16): 6130–6138.

99 Revercomb, H.E. and Mason, E.A. (1975). Theory of plasma chromatography/ gaseous electrophoresis. *Rev. Anal. Chem.* 47 (7): 970–983.

100 Kemper, P.R., Dupuis, N.F., and Bowers, M.T. (2009). A new, higher resolution, ion mobility mass spectrometer. *Int. J. Mass Spectrom.* 287 (1–3): 46–57.

101 Merenbloom, S.I., Glaskin, R.S., Henson, Z.B., and Clemmer, D.E. (2009). High-resolution ion cyclotron mobility spectrometry. *Anal. Chem.* 81 (4): 1482–1487.

102 Deng, L., Ibrahim, Y.M., Baker, E.S. et al. (2016). Ion mobility separations of isomers based upon long path length structures for lossless ion manipulations combined with mass spectrometry. *ChemistrySelect* 1 (10): 2396–2399.

103 Deng, L., Ibrahim, Y., Hamid, A.M. et al. (2016). Ultrahigh resolution ion mobility separations utilizing traveling waves in a 13-m serpentine path length structures for lossless ion manipulations module. *Anal. Chem.* 88: 8957–8964.

104 Deng, L., Webb, I.K., Garimella, S.V.B. et al. (2017). Serpentine ultralong path with extended routing (SUPER) high resolution traveling wave ion mobility-MS using structures for lossless ion manipulations. *Anal. Chem.* 89 (8): 4628–4634.

105 Nagy, G., Attah, I.K., Conant, C.R. et al. (2020). Rapid and simultaneous characterization of drug conjugation in heavy and light chains of a monoclonal antibody revealed by high-resolution ion mobility separations in SLIM. *Anal. Chem.* 92 (7): 5004–5012.

106 Deng, L., Garimella, S.V.B., Hamid, A.M. et al. (2017). Compression ratio ion mobility programming (CRIMP) accumulation and compression of billions of ions for ion mobility-mass spectrometry using traveling waves in structures for lossless ion manipulations (SLIM). *Anal. Chem.* 89 (12): 6432–6439.

107 Ibrahim, Y.M., Hamid, A.M., Cox, J.T. et al. (2017). Ion elevators and escalators in multilevel structures for lossless ion manipulations. *Anal. Chem.* 89 (3): 1972–1977.

108 Hollerbach, A.L., Li, A., Prabhakaran, A. et al. (2020). Ultrahigh resolution ion mobility separations over extended path lengths and mobility ranges achieved using a multilevel structures for lossless ion manipulations (SLIM) module. *Anal. Chem.* 92 (11): 7972–7979.

109 Wojcik, R., Webb, I.K., Deng, L. et al. (2017). Lipid and glycolipid isomer analyses using ultrahigh resolution ion mobility spectrometry separations. *Int. J. Mol. Sci.* 18 (1): 183.

110 Chouinard, C.D., Nagy, G., Webb, I.K. et al. (2018). Rapid ion mobility separations of bile acid isomers using cyclodextrin adducts and structures for lossless ion manipulations. *Anal. Chem.* 90 (18): 11086–11091.

111 Nagy, G., Chouinard, C.D., Attah, I.K. et al. (2018). Distinguishing enantiomeric amino acids with chiral cyclodextrin adducts and structures for lossless ion manipulations. *Electrophoresis* 39 (24): 3148–3155.

112 Chouinard, C.D., Nagy, G., Webb, I.K. et al. (2018). Improved sensitivity and separations for phosphopeptides using online liquid chromotography coupled with structures for lossless ion manipulations ion mobility–mass spectrometry. *Anal. Chem.* 90 (18): 10889–10896.

113 Dou, M., Chouinard, C.D., Zhu, Y. et al. (2019). Nanowell-mediated multidimensional separations combining nanoLC with SLIM IM-MS for rapid, high-peak-capacity proteomic analyses. *Anal. Bioanal. Chem.* 411 (21): 5363–5372.

114 Nagy, G., Kedia, K., Attah, I.K. et al. (2019). Separation of β-amyloid tryptic peptide species with isomerized and racemized l-aspartic residues with ion mobility in structures for lossless ion manipulations. *Anal. Chem.* 91 (7): 4374–4380.

115 Nagy, G., Veličković, D., Chu, R.K. et al. (2019). Towards resolving the spatial metabolome with unambiguous molecular annotations in complex biological systems by coupling mass spectrometry imaging with structures for lossless ion manipulations. *Chem. Commun.* 55 (3): 306–309.

116 Wojcik, R., Nagy, G., Attah, I.K. et al. (2019). SLIM ultrahigh resolution ion mobility spectrometry separations of isotopologues and isotopomers reveal mobility shifts due to mass distribution changes. *Anal. Chem.* 91 (18): 11952–11962.

117 May, J.C., Leaptrot, K.L., Rose, B.S. et al. (2021). Resolving power and collision cross section measurement accuracy of a prototype high-resolution ion mobility platform incorporating structures for lossless ion manipulation. *J. Am. Soc. Mass Spectrom.* 32 (4): 1126–1137.

118 Giles, K., Ujma, J., Wildgoose, J. et al. (2019). A cyclic ion mobility-mass spectrometry system. *Anal. Chem.* 91 (13): 8564–8573.

119 Poad, B.L.J., Zheng, X., Mitchell, T.W. et al. (2018). Online ozonolysis combined with ion mobility-mass spectrometry provides a new platform for lipid isomer analyses. *Anal. Chem.* 90 (2): 1292–1300.

120 Pellegrinelli, R.P., Yue, L., Carrascosa, E. et al. (2022). A new strategy coupling ion-mobility-selective CID and cryogenic IR spectroscopy to identify glycan anomers. *J. Am. Soc. Mass Spectrom.* 33 (5): 859–864.

7

Differential Mobility Spectrometry and Its Application to High-Throughput Analysis

Bradley B. Schneider[1], *Leigh Bedford*[1], *Chang Liu*[1], *Eva Duchoslav*[1], *Yang Kang*[1], *Subhasish Purkayastha*[2], *Aaron Stella*[2], *and Thomas R. Covey*[1]

[1] *SCIEX, Concord, ON, Canada*
[2] *SCIEX, Framingham, MA, USA*

7.1 Introduction

It is fair to say that the developments toward ultrahigh-throughput mass spectrometry over the past three decades have largely been driven by the evolving requirements of the drug discovery process. Technologies were developed in the pharmaceutical industry to dramatically increase the rate at which chemical entities could be synthesized, chemical libraries grew into the tens of millions of compounds, and the numbers of validated biological targets as well as physical–chemical tests predictive of a chemical's efficacy rapidly expanded. In order to screen these libraries with the multitude of tests being developed, quantitative measurements on tens to hundreds of thousands of samples per day are in order.

HPLC emerged over this time as the premier method of sample introduction into the mass spectrometer and will continue to be so. This is largely because it separates the components of a sample on the basis of their chemical properties, enabling substances of identical mass to be distinguished to the extent that isobaric isomers and even enantiomers can be distinguished with HPLC. Because HPLC and mass spectrometry have orthogonal separation mechanisms, one based on chemistry, one based on nuclear mass, the universe of chemical space that can be interrogated has been enormously expanded compared to what was possible before the big bang in 1989 when it first became commercially available.

However, because of the physical principles HPLC employs, it presents limitations to analysis speed. Diffusion rates of compounds from the surfaces and pores of stationary phases lie at the core of this limitation. Despite the myriad of

High-Throughput Mass Spectrometry in Drug Discovery, First Edition. Edited by Chang Liu and Hui Zhang.
© 2023 John Wiley & Sons, Inc. Published 2023 by John Wiley & Sons, Inc.

advancements in HPLC technology over this time frame, discussed in Chapter 1 of this book, it remains as the bottleneck to achieving ultrahigh-throughput analyses where sampling rates ≥1 Hz are required.

Ion mobility is a separation technique that has the speed to accommodate the duty cycle of a sample introduction technology that operates at this frequency. It has also been shown to be relatively straightforward to interface to mass spectrometry. Ion mobility is a reasonable candidate to substitute for the functions of the HPLC in situations where ultrahigh-throughput is the goal.

There are currently two different types of ion mobility in the commercial sphere that have shown utility coupled to mass spectrometry. The first and oldest form is classical low field mobility based on the time of flight of an ion over a distance and having several variations on the design of the drift tube and electric fields, which is detailed in Chapter 6 of this book. The second is differential mobility based on the difference in an ion's mobility in high and low electric fields. There are two variants, high field asymmetric waveform ion mobility spectrometry (FAIMS) and differential mobility spectrometry (DMS).

The theme of this chapter is an analysis of the speed and selectivity characteristics of the three mobility-MS approaches and demonstration that DMS is uniquely suited to substitute for the HPLC in ultrahigh-throughput-targeted quantitation situations. The sample introduction technology used here is acoustic ejection mass spectrometry (AEMS) which has shown that a 1 Hz sampling frequency is routinely achievable (more details about AEMS system are provided in Chapter 5 of this book) [1–3] but has the potential to be able to exceed those speeds by over 10-fold [4].

7.2 Separation Speed

7.2.1 Classical Low Field Ion Mobility

At high pressure, ions or charged particles in the presence of an electric field (E) will drift with a speed given by $v = KE$, where K is the ion mobility. Under the effect of low electric fields, the ion mobility is a constant value for a given ion that can be estimated from the Mason–Schamp equation. Classical ion mobility spectrometry (IMS) devices include an ion source and a gating device to pulse ions into a drift tube with constant electric field [5, 6]. Ions with differences in their mobility drift down the flight tube at different speeds, resulting in a range of arrival times at the exit.

A number of variants of IMS devices have been developed over the years based upon static or dynamic electric fields and include different combinations of electric field and gas flow vectors [7–9]. Various configurations include devices with electric field and gas flow vectors opposing each other or oriented in the same direction, as well as devices where the electric field and gas flow are oriented orthogonal to each other. One similarity of all of these devices is that the separation mechanism depends upon differences in ion mobility for charged species, and the ion mobility is related to mass. This means that IMS devices separate charged species in a similar manner to a low-resolution mass spectrometer, where ions with very different m/z can be separated easily, while isobaric species with the same charge state have similar ion mobility values and therefore drift with similar speeds. To alleviate this issue, ion mobility devices with very long path lengths [10] have been designed, and these devices can include extended drift tubes [11] or cyclic mobility analyzers [12] where ions can make multiple cycles to improve separation.

The separation capability for an IMS is generally referred to as resolving power, and this can be defined as the ratio of drift time/FWHM of the mobility peak. Longer drift tubes improve resolving power by extending the drift time; however, the gain is generally not a linear function of drift tube length because diffusional broadening may also widen the mobility peak. For species with very different m/z, IMS devices with low resolving power are generally sufficient, and separations can be achieved in a few milliseconds. However, isobaric species have more similar ion mobility values and the separation time requirements can extend to 10s or 100s of milliseconds. In the case of structural isomers, such as diastereomers, the necessary separation time can extend to seconds or 10s of seconds, if possible at all. The pulsed nature of IMS devices imparts additional difficulties for coupling with continuous mass analyzers such as triple quadrupole instruments.

7.2.2 Differential Mobility Spectrometry

DMS is an alternate approach for ion mobility separations where the separation mechanism depends upon the difference in mobility for a given ion under different electric fields [13, 14]. As described previously, the ion mobility is a constant for a given ion at low electric fields. However, as the electric field increases to approximately 10 Td or greater, the mobility becomes field-dependent, and is no longer constant. Differential mobility devices exploit this effect by applying a custom separation waveform between two equally spaced plates to cause ions to oscillate in a direction perpendicular to a transport gas flow between the plates. The separation waveform includes a short high field component and a longer low field component of the opposite polarity. An ion within the space between the two plates will drift a distance ($d1$) toward one of the plates during the high field portion of the waveform and a distance ($d2$) toward the other plate during the low field portion of the waveform. This induces a zig-zag trajectory for a given ion between the two plates with a net displacement toward one or the other of the plates. A small DC potential (referred to as a compensation voltage, CoV) applied between the plates can correct the trajectory for a given ion to allow it to pass through the analyzer. Under these conditions, the trajectory for a given ion depends upon the difference in mobility between the high and low field portions of the waveform, and the normalized difference between the high and low field mobility $(K(E)-K(0))/K(0)$ is referred to as alpha. While low field mobility is related to mass, the difference between an ion's high and low field mobilities depends to a greater extent upon factors such as chemistry of the ion/neutral interaction. As a result, differential mobility separations are far more orthogonal to mass than traditional IMS separations.

The concept of differential mobility separations was originally conceived in the former Soviet Union in the early 1980s [15] and the work diverged into two different analyzer designs [14, 16, 17]. The first involves application of the separation voltage to electrodes with curved surfaces and this approach has been described as FAIMS, and the second involves application of the separation voltage to flat parallel plates and has been described as DMS. The electrode structure (curved versus flat) imparts significant differences to the separation characteristics of the analyzer and this will be discussed in more detail in Sections 7.2.2.1 and 7.2.2.2.

7.2.2.1 FAIMS

The term FAIMS generally denotes differential mobility separations using curved electrodes with either concentric cylinders or spherical electrodes [18]. In either case, the curved electrode structure gives inhomogeneous fields between the two electrodes which can result in focusing or defocusing effects depending upon the polarity of the ion, the separation waveform and the CoV. These effects have been described in detail by Krylov [19]. The concept of applying differential mobility separation waveforms to curved surfaces originated in the former Soviet Union and the technology migrated to North America as a detector for dangerous chemicals in the mining industry, with the commercialization of stand-alone sensors by a company known as Mine Safety Appliances. While the original Russian work did include combination of the technology with electrospray ionization and mass spectrometry, the work of Roger Guevremont and his colleagues at the National Research Council in Ottawa, Canada, advanced the technology for routine ESI-MS [20–22]. The original hardware developed by Roger Guevremont and his subsequent company (Ionalytics Corporation) included cylindrical FAIMS hardware designed to capitalize on the focusing effect of nonhomogeneous fields. One of the drawbacks with this earlier design was the ratio of the inner volume of the cylindrical device to the transport gas flow rate, resulting in very long residence times on the order of 200 ms [22]. When analyzing multiple compounds within a single method, the transit time requirement to equilibrate between alternating analysis times rendered these devices unsuitable for high-throughput MS.

In 2005, the Ionalytics Corporation and technology was purchased by Thermo, and new devices were engineered with a goal to reduce residence time. The new geometry transported ions around the center of the cylinder, rather than down the entire length, and this reduced the residence time into the ballpark of 100 ms [23, 24]. The new geometry also included means to heat the two cylindrical electrodes separately, to offer a new mode of operation to reduce the inhomogeneity of the separation field. This reduced the focusing effect and enabled the use of chemically modified transport gases, albeit with substantially reduced resolution relative to planar DMS devices that have true homogeneous field distributions [25]. While these devices achieved a substantial improvement in residence time, the requirement for 100 ms pause between measuring two MRM transitions was still too long for high-throughput MS. Further work has improved the gas flow dynamics for cylindrical devices [26]; however, the necessary residence time is still substantially longer than typical planar DMS devices, and approximately 2 orders of magnitude longer than micromachined devices [27].

7.2.2.2 DMS

Research work on differential mobility separations in the former Soviet Union diverged in two directions with different groups in Tashkent and Novosibirsk [28]. In addition to the cylindrical FAIMS work, research was also conducted using planar electrodes, and this innovation provided significant advantages in comparison to using curved electrodes. In the planar variant, the separation field between the two plates is homogeneous and therefore complications arising from focusing and defocusing effects are negligible. There is no requirement for a specific polarity to the waveform and the peak broadening effects that limit resolution with the cylindrical variant are nonexistent. The absence of focusing effects imparts ion losses in planar devices as a result of diffusion to the electrode surfaces, but these effects can be minimized by designing devices with very short residence time. Planar DMS technology and some of the original Russian developers migrated to North America through the laboratory of Professor Gary Eiceman at New Mexico State University. This group worked on the development of DMS sensors for field-based applications and eventually collaborated with others on combined DMS/MS interfaces [29–32].

There is a high degree of flexibility in DMS design; the spacing between the electrodes (defined as gap height) dictates the magnitude of the CoV that is necessary to restore the trajectory of an ion and allow it to pass. The larger the DMS gap height, the bigger the spread of peaks in CoV space, contributing directly to improved peak capacity [33]. The other factor in the peak capacity equation is the width of the peaks. In DMS, there is a 1/X relationship between the peak width observed when ramping the CoV and the residence time [34]. Therefore, the residence time can be optimized specifically for either high peak capacity or high-throughput. At one extreme of this spectrum are micromachined DMS devices with gap heights down to 35 μm [27, 35], and residence times in the 100–300 μs time frame. At the other extreme are DMS devices with gap heights between 1 and 2 mm with residence time ranging from about 5 to 600 ms [36–38]. For such devices, the residence time can be directly controlled by sealing the DMS cell onto a mass spectrometry inlet and providing a port for gas flow modification in the region between the DMS and mass spectrometer inlet. Generally, there is a trade-off between peak capacity and time when designing a DMS device for high-throughput analysis, and typical residence times for commercial devices are on the order of 1–7 ms.

7.3 Separation Selectivity

7.3.1 Classical Low Field Ion Mobility

When adding additional means for selectivity improvement to a mass spectrometry system for high-throughput drug discovery, it is desirable to provide as orthogonal of a separation mechanism as possible to mass. Low field ion mobility devices separate ions based upon differences in the low field mobility constant; however, as described above, the mobility constant is related to physical size and shape. As a result, low field IMS devices provide separations with very little orthogonality to mass spectrometry separations for ions of the same charge state. For isobaric species with different charge states, additional separation orthogonality can be achieved. In addition, the pulsed nature of IMS devices that require drift times makes coupling with continuous analyzers such as triple quadrupole mass spectrometers tricky. For these reasons, as well as the extended flight times necessary to separate isobaric species, low field IMS devices are rarely coupled with high-throughput MS systems.

7.3.2 Differential Mobility Spectrometry

7.3.2.1 FAIMS

Both FAIMS and DMS devices provide separations that are more orthogonal to mass than IMS devices and this makes them both better choices for coupling with high-throughput MS systems. In general, hard sphere separations can be quite effective in FAIMS devices, particularly for peptides or multiply charged species. Separations that depend upon clustering chemistry are less effective than separations with the planar design because of the heterogeneous fields present in cylindrical analyzers. It is common practice to add clustering agents to the transport gas in planar DMS devices to induce additional chemical selectivity and spread peaks across a wider range of CoV values [39–41]. In some cases, peaks that cluster heavily will shift to very large negative CoV values while other chemical entities that do not undergo clustering can have positive CoV values. This is problematic for FAIMS devices because the strength of the focusing depends to some extent on the magnitude of the CoV value. Under conditions that provide maximum focusing for some ions (e.g. the ones with the largest negative CoV values), other ions can be defocused and lost (e.g. the ones with the largest positive CoV values) and others will exhibit only weak focusing effects (e.g. the ones with small negative CoV values). In addition to these effects, compounds that exhibit strongest field focusing and therefore have the maximum signal will exhibit peak broadening as a result of the focusing effect. This has been demonstrated for a group of four small molecules previously [25]. As a result of these compromises, FAIMS devices have rarely been used with chemical modifiers in the transport gas flow.

7.3.2.2 DMS

In addition to the benefit of very short residence times, the homogeneous fields present in the planar DMS design provide unique advantages for maximizing chemical selectivity with the use of chemical modifiers in the transport gas. Because there are no focusing/defocusing issues to contend with, the major determinant for ion transmission is diffusion or chemistry losses in the analyzer. In addition, chemical modifiers can be added to the transport gas to augment separations via the dynamic cluster/decluster mechanism whereby ions cluster during the low field portion of the waveform (decreases the apparent low field mobility) and decluster under the intense electric fields of the high field portion of the waveform. This leads to large shifts in the CoV value for compounds that cluster and smaller shifts for compounds that do not. The lack of focusing effects means that the peak widths with and without chemical modifier remain relatively constant and all of the ions transmit with a single waveform polarity. The chemical separations imparted in the presence of modifiers or clustering agents provide additional orthogonality relative to mass separation and even relative to other chemical modifiers [42]. Figure 7.1 shows an example of separation orthogonality for a mixture of about 40 compounds. The y-axis displays the CoV for each of the compounds and the x-axis plots the m/z. The data were taken with three different chemical modifiers in the transport gas: isopropanol (black dots), acetonitrile (light gray dots), and ethylacetate (dark gray dots). In all cases, the separations that are achieved with DMS show a high degree of orthogonality to m/z. At any particular mass, it is clear that isobars are separated across a wide range of CoV values. The spread in CoV space is greatest at low m/z and compresses to some extent at higher m/z. This is not surprising because the change in mobility due to clustering is greatest when the ion of interest is smaller.

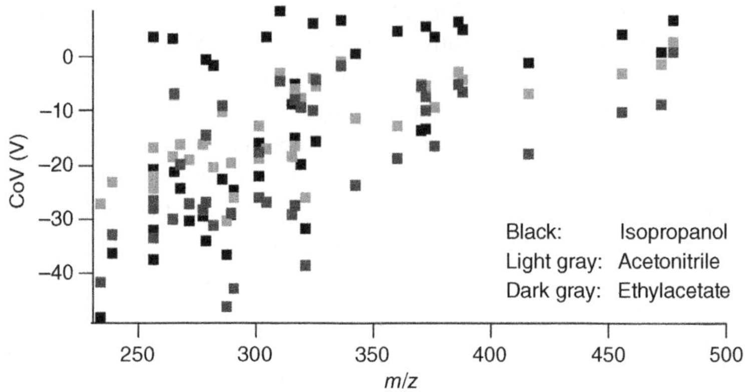

Figure 7.1 Orthogonality of DMS separations versus *m/z* for three different chemical modifiers in nitrogen transport gas.

The results presented in Figure 7.1 suggest that the addition of DMS to a high-throughput MS system can at least to some extent compensate for the elimination of the LC component by providing an orthogonal separation mechanism. An example is provided for the protonated compounds chlordiazepoxide (*m/z* 300) and temazepam (*m/z* 301). These two compounds are not isobaric (structures provided in Appendix 7.A); however, there is an isobaric overlap of the first higher isotopic peak of chlordiazepoxide and protonated temazepam as shown in Figure 7.2. The most intense daughter ion for temazepam is the *m/z* 255 fragment ion and this ion is also generated by MS/MS of the first higher isotopic peak for chlordiazepoxide. Therefore, a triple quadrupole mass spectrometer cannot

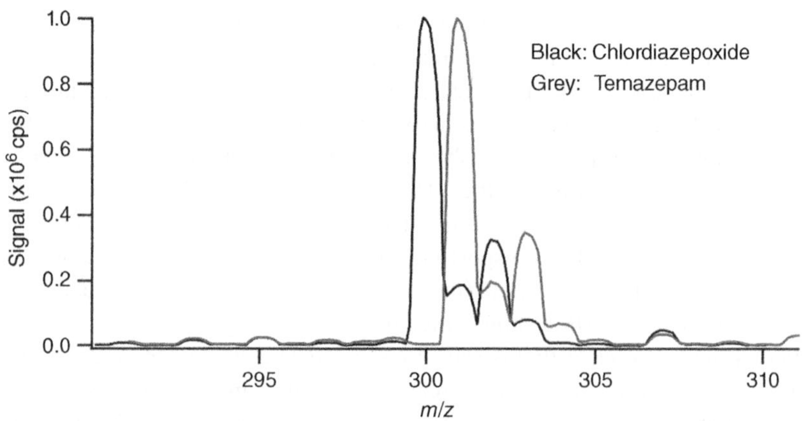

Figure 7.2 Q1 scans for chlordiazepoxide and temazepam.

distinguish between a real temazepam signal and an interference resulting from injection of chlordiazepoxide.

Figure 7.3 shows an LC separation of chlordiazepoxide and temazepam using a standard C18 column (top pane), where the two compounds are easily separated. The middle pane of Figure 7.3 shows the interference peak present in the MRM trace for temazepam at a retention time of approximately 7.15 minutes. For the data presented in the middle pane, only chlordiazepoxide was injected into the LC/MS system, and this can be clearly determined by looking at the retention

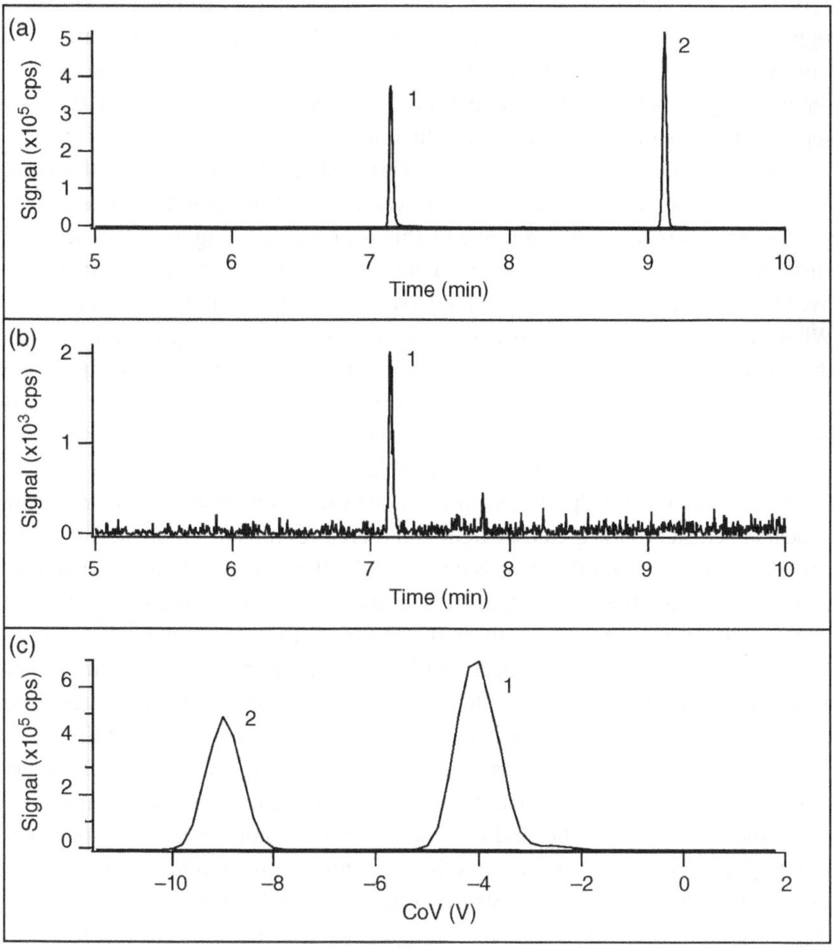

Figure 7.3 (a) LC separation of 1. chlordiazepoxide and 2. temazepam. (b) MRM signal for temazepam showing the presence of an interference peak from injection of chlordiazepoxide. (c) DMS separation of 1. chlordiazepoxide and 2. temazepam.

time for the signal. When switching to a high-throughput analysis system without LC separation, the retention time information is eliminated, and it would not be possible to tell which analyte or analytes actually contributed to the signal. The bottom pane of Figure 7.3 shows how DMS separation can replace the LC for these two compounds. These data were acquired using 3% ethylacetate modifier in the nitrogen transport gas, and under these conditions the two compounds were baseline separated. Temazepam appeared to cluster to the greatest extent, and therefore was shifted to the most negative CoV value (~−9 V), while chlordiazepoxide was shifted to a lesser extent (CoV ~ −4 V). By setting the appropriate CoV value for either of the compounds, it is possible to completely filter out signals from the other, resulting in an interference-free signal for either compound. Comparing the top and bottom panes of Figure 7.3, it is apparent that the LC separation gave better peak capacity than the DMS separation; however, the DMS separation was sufficient to eliminate the interference.

In addition to separation of isobaric species, DMS devices are capable of separating structural isomers, and this can be critical for high-throughput analysis in areas such as drug screening. An example of this is provided in Figure 7.4 for a trio of structural isomers that are difficult to separate using ion mobility devices: morphine, hydromorphone, and norhydrocodone (structures provided in Appendix 7.A). When using nitrogen transport gas in the absence of clustering modifiers, these three compounds are not separated in a standard DMS device as shown in the top pane. The three peaks are overlapping to a sufficient extent that it would not be possible to differentiate them. The second pane from the top shows an improved separation of these three structural isomers achieved with the addition of 1.5% isopropanol to the transport gas flow. Under these conditions, norhydrocodone appears to cluster to the greatest extent and therefore shifts to a more negative CoV value compared to the other two isomers. While this is better than the separation with pure nitrogen transport gas, it still does not provide differentiation of the three compounds. The second pane from the bottom shows the separation of these isomers with the addition of 3% acetonitrile to the transport gas. The chemical interactions between these ions and acetonitrile modifier are significantly different than with the alcohol and the distribution of the peaks in CoV space reflects this.

Norhydrocodone is easily separated from the other two isomers; however, morphine and hydromorphone are not. The bottom pane shows the separation of the same species with the addition of 3% ethylacetate to the transport gas. Ethylacetate clustering provides a sufficiently unique interaction with each of the isomers that the optimized CoV values differ sufficiently to separate them. These results demonstrate the chemical orthogonality that can be realized for DMS separations by changing the chemical modifier. For these results, ethylacetate is the preferred modifier; however, for other systems, better separations can be achieved using different modifiers.

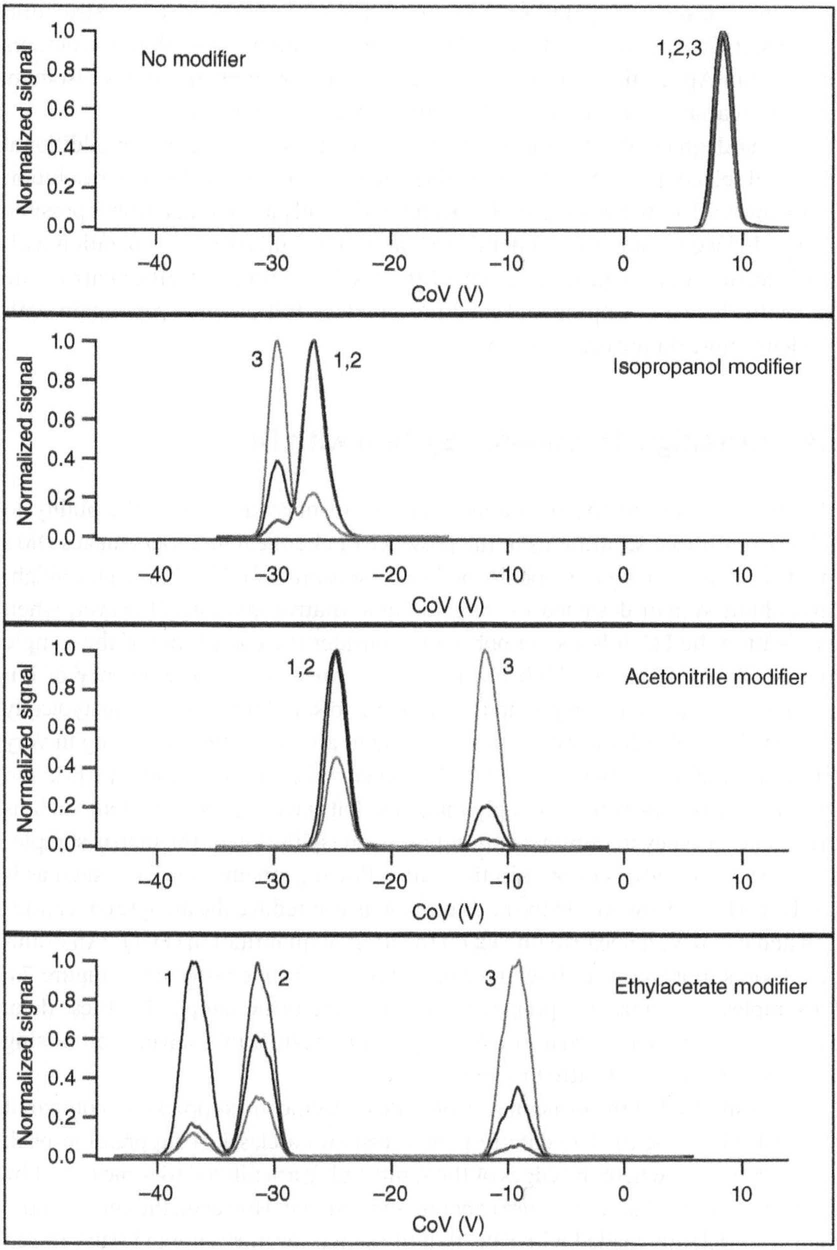

Figure 7.4 Separation of the structural isomers: 1. morphine, 2. hydromorphone, and 3. norhydrocodone under different DMS transport gas conditions.

Figure 7.5 shows separation data for a different group of three structural isomers, including oxymorphone, dihydrocodeine, and noroxycodone (structures provided in Appendix 7.A). Similar to the previous example, these three isomers are not separated in the absence of chemical modifiers (top pane).

Varying degrees of separation are possible for these isomers when additional chemical selectivity is imparted by adding chemical modifiers. Baseline resolution is possible with either isopropanol or acetonitrile, and partial separation is possible using ethylacetate modifier under these conditions. Full baseline separation with ethylacetate would require extension of the residence time to further narrow the peaks. In this case, isopropanol modifier provides full baseline separation with residence time on the order of 5 ms.

7.4 Ultrahigh-Throughput System with DMS

The orthogonality of the separations relative to mass as well as the ability to achieve optimized separations in the presence of chemical modifiers makes DMS uniquely suited among the mobility options to substitute for HPLC in an ultrahigh-throughput system designed for targeted quantitative analysis. However, when eliminating the LC, it is also important to consider the complexity of the sample matrix. Matrices that are high in salt content can reduce the efficiency of ion generation for a given compound as compared to standards. These salts typically elute in the void volume rather than binding to an LC column, resulting in very effective desalting when using LC. However, when eliminating the LC, ion suppression is common for sample matrices that have high salt content such as urine. One strategy to minimize ion suppression is dilution of the matrix samples either prior to analysis or by injection into a flowing stream of solvent such as is used for AEMS. However, extreme dilution will also reduce the analyte concentration and can have a negative impact on the limit of quantitation (LOQ). An example of ion suppression for flow injection analysis (FIA) is provided in Figure 7.6 for samples of mirtazapine prepared in a controlled urine sample. For these data, 5 μL of sample was injected into a flowing stream of 50 μL/min solvent comprising 50/50 water/methanol with 0.1% formic acid.

Pane b shows that the urine matrix provides extreme ion suppression compared to standards (Pane a). The data in pane b displays a classical suppression peak shape from FIA, where the edges of the sample plug are diluted to some extent by the flowing solvent and therefore generate some signal. However, the center of the plug is sufficiently loaded with salts that the signal for mirtazapine is suppressed to a level lower than the chemical background in the MRM trace. This represents 100% suppression as the peak cannot be integrated to any reliable extent. For comparison, the urine sample was desalted using a standard procedure with a

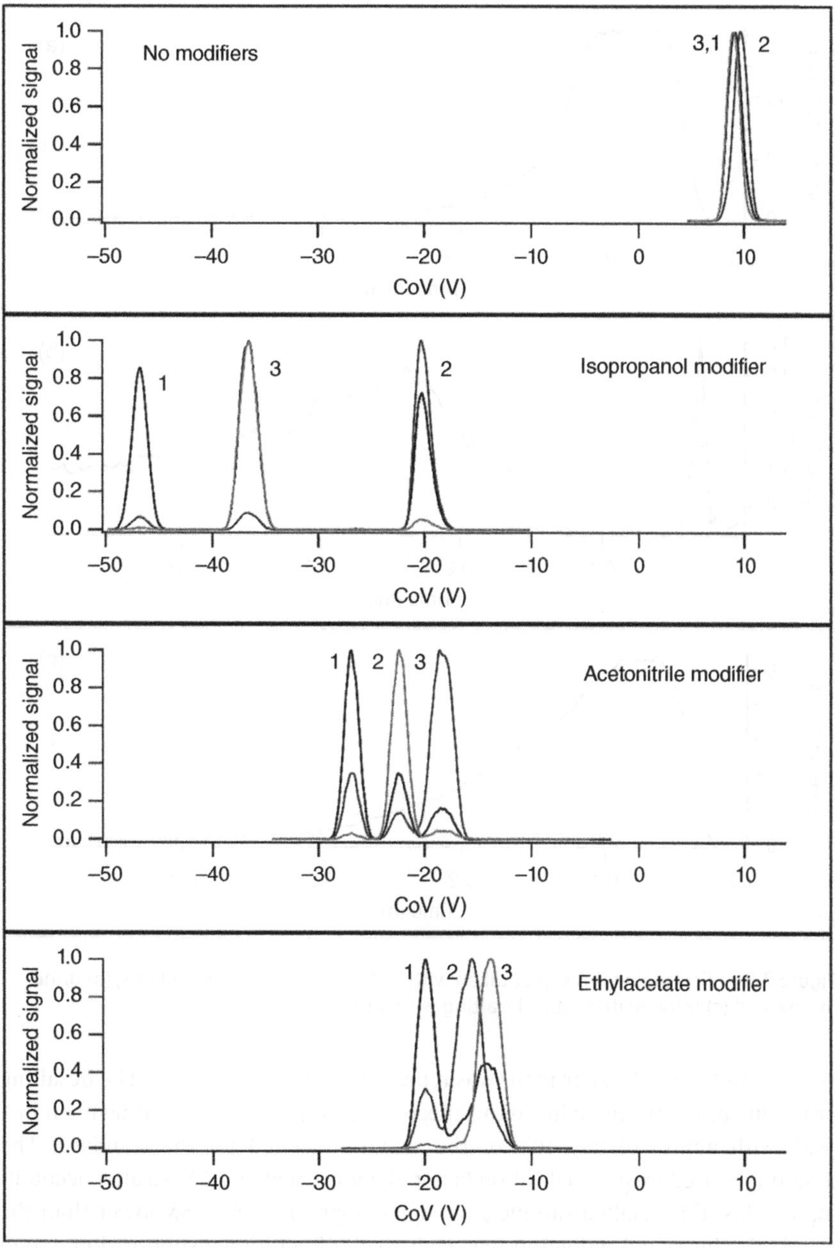

Figure 7.5 Separation of the structural isomers: 1. oxymorphone, 2. dihydrocodeine, and 3. noroxycodone under different DMS transport gas conditions.

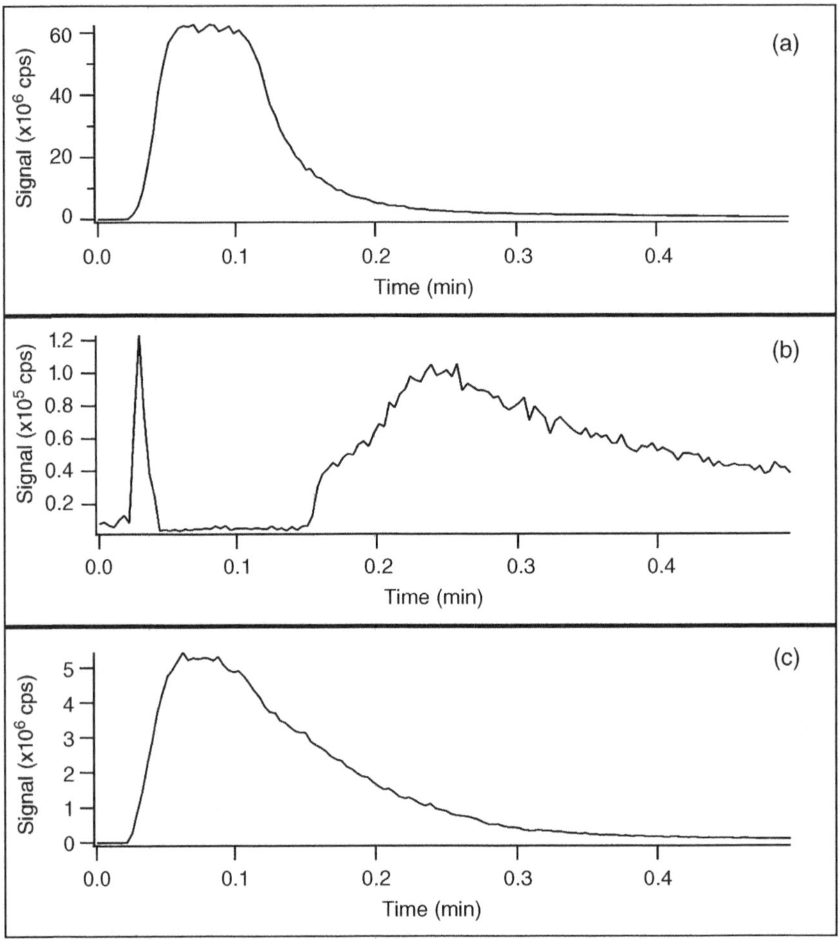

Figure 7.6 Mirtazapine FIA signal taken with 5 μL injection for (a) standards, (b) urine matrix, and (c) urine matrix after desalting with a DPX tip.

DPX tip and the FIA experiment was repeated, as shown in pane c. The desalting procedure dramatically reduced the effects of ion suppression, resulting in an FIA peak with approximately 10% of the signal generated for the standard. The desalting procedure gave a dilution factor of approximately 2.2X, so after accounting for this, the desalted sample gave signal approximately 4.5X lower than the standard. This signal reduction may be a result of additional suppression due to sample complexity, or lack of binding of the analyte to the desalting tip. Regardless, the data clearly show that when eliminating the LC system, desalting can be a critical aspect of sample preparation.

Dilution can also help to limit ion suppression effects in the absence of LC. The same mirtazapine sample prepared in urine matrix with and without desalting was flow injected at smaller volumes into a flowing liquid stream to provide greater dilution. For the data presented in Figure 7.7, 50 nL of sample was injected into a flowing stream of 300 μL/min solvent.

This represents approximately 100-fold dilution compared to the results of Figure 7.6, consistent with the approximate 100X decreased peak height for the desalted peak (black trace). The gray trace in Figure 7.7 shows the non-desalted urine peak, which still shows significant characteristics of peak suppression; however, the extent of signal reduction is substantially reduced compared to the 5 μL injection. Further dilution can reduce the extent of suppression to a greater extent as shown in Table 7.1, where the ratio of mirtazapine signal for a standard

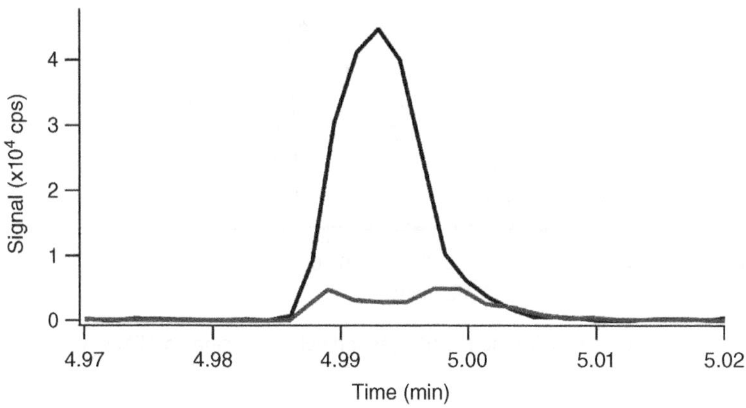

Figure 7.7 Zoom of AEMS/DMS/MS data for Mirtazapine samples prepared in urine (gray trace) and desalted urine (black trace).

Table 7.1 Ratio of signal intensity for standards of mirtazapine to the same concentration prepared in urine.

Injection vol (nL)	Std/urine
2.5	9.27
5	17.67
10	22.86
20	31.88
50	54.97

sample/mirtazapine prepared in urine decreased from about 55X for injections of 50 nL of sample down to about 9X for injection of 2.5 nL of sample.

This effect can be visualized by injecting increasing volumes for a sample as shown in Figure 7.8 for desmethyldoxepin prepared by DPX desalting of a urine sample.

After desalting, the expected linear signal increase is observed with injection volume as shown in the black trace of Figure 7.9. However, when injecting raw urine, significant gains with injection volume are not observed. In fact, the signal decreases somewhat with increased injection volume from 20 to 50 nL, suggesting extreme suppression.

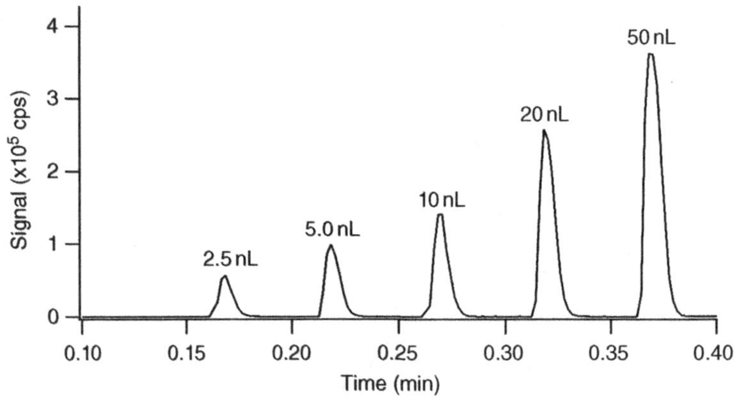

Figure 7.8 Analytical signal with increased injection volumes using an AEMS system for a desalted urine sample.

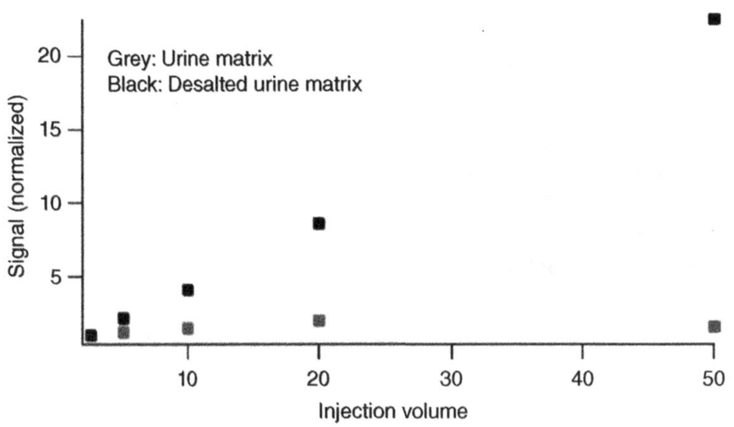

Figure 7.9 Suppression effects for samples prepared in urine versus desalting the urine matrix.

7.4.1 AEMS Data

Figure 7.10 shows a schematic of an AEMS/DMS/MS system that is suitable for high-throughput analysis. The system includes an acoustic transducer for applying acoustic energy to analyte samples contained in various wells of a sample plate. The acoustic energy causes ejection of one of more droplets from a given well, and the droplets are captured into an Open Port Interface [43], as has been previously described. The resulting sample plugs are drawn through a transport capillary to the tip of an electrospray ionization probe, and the liquid flow is established by the venturi force from a nebulizer at the tip of the probe. Ions and charged species travel through a curtain gas [36] prior to entering a curtain chamber with a DMS analyzer. The DMS analyzer is sealed to the vacuum inlet of the mass spectrometer such that the gas flow into the vacuum system draws a portion of the total curtain gas flow through the DMS device and into the mass spectrometer. The AEMS system has been described in more detail in Chapter 5 of this book.

Ultrahigh-throughput analysis systems based upon AEMS impart significant challenges for ion mobility devices. In particular, AEMS devices generate very short signal transients that impose speed restrictions to gas-phase separations. To properly define the shape of the signal transient, it is necessary to measure sufficient points across the width of the peak, in an analogous fashion to sharp chromatography peaks. If the signal transient is poorly defined due to lack of sufficient measurements, the signal coefficient of variation (CV) will suffer. Figure 7.11 shows an example of data taken using an AEMS device with a DMS/MS. The *y*-axis shows the CV for sets of 30 replicate measurements for penbutolol ions taken under a series of different conditions including measurement dwell

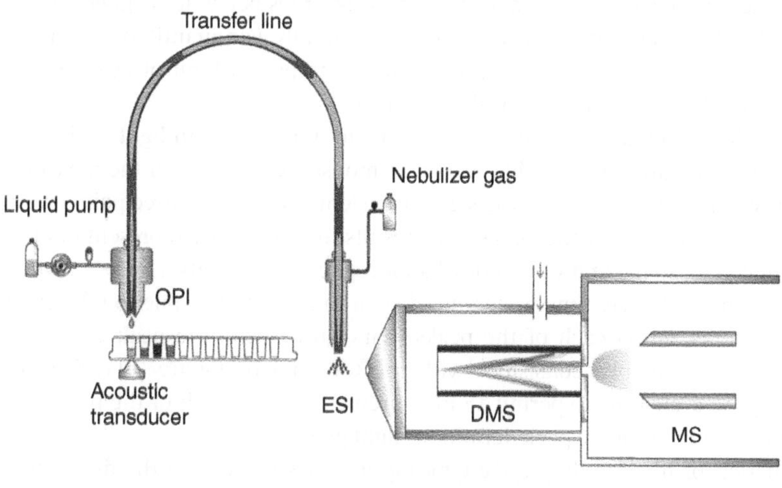

Figure 7.10 Schematic of an AEMS system with DMS system.

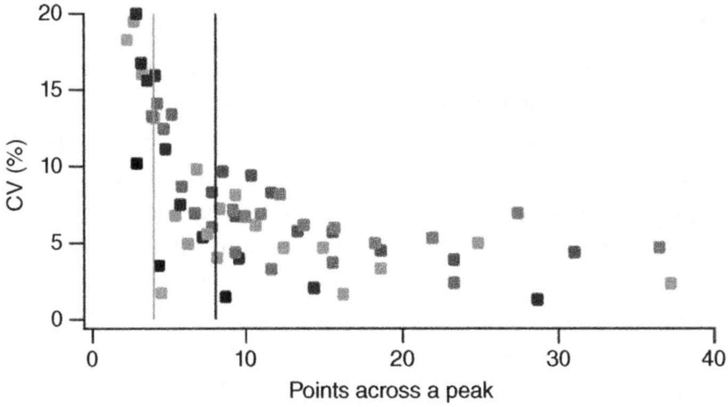

Figure 7.11 Coefficient of variation (CV) versus points across a peak for penbutolol measured from a complex sample with various dwell time settings from 1 to 200 ms ($N = 30$ injections). The data show that to achieve CV better than 15%, we need to measure at least four points across a peak. Once we have at least eight points across a peak, there is no significant further improvement in CV. It is possible to achieve better than 10% CV with at least eight points across a peak.

times from 1 to 200 ms and a variable number of MRM transitions monitored simultaneously (1–12). The *x*-axis shows the number of points measured across the signal transient peak from more than 90 different measurements. The CV of the measurement improves significantly as the number of points recorded across a peak increases to approximately 8 (black vertical line in the plot). Further increases in the number of points measured across the peak provide minimal additional improvements in the peak area reproducibility. When the number of points measured across the signal transient peak drops below four (gray vertical line in the plot), the percent CV increases exponentially. For quantitative analysis, the desirable CV is less than 15%, which would impose a limit of at least four measurement points across the peak of interest.

The data presented in Figure 7.11 can be better understood in light of the raw data plots in Figure 7.12. The black trace demonstrates that when the measurement cycle time is too slow for the signal transient width, only three points were measured across each of the peaks. This results in a very poor representation of the actual peak shape and very poor CVs for replicate injections. However, when the measurement cycle time was reduced such that it was possible to take eight measurements across each of the peaks (dark gray trace), the peak definition improved considerably, consistent with the CV data of Figure 7.11. Further reducing the cycle time to permit 15 measurements across each of the peaks did not substantially improve peak definition (light gray trace).

Reduction of method duty cycle typically includes decreasing the dwell time during which measurements are taken for each compound. Excessively short

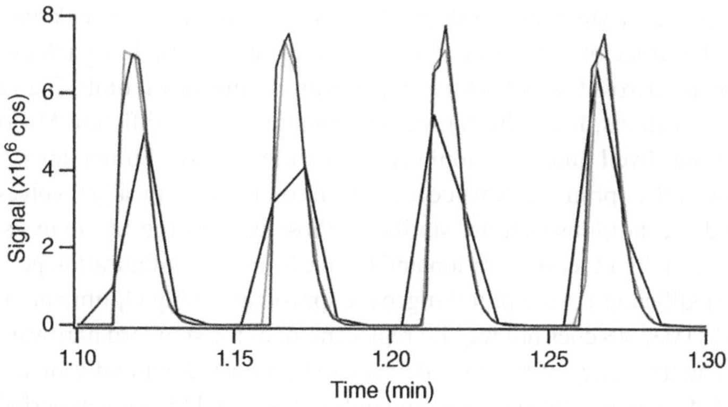

Figure 7.12 Raw data for four replicate AEMS injections using instrument conditions to achieve 3 measurements across the peaks (black trace), 8 measurements across the peaks (dark gray trace), and 15 measurements across the peaks (light gray trace).

dwell times limit the total number of ions counted during each measurement and can therefore have a detrimental effect on count statistics for a given ion current. For this reason, it is generally beneficial to adjust the cycle time to achieve the minimum number of points across the signal transient peak to properly define the peak, e.g. set the conditions to achieve approximately eight points across the peaks. The number of points across a peak can be defined by Eq. (7.1).

$$\text{Number of Measurements} = \text{Peak Width/Cycle Time} \tag{7.1}$$

For a standard analysis method using a triple quadrupole mass spectrometer, the cycle time depends upon the sum of the dwell time and pause time, multiplied by the total number of MRM transitions to be monitored (N) as defined in Eq. (7.2).

$$\text{Cycle Time} = N(\text{Dwell} + \text{Pause}) \tag{7.2}$$

The peak width in AEMS depends upon a number of factors including the transport solvent composition, the specific design of the open port probe and the transfer tubing, and the volume of liquid sampled by the device. While the relationship of these parameters to the peak width is difficult to predict a priori, the experimental determination of the peak width is trivial. From the peak width information and the desired number of measurements across a peak (i.e. 8), the cycle time can be determined and hence the optimal dwell time for any total number of MRM transitions. This entire process can be automated to eliminate the need for manual dwell time optimization. In the case of a standard DMS separation, typical pause times are on the order of 15 ms to account for 5–7 ms transit time in the DMS, 5 ms fill time for the inlet optics of the mass spectrometer, and a few milliseconds to stabilize the SV and CoV. As an example, for a peak with base width on the order of approximately 1.8 seconds, it is possible to achieve eight measurements across the peaks by using 2 ms dwell times for monitoring 12 different MRM transitions, 10 ms dwell times for monitoring eight different MRM transitions, or 25 ms dwell times for monitoring five different MRM transitions.

Determination of the optimal CoV value for a given compound typically involves tee infusion under conditions that are similar to those used in the LC analysis. Conversion between the LC configuration and the tee infusion configuration generally requires modification of the plumbing to the sprayer assembly. Optimization of CoVs in ADE/DMS/MS does not require replacement of the standard hardware configuration. The frequency of the ADE device can be adjusted to rapidly switch back and forth between generating a pseudo-continuous signal for a given period of time and generating very short signal transients as shown in Figure 7.13 for an acoustic device that dispenses droplets with 2.5 nL volume. In a first mode of operation, a series of individual droplets were dispensed with a constant frequency of 10 Hz as shown on the left side of pane a. Transport of the sample plug through the transfer line yields peak widths on the order of one second, resulting in substantial overlap of the signal for the various individual injections as shown on the left side of pane b. The net result is a pseudo-continuous ion signal similar to a standard infusion experiment (left side of pane c). During a second time period (1.3–1.8 minutes), the ejection frequency was increased to 400 Hz to eject a series of six droplets over a time period of approximately 2.5 ms to generate a discrete peak comprising 15 nL of analyte diluted in the transport flow stream (right side of pane b). The resulting peak resembles a standard flow injection peak with reduced tailing. A series of nine replicate injections were conducted with a three seconds delay time between each of them. Pane c shows analytical data acquired for a sample of reserpine switching from pseudo-continuous mode to pulsed mode. The transport liquid composition and flow rate, as well as all of the ion source voltages, gas flows, and temperature remain constant while switching from pseudo-continuous mode to discontinuous mode and therefore DMS conditions optimized in pseudo-continuous mode are also optimal in discontinuous mode. This greatly simplifies the process of DMS voltage optimization in ADE/DMS/MS.

The transport liquid flow in AEMS with DMS analysis typically comprises pure organic species such as methanol. Conversely, LC/DMS/MS analysis typically includes solvent with some portion of aqueous content. Previous results suggested that the presence of a curtain gas flow prior to the DMS cell prevented source solvents from penetrating into the DMS region and affecting optimal CoVs [39]. This is particularly important in the case of AEMS with DMS, where the transport solvent may comprise exclusively organic solvent. Figure 7.14 shows a comparison of the separation of the two isobaric species clozapine and midazolam

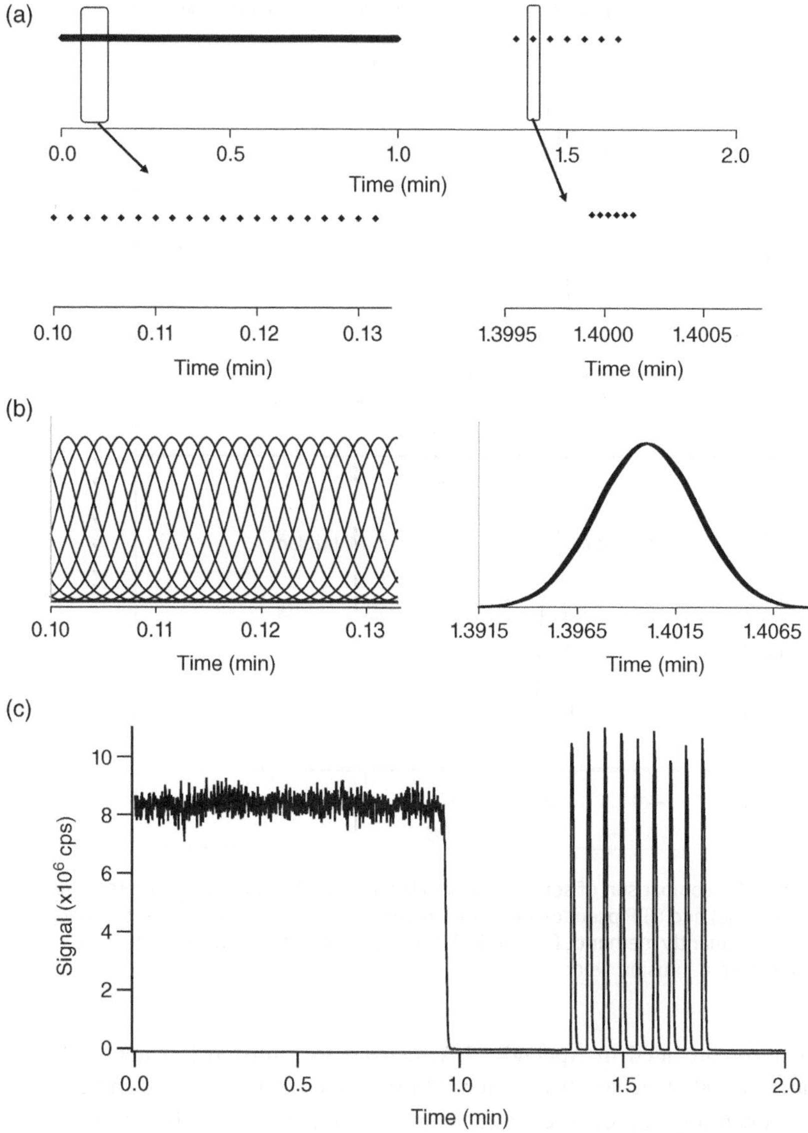

Figure 7.13 Switching between pseudo-continuous signal mode and pulsed mode. (a) Marks for time stamps when each individual droplet is acoustically dispensed. In the first period (up to ~1 min), droplets were ejected at the constant frequency of 10 Hz. The second period (1.35–1.65 min) contains eight groups of ejections, and each group contains six droplets ejected with the internal time of 2.5 ms (400 Hz). There are 3 s delay time between each ejection groups. The time periods of 0.10–0.13 and 1.3995–1.4008 min were expanded for better visualization. (b) The overlay of simulated MS signal from each individual acoustic ejected droplet with the ejection pattern described in (a), at the time periods 0.10–0.13 and 1.3915-1.4010 min. (c) The summation of the MS signal as described in (b) as the illustration of the observed MS signal from the described ejection pattern.

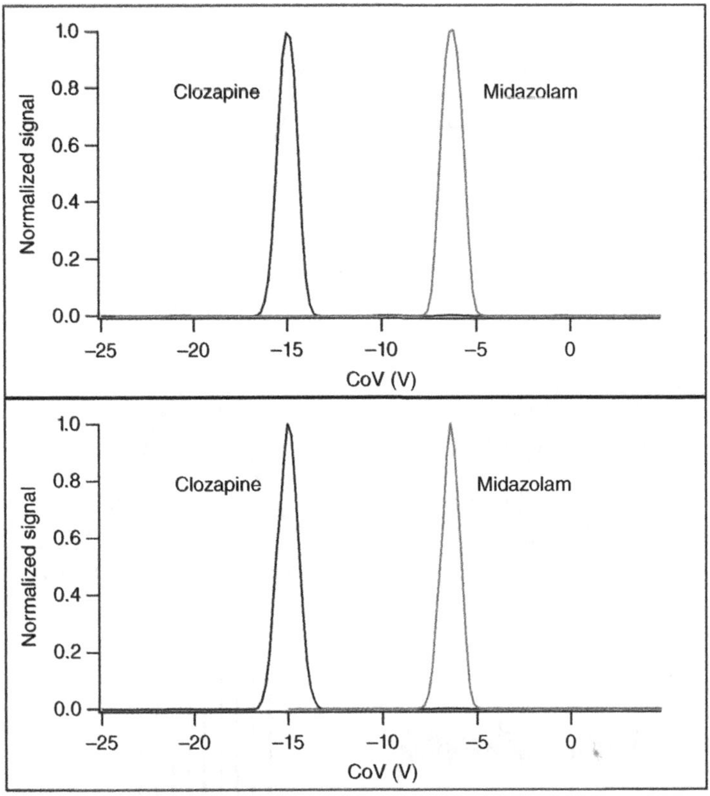

Figure 7.14 Comparison of separations for clozapine and midazolam using tee infusion ESI (top pane) and OPP pseudo-continuous mode (bottom pane). The solvent composition was predominantly methanol for the OPP work and 90% acetonitrile with 0.1% formic acid for the tee infusion work.

(structures provided in Appendix 7.A) using a tee infusion ESI approach (top pane) with 90/10 acetonitrile/water with 0.1% formic acid solvent and an ADE device operating in pseudo-continuous mode (bottom pane) with 100% methanol solvent. In both cases, the transport liquid flow rate was 500 μL/min, the DMS transport gas temperature was approximately 105 °C, and the ion source heaters were set to 300 °C. As demonstrated in Figure 7.14, baseline separation was achieved for the clozapine peak (black trace) and the midazolam peak (gray trace). The optimal CoV values generated by the ADE device were the same as those from the tee infusion approach and the peak widths were approximately equivalent. These data show that the high-throughput approach does not affect the separation capability of the DMS device.

7.4.2 DMS Sensitivity (Ion Transmission)

Elimination of chemical background can dramatically improve the detection limit in mass spectrometry; however, the extent of the gain is limited by the transmission of the mobility device. As a general rule of thumb, detection limit increases are proportional to the square root of background reduction. Conversely, detection limits scale directly with signal intensity. As an example, a 100-fold background reduction would be canceled by a 10-fold signal decrease, resulting in negligible S/N improvement. For this reason, ion transmission is a critical aspect of an ion mobility analyzer designed for mass spectrometers. In the case of planar DMS devices, the maximum transmission is limited by diffusion losses within the ion mobility cell. Additional loss mechanisms can further reduce transmission and must therefore be accounted for when designing a DMS analyzer [34, 44]. The amplitude of ion oscillations between the DMS electrodes is inversely proportional to the waveform frequency, and therefore it is desirable to maximize the waveform frequency such that radial oscillations comprise only a small portion of the DMS gap height. Ion heating and fragmentation is an additional loss mechanism that can reduce transmission; however, this phenomenon is mainly observed with micromachined devices that include very small gap heights and separation fields up to approximately 300 Td. Losses can also occur at the inlet and outlet of DMS devices due to detrimental fringing fields. As ions approach the DMS cell inlet, the DC offset between the two electrodes diverts ions prior to experiencing the effects of the separation field and ions can enter the gap offset from the central axis. This issue can be resolved by using a directed gas stream to sweep the ions through the fringing field region and direct them into the center of the DMS gap. When losses due to fringing fields, ion fragmentation, and radial oscillations are minimized, ion transmission approaching the diffusion limits can be achieved using planar DMS devices. The situation is more complicated when using FAIMS devices with curved electrodes because focusing and defocusing effects due to the inhomogeneous fields must also be accounted for. The FAIMS literature frequently describes the benefits of inhomogeneous fields and ion focusing effects that can minimize ion losses, particularly for multiply charged peptides where the focusing effects can be strong [45]. However, these publications rarely describe the drawbacks of simultaneous defocusing for some ions or weak focusing for others [19], and this can lead to confusion in the scientific literature, where some publications describe near lossless performance [23, 46] and others show losses that can be orders of magnitude [25].

The relative transmission of DMS devices to optimized ESI/MS systems with the mobility cell removed has been described in the scientific literature over the past decade. Early systems did not account for fringing field losses at the inlet of the cell and therefore the losses for analytical flows were on the order of 4–8X [39].

A subsequent design improvement with a jet injector lens to reduce inlet losses improved transmission by a factor of approximately 2 [44]. It is important to note that the comparison data was a fully optimized ESI/MS with source temperatures of 750 °C. AEMS devices use transport solvent flow rates similar to high flow ESI/MS systems; however, the maximum source temperature can be limited as compared to standard ESI experiments, particularly when using transport flows that are comprised predominantly of organic solvents. It is important to avoid boiling effects at the sprayer tip because the nebulizer venturi establishes the liquid pull from the OPP port through the transport tubing, and therefore source temperatures are normally restricted to approximately 300 °C. The source temperature restriction lowers the baseline signal for an AEMS system in the absence of DMS. The DMS hardware includes an additional heat exchanger to elevate the transport gas temperature up to approximately 200 °C. This additional heating can compensate for compromised source heating and is sufficiently downstream from the spray electrode that it does not contribute to spray instability. The net result is a general normalization of the data with and without DMS installed. Figure 7.15 shows an example of optimized data for replicate injections of reserpine with an AEMS system on a mass spectrometer equipped with a jet injector DMS cell (black trace) and with the DMS removed (gray trace). The average signal reduction due to the presence of the DMS cell was 1.28X for this data set, as shown in Figure 7.15.

At the same time, there was approximately a 10-fold reduction in the solvent background continuum as shown in Figure 7.16, with the net effect being an improved signal/noise ratio (S/N) as a result of DMS filtering. Typical losses for this compound for standard high flow rate ESI are on the order of 2X with the present configuration.

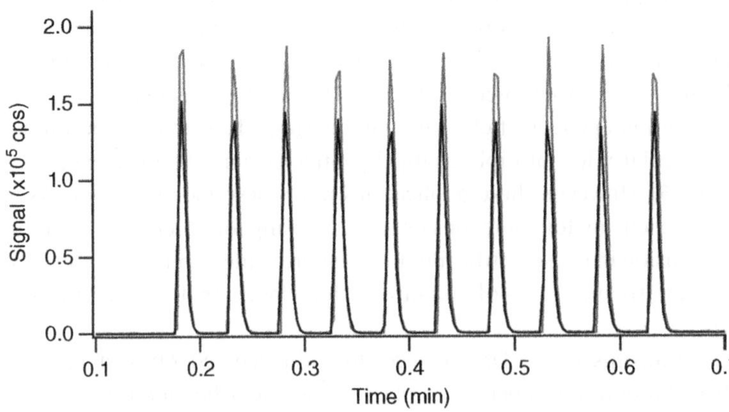

Figure 7.15 Replicate measurements for reserpine injections with a DMS installed (black trace) and removed (gray trace) from a triple quadrupole mass spectrometer.

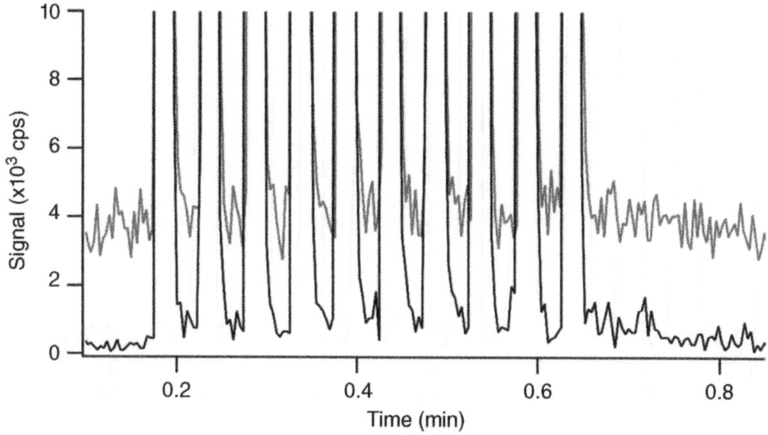

Figure 7.16 Zoomed view of the plot of reserpine signal intensity with a DMS installed (black trace) and removed (gray trace) from a triple quadrupole mass spectrometer.

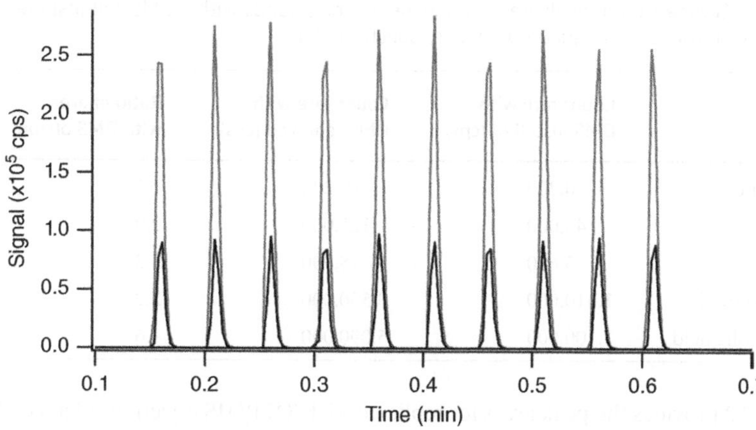

Figure 7.17 Replicate measurements for minoxidil injections with a DMS installed (black trace) and removed (gray trace) from a triple quadrupole mass spectrometer.

Figures 7.17 and 7.18 show a similar peak intensity and S/N comparison for minoxidil ions, where the signal reduction due to DMS was somewhat greater.

For the minoxidil comparison, there was a 2.9-fold signal reduction as a result of the installation of the DMS cell. Minoxidil ions are much smaller than reserpine ions and therefore the expected losses due to diffusion within the DMS cell are greater. Despite the signal reduction, the S/N ratio was approximately 2-fold greater with the DMS in place as a result of the background reduction as shown in Figure 7.18.

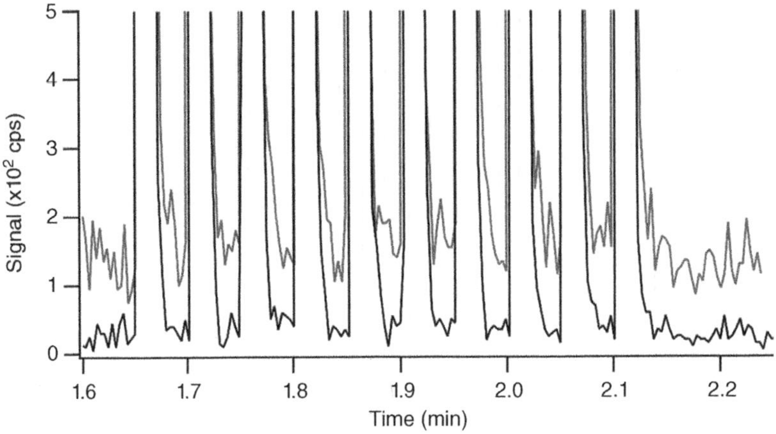

Figure 7.18 Zoomed view of the plot of minoxidil signal intensity with a DMS installed (black trace) and removed (gray trace) from a triple quadrupole mass spectrometer.

Table 7.2 Comparison of peak areas for a panel of compounds with a DMS cell installed and removed from a triple quadrupole mass spectrometer.

Analyte	Count rate with DMS installed (cps)	Count rate with DMS removed (cps)	Ratio of area with DMS off/on
Reserpine	70,200	90,200	1.3
Minoxidil	43,000	125,000	2.9
Proline	5,680	18,900	3.3
5-Fluorouracil	1,610,000	3,530,000	2.2
Taurocholic acid	3,100,000	5,090,000	1.6

Table 7.2 provides the peak areas for replicate ADE/OPP/MS injections of a panel of compounds comprising reserpine, minoxidil, proline, 5-FU, and taurocholic acid with a DMS cell installed and removed from the same mass spectrometer.

The signal intensity losses due to installation of the DMS ranged from about 1.3–3.3X, whereas the signal losses were 2.0–4.8X for the same compounds when using high flow ESI for FIA. The losses were slightly reduced when using AEMS because of the compromised source heating.

7.4.3 Examples of AEMS Analyses with DMS

7.4.3.1 Example 1. DMS to Eliminate Interferences from Isobaric Species

The AEMS process samples directly from wells of a sample plate and therefore does not include provisions for LC separation. As a result, the general applicability of this approach is limited by the presence of chemical interferences. The benzodiazepine drugs mirtazapine and desmethyldoxepin can be used to demonstrate this effect. These drugs have similar chemical structures, as shown in Appendix 7.A, and are isobaric in Q1. They also share a significant number of daughter ion *m/z* ratios, such that the MRM transitions that yield the highest signal intensity have interferences from the presence of the other drug. Figure 7.19 shows AEMS data for injection of a mirtazapine standard sample while monitoring the MRM transition for mirtazapine (top pane) and desmethyldoxepin (bottom pane). As shown in Figure 7.19, significant peak intensities were measured in each of the MRM channels, corresponding to the real mirtazapine signal (top pane), and a false positive interference peak in the bottom pane.

A similar effect was observed when injecting a desmethyldoxepin standard solution (Figure 7.20). In this case, the expected desmethyldoxepin signal is shown in the bottom pane, and a false positive interference peak is shown in the top pane corresponding to the mirtazapine MRM response.

The data presented in Figures 7.19 and 7.20 are problematic because it is not possible to differentiate the source of signals measured for these isobaric species. The gas-phase separation capability of the DMS is critical for analyzing these compounds. Figure 7.21 shows ionograms acquired for mirtazapine and desmethyldoxepin using DMS, demonstrating that it is possible to baseline separate these isobaric species when adding the DMS to the AEMS workflow.

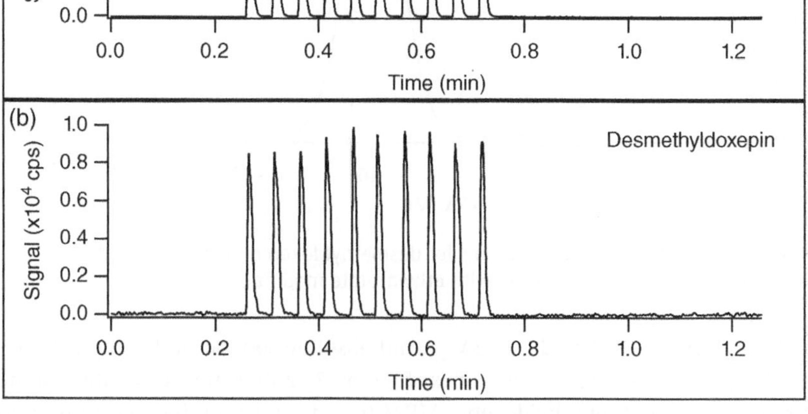

Figure 7.19 AEMS data for injection of a mirtazapine standard while measuring the optimal MRM transitions for mirtazapine (a) and desmethyldoxepin (b).

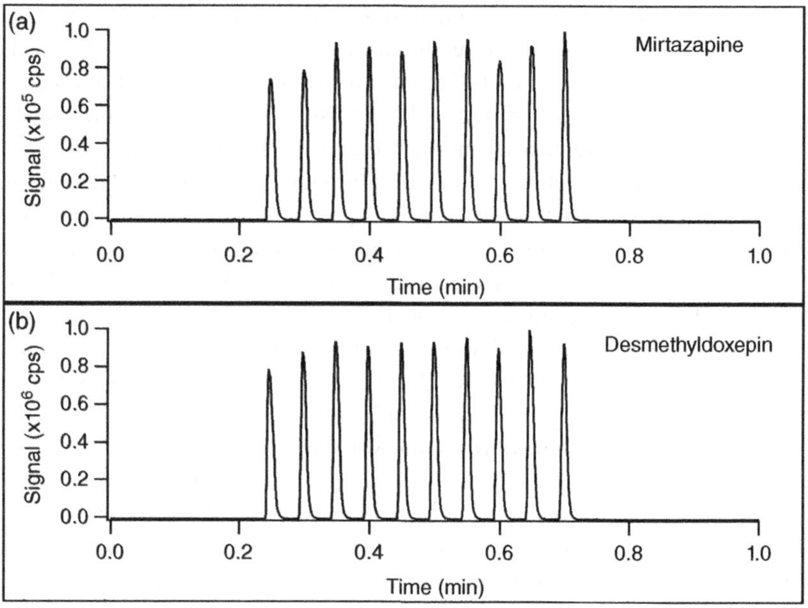

Figure 7.20 AEMS data for injection of a desmethyldoxepin standard while measuring the optimal MRM transitions for mirtazapine (a) and desmethyldoxepin (b).

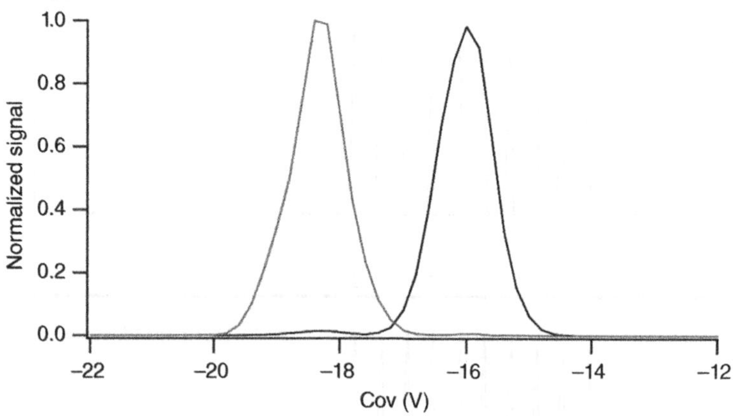

Figure 7.21 Overlay of DMS ionograms for desmethyldoxepin (gray trace) and mirtazapine (black trace) using DMS with ethylacetate modifier.

CoV settings of −15.9 and −18.3 V permit absolute separation for the elimination of interferences. The top two panes of Figure 7.22 show the elimination of the interference in the desmethyldoxepin MRM trace for mirtazapine injections, and the bottom two panes show the elimination of the interference in the mirtazapine

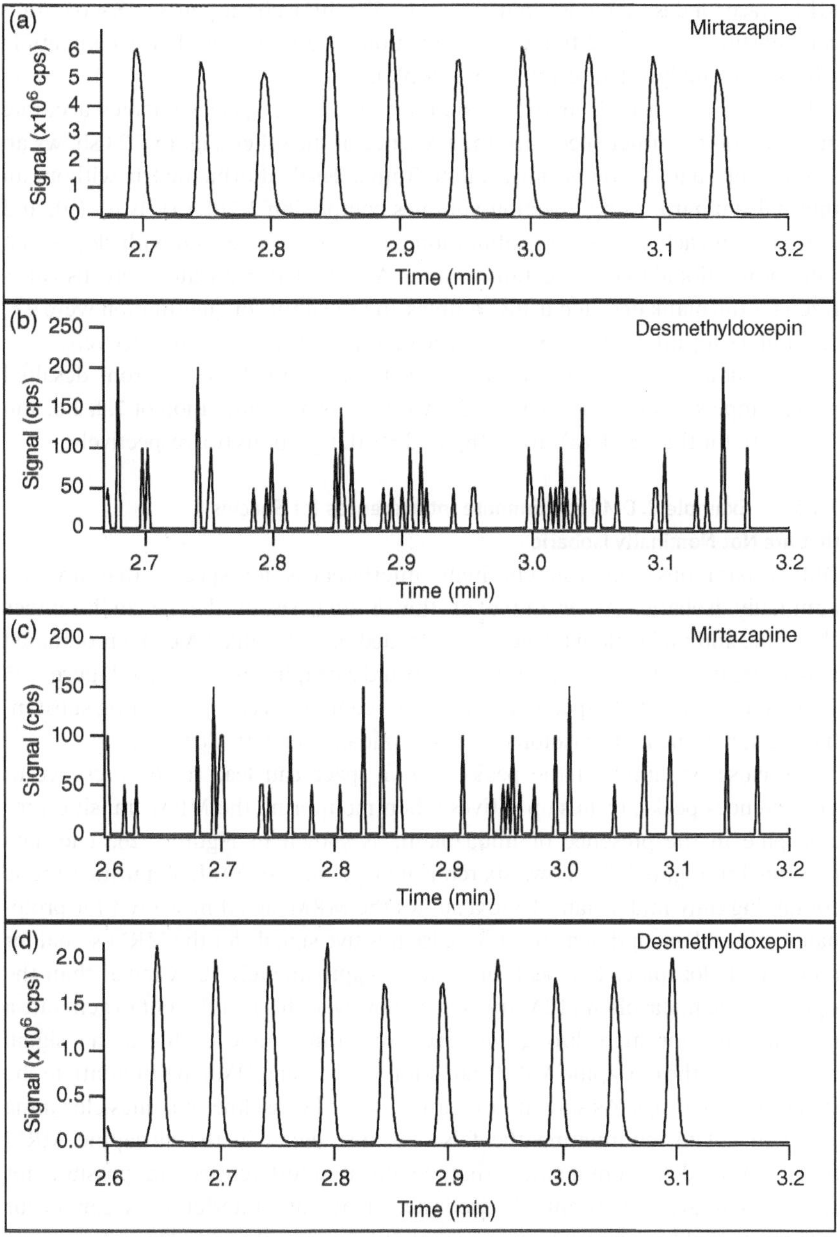

Figure 7.22 AEMS data for mirtazapine and desmethyldoxepin with a DMS installed for isobaric separation: (a) mirtazapine signal when injecting a mirtazapine standard; (b) desmethyldoxepin signal when injecting a mirtazapine standard; (c) mirtazapine signal when injecting a desmethyldoxepin standard; and (d) desmethyldoxepin signal when injecting a desmethyldoxepin standard.

MRM trace for desmethyldoxepin injections. For this isobaric pair, DMS separation can substitute for a traditional LC separation on a timescale that is compatible with AEMS analysis (~1 seconds per sample).

The elimination of isobaric interferences for these species enables accurate quantitation for either species in the presence of the other. Figure 7.23 shows an example of quantitation of mirtazapine from a desalted urine matrix without an internal standard, using injection volumes comprising 2.5 nL (a), 10 nL (b), and 50 nL (c). In each case, linear calibration curves were generated with slopes that were proportional to the injection volume. A total of 10 replicates were also analyzed for the blank desalted urine samples and the limits of quantitation were 9.6, 1.9, and 0.6 ng/mL for the 2.5, 10, and 50 nL injection volumes, respectively.

High quality data were also generated for desmethyldoxepin from desalted urine samples as shown in Figure 7.24, with limits of quantitation of 7.2, 2.6, and 0.4 ng/mL for the 2.5 nL (a), 10 nL (b), and 50 nL injections (c), respectively.

7.4.3.2 Example 2. DMS to Eliminate Interferences for Species that are Not Nominally Isobaric

DMS separations can also eliminate interferences for species that are not nominally isobaric. An example of this is analysis of the benzodiazepines clozapine and midazolam (structures provided in Appendix 7.A). The protonated masses are *m/z* 326 and 327 for midazolam and clozapine, respectively. Figure 7.25 shows an overlay of Q1 spectra for these two compounds demonstrating substantial overlap of the higher isotopic peaks of midazolam with clozapine.

The close proximity of the peaks in *m/z* space can lead to the presence of interferences peaks, or false positives when monitoring the MRM transition for clozapine in the presence of midazolam, as shown in Figure 7.26. The data presented in Figure 7.26 shows six replicate injections of 50 nL of a urine sample containing only midazolam. Pane a shows the MRM signal measured for protonated midazolam, and pane b shows false positive signals for the MRM signal for protonated clozapine. The peak intensity is approximately 100X lower than the signal for the midazolam MRM transition; however, this is sufficient to generate a false positive for high-throughput drug screening. Pane c shows the signal measured for the clozapine MRM transition when using DMS to differentiate the signals for the drugs. As shown in Figure 7.14, DMS baseline separates clozapine and midazolam, eliminating the false positive signal in the clozapine MRM trace. These data demonstrate that chemical interferences are possible for species that are not nominally isobaric. Significant interferences can occur for species that have nominal masses that differ by less than 3–4 Da.

In some cases, interferences can also be observed for species with even larger mass differences due to fragmentation in the atmosphere to vacuum interface region. An example of this occurs in screening assays of cytochrome P450 enzyme CYP1-A2 [47]. The marker metabolite is acetaminophen for the probe substrate

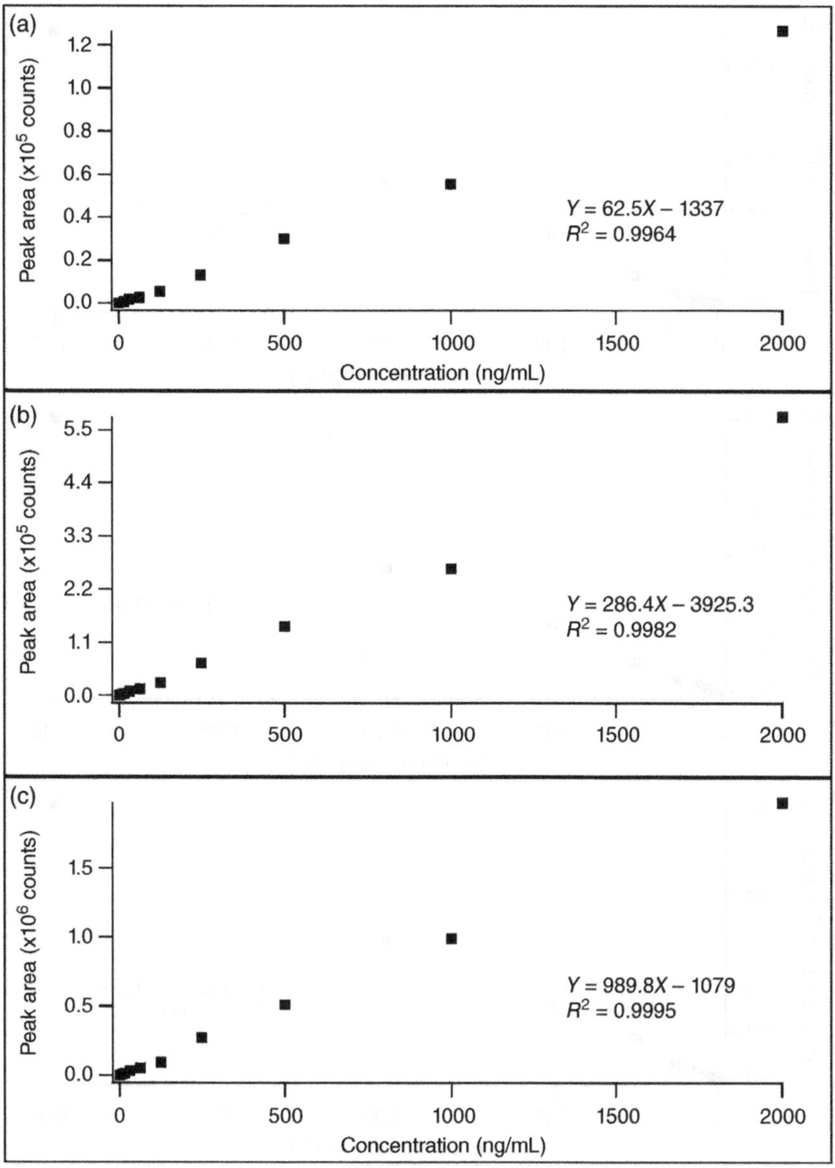

Figure 7.23 Calibration curves for mirtazapine in desalted urine. The injection volume was (a) 2.5, (b) 10 , and (c) 50 nL. The LOQs were 9.6, 1.9, and 0.6 ng/mL.

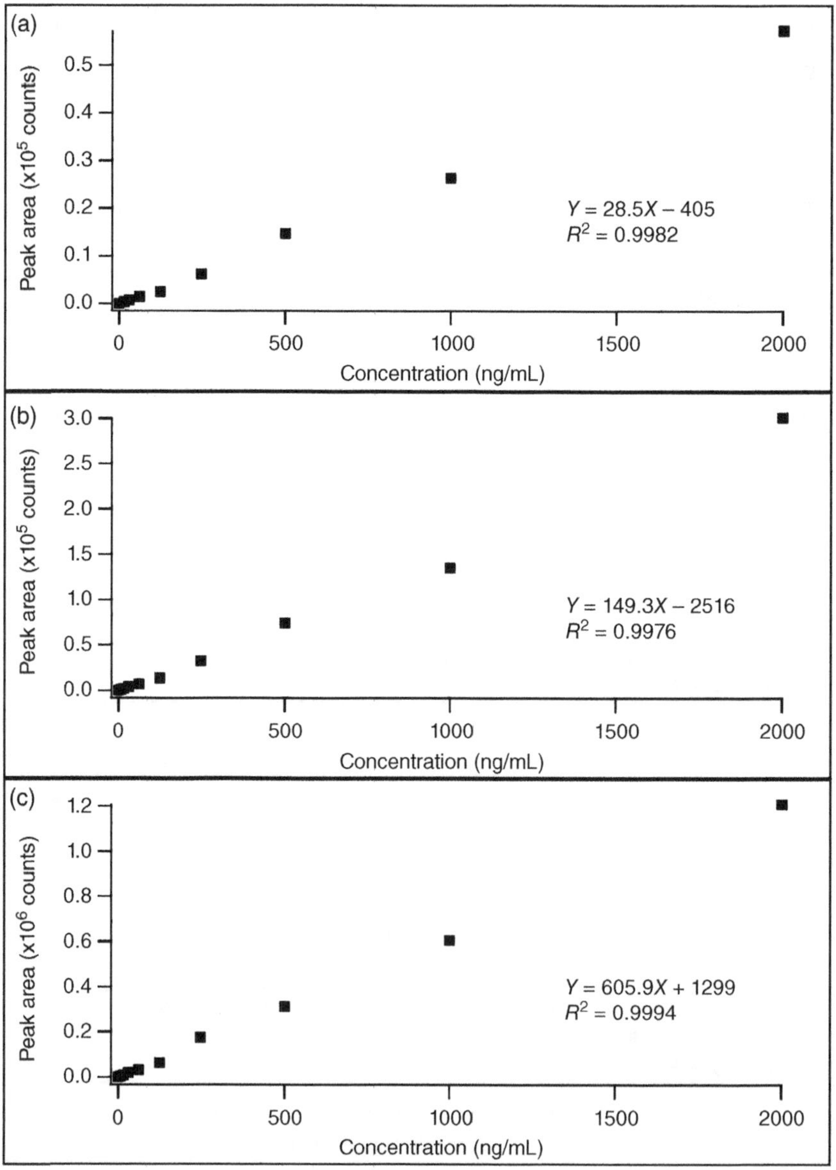

Figure 7.24 Calibration curves for desmethyldoxepin in desalted urine. The injection volume was (a) 2.5, (b) 10, and (c) 50 nL. The LOQs were 7.2, 2.6, and 0.4 ng/mL.

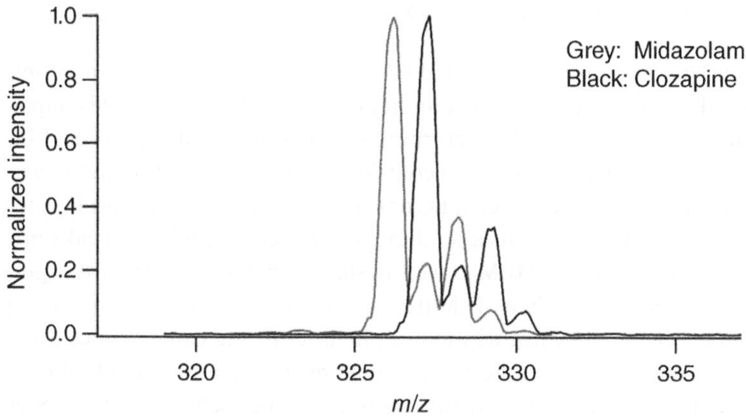

Figure 7.25 Overlay of Q1 spectra for midazolam (gray trace) and clozapine (black trace).

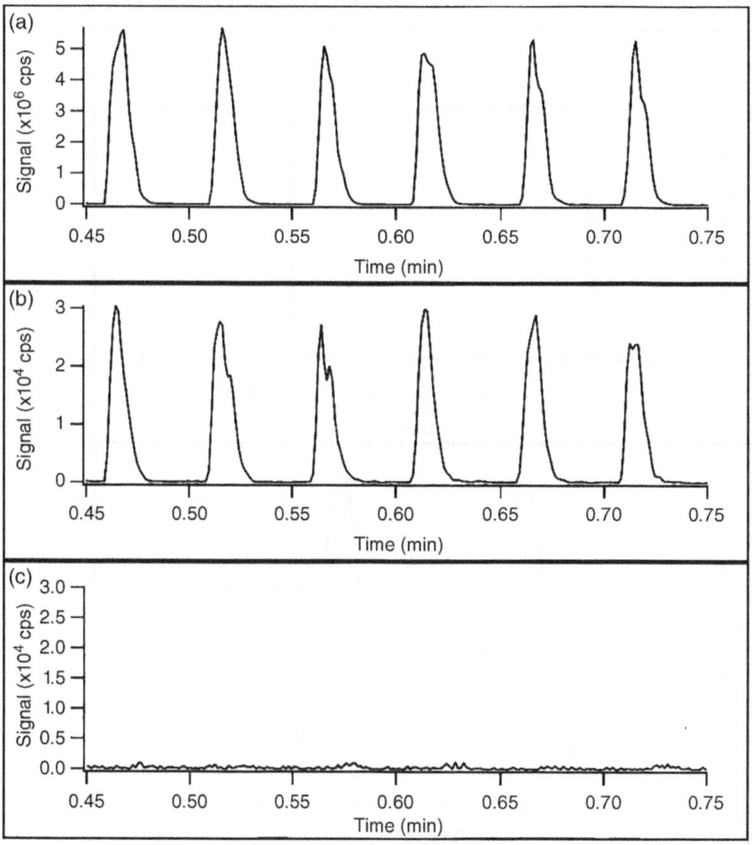

Figure 7.26 Midazolam and clozapine example showing midazolam injection (a) and the clozapine interference with DMS in transparent mode (b) and the filtered Clozapine signal with DMS (c).

phenacetin. This analysis is complicated by fragmentation of phenacetin to yield an ion with the same structure as acetaminophen [48, 49]. Therefore, it is important to differentiate between true acetaminophen signal and signal resulting from conversion of phenacetin due to collision-induced dissociation in the vacuum stages of the mass spectrometer. LC normally separates the two species such that the retention time can be used to differentiate between the two peaks present in the acetaminophen MRM trace, as shown in Figure 7.27. Any signal measured in the acetaminophen MRM trace at a retention time of approximately 7.4 minutes is generated by degradation of phenacetin, and can be disregarded. The issue becomes more complicated for analyses that do not include the LC separation due to the requirement to differentiate between actual acetaminophen signal and signal resulting from degradation of phenacetin. DMS is also capable

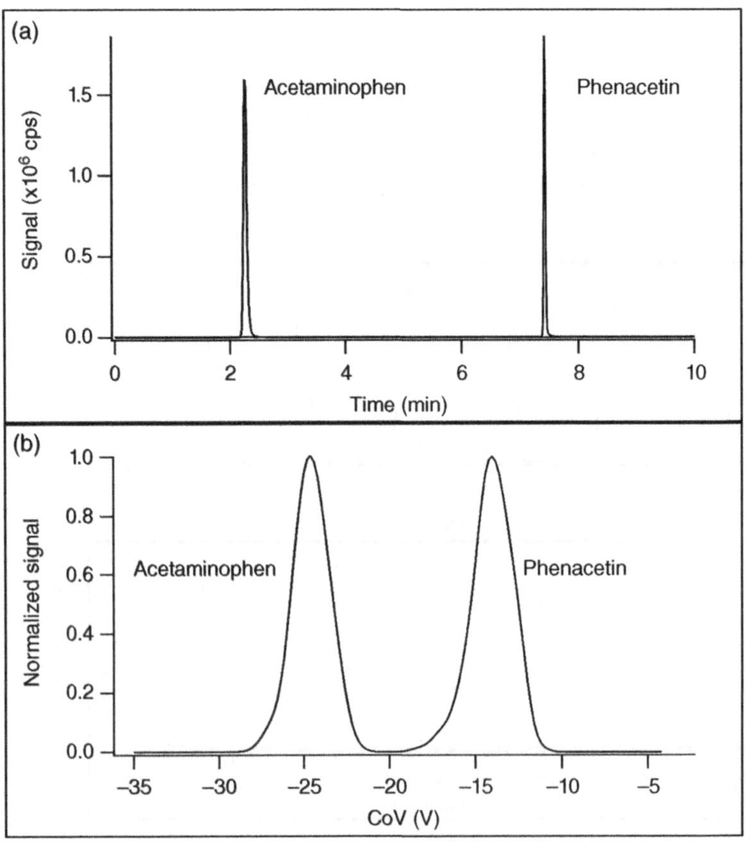

Figure 7.27 Separation of acetaminophen and phenacetin using (a) LC/MS and (b) DMS/MS.

of separating these two compounds when adding acetonitrile modifier to the transport gas, as shown in the bottom pane of Figure 7.27. Despite a substantially lower peak capacity than LC, the compounds of interest are baseline separated, and this suggests that DMS would be a suitable replacement for LC to enable high-throughput analysis.

The baseline separation afforded by DMS enables accurate determination of the peak area for acetaminophen, free from any contribution resulting due to downstream conversion of phenacetin. Figure 7.28 shows an example of this for injection of phenacetin, while monitoring the signal for the phenacetin MRM transition (a) and the acetaminophen MRM transition (b). LC was used for this demonstration to help visualize the interference from phenacetin (pane b, black trace), and its removal through the use of DMS (pane b, gray trace). In addition

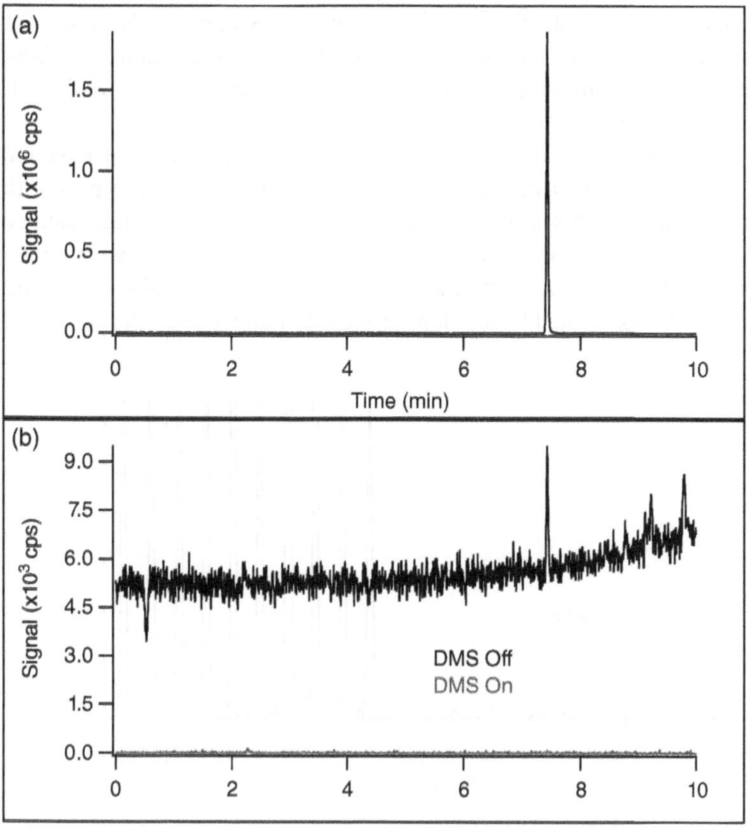

Figure 7.28 LC/MS data for injection of a phenacetin sample: (a) MRM signal for phenacetin and (b) MRM signal for acetaminophen.

to elimination of the signal due to fragmentation of phenacetin at a retention time of approximately 7.4 minutes, the general MRM background for the acetaminophen MRM transition decreased by more than an order of magnitude.

7.4.3.3 Example 3. DMS to Eliminate Unknown Interferences from Species Endogenous to the Solvent Matrix

In the previous examples, the potential interferences were well understood, as they were generated by alternate analytes of interest in the analyses. The situation is more complicated when it is not possible to predict them apriori. An example of this is endogenous species that may be present in a given sample matrix, as in many cases, the identity of these interferences may be completely unknown. One example of this relates to drug screening of fentanyl and its metabolite norfentanyl from urine matrices (structures provided in Appendix 7.A). Figure 7.29 shows ADE/OPP data taken for 10 repeat injections of desalted urine matrix blanks and 10 injections for a desalted urine matrix containing 7.8 ng/mL of fentanyl. For these data, the injection volume was 50 nL and the data were acquired without DMS separation. The blanks were relatively clean, with general background levels in the 100–200 cps range.

Figure 7.30 shows calibration curve data taken with three different injection volumes; 50 nL (a), 10 nL (b), and 2.5 nL (c) for desalted urine samples with concentrations down to 7.8 ng/mL. The fentanyl signal was normalized using a deuterated internal standard. In each case, linear calibration curves were generated, with good linearity. The LOQ was determined from 10 times the standard deviation of the blank divided by the slope of the calibration curve. The calculated

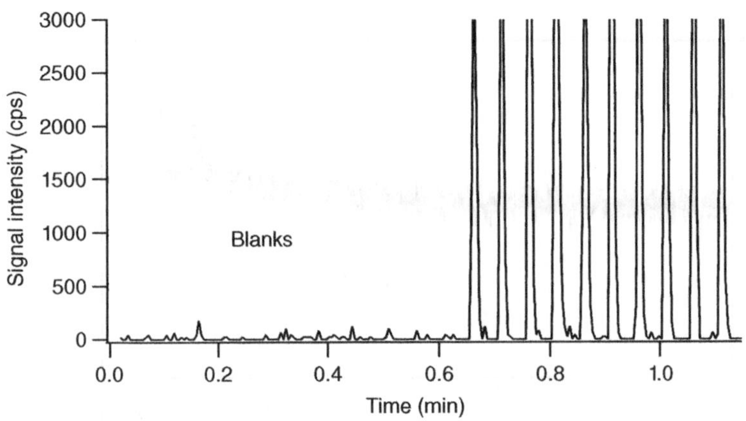

Figure 7.29 Raw data for injections of fentanyl samples prepared in desalted urine. The data show 10 blank injections followed by 10 injections for a desalted urine sample comprising 7.8 ng/mL fentanyl.

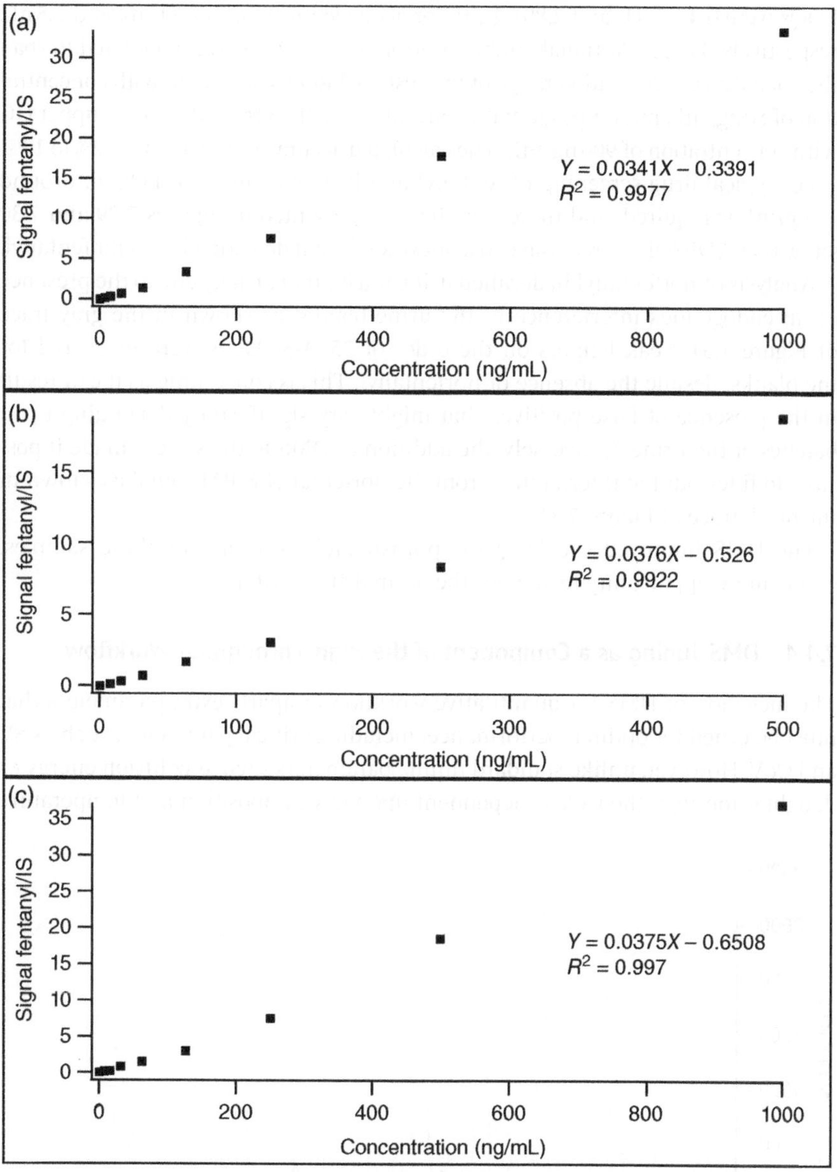

Figure 7.30 Calibration curves for fentanyl from a desalted urine sample acquired by injecting (a) 50 nL volume, (b) 10 nL volume, and (c) 2.5 nL volume.

LOQs were 0.46, 1.41, and 2.89 ng/mL for acoustic injections of 50, 10, and 2.5 nL, respectively. Three additional quality control samples (QCs) were prepared to span the analytical concentration range of interest, including a lower QC with concentration of 50 ng/mL, midrange QC with concentration of 350 ng/mL, and an upper QC with concentration of 900 ng/mL. The calculated accuracies ranged from 2% to 13%.

For typical drug screening of fentanyl and its metabolites, an LOQ of around 10 ng/mL is required, and therefore, the data presented in Figures 7.29 and 7.30 show that ADE/OPP screening can achieve sufficient detection limits for fentanyl.

Analysis of norfentanyl in desalted urine is a bit more tricky due to the presence of an endogenous interference in the urine blanks, as shown in the gray trace of Figure 7.31. Peak heights on the order of 2500–3000 cps were measured for the blanks, despite the absence of norfentanyl. This is concerning as it can result in the presence of false positives that might vary significantly depending upon batches of the urine. Conversely, the addition of DMS to the system made it possible to filter out the interference from the norfentanyl MRM signal as shown in the black trace of Figure 7.31.

The DMS improved the LOQ to approximately 5 ng/mL for these samples, giving more opportunity to achieve the desired 10 ng/mL LOQ.

7.4.4 DMS Tuning as a Component of the High-Throughput Workflow

The inclusion of DMS for quantitative workflows imparts extra parameters that must be tuned for optimal performance, including critical parameters such as SV and CoV. However, unlike standard tuning parameters such as collision energy or daughter ion *m/z*, the CoV is dependent upon gas composition and temperature

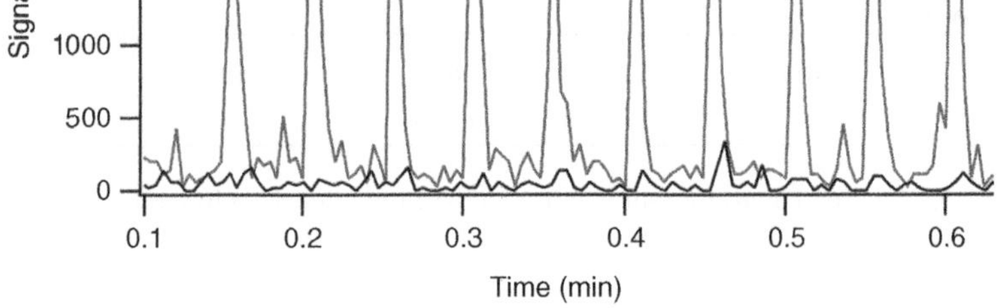

Figure 7.31 Blank injections for norfentanyl from a desalted urine sample with DMS separation (black trace) and without DMS separation (gray trace).

conditions of the DMS cell and transport gas flow. A consequence of this is that the optimal CoV must be determined under the specific conditions that are used for subsequent quantitative analysis. For standard LC/MS approaches with ESI, it is necessary to either inject a series of replicate injections under standard operational conditions, or the sample inlet line is replaced with a tee infusion set up to provide a continuous analyte signal for tuning while maintaining the standard mobile-phase flow rate and source temperature conditions. The conversion between different methods offers potential pitfalls for CoV determination, including lack of sufficient temperature equilibration or errors in flow rate. Difficulties determining the optimal DMS parameters for quantitative workflows are believed to be a main factor limiting general DMS adoption.

The combination of DMS with OPP reduces the complexity for CoV tuning and optimization. By changing the frequency of the acoustic droplet ejector while maintaining a constant total flow into the OPP interface, it is possible to maintain near-constant conditions within the source and DMS cell while generating either pulses of signal for an ion of interest or a continuous signal that allows for CoV ramping for optimization purposes. Conversion between pseudo-continuous mode for CoV optimization and pulsed signal mode for high-throughput quantitative analysis is trivial and the constant conditions ensure that the CoV values determined in pseudo-continuous mode remain ideal for pulsed mode.

Parameter optimization in mass spectrometry is not new. Automated programs have been developed to establish optimal MS conditions and store those conditions in databases [50]. The same is currently happening for DMS parameter optimization. Automated parameter optimization eliminates human error and will be critical for optimization of DMS parameters for hundreds or thousands of compounds that may be run using high-throughput quantitation platforms.

7.4.5 Automation of the Tuning Process

The relationship between the compensation field (C) and separation field (S) is presented in Eq. (7.3).

$$C = \frac{-\langle S\alpha \rangle}{1+\langle \alpha \rangle} \tag{7.3}$$

There are two direct consequences of Eq. (7.3). The first is that by measuring the compensation field at a series of different separation field settings, it is possible to determine the alpha function for a given analyte [34], and the second is that when the alpha function is known for a given compound, it is possible to calculate the CoV or field for any separation voltage or field setpoint. In the case of the former, the determination of alpha requires collection of a series of ionograms for a compound of interest under a series of different electric field conditions. One approach that has been described in the literature includes collection of 18 different ionograms with a range of SV settings from 0 V to the maximum setting, as shown in Figure 7.32.

Each ionogram is smoothed and centroided to determine the optimal CoV value and these values are fed into an automated calculator to determine the alpha curve. The mathematical calculations to determine the alpha function for a given ion are not trivial and a component of the calculation involves curve fitting. However, the approach uses the same fundamental calculations for each compound, so can be repeated rapidly once the basic formulas are included in an automated calculator. Since alpha is a fundamental intrinsic property for an ion under a given transport gas composition, the data are ideal for library storage. The time requirements for collecting the ionogram ramps for alpha determination are compatible with AEMS technology. A convenient volume for a single AEMS reservoir on a 384-well plate is 40 μL of sample. Typical dwell and pause times for MRM analysis are on the order of 20–200 and 15 ms, respectively. Using the 20 ms dwell time, it is possible to step across a 40 V CoV window with 0.1 V step size in approximately 14 seconds. The peak fitting algorithm for alpha determination requires sufficient coverage of SV space, and typically a total of 18 ionogram ramps is generated from SV = 0 V to high SV. This corresponds to a total time on the order of 252 seconds or 4.2 minutes per alpha acquisition. For the AEMS infusion method used in these studies, it was possible to generate continuous signal for 15–20 minutes prior to depleting the 40 μL of sample in a given well. While not absolutely necessary, it is possible to further reduce the time requirement for alpha curve generation because the first separation voltage steps have values

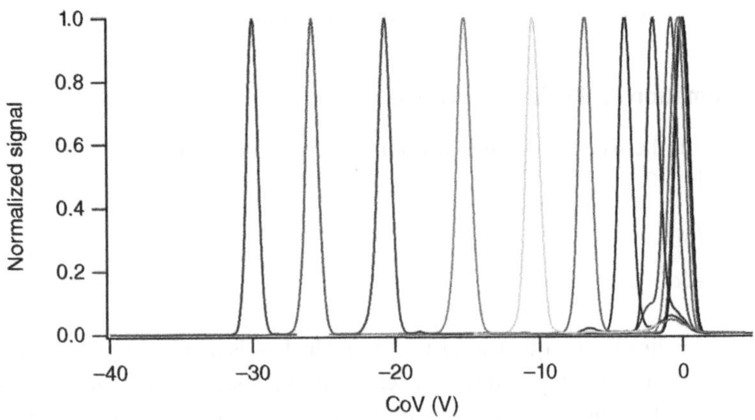

Figure 7.32 Overlay of a series of ionorams for determining the alpha curve for amobarbital.

below 10 Td and give negligible CoV shift. For these ionogram ramps, it would be possible to restrict the total ramping range to 10 V. It is also possible to increase the CoV step size to 0.2 V under some conditions to reduce the total time requirement by an additional factor of 2.

Current automated DMS optimization schemes involve collecting ionogram data for a series of different SV settings and determining the SV value that gives maximum ion signal. However, the alpha curve data described above opens the door to a new automated optimization process that can be realized by comparing alpha curves for different compounds. Once the alpha curve is known for a given compound, the fundamental DMS behavior can be determined for the compound under the specific transport gas conditions that were used for alpha determination. If the alpha curves are known for two compounds of interest, it is possible to use the curves to determine the optimal SV conditions for separating those compounds. An example of how this is done is provided for the structural isomers amobarbital and pentobarbital that have the structures shown in Appendix 7.A.

Figure 7.33 shows the alpha curves for amobarbital and pentobarbital determined under two different transport gas conditions: nitrogen transport gas and nitrogen with 3% acetonitrile modifier.

Under nitrogen transport gas conditions, the alpha curves for the two barbitals almost directly overlay, demonstrating classical Type C behavior [51]. The near-identical alpha functions ensure that the CoV measured for these two compounds are almost identical under any separation field tested during the generation of this plot, and therefore the DMS cannot separate them. The addition of acetonitrile dramatically changes the alpha functions for both barbital compounds to yield Type B behavior, where the alpha function becomes more positive with increasing

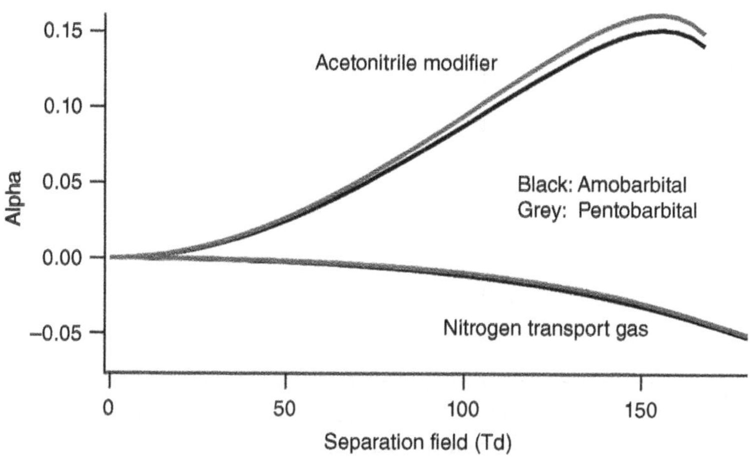

Figure 7.33 Alpha curves for amobarbital and pentobarbital.

separation field until a maximum value and then shifts to becoming more negative with further field increase. When acetonitrile is added to the transport gas, the alpha curves for the two barbitals diverge to some extent. A difference in the alpha value for a given field presents the opportunity to separate the two compounds in a DMS cell, and the greater the difference in alpha values, the greater the difference in optimal CoV. Generally, when the alpha value difference is greater than about 0.01, the two compounds can be baseline separated when operating the DMS cell with reasonable residence times. Referring to Figure 7.33, it is clear that the two compounds are separable when the separation field is set between about 125 and 160 Td. By comparing the two alpha curves, it is possible to determine the maximum difference in alpha across the entire range of the separation field. This can also be done in an automated fashion by simply subtracting one of the alpha curves from the other to generate a difference function as shown in Figure 7.34.

The maximum of the difference function provides the optimal separation field for separation of the two compounds. As shown in Figure 7.34, this corresponds to approximately 148 Td for the separation of amobarbital and pentobarbital in nitrogen with 3% acetonitrile transport gas. This approach enables an automated optimization approach where a user may specify maximum separation for two compounds, rather than simply maximum signal for a given compound. Determination of the separation voltage and CoV settings for the DMS is trivial once the desired field values and transport gas temperature are known. Figure 7.35 shows baseline separation of a mixture of amobarbital and pentobarbital using the automated optimization approach described above.

Thus far, the data described in this section have covered a novel approach for automated optimization of separations for two compounds of interest. However, there are many situations where more than just one isobaric interferences may be present

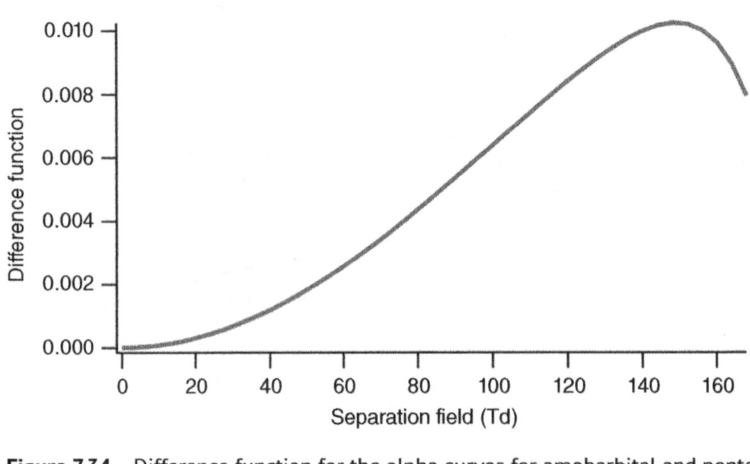

Figure 7.34 Difference function for the alpha curves for amobarbital and pentobarbital taken with 3% acetonitrile in the transport gas.

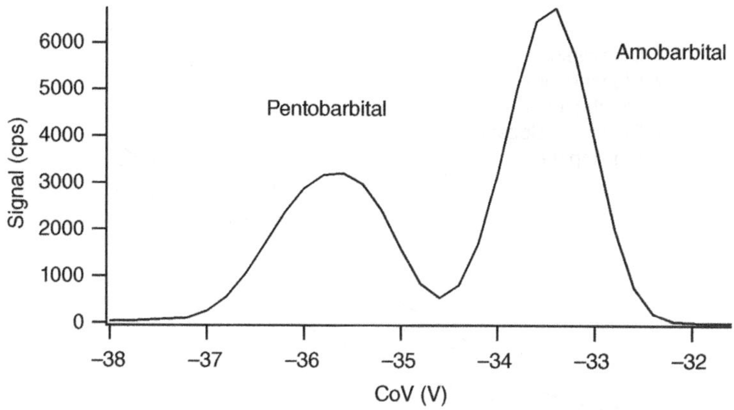

Figure 7.35 DMS separation for a mixture containing pentobarbital and amobarbital.

for a given compound. The alpha curve approach can also be used in these more complex situations. Figure 7.36 provides an example of alpha curves generated for five different isobaric or near-isobaric benzodiazepines taken with 3% ethylacetate modifier (structures provided in Appendix 7.A). The compounds span an m/z range of 313–316 and have significant isobaric interferences from the protonated ions and isotopes. Chemical clustering with the ethylacetate modifier is observed for all five compounds as indicated by the positive slope of the alpha curves. The alpha curve for flunitrazepam has an unconventional shape, overlapping with the curves for clonazepam and desmethylclozapine at different separation fields. In addition, the alpha curves for olanzapine and clonazepam diverge at relatively low separation field but converge at the highest measured separation fields. The data in Figure 7.36 show the difficulty with trying to determine the optimal separation field by trial and error methods using a few different tested SV values. The alpha curve data of Figure 7.36 provide very useful information for optimizing separations. For instance, if the goal is to separate clonazepam and amoxapine, it is desirable to set the highest possible separation field, or about 200 Td in this case. Conversely, if the goal were to separate olanzapine and clonazepam, 200 Td would be a poor choice of separation field. For the data presented below, the goal was to separate all of the five compounds to ensure that there would be no issue with isobaric interferences. Under these conditions, an optimal separation field would be around 148 Td.

Figure 7.37 shows an example of the DMS separation that is possible when using the optimized value from the plot of Figure 7.36 with the DMS peak width set to approximately 3 V at the base.

It is possible to set specific CoV values to selectively transmit each of the different compounds. Automated alpha determination, creation of alpha libraries, and automated determination of the best DMS settings for various separations are all possible today, and this will be critical for enabling DMS for high-throughput analysis.

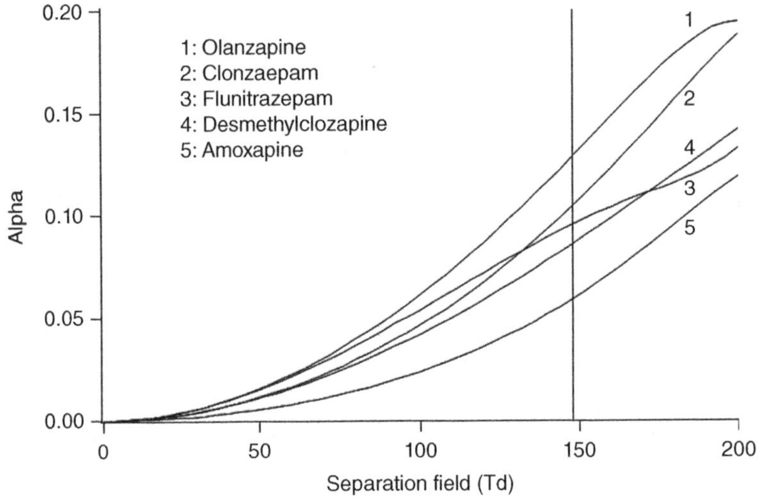

Figure 7.36 Alpha curves for a group of five isobaric and near-isobaric benzodiazepines with *m/z* from 313 to 316, measured with 3% ethylacetate in the nitrogen transport gas. Optimal separations for these five compounds can be achieved with the separation field set between 140 and 150 Td.

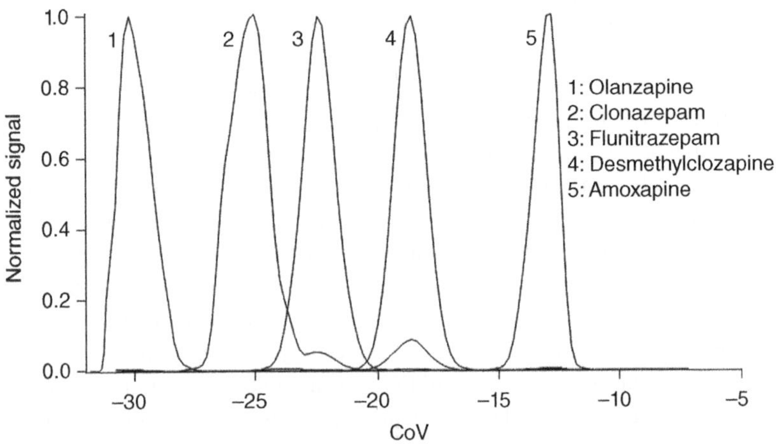

Figure 7.37 Alpha example of the separation with the separation field set to 143 Td.

7.5 Conclusions

High-throughput screening approaches that target analysis of tens of thousands of samples per day with throughput on the order of 1 Hz or greater are not presently compatible with liquid chromatography separations. The elimination of LC significantly reduces analytical selectivity, substantially increasing the likelihood of isobaric interferences in a tandem mass spectrometry measurement from species endogenous or exogenous to the sample matrix. While its peak capacity is generally lower than LC approaches, DMS presents an alternative method to augment selectivity for high-throughput screening. When compared to other ion mobility approaches, DMS provides a unique combination of fast separations and orthogonality to mass separation, and this makes it ideal for coupling to high-throughput MS workflows such as those based upon the open port interface. DMS theoretical understandings have progressed to the point that it is now possible to run fully automated processes to not only determine optimal DMS set points, but to define the best separation conditions for two or more compounds. Moving forward, this will enable the creation of libraries of DMS information such as alpha curves for defining peak positions and relative separations. More work is needed on the software and processing; however, the seeds are in place for development of fully automated high-throughput MS/MS screening systems that employ DMS separations in place of liquid chromatography.

7.A Chemical Structures

Figures 7.2 and 7.3: Temazepam and Chlordiazepoxide

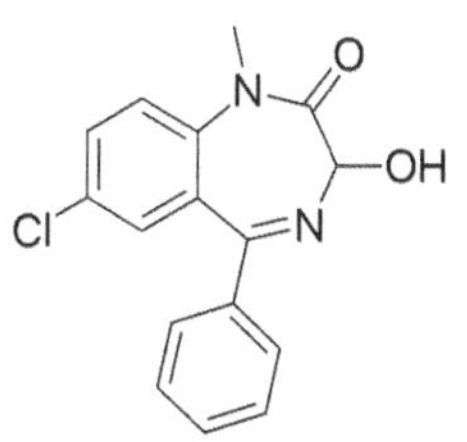

Temazepam

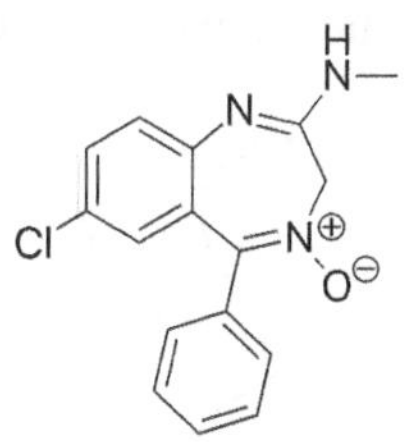

Chlordiazepoxide

Figure 7.4: Morphine, Hydromorphone, and Norhydrocodone

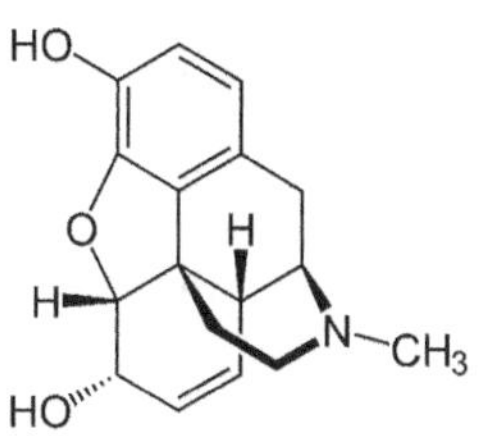

Morphine

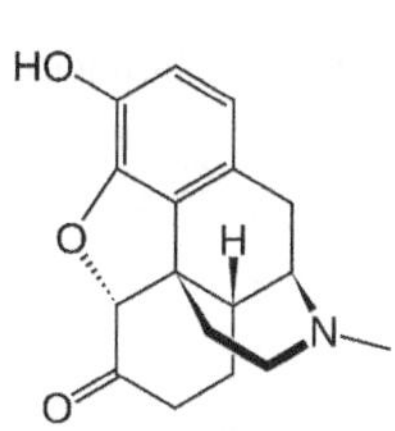

Hydromorphone

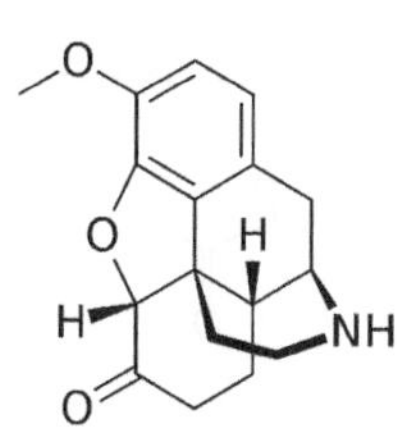

Norhydrocodone

Figure 7.5: Oxymorphone, Dihydrocodeine, and Noroxycodone

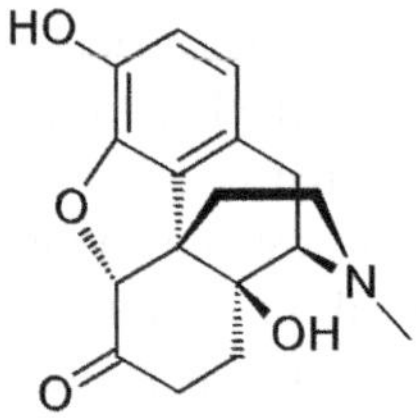
Oxymorphone

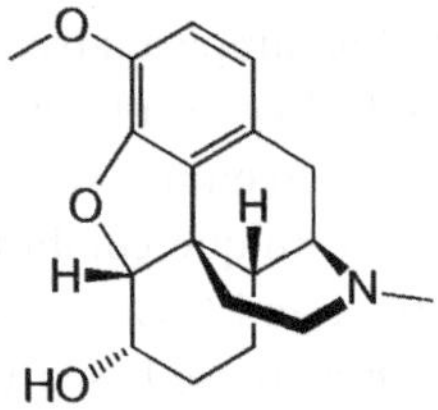
Dihydrocodeine

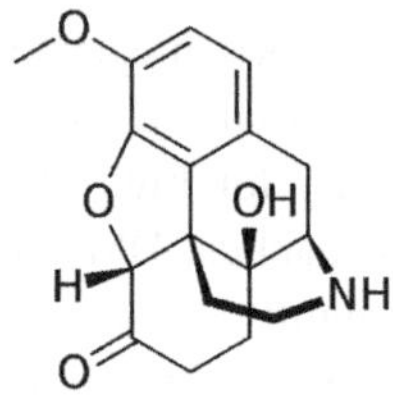
Noroxycodone

Figures 7.14 and 7.25, 7.26: Clozapine and Midazolam

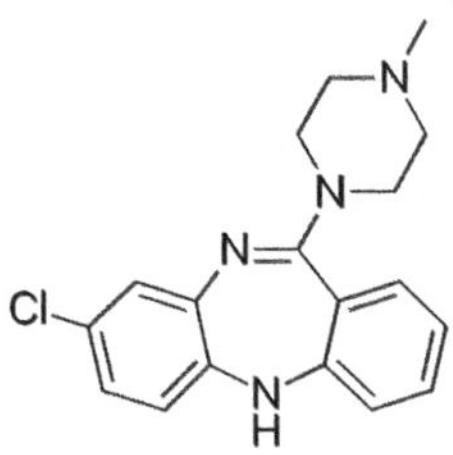
Clozapine

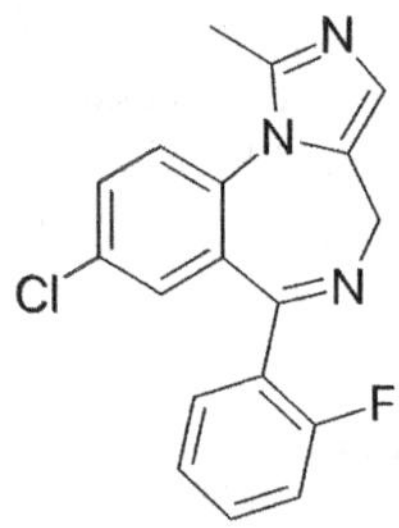
Midazolam

Figures 7.19–7.24: Mirtazapine and Desmethyldoxepin

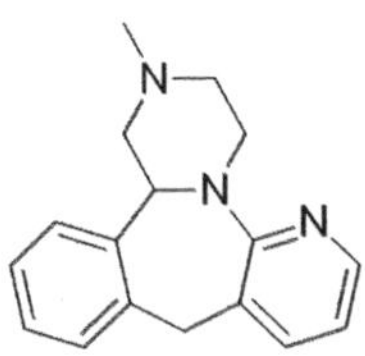
Mirtazapine

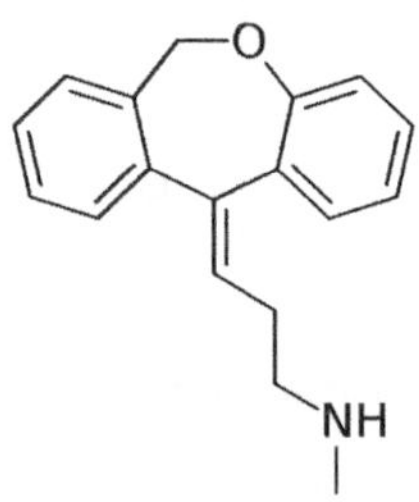
Desmethyldoxepin

Figures 7.27 and 7.28: Acetaminophen and Phenacetin

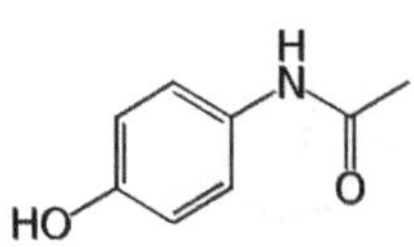
Acetaminophen

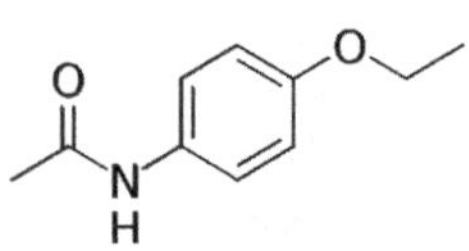
Phenacetin

Figures 7.29–7.31: Fentanyl and Norfentanyl

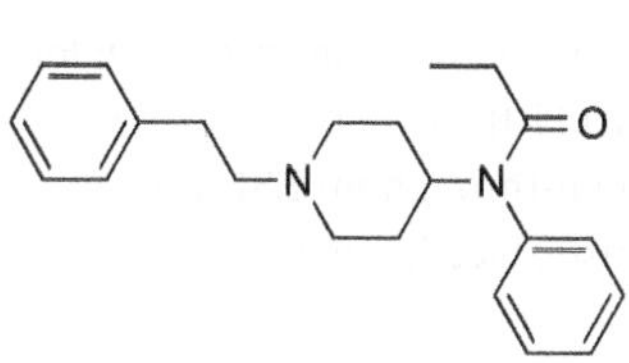

Fentanyl

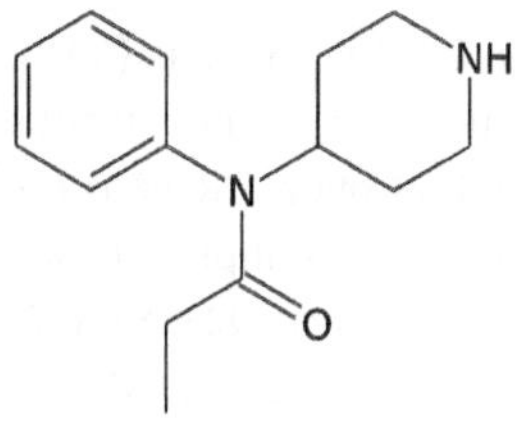

Norfentanyl

Figures 7.32–7.35: Amobarbital and Pentobarbital

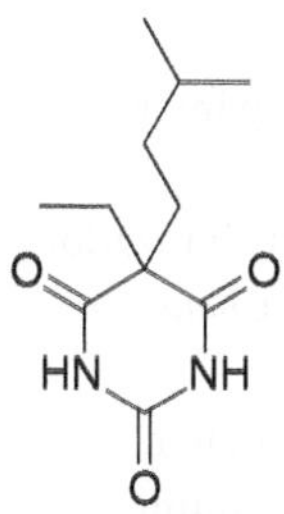

Amobarbital

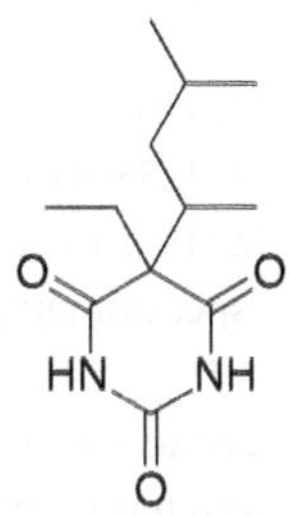

Pentabarbital

Figures 7.36–7.37: Olanzapine, Clonazepam, Flunitrazepam, Desmethylclozapine, and Amoxapine

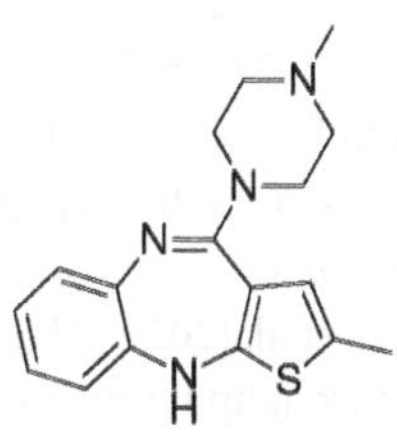

Olanzapine

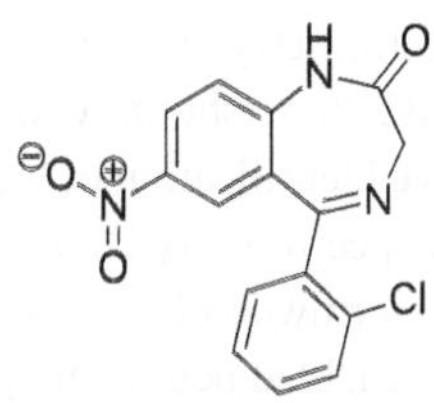

Clonazepam

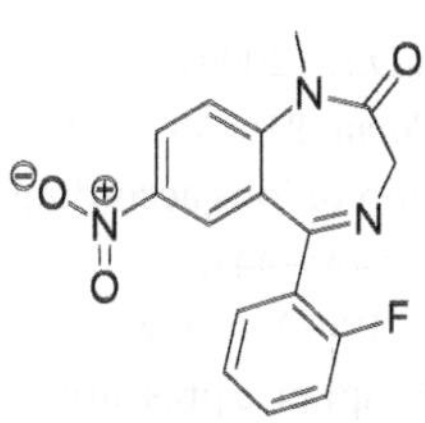

Flunitrazepam

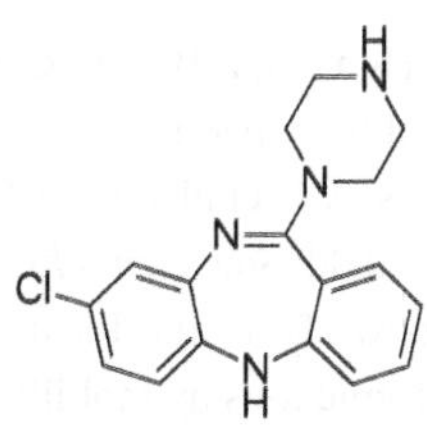

Desmethylclozapine

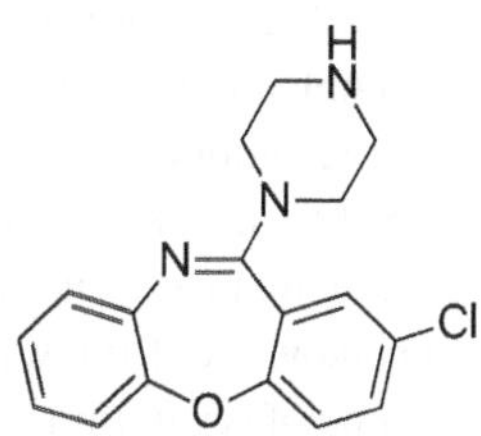

Amoxapine

References

1 Zhang, H., Liu, C., Hua, W. et al. (2020). Acoustic ejection mass spectrometry for high-throughput analysis. *bioRxiv* https://doi.org/10.1101/2020.01.28.923938.

2 Häbe, T.T., Liu, C., Covey, T.R. et al. (2020). Ultrahigh-throughput ESI-MS: sampling pushed to six samples per second by acoustic ejection mass spectrometry. *Anal. Chem.* 92 (18): 12242–12249.

3 Liu, C., Van Berkel, G.J., Cox, D.M., and Covey, T.R. (2020). Operational modes and speed considerations of an acoustic droplet dispenser for mass spectrometry. *Anal. Chem.* 92 (24): 15818–15826.

4 Liu, C., Van Berkel, G.J., Kovarik, P. et al. (2021). Fluid dynamics of the open port interface for high speed nanoliter volume sampling mass spectrometry. *Anal. Chem.* 93 (24): 8559–8567.

5 Hoaglund, C.S., Valentine, S.J., and Clemmer, D.E. (1997). An ion trap interface for ESI-ion mobility experiments. *Anal. Chem.* 69: 4156–4161.

6 Wu, C., Siems, W.F., Asbury, R., and Hill, H.H. Jr. (1998). Electrospray ionization high-resolution ion mobility spectrometry-mass spectrometry. *Anal. Chem.* 70: 4929–4938.

7 De La Mora, J.F., Thomson, B.A., and Gamero-Castano, M. (2005). Tandem mobility mass spectrometry study of electrosprayed tetraheptyl ammonium bromide clusters. *J. Am. Soc. Mass Spectrom.* 16: 717–732.

8 Giles, K., Pringle, S.D., Worthington, Ḳ.R. et al. (2004). Applications of a traveling wave-based radio-frequency-only stacked ring ion guide. *Rapid Commun. Mass Spectrom.* 18: 2401–2414.

9 Silveira, J.A., Ridgeway, M.E., and Park, M.A. (2014). High resolution trapped ion mobility spectrometry of peptides. *Anal. Chem.* 86: 5624–5627.

10 Ibrahim, Y.M., Shvartsburg, A.A., Smith, R.D., and Belov, M.E. (2011). Ultrasensitive identification of localization variants of modified peptides using ion mobility spectrometry. *Anal. Chem.* 83: 5617–5623.

11 Arndt, J.R., Wormwood Moser, K.L., Van Aken, G. et al. (2021). High-resolution ion-mobility-enabled peptide mapping for high-throughput critical quality attribute monitoring. *J. Am. Soc. Mass Spectrom* 32 (8): 2019–2032. https://doi.org/10.1021/jasms.0c00434.

12 Merenbloom, S.I., Glaskin, R.S., Henson, Z.B., and Clemmer, D.E. (2009). High-resolution ion cyclotron mobility spectrometry. *Anal. Chem.* 81: 1482–1487.

13 Buryakov, I.A., Krylov, E.V., Makas, A.L. et al. (1991). Separation of ions according to mobility in a strong AC electric field. *Sov. Tech. Phys. Lett.* 17: 446–447.

14 Buryakov, I.A., Krylov, E.V., Nazarov, E.G., and Rasulev, U.K. (1993). A new method of separation of multi-atomic ions by mobility at atmospheric pressure using a high-frequency amplitude-asymmetric strong electric field. *Int. J. Mass Spectrom. Ion Process.* 128: 143–148.

15 Gorshkov, M.P. (1982). Inventor's certificate of USSR No. 966583, G01N27/62.

16 Carnahan, B., Day, S., Kouznetsov, V., and Tarassov, A. (1995). Development and applications of transverse field compensation ion mobility spectrometer. *Proceeding of an Fourth International Workshop on Ion Mobility Spectrometry*, Cambridge, UK.

17 Carnahan, B., Day, S., Kouznetsov, M., et al. (1996). Field ion spectrometry – a new analytical technology for trace gas analysis. *Proceedings of the 41st Annual ISA Analysis Division Symposium* (21–24 April 1996), vol. 51 (1), pp. 87–96. Framingham, MA: ISA.

18 Purves, R.W. and Guevremont, R. (1999). Electrospray ionization high-field asymmetric waveform ion mobility spectrometry-mass spectrometry. *Anal. Chem.* 71: 2346–2357.

19 Krylov, E.V. (2003). Comparison of the planar and coaxial field asymmetrical waveform ion mobility spectrometer (FAIMS). *Int. J. Mass Spectrom.* 225: 39–51.

20 Barnett, D.A., Ells, B., Guevremont, R., and Purves, R.W. (1999). Separation of leucine and isoleucine by electrospray ionization-high field asymmetric waveform ion mobility spectrometry-mass spectrometry. *J. Am. Soc. Mass Spectrom.* 10: 1279–1284.

21 Ells, B., Barnett, D.A., Froese, K. et al. (1999). Detection of chlorinated and brominated byproducts of drinking water disinfection using electrospray ionization-high-field asymmetric waveform ion mobility spectrometry-mass spectrometry. *Anal. Chem.* 71: 4747–4752.

22 Barnett, D.A., Ells, B., Guevremont, R. et al. (2000). Evaluation of carrier gases for use in high-field asymmetric waveform ion mobility spectrometry. *J. Am. Soc. Mass Spectrom.* 11: 1125–1133.

23 Barnett, D.A., Belford, M., Dunyach, J.J., and Pruves, R.W. (2007). Characterization of a temperature-controlled FAIMS system. *J. Am. Soc. Mass Spectrom.* 18: 1653–1663.

24 Wu, S.T., Xia, Y.Q., and Jemal, M. (2007). High-field asymmetric waveform ion mobility spectrometry coupled with liquid chromatography/electrospray ionization tandem mass spectrometry (LC/ESI-FAIMS-MS/MS) multi-component bioanalytical method development, performance evaluation and demonstration of the constancy of the compensation voltage with change of mobile phase composition or flow rate. *Rapid Commun. Mass Spectrom.* 21: 3667–3676.

25 Purves, R.W., Ozog, A.R., Ambrose, S.J. et al. (2014). Using gas modifiers to significantly improve sensitivity and selectivity in a cylindrical FAIMS device. *J. Am. Soc. Mass Spectrom.* 25: 1274–1284.

26 Purves, R.W., Prasad, S., Belford, M. et al. (2017). Optimization of a new aerodynamic cylindrical FAIMS device for small molecule analysis. *J. Am. Soc. Mass Spectrom.* 28: 525–538.

27 Shvartsburg, A., Smith, R.D., Wilks, A. et al. (2009). Ultrafast differential ion mobility spectrometry at extreme electric fields in multichannel microchips. *Anal. Chem.* 81 (15): 6489–6495.

28 Nazarov, E.G. (2012). A journey into DMS/FAIMS technology. *Int. J. Ion Mobil. Spectrom.* 15: 83–84.

29 Eiceman, G.A., Nazarov, E.G., and Miller, R.A. (2000). A micro-machined ion mobility spectrometer-mass spectrometer. *Int. J. Ion Mobil. Spectrom.* 3: 15–27.

30 Eiceman, G.A., Tadjikov, B., Krylov, E. et al. (2001). Miniature radio-frequency mobility analyzer as a gas chromatographic detector for oxygen-containing volatile organic compounds, pheromones and other insect attractants. *J. Chromatogr. A* 917: 205–217.

31 Krylov, E., Nazarov, E.G., Miller, R.A. et al. (2002). Field dependence of mobilities for gas-phase-protonated monomers and proton-bound dimers of ketones by planar field asymmetric waveform ion mobility spectrometer (PFAIMS). *J. Phys. Chem. A* 106: 5437–5444.

32 Krylova, N., Krylov, E., Eiceman, G.A., and Stone, J.A. (2003). Effect of moisture on the field dependence of mobility for gas-phase ions of organophosphorus compounds at atmospheric pressure with field asymmetric ion mobility spectrometry. *J. Phys. Chem. A* 107: 3648–3654.

33 Schneider, B.B., Nazarov, E.G., Londry, F., and Covey, T.R. (2015). Comparison of the peak capacity for DMS filters with various gap height: experimental and simulations results. *Int. J. Ion Mobil. Spectrom.* 18: 159–170.

34 Schneider, B.B., Nazarov, E.G., Londry, F. et al. (2016). Differential mobility spectrometry/mass spectrometry history, theory, design optimization, simulations, and applications. *Mass Spectrom. Rev.* 35: 687–737.

35 Shvartsburg, A.A., Tang, K., Smith, R.D. et al. (2009). Ultrafast differential ion mobility spectrometry at extreme electric fields coupled to mass spectrometry. *Anal. Chem.* 81: 8048–8053.

36 Schneider, B.B., Covey, T.R., Coy, S.L. et al. (2010). Planar differential mobility spectrometer as a pre-filter for atmospheric pressure ionization mass spectrometry. *Int. J. Mass Spectrom.* 298: 45–54.

37 Shvartsburg, A.A., Prior, D.C., Tang, K., and Smith, R.D. (2010). High-resolution differential ion mobility separations using planar analyzers at elevated dispersion fields. *Anal. Chem.* 82: 7649–7655.

38 Shvartsburg, A.A. and Smith, R.D. (2011). Ultrahigh-resolution differential ion mobility spectrometry using extended separation times. *Anal. Chem.* 83: 23–29.

39 Schneider, B.B., Covey, T.R., Coy, S.L. et al. (2010). Control of chemical effects in the separation process of a differential mobility mass spectrometer system. *Eur. J. Mass Spectrom.* 16: 57–71.

40 Schneider, B.B., Covey, T.R., Coy, S.L. et al. (2010). Chemical effects in the separation process of a differential mobility/mass spectrometer system. *Anal. Chem.* 82: 1867–1880.

41 Rorrer, L.C. III and Yost, R.A. (2011). Solvent vapor effects on planar high-field asymmetric waveform ion mobility spectrometry. *Int. J. Mass Spectrom.* 300: 173–181.

42 Schneider, B.B., Covey, T.R., and Nazarov, E.G. (2013). DMS-MS separations with different transport gas modifiers. *Int. J. Ion Mobil. Spectrom.* 16: 207–216.

43 Van Berkel, G.J. and Kertesz, V. (2015). An open port sampling interface for liquid introduction atmospheric pressure ionization mass spectrometry. *Rapid Commun. Mass Spectrom.* 29 (19): 1749–1756.

44 Schneider, B.B., Londry, F., Nazarov, E.G. et al. (2017). Maximizing ion transmission in differential mobility spectrometry. *J. Am. Soc. Mass Spectrom.* 28: 2151–2159.

45 Barnett, D.A. and Ouellette, R.J. (2011). Elimination of the helium requirement in high-field asymmetric waveform ion mobility spectrometry (FAIMS): beneficial effects of decreasing the analyzer gap width on peptide analysis. *Rapid Commun. Mass Spectrom.* 25: 1959–1971.

46 Barnett, D.A., Ells, B., Guevremont, R., and Purves, R.W. (2002). Application of ESI-FAIMS-MS to the analysis of tryptic peptides. *J. Am. Soc. Mass Spectrom.* 13: 1282–1291.

47 Wang, J.J., Guo, J.J., Zhan, J. et al. (2014). An in-vitro cocktail assay for assessing compound-mediated inhibition of six major cytochrome P450 enzymes. *J. Pharm. Anal.* 4: 270–278.

48 Lim, K.B., Ozbal, C.C., and Kassel, D.B. (2010). Development of a high-throughput online-phase extraction/tandem mass spectrometry method for cytochrome P450 inhibition screening. *J. Biomol. Screening* 15: 447–452.

49 Weaver, R., Graham, K.S., Beattle, I.G. et al. (2003). Cytochrome P450 inhibition using recombinant proteins and mass spectrometry/multiple reaction monitoring technology in a cassette incubation. *Drug Metab. Dispos.* 31: 955–966.

50 Laycock, J. (2010). Integration and streamlining of robotics, automated tuning LC-MS/MS and LIMS workflows for high-throughput in vitro ADME bioanalysis. *Presented at: 58th American Society of Mass Spectrometry Conference on Mass Spectrometry and Allied Topics*, Salt Lake City, UT (May 23–25 2010).

51 Guevremont, R. and Purves, R.W. (1999). Atmospheric pressure ion focusing in a high-field asymmetric waveform ion mobility spectrometer. *Rev. Sci. Instr.* 70: 1370–1383.

Section 4

Special Sample Arrangement

8

Off-Line Affinity Selection Mass Spectrometry and Its Application in Lead Discovery

Christopher F. Stratton, Lawrence M. Szewczuk, and Juncai Meng

Discovery Technologies and Molecular Pharmacology (DTMP), Janssen Research & Development, LLC, Spring House, PA, USA

8.1 Introduction to Off-Line Affinity Selection Mass Spectrometry

Affinity selection mass spectrometry (ASMS) is an unbiased biophysical assay that connects a binding event with compound identity via accurate mass. ASMS is widely applied in hit identification and hit validation during early small-molecule drug discovery and is established as a high-throughput screening (HTS) approach complementary to traditional HTS or DNA-encoded library (DEL) strategies [1]. With the recent emergence of targeted protein degradation as a novel therapeutic modality, ASMS has gained popularity as a rapid method to identify small-molecule binders of E3 ligases or target proteins to potentially form heterobifunctional PROTAC molecules [2]. In the most simplistic form of ASMS, compounds (either in mixtures or as single entities) are incubated with a target protein and then an affinity selection step is used to separate the bound from unbound fraction of compounds. Compounds retained in the bound fraction are identified by traditional liquid chromatography-mass spectrometry (LC-MS) or matrix-assisted laser desorption ionization (MALDI) mass spectrometry. Although a variety of ASMS workflows have been reported, these approaches can generally be divided into two main categories: "on-line" and "off-line." In the on-line methodology, the affinity selection step is coupled to the identification step using two-dimensional LC coupled with MS; by contrast, off-line ASMS methods implement affinity selection and compound identification in two discrete steps (Figure 8.1). There are advantages and disadvantages to each of these methods, namely on-line

High-Throughput Mass Spectrometry in Drug Discovery, First Edition. Edited by Chang Liu and Hui Zhang.
© 2023 John Wiley & Sons, Inc. Published 2023 by John Wiley & Sons, Inc.

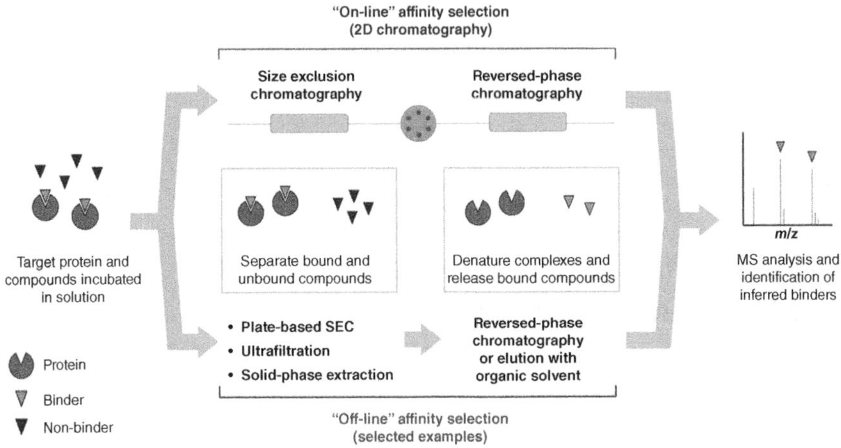

Figure 8.1 Schematic diagram of on-line versus off-line affinity selection mass spectrometry.

platforms are easy to automate, but lack the ability to execute parallel affinity selections; whereas the parallel nature of off-line methods provides higher throughput, but introduces challenges to full automation of the process. In the following chapter, we seek to introduce several approaches for off-line ASMS and discuss the pros and cons for these methods in the context of small-molecule lead discovery. On-line ASMS will be discussed in Chapter 9.

8.2 Selected Off-Line Affinity Selection Technologies and Its Application in Lead Discovery

8.2.1 Membrane Ultrafiltration-Based Affinity Selection

8.2.1.1 Introduction of Membrane Ultrafiltration-Based ASMS

Membrane ultrafiltration was first developed in early 1980s for serum protein binding determinations [3, 4]. As a solution-based affinity selection approach, ultrafiltration takes advantage of the pore size of a filter membrane, allowing separation of large molecules from small molecules. The workflow of the ultrafiltration-based ASMS involves: incubation of a library of small molecules with a therapeutic protein target leading to the formation of a protein–ligand complex; separation of the protein–ligand complex from unbound compounds using a molecular weight cutoff ultrafiltration membrane; and identification of bound or unbound compounds via mass spectrometry. During the affinity selection step, an assay solution is transferred into a plate or a tube containing

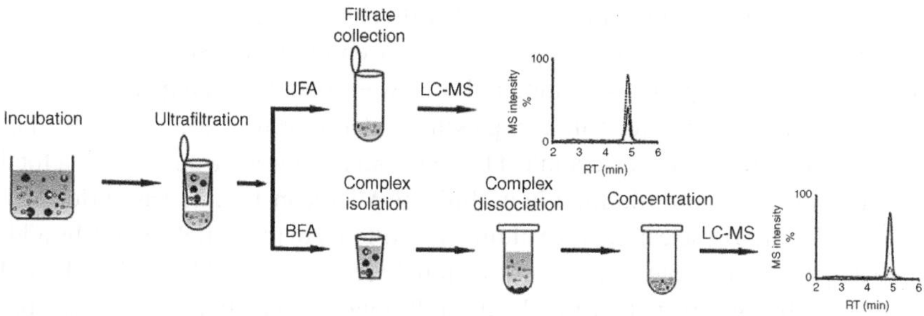

Figure 8.2 Membrane ultrafiltration-based affinity selection mass spectrometry on unbound fraction analysis (UFA, top) versus bound fraction analysis (BFA, bottom). *Source:* Reproduced with permission from Qin et al. [5]/Elsevier.

ultrafiltration membrane. When a pressure is applied using a pump, vacuum, or centrifugal force, free protein and protein–ligand complex exceeding the mass cutoff of the membrane are retained while buffer and unbound compounds pass through the membrane, allowing effective separation of the bound fraction from the unbound fraction (Figure 8.2).

The fraction containing protein–ligand complex can be collected and treated with organic solvents or pH change to facilitate dissociation of bound ligands from the complex, followed by direct characterization of small-molecule binders in LC-MS. To reduce the time required for liquid chromatography, MALDI can be used for direct analysis of the protein–ligand complex fraction at sampling rate of one to three seconds per well [6]. Ligands can be dissociated from the protein–ligand complex on MALDI plate surface by denaturation in matrix or under impact of laser irradiation, followed by detection in MALDI mass spectrometer.

Alternatively, the protein–ligand binding event can be indirectly measured by analysis of the unbound compounds that pass through the membrane. By careful comparison of unbound small molecules in LC-MS to that of the control assay using denatured protein target, small molecules that exhibit significant decrease in recovery indicate potential binding against protein target of interest. While both direct and indirect analysis approaches have been reported for hit identification in the literature, indirect analysis using unbound fraction usually involves fewer steps, but requires more complex analyses of the affinity selection data [5, 7].

8.2.1.2 Application of Membrane Ultrafiltration-Based ASMS in Lead Discovery

Although the general principle and workflow of membrane ultrafiltration is straightforward, it was not until the late 1990s that membrane ultrafiltration-based ASMS was implemented in target-based drug discovery as a means to identify small-molecule binders [8–10]. Researchers at Abbot have used membrane ultrafiltration-based affinity selection for screening of 15 targets against a library of

230,000 small molecule compounds [11–15]. The screens were carried out in mixtures of 700–3000 compounds per well at concentration of 1.5 μM each member against 10 μM target proteins. Compounds that were identified in duplicate primary screens and met the quantitative picking criteria were selected for further confirmation. To reduce the chemical background and enrich true binders, a total of three rounds of ultrafiltration in a Millipore Microcon-10 unit was performed. After each round of selection, the volume was restored to the initial volume by adding back the target protein at 10% of the initial concentration. Although unbound ligands are being depleted during the ultrafiltration, with adding protein back the initial equilibrium quantity of bound ligand is maintained. This is particularly important for discovery of weak binders as adding back protein solution allows the enrichment to be achieved based on equilibrium rather than dissociation rate. The final retentate containing free protein and protein–ligand complex was collected and extracted using organic solvents to remove buffers and proteins. The organic extracts were subsequently subjected to LC-MS analysis for characterization of small-molecule binders. Using membrane ultrafiltration-based affinity selection mass spectrometry, small-molecule hits were successfully identified for oncology targets Bcl-xL [15], mammalian checkpoint kinase CHK1 [13], anti-infective target *Streptococcus* enzyme MurF, and the oncology target MetAP2 [12].

Qian et al. reported screening of Bcl-xL, a member of the Bcl-2 family of proteins, that plays a key role in the mechanism of apoptosis. A library of 263,382 compounds was pooled into mixtures of approximately 2400 compounds per well for ASMS screen [15]. High protein concentration is selected to increase the chances of finding weak binders. A screening strategy using two sets of orthogonal mixtures was implemented. Out of ~263k compounds screened in duplicates, 29 small molecules were identified as Bcl-xL binders with affinity below 100 μM, representing an overall hit rate of 0.012%. Further characterization of the screening hits (estimated K_d 1–4 μM in ASMS) in both a fluorescence polarization (FP) assay and NMR assay led to the confirmation of two classes of novel binders to Bcl-xL with IC_{50} 2–8 μM in FP assay, which could potentially serve as chemistry starting points for the development of anticancer drugs. It is worth noting that in the Bcl-xL primary screen, compounds providing strong MS intensities were often not true binders, although they still met empirically established criteria for selection of ligands.

Chen et al. reported a screen of a fragment-based combinatorial library against the antibiotic-resistance target New Delhi metallo-lactamase1 (NDM-1) using membrane ultrafiltration-based ASMS [16]. Membrane Amicon Ultra-0.5 mL with mass cutoff at 10 kDa from Millipore (Bedford, MA) was used in this study. Relatively high concentrations of protein (25 μM) and compounds (25–50 μM) were used in the screen to allow for identification of weak fragment binders against NDM-1. As a negative control, glutathione S-transferase (GST) was introduced as a blocking protein to reduce nonspecific binding of compounds to the filter membrane in ultrafiltration step. It was found that this strategy could remarkably reduce false positives and enhance the selectivity and accuracy of the NDM-1 ASMS screen. Membrane ultrafiltration-based ASMS has also been employed to screen and study the interaction between biological targets and active components from medicinal plants [17–21].

8.2.1.3 Pulse Ultrafiltration-Based ASMS Technology

Different from traditional off-line ultrafiltration ASMS, van Breemen and coworkers reported pulse ultrafiltration mass spectrometry for identification of ligands in library mixtures that bind to solution-phase receptors [22]. This approach can also be considered as an on-line ASMS method, but since it employs ultrafiltration membrane for affinity selection with same principle to any other membrane ultrafiltration-based ASMS, it is included here for discussion. The system consists of an ultrafiltration chamber coupled with a mass spectrometer. The chamber is fitted with a Teflon-coated magnetic stir bar and a methylcellulose ultrafiltration membrane. When a pulse of analyte solution containing a mixture of compounds and a protein target in assay buffer is injected into the chamber, the formed protein–ligand complex is retained on the ultrafiltration membrane while unbound small molecules pass through along with mobile phase. Then an organic solvent or acid is added into the pulse to dissociate ligands from the complex, and the released ligands are characterized by LC-MS (Figure 8.3). van Breemen's lab has successfully implemented this method for screening of xenobiotic compounds [23], study of metabolic stability of drugs [24], characterization of inhibitors of cyclooxygenase-2 [25], and identification of binders for human retinoid X receptor [26] as well as inhibitors of dihydrofolate reductase (DHFR) [27]. Similar to other ultrafiltration-based ASMS methods, the rate-limiting step of this method is mainly from filtration step where lengthy wash time is needed to remove unbound ligands completely. It should be noted that nonspecific binding of small molecules to the ultrafiltration membrane is observed. However, one of the advantages of pulse ultrafiltration ASMS is that the protein can potentially be used repeatedly by careful treatment of the protein–ligand complex in a manner where bound ligands can be released without irreversibly denaturing the protein. Unfortunately, due to the requirement of a specialized operation unit that consists of an ultrafiltration chamber, this approach has not been adopted by other laboratories where traditional ultrafiltration or other affinity selection methods dominate.

8.2.1.4 Affinity Rank-Ordering Using Pulse Ultrafiltration-Based ASMS

In addition to screening large numbers of compounds for hit identification, several ASMS strategies have been developed for affinity rank-ordering or measurement of relative binding affinities ACE50 during hit validation [28–30]. van Breemen and coworker reported a method to rank order compounds that

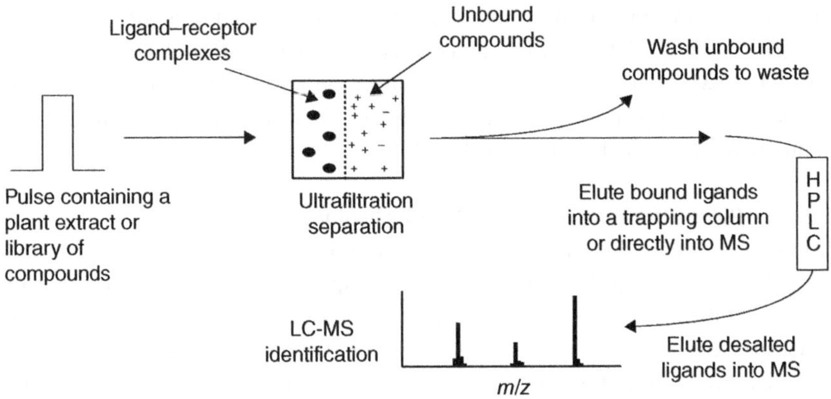

Figure 8.3 Pulsed ultrafiltration-mass spectrometry (PUF-MS) for screening of a target protein against chemical mixtures of compounds or plant extracts. *Source:* Reproduced with permission from Johnson et al. [22]/John Wiley & Sons.

inhibit the aggregation of amyloid β protein 1-40 (Aβ1-40) [31]. Aβ1-40 was incubated with a mixture of small molecules and then subjected to membrane ultrafiltration and mass spectrometric analysis. The concentration of monomer Aβ1-40 in the filtrate was measured based on the linear calibration curve of the analyte established. Compounds that inhibit the Aβ1-40 aggregation can be rank ordered based on the percentage of monomer Aβ1-40 present in the filtrate. A similar strategy was employed to measure binding percentage of screening hits against DHFR, where a good correlation between the inhibition constants (K_i values) and the extent of ASMS binding of inhibitors was observed [32].

Qin et al. further developed this approach for affinity measurement of fragment mixtures against target proteins using ultrafiltration-based ASMS [5]. The concentration of unbound compounds in the filtrate was determined based on a calibration curve prepared using the mixture of the compounds. Given the protein target is usually in excess of ligands in the screen with negligible competition, the binding affinity of the compounds can be then calculated using a single-point calculation equation. This workflow was implemented for screening and binding affinity measurement of fragments in mixtures against the HCV RNA polymerase NS5B, and the affinity values measured in this approach were comparable to what was determined in a conventional SPR assay. Comparing to other ASMS approaches described above, it is noteworthy that this approach does not require titration of compounds in multiple concentrations. Another advantage is that by analyzing the unbound fraction of compounds in the filtrate, it allowed detection and K_d measurement of low-affinity ligands, such as fragments, without compromising throughput. More reviews on membrane ultrafiltration affinity selection mass spectrometry have been published [7, 20].

8.2.1.5 Advantages and Disadvantages of Membrane Ultrafiltration-Based ASMS

Off-line ultrafiltration-based ASMS requires no sophisticated procedures and has been adapted by scientists for hit identification. The advantages of this technology include minimal requirement on instrumentation, high flexibility, in solution protein–ligand complex formation, as well as no tagging of proteins or compounds. However, in practice, poor consistency between individual wells is observed. Leakages and nonspecific binding of small molecules to ultrafiltration membrane are common problems. It was reported that during ultrafiltration step, small molecules may nonspecifically bind to the polymer-constructed materials of devices, resulting in inaccurate measurement of small molecule binders. An additional counter screen is usually implemented to address this problem. In some cases, adding back of protein solution in multiple rounds of ultrafiltration workflow is needed to maintain equilibrium condition during separation process and improve signal to noise ratio [33]. In addition, ultrafiltration-based ASMS generally requires large assay volumes (~40–400 μL). These factors lead to the consumption of relatively large amount of protein material. Furthermore, the concentration of the small molecules in unbound fraction can be further diluted through multiple rounds of ultrafiltration, which may compromise the detectability of compounds in mass spectrometer in cases where the protein–ligand binding event is indirectly measured by analysis of the unbound fraction. As a consequence of the disadvantages described above, the use of this technology in pharmaceutical industry is limited.

8.2.2 Plate-Based Size Exclusion Chromatography

8.2.2.1 Introduction of SpeedScreen: A Plate-Based SEC ASMS Technology

The concept of separating compounds by size was developed in 1960s where a cross-linked dextran gel was used as the column material for the analysis of high molecular weight polymers. The porosity of the gel determines its separation properties. Small molecules can diffuse into these pores and larger ones cannot, resulting in longer retention time of small molecules. With the development of more advanced gels and other size-exclusion materials, the separation power of size-exclusion-based methods has steadily increased. A detailed review of this approach as practiced by Amgen, Novartis, and Wyeth Pharmaceutical has been reported, where gel permeation chromatography (GPC) spin columns are used to separate protein–ligand complexes from unbound ligands using off-line ASMS [34].

A popular in-plate size-exclusion chromatography ASMS technique, termed SpeedScreen, was developed at Novartis and implemented for industrial-scale screening of protein targets, particularly orphan target proteins or those that may not be amenable to conventional screening methodologies [35, 36]. The SpeedScreen "sandwich" can be assembled in either 96- or 384-well plate format (Figure 8.4). Separation of the protein–ligand complex from unbound compounds

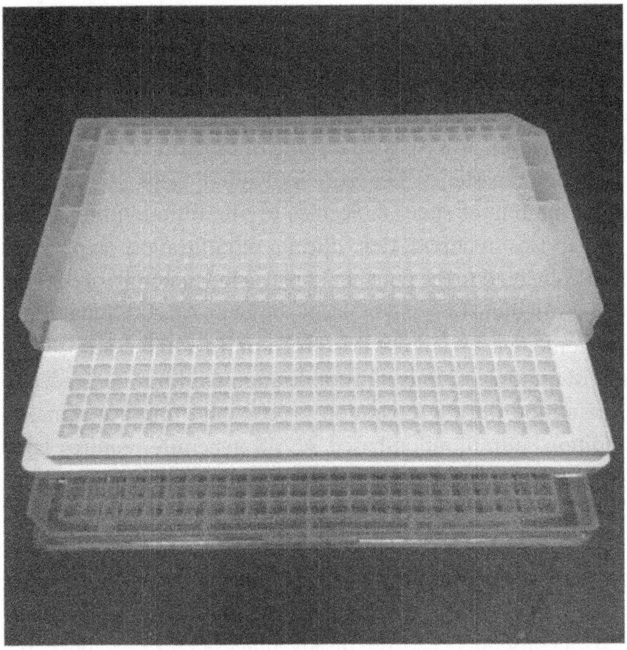

Figure 8.4 "SpeedScreen sandwich" assembly of three microtiter plates (in either 96- or 384-well plate format). On top is the assay plate with a pinhole at bottom to hold assay solution. In the middle is the size exclusion plate containing gel in each well. At the bottom is the eluates collection plate.

in the 96-well plate format assembly was carried out in a filter plate packed with gel (MultiScreen-HA, MAHAN4510, Millipore). To make a gel plate, usually 45 μL of Sephadex-G50 was added in each well. Hydration of the gel, followed by centrifugation at 910× g at 4 °C for three minutes, leads to the formation of homogeneously packed gel columns. Sephadex-G50 gel plates need to be used immediately for the separation process as the gel dries rapidly and should be handled with extreme care to avoid any cracking of the gel structure. To prepare affinity selection, a "sandwich" three-plate assembly was established including a top plate with small-sized pinholes in the bottom of each well that can hold assay solution without leakage until a centrifugal force is applied, a middle plate containing gel columns preconditioned with target-specific assay buffer, and a bottom plate to collect eluates. Assay solution (25 μL) was loaded onto the sandwich-like assembly of SEC plate, and a fast separation of the protein–ligand complex from unbound compounds is achieved through two minutes centrifugation of the three-plate assembly at low temperature. The estimated separation time on gel column in each well is around 10 seconds. During the separation process, ligand continuously dissociates from the protein–ligand complex, therefore a short separation time at low temperature is critical to identify weak binders (with relatively fast off-rate). The eluate is then injected onto a LC-MS system, where bound ligands were dissociated from the protein–ligand complex under acidic condition at high temperature on reverse phase column, and then enriched and eluted into a mass spectrometer for identification.

8.2.2.2 Application of SpeedScreen in Lead Discovery

The SpeedScreen technology was implemented for screening of various targets at Novartis, such as transcription factors, adapter molecules, regulatory subunits, heat shock proteins, metal binding proteins, RNA binding proteins, transferases, dehydrogenases, kinases, isomerases, proteases, phosphatases, and oxidoreductases [35–37]. Usually, affinity selection of about 500,000 compounds as mixtures of 400 against a target using "sandwich" assembly can be completed within a day. However, it takes about seven to nine days to complete analysis of the elutes using one LC-MS instrument. Of course, the throughput can be significantly increased by scaling up LC-MS capacity. The hit rate of SpeedScreen assays is typically low when compared to traditional HTS. While it varies from target to target, the hit rates of primary screens typically range between 0.05% and 1.5%.

About 26 targets were screened using the SpeedScreen approach and the chemical nature of the hits identified was studied [38]. It was found that the hits generated from SpeedScreen in general are drug-like compounds, albeit with greater lipophilicity and relatively higher molecular weights. This may be associated with the assay format employed for hit identification, or in some extent related to the protein target itself. Similar to traditional HTS workflows, hits identified from SpeedScreen ASMS should be triaged through various orthogonal assays to eliminate promiscuous binders, false positives, and for further validation of true binders.

The SpeedScreen approach was also implemented to assess the structural continuity of biologics in a high-throughput manner [39]. In the biologics development paradigm, it is critical to constantly assess the quality and the consistency of the tertiary and quaternary structure of biologics, because lacking of structural continuity between discrete samples can have dramatic physiological consequences. Any changes in the higher-order structure of biologics could impact efficacy and safety. However, due to the complexity of the structure, physicochemical characterization of biologics is usually resource intensive and time-consuming. Utilizing SpeedScreen-based ASMS, Musetti et al. reported a high-throughput chemoprinting platform that is capable of assessing the consistency of structure of protein biologics. Several protein targets that cover a range of functional and structural classes, including bacterial acetyl coenzyme-A carboxylase (carboxy transferase domain, BACC), human focal adhesion kinase

(FAK), human SmyD3 histone methyltransferase (SmyD3), and human Sirtuin3 (Sirt3), were selected and incubated with a library of 100 compounds, followed by SpeedScreen ASMS. The library was assembled based on available data sets obtained from high-throughput biochemical assay and covers a wide range of potency for each target (pIC_{50} 4.5–8.9).

Following the initial screen, a relative binding affinity percentage (%-RBA) was generated based on the ratio between MS signal of bound ligands in the presence of protein target and MS signal of ligands in the absence of protein target (control samples). In general, any perturbations of target proteins, such as amino acid transpositions or higher-order structural changes will potentially lead to changes of %-RBA (Figure 8.5). The authors demonstrated that this method is sensitive enough to distinguish single-point mutations of biologics. This is the first report utilizing ASMS-based high-throughput chemoprinting method to assess gross structural changes within the same target.

In addition to hit identification in target-based screening, Salcius et al. has utilized SpeedScreen to assess binding of bioactive small molecules to approximately 1000 individually purified proteins for target identification, named Size-Exclusion Chromatography for Target Identification (SEC-TID) [40]. The 384-well SEC plate is a low-protein-binding PVDF filter plate loaded with G50 size exclusion resin, which allows separation of protein–ligand complexes from unbound small molecules under centrifugation. The amount of small molecules bound to the target protein is then quantified using LC-MS. Since a collection of recombinant proteins were used, SEC-TID enables rapid identification of protein target against bioactive small molecules. As a proof of concept, bromodomain profiling with JQ1 and bromosporine was performed. Bromodomains were chosen because there are less than 50 proteins that contain this domain in human proteome, and small molecules that specifically target these domains have been identified. The authors expressed and purified a panel of 31 bromodomain proteins, and then profiled against known binders JQ1 and bromosporine using SEC-TID (Figure 8.6). In general, the interactions reported by SEC-TID are in good agreement with the interactions observed by differential scanning fluorimetry (DSF). This approach is particularly useful in target deconvolution of a phenotypic screen. Although purification of large number of proteins can be resource intensive, SEC-TID screen does provide an alternative way of identification of novel targets (or novel mechanism of action), which may potentially lead to the development of first-in-class drugs.

8.2.2.3 Advantages and Considerations of SpeedScreen

As a homogeneous solution-phase assay, SpeedScreen rapidly separates the protein–ligand complex from unbound compounds, which enables the detection of ligands with weak affinity in a double-digit micromolar range against the target.

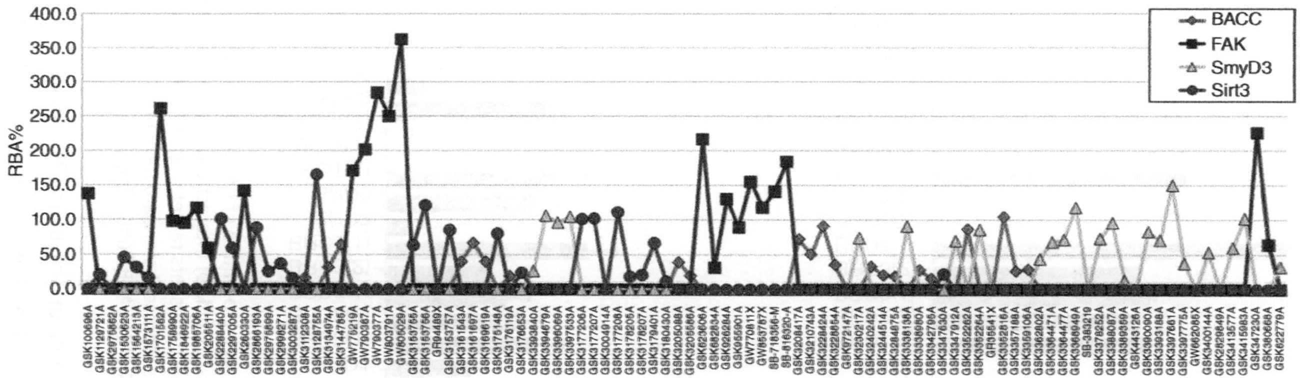

Figure 8.5 Relative binding affinity percentage %-RBA (*y*-axis) of 100 small molecules (*x*-axis) versus BACC (diamonds), FAK (squares), SmyD3 (triangles), and Sirt3 (dots). *Source:* Reproduced with permission from Musetti et al. [39]/American Chemical Society.

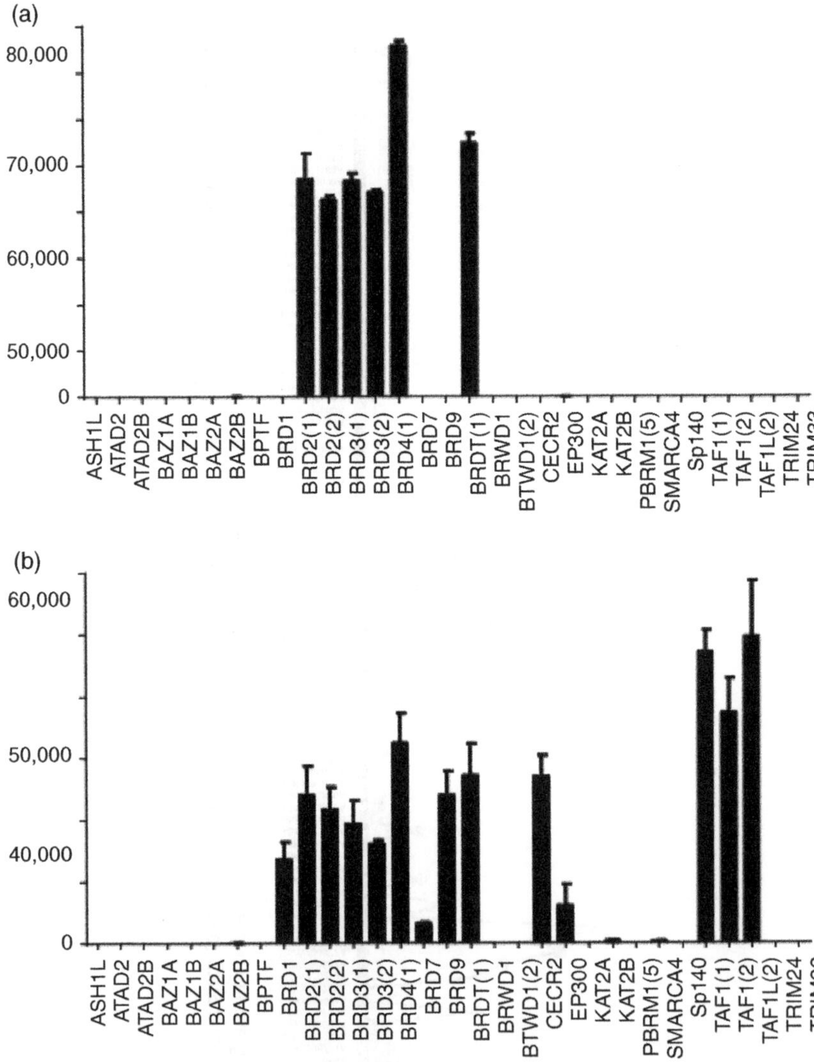

Figure 8.6 JQ1 and bromosporine profiling of the bromodomain panel. (a) Interactions observed after profiling JQ1 at 10 μM. (b) Interactions observed after profiling bromosporine at 10 μM. *X*-axis represents target protein; *Y*-axis represents MS signal of bound compounds; Error bars represent standard deviation. *Source:* Reproduced with permission from Salcius et al. [40]/SAGE Publishing.

No reagent (protein or compound) labeling is required and this technology is applicable for screening of potentially any classes of soluble targets against nearly all compound sources, such as compounds from medicinal chemistry, combinatorial chemistry, or natural products. A few limitations of SpeedScreen were also reported: First, a relatively large volume of assay solution (>25 μL) is needed, resulting in high protein consumption. Second, different proteins behave differently on the in-plate Sephadex G50 gel column, with recoveries on average ranging from as low as 12% to up to 90%. For those target proteins with poor recovery, different type of gel material may need to be evaluated during assay development. Finally, certain compounds may form aggregates at screening concentration and pass through the SEC gel columns in the absence of binding to the protein. However, in SpeedScreen at Novartis, it was found that the percentage of compounds passing through the gel column is relatively low <0.001% [38]. These compounds were identified as false positive hit and removed from hit confirmation list. It is noteworthy that O'Connell et al. conducted similar SpeedScreen using fresh 96-well MicroSpin P-2 SEC plate from The Nest Group (Southborough, MA) [41] and a significant number of gel wells in the MicroSpin P-2 SEC plate had compounds breakthrough potentially due to gel degradation, which yields irreproducible data. This highlights how critical it is to evaluate the gel material against screening reagents as part of the assay development in SpeedScreen.

8.2.3 Bead-Based Affinity Selection

8.2.3.1 Introduction to Bead-Based Affinity Selection

Several methods for off-line ASMS have been developed using solid-support beads, which are employed to partition the bound and unbound fraction of ligand through a solid-phase extraction. This strategy has been leveraged in hit identification efforts [42–46] and recently used to validate the binding of hits from DEL screening prior to off-DNA synthesis [47]. In these approaches, a protein of interest is first immobilized on solid-support beads and the protein-bound beads are then incubated with a mixture of ligands or screening compounds. The compounds that bind to the target protein will remain associated with the beads, which then undergo a series of gentle wash steps to remove the unbound fraction of ligand that remain in the supernatant. Ligands that bind the immobilized protein are then released from the beads by treatment with an organic solvent or heat. When coupled with LC-MS as a method for ligand detection, bead-based affinity selection provides an efficient format for off-line ASMS and the identification of inferred binders. Approaches for ASMS have been implemented with various bead types, which dictate the strategy for solid-phase extraction. For example, if the protein target is immobilized on agarose beads, centrifugation can be used to pellet the beads at the bottom of the reaction vessel during the wash steps [43]. More commonly, magnetic beads are used as the solid-support to provide a simple and effective medium for solid-phase extraction in a variety of assay structures ranging from individual tubes to plate-based formats.

8.2.3.2 Application and Discussion of Bead-Based Affinity Selection in Lead Discovery

Magnetic microbead affinity selection screen (MagMASS) is an ASMS strategy initially developed to increase the throughput of screening natural product extracts [42, 46]. In the MagMASS approach, magnetic beads are used to enable affinity selection via solid-phase extraction (Figure 8.7). The beads bearing the immobilized target of interest are incubated with the mixture of compounds to allow for compound binding. The original workflow for the MagMASS system reports the workflow in 96-well plates; however, 384-well formats can be envisioned. Following the incubation with compounds, the beads are trapped at the bottom of the well using a magnetic plate consisting of neodymium rare earth magnets [46]. The unbound fraction of compound is then removed from the system through a series of wash steps with buffer and water. The bound fraction of compound is then released from the protein with a methanol extraction and binders are identified via UHPLC/MS (Figure 8.7). Nonspecific binding is measured by parallel experiments with heat-denatured protein. Data are reported as a binding enrichment, which is the ratio of MS signal area under the curve (AUC) for the compound of interest with the native protein versus that of the denatured protein [46].

Rush et al. report a proof-of-concept study [46] in which MagMASS is used to assess the binding of retinoid X receptor-α (RXRα) with its endogenous ligand 9-*cis*-retinoic acid (K_d = 15.7 nM) [48] and a tool compound LG100268 (K_d = 3 nM) [49]. A variety of conditions were contrived to approximate the challenges of implementing the MagMASS workflow on complex systems, including botanical extracts and mixtures of combinatorial compounds. The botanical extracts were prepared in methanol and spiked with 10 pmol of 9-*cis*-retinoic acid and LG100268. The control compounds were accurately identified as binders of

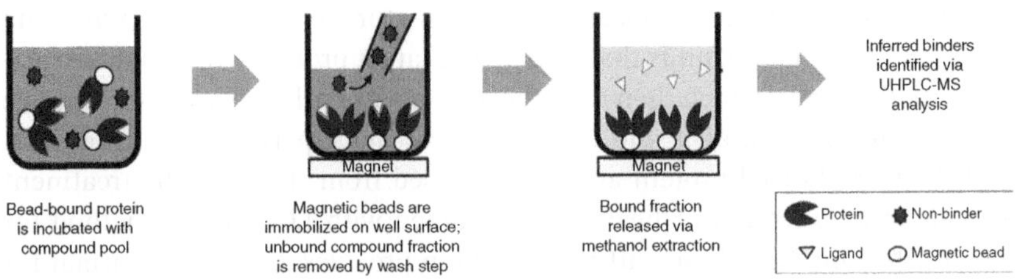

Figure 8.7 Schematic for the MagMASS approach for off-line affinity selection. The target protein is immobilized on magnetic beads and incubated with the compound pool. The complex of bead-bound protein and affinity-bound compounds is trapped at the bottom of the vessel using an external magnet. Unbound compounds are removed by washing the beads with buffer. Bound compounds are released from the protein complex by extracting with an organic solvent (e.g. methanol), and compound identity is confirmed via MS analysis. *Source:* Reproduced with permission from Rush et al. [46], (Figure 1)/ American Chemical Society.

RXRα (immobilized at 100 pmol) demonstrating that the extract matrix did not interfere with the MagMASS analysis. Additionally, 9-*cis*-retinoic acid and LG100268 were also accurately identified as binders of RXRα when spiked into a mixture of 20 random compounds (10 μM each compound) meant to mimic the effects of screening combinatorial compound libraries [46].

Whereas the affinity selection step for other off-line ASMS approaches such as SpeedScreen occurs in solution, a key distinction with bead-based approaches is the requirement for immobilizing the target protein on solid support. The immobilization strategy in the proof-of-concept study with RXRα used N-hydroxysuccinimide (NHS) ester chemistry to tether the protein to the magnetic beads in a nonspecific, covalent fashion [46]. However, bead-based ASMS is amenable to a range of common protein immobilization strategies, such as the use of biotin and polyhistidine affinity tags. These approaches have been employed for ASMS screens to identify high-affinity peptide binders using streptavidin-coated magnetic beads [45] and ASMS screens of fragment-based libraries using nickel agarose beads [44], respectively. Although the nonspecific immobilization is most flexible with respect to protein construct design, the use of a site-specific affinity tag can provide a more uniform protein surface on the bead for compound capture and avoid the potential of covalent immobilization impacting protein activity or function [50]. Overall, bead-based approaches for affinity selection provide a simple and rapid means of implementing ASMS workflows. The approach is both amenable to automation and accessible to laboratories without the need for specialized affinity selection infrastructure.

8.2.4 Self-Assembled Monolayers and Matrix-Assisted Laser Desorption Ionization (SAMDI)

8.2.4.1 Introduction to SAMDI Technology

SAMDI was developed in the laboratory of Prof. Milan Mrksich at the University of Chicago [51] and is currently implemented as a proprietary platform technology by SAMDI Tech, LLC (Webpage available from: www.samditech.com). SAMDI couples rapid sample preparation using self-assembled monolayers on gold surface arrays with the high-throughput sampling rate of matrix-assisted laser desorption ionization mass spectrometry (MALDI-MS) [52]. High-throughput MS activity assays have been developed for multiple enzyme classes using SAMDI [53–55], and more recently this technology has been applied to detect direct ligand binding via an off-line affinity selection approach [56].

An illustration depicting how SAMDI technology is adapted to an ASMS workflow is shown in Figure 8.8. The protein of interest in this scheme requires a biotin affinity tag to enable immobilization on the SAMDI gold surface array; however, other protein capture strategies are possible [54]. The target protein is incubated

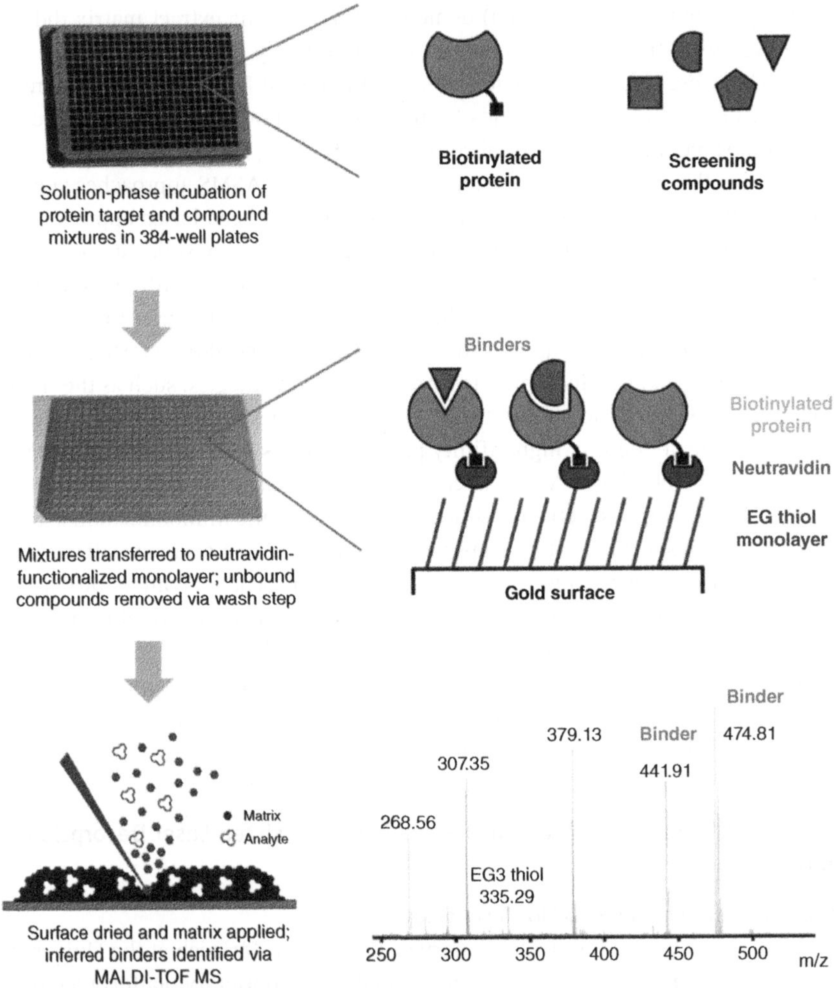

Figure 8.8 Schematic for ASMS workflow using self-assembled monolayers and matrix-assisted laser desorption ionization (SAMDI). A biotinylated target protein and pool of screening compounds are mixed and incubated in solution using 384-well plates. The mixtures are then transferred to a gold surface SAMDI array functionalized with neutravidin, which allows for affinity capture of the target protein. Compounds that bind to the target protein remain associated with the monolayer surface, while unbound compounds are removed by successive wash steps. Compound masses are identified using MALDI-TOF MS and data analysis infers binders. EG thiol monolayer = mixture of (biotin-ethylene glycol)$_6$-SH and 97% (ethylene glycol)$_3$-SH (EG3 thiol), MALDI-TOF, matrix-assisted laser desorption/ionization – time of flight; MS, mass spectrometry. *Source:* [56] / with permission from Elsevier and [53] / with permission from John Wiley & Sons.

with compound mixtures in solution using 384-well plates. The solutions of protein and compounds are then spotted onto a gold surface SAMDI array. The SAMDI plate presents a self-assembled monolayer of alkanethiolates capped with polyethylene glycol and functionalized with an appropriate affinity capture reagent [51, 52]. For example, when capturing biotinylated proteins (as represented in Figure 8.8), an array comprising immobilized neutravidin on a self-assembled monolayer of 3% (biotin-ethylene glycol)$_6$-SH and 97% (ethylene glycol)$_3$-SH (EG3 thiol) has been shown to be effective [56]. The protein and compound mixtures are incubated on the neutravidin-coated array, followed by a rapid wash with water to extract unbound compounds. The plate is then dried with compressed air, matrix is applied to the array, and samples are analyzed via MALDI-MS. Peaks for the small molecules are quantitated via AUC and normalized against the AUC for the (ethylene glycol)$_3$-SH monolayer component. Inferred binders are identified by matching observed mass-to-charge values with calculated monoisotopic masses from molecular formulas [56].

Similar to bead-based ASMS approaches, SAMDI uses a solid-phase approach to enable affinity selection, which requires immobilization of the protein target. Many reports on bead-based ASMS pre-immobilize the protein target on solid support before compound capture, even in cases where site-specific affinity tags are employed (e.g. biotin and His-Tag) [44, 45, 47]. By contrast, a key feature of the SAMDI workflow is that the initial compound binding event occurs through a solution-phase incubation (Figure 8.8). This allows for mechanistic elements of target engagement such as conformational changes or other ligand-induced effects to occur before the protein is immobilized on the SAMDI surface. As an alternative approach, solid-phase compound incubation is possible, wherein the target protein is immobilized on the monolayer surface first, followed by incubation with the compound pools. Such strategies are useful in situations when protein immobilization is not possible via affinity capture tags and nonspecific covalent protein immobilization must be used (e.g. maleimide or NHS ester chemistries) [53, 54]. Solid-phase compound capture affords SAMDI ASMS flexibility with respect to protein construct design and affinity tag requirements. However, solution-phase compound capture is likely the best approach for SAMDI ASMS with most systems, as it provides higher overall signal and decreased sample variability relative to corresponding solid-phase incubations [56]. Independent of the capture strategy, inferred binders identified via SAMDI ASMS must be evaluated for nonspecific binding with the monolayer array itself. To gauge nonspecific binding, compound pools are evaluated in the SAMDI ASMS workflow both in the presence and absence of the protein target. Compounds that display comparable signals both with and without protein are identified as nonspecific binders of the SAMDI surface, and not specific hits for the target protein [56].

8.2.4.2 Discussion and Proof-of-Concept of SAMDI Technology for Off-Line ASMS

As with other affinity selection methods such as off-line SEC or membrane ultrafiltration, SAMDI ASMS can be implemented in single-point or dose–response formats. The high-throughput sampling rate of SAMDI ASMS makes it feasible to run full-deck primary screens, and the ability to follow up with dose–response analyses provides an early read into compound behavior and relative binding potencies. Importantly, binding data from SAMDI ASMS screen of β-secretase (BACE) has been shown to correlate well with orthogonal biochemical binding assays. For example, single-point AUC data acquired with a pool of known binders for BACE show conserved rank ordering with IC_{50} values from an established time-resolved fluorescence energy transfer (TR-FRET) assay (Figure 8.9a) [57]. Additionally, when compounds were individually evaluated against BACE in dose response, EC_{50}s generated from SAMDI ASMS data were consistent with inhibition values measured in the HTRF assay (Figure 8.9b) [56].

In off-line ASMS formats such as the SpeedScreen approach, equilibria for ligand binding are generally driven by using protein concentrations in excess of any individual compound in the pool. By contrast, binding equilibria in SAMDI ASMS are driven by using compound concentrations (roughly 10–20 μM) in excess of the protein target. A key benefit to this configuration is relatively low protein consumption, with assays generally using nM to single-digit μM protein concentrations in low reaction volumes (as low as 300 nL) [56]. However, the requirement for high ligand concentrations limits the compound pool size for SAMDI ASMS. Whereas other ASMS formats may use mixtures containing hundreds or thousands of compounds, SAMDI ASMS will generally use mixtures of approximately 10 compounds per well. The requirement of screening smaller compound mixtures is compensated for by the rapid sampling rate of MALDI-MS (1–2 second per sample), which provides higher throughput than traditional chromatography-coupled MS approaches (e.g. LC-ESI-MS) [58]. Relative to ESI-MS, MALDI-MS can also be less susceptible to ion suppression and source contamination from assay buffer components such as surfactants and salts [59]. The combination of using a pooled compound library, sensitive MALDI instrumentation, and automation provides an estimated throughput for SAMDI ASMS at approximately >100,000 compounds per day [56].

8.2.5 Ultracentrifugation Affinity Selection

8.2.5.1 Introduction to Ultracentrifugation Affinity Selection

A method for off-line affinity selection using ultracentrifugation was reported by Harlan et al. [60]. In this approach, a macromolecular target is incubated with a compound (or compound mixture) and ultracentrifugation is used to establish a

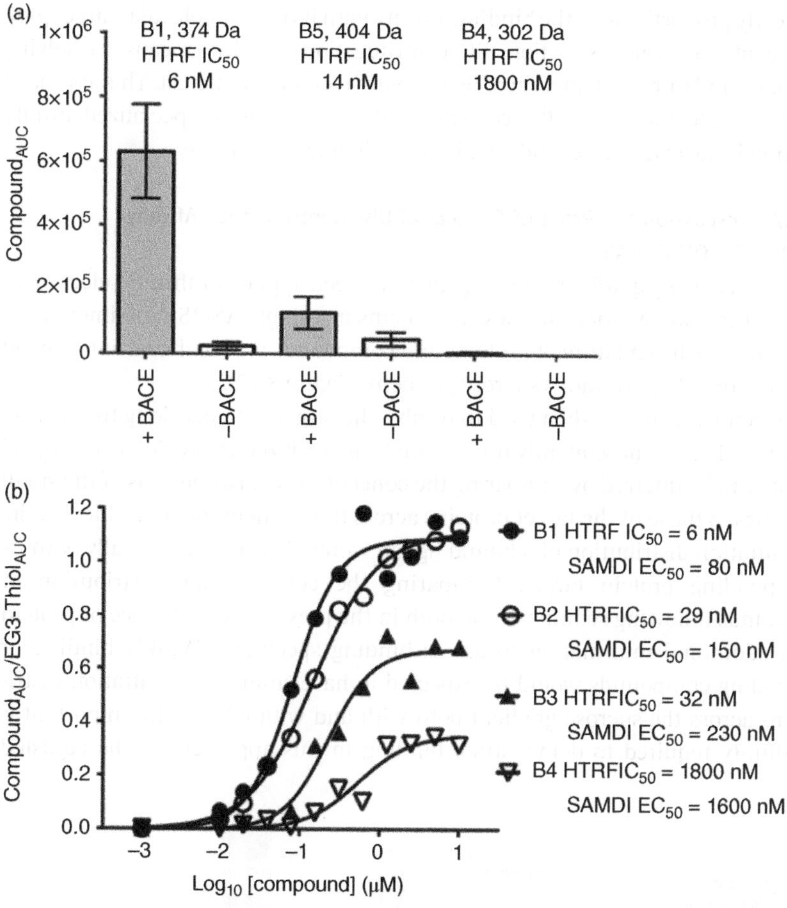

Figure 8.9 Correlation of SAMDI ASMS data with an orthogonal biochemical binding assay. (a) SAMDI ASMS evaluation using a pool of three compounds (B1, B5, and B4; 10 μM each) and 1 μM BACE. Signal from SAMDI ASMS (AUC) correlates with potencies from an orthogonal HTRF assay. (b) Dose–response curves for individual compounds (B1, B2, B3, and B4) using SAMDI ASMS. Relative potencies (EC_{50}) by SAMDI ASMS are consistent with potencies in an orthogonal HTRF assay. AUC, area under the curve; HTRF, homogenous time-resolved fluorescence; BACE, β-secretase, SAMDI, self-assembled monolayers and matrix-assisted laser desorption ionization; EG3-thiol, (ethylene glycol)$_3$-SH. *Source:* Reproduced with permission from VanderPorten et al. [56], (Figures 2 and 3)/SAGE Publishing.

differential concentration gradient between the macromolecular target and potential small-molecule binders. This strategy exploits the hydrodynamic principle that molecules with different molecular weights display different sedimentation rates when exposed to a centrifugal driving force. It is proposed that the concentration gradient of a small-molecule ligand upon sedimentation should be directly proportional to the binding affinity with its macromolecular target [60]. In the context of affinity selection, ultracentrifugation provides a means of resolving the bound and unbound fraction of ligand with a protein of interest. This approach for affinity selection is broadly accessible and does not require specialized affinity selection infrastructure beyond standard biochemical laboratory.

8.2.5.2 Discussion and Proof-of-Concept of Ultracentrifugation Affinity Selection for Off-line ASMS

Rate-zonal density gradient centrifugation has been applied to affinity selection in a manner that allows for multiplexing proteins in a single ASMS experiment [61]. In this approach, a mixture of proteins and compounds are incubated and applied to the top of a discontinuous sucrose gradient (Figure 8.10).

Ultracentrifugation is then used to resolve the target proteins along the sucrose gradient, which is subsequently fractionated and analyzed via LC-MS [61]. Ligand binding can be inferred by comparing the concentration distributions of the small molecules to those of the target proteins across the gradient fractions. Ideally, the concentration distribution of a bound ligand would distribute identically with its corresponding protein target. Comparing the concentration distribution of compounds following sedimentation both in the presence and absence of protein offers additional information to assess binding specificity. Weakly binding or nonbinding compounds would be expected to have similar concentration distributions across the sucrose gradient both with and without protein. Importantly, the affinity required to demonstrate binding in this approach can be adjusted

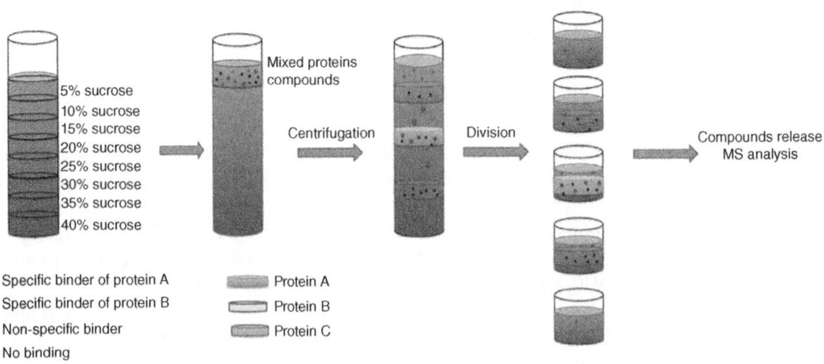

Figure 8.10 Schematic of ultracentrifugation-based strategy for multiplexed affinity selection. A mixture of multiple proteins and multiple compounds is placed on top of a discontinuous sucrose gradient. The proteins are resolved through the sucrose gradient via centrifugation and then the gradient solution is divided into five equivolume fractions. Compounds are released from the protein complexes using reversed-phase chromatography and analyzed by mass spectrometry. *Source:* Reproduced with permission from Jin et al. [61]/Royal Society of Chemistry.

through the incubation conditions for the proteins and ligands, as well as the centrifugal force applied during the affinity selection step [60, 61].

Proof-of-concept studies with ultracentrifugation affinity selection have demonstrated the potential for multiplexing ASMS with three proteins of varying sedimentation coefficients: bovine serum albumin (BSA, 66.4 kDa), tyrosine phosphatase 1B (PTP1B, 37.5 kDa), and low-density lipoprotein receptor (LDLR, 11.5 kDa) [61]. Distribution of the three proteins in a sucrose gradient was assessed following sedimentation for the proteins as individual samples (concentration quantification via Bradford assay, Figure 8.11a) and as a mixture (concentration quantification via LC-MS, Figure 8.11b). Tool compounds were then used to demonstrate the identification of known binders with the target proteins both in single-protein and multiplexed formats. Sedimentation experiments show that a known small-molecule binder of PTP1B has a concentration distribution consistent with that of the PTP1B protein, in both individual and multiplexed ASMS formats (Figure 8.11c) [62]. The concentration distribution of the tool compound following sedimentation with PTP1B was differentiated from those in individual experiments with LDLR and BSA, as well as a no-protein control. Warfarin, an anticoagulant that binds to BSA with a K_d in the μM range, was shown to have a similar concentration distribution following sedimentation with BSA, or with the protein mixture (Figure 8.11d). This profile was distinct from the profile for warfarin with LDLR and PTP1B, and in the no-protein control, suggesting a specificity to the engagement with BSA over the other proteins. A negative control compound that has no affinity for LDLR, PTP1B, or BSA was shown to have similar concentration distributions when evaluated with the individual proteins, the protein mixture, or the no-protein control (Figure 8.11e). These data are consistent with a lack of engagement between the negative control compound and the three protein targets.

A clear advantage offered by the ultracentrifugation approach is the potential to multiplex ASMS with several proteins in a single experiment. This complements many other off-line affinity selection formats, which are generally optimized to satisfy the screening of a single protein. However, multiplexed ultracentrifugation affinity selection requires that the target proteins are resolved through the sedimentation process. Accordingly, the target proteins must have different sedimentation coefficients or otherwise migrate differently through the centrifugal medium. Such a constraint is easier to accommodate in idealized or proof-of-concept systems but may be less practical as applied in actual drug discovery programs, especially if the end user would like to multiplex highly homologous proteins. As with any multiplexed affinity selection approach, experimental design would need to consider and address the potential for internal competition of ligands between the target proteins in the mixture. Again, this consideration is easier to address in simplified systems, but may become more challenging when

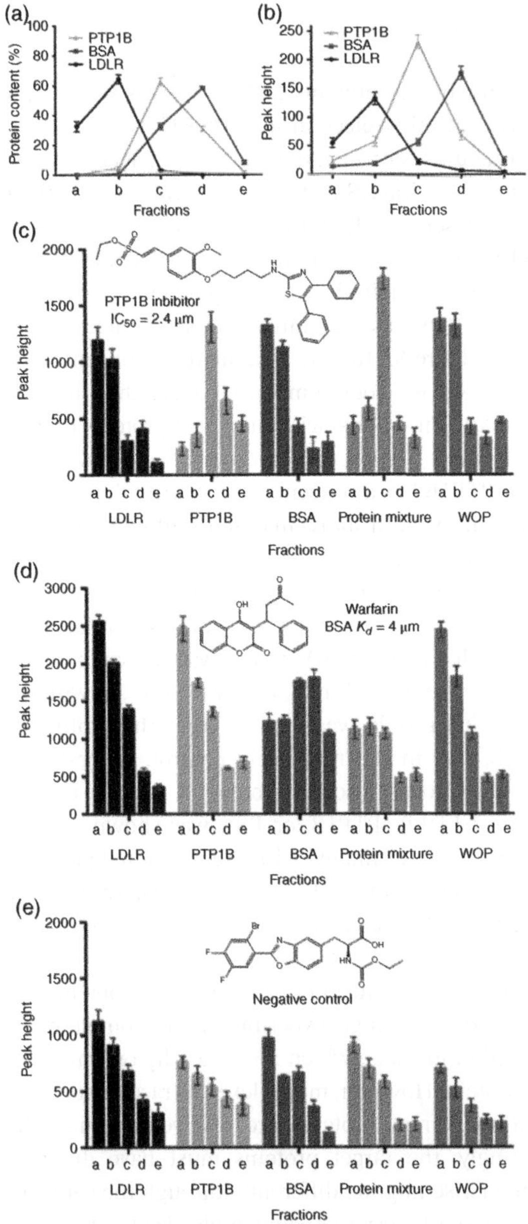

Figure 8.11 Concentration distributions for proteins and tool compounds following ultra-centrifugation sedimentation in single-protein and multiplexed ASMS formats. (a) Concentration distribution of individual proteins following 80 minutes centrifugation as determined by Bradford assay. (b) Molecular weight distributions of each protein in a protein mixture as determined by MS. Concentration distributions of (c) a PTP1B inhibitor, (d) warfarin, and (e) a negative control in the compound mixture treated with single proteins or a 3-protein mixture following 80 minutes of centrifugation. PTP1B, tyrosine phosphatase 1B; BSA, bovine serum albumin; LDLR, low-density lipoprotein receptor; WOP, without protein. *Source:* Reproduced with permission from Jin et al. [61], (Figures 2 and 3)/Royal Society of Chemistry.

implementing the ultracentrifugation ASMS on the scale required to support lead discovery efforts. To this end, initial experiments do show promise for the ultracentrifugation approach with multiple proteins and larger mixtures of up to 45 compounds [61]. However, practical limitations may make the ultracentrifugation approach for ASMS more suitable for lower throughput applications, such as hit confirmation or mechanistic studies, as opposed to hit identification campaigns.

8.3 Future Perspectives

ASMS is a promising technology that has become increasingly popular as a platform that complements traditional small-molecule HTS and other affinity selection platforms (e.g. DEL and mRNA display). Relative to on-line approaches, off-line ASMS is typically more accessible because it eliminates the need for complex liquid chromatography systems and provides more versatility in how the affinity selection step can be executed. In addition, the parallel nature of off-line affinity selection drastically increases throughput while simultaneously offering a way to nullify differences in compound-protein incubation time domain, which allows for every sample to be treated identically during the affinity selection step. Recent advances in removing the LC components completely by using SAMDI/MALDI-based detection methods offer a framework to increase throughput of the platform to levels often achieved by more traditional biochemical and cellular HTS approaches (i.e. >100,000 compounds per day). Alternatively, acoustic ESI-MS allows sample analysis at speed of ~1–3 seconds per sample, more than an order of magnitude faster than the conventional "gold standard" LC-MS technology. When coupled with off-line affinity selection methods, acoustic ESI-MS potentially enables affinity selection screening of millions of compounds in a few days. Due to its high sample readout speed, compounds can be screened in smaller pool size but still achieve ultrahigh-throughput. As off-line ASMS platforms continue to evolve and automation solutions continue to improve, we may see a migration from on-line ASMS platforms toward the off-line counterparts to reduce variability, increase throughput, and decrease cycle times for early small molecule drug discovery.

In addition to providing a fast, unbiased screening approach complementary to conventional HTS for hit identification, off-line ASMS can play an important role in hit triage, where target engagement of HTS or DEL hits is desired to prioritize promising compounds for further follow up. Certain targets, such as protein complexes, may not behave well in surface-based biophysical assays. Off-line ASMS, as an in-solution binding assay, provides an alternative approach for target engagement study of complex targets without large investments in instrumentation. Given the high-throughput and minimal assay development needed, off-line

ASMS could potentially become a routine target engagement assay for small molecule hit triage in high-throughput manner. Although off-line ASMS technology was established decades ago, it regained popularity in early drug discovery recently and many of the unique applications have just begun to be demonstrated. We expect that many more innovations and applications of off-line AS-MS will occur during the next 10–20 years.

References

1 Prudent, R., Annis, D.A., Dandliker, P.J. et al. (2021). Exploring new targets and chemical space with affinity selection-mass spectrometry. *Nat. Rev. Chem.* 5 (1): 62–71.

2 Petrilli, W.L., Adam, G.C., Erdmann, R.S. et al. (2020). From screening to targeted degradation: strategies for the discovery and optimization of small molecule ligands for PCSK9. *Cell Chem. Biol.* 27 (1): 32–40.e3.

3 Whitlam, J.B. and Brown, K.F. (1981). Ultrafiltration in serum protein binding determinations. *J. Pharm. Sci.* 70 (2): 146–150.

4 Zhirkov, Y.A. and Piotrovskii, V.K. (1984). On the usefulness of ultrafiltration in drug-protein binding studies. *J. Pharm. Pharmacol.* 36 (12): 844–845.

5 Qin, S., Ren, Y., Fu, X. et al. (2015). Multiple ligand detection and affinity measurement by ultrafiltration and mass spectrometry analysis applied to fragment mixture screening. *Anal. Chim. Acta* 886: 98–106.

6 Hannewald, P., Maunit, B., and Muller, J.F. (2006). Tubulin-binding drug screening by MALDI-TOFMS. *Anal. Chem.* 78 (13): 4390–4397.

7 Wei, H., Zhang, X., Tian, X., and Wu, G. (2016). Pharmaceutical applications of affinity-ultrafiltration mass spectrometry: recent advances and future prospects. *J. Pharm. Biomed. Anal.* 131: 444–453.

8 Cloutier, T.E. and Comess, K.M. (2007). Library screening using ultrafiltration and mass spectrometry. In: *Mass Spectrometry in Medicinal Chemistry* (ed. K.T. Wanner and G. Höfner), 157–183. Wiley.

9 van Breemen, R.B., Huang, C.R., Nikolic, D. et al. (1997). Pulsed ultrafiltration mass spectrometry: a new method for screening combinatorial libraries. *Anal. Chem.* 69 (11): 2159–2164.

10 Wieboldt, R., Zweigenbaum, J., and Henion, J. (1997). Immunoaffinity ultrafiltration with ion spray HPLC/MS for screening small-molecule libraries. *Anal. Chem.* 69 (9): 1683–1691.

11 Comess, K.M. and Schurdak, M.E. (2004). Affinity-based screening techniques for enhancing lead discovery. *Curr. Opin. Drug Discovery Dev.* 7 (4): 411–416.

12 Comess, K.M., Schurdak, M.E., Voorbach, M.J. et al. (2006). An ultraefficient affinity-based high-throughput screening process: application to bacterial cell wall biosynthesis enzyme MurF. *J. Biomol. Screening* 11 (7): 743–754.

13 Comess, K.M., Trumbull, J.D., Park, C. et al. (2006). Kinase drug discovery by affinity selection/mass spectrometry (ASMS): application to DNA damage checkpoint kinase Chk1. *J. Biomol. Screening* 11 (7): 755–764.

14 Kawai, M., BaMaung, N.Y., Fidanze, S.D. et al. (2006). Development of sulfonamide compounds as potent methionine aminopeptidase type II inhibitors with antiproliferative properties. *Bioorg. Med. Chem. Lett.* 16 (13): 3574–3577.

15 Qian, J., Voorbach, M.J., Huth, J.R. et al. (2004). Discovery of novel inhibitors of Bcl-xL using multiple high-throughput screening platforms. *Anal. Biochem.* 328 (2): 131–138.

16 Chen, X., Li, L., Chen, S. et al. (2013). Identification of inhibitors of the antibiotic-resistance target New Delhi metallo-β-lactamase 1 by both nanoelectrospray ionization mass spectrometry and ultrafiltration liquid chromatography/mass spectrometry approaches. *Anal. Chem.* 85 (16): 7957–7965.

17 Chen, G., Huang, B.X., and Guo, M. (2018). Current advances in screening for bioactive components from medicinal plants by affinity ultrafiltration mass spectrometry. *Phytochem. Anal.* 29 (4): 375–386.

18 Huai, J., Zhao, X., Wang, S. et al. (2019). Characterization and screening of cyclooxygenase-2 inhibitors from Zi-Shen pill by affinity ultrafiltration-ultra performance liquid chromatography mass spectrometry. *J. Ethnopharmacol.* 241: 111900.

19 Jin, Y., Cheng, X., Jiang, F. et al. (2015). Application of the ultrafiltration-based LC-MS approach for screening PTP1B inhibitors from Chinese red yeast rice. *Anal. Methods* 8 (2): 353–361.

20 Muchiri, R.N. and van Breemen, R.B. (2021). Affinity selection-mass spectrometry for the discovery of pharmacologically active compounds from combinatorial libraries and natural products. *J. Mass Spectrom.* 56 (5): e4647.

21 Tao, Y., Yan, J., and Cai, B. (2021). Label-free bio-affinity mass spectrometry for screening and locating bioactive molecules. *Mass Spectrom. Rev.* 40 (1): 53–71.

22 Johnson, B.M., Nikolic, D., and van Breemen, R.B. (2002). Applications of pulsed ultrafiltration-mass spectrometry. *Mass Spectrom. Rev.* 21 (2): 76–86.

23 Nikolic, D., Fan, P.W., Bolton, J.L., and Van Breemen, R.B. (1999). Screening for xenobiotic electrophilic metabolites using pulsed ultrafiltration-mass spectrometry. *Comb. Chem. High-Throughput Screening* 2 (3): 165–175.

24 Geun Shin, Y., Bolton, J.L., and van Breemen, R.B. (2002). Screening drugs for metabolic stability using pulsed ultrafiltration mass spectrometry. *Comb. Chem. High-Throughput Screening* 5 (1): 59–64.

25 Nikolic, D., Habibi-Goudarzi, S., Corley, D.G. et al. (2000). Evaluation of cyclooxygenase-2 inhibitors using pulsed ultrafiltration mass spectrometry. *Anal. Chem.* 72 (16): 3853–3859.

26 Liu, D., Guo, J., Luo, Y. et al. (2007). Screening for ligands of human retinoid X receptor-alpha using ultrafiltration mass spectrometry. *Anal. Chem.* 79 (24): 9398–9402.

27 Nikolic, D. and van Breemen, R.B. (1998). Screening for inhibitors of dihydrofolate reductase using pulsed ultrafiltration mass spectrometry. *Comb. Chem. High-Throughput Screening* 1 (1): 47–55.

28 Adam, G.C., Meng, J., Athanasopoulos, J. et al. (2007). Affinity-based ranking of ligands for DPP-4 from mixtures. *Bioorg. Med. Chem. Lett.* 17 (9): 2404–2407.

29 Annis, D.A., Nazef, N., Chuang, C.C. et al. (2004). A general technique to rank protein-ligand binding affinities and determine allosteric versus direct binding site competition in compound mixtures. *J. Am. Chem. Soc.* 126 (47): 15495–15503.

30 Annis, D.A., Shipps, G.W., Deng, Y. et al. (2007). Method for quantitative protein-ligand affinity measurements in compound mixtures. *Anal. Chem.* 79 (12): 4538–4542.

31 Cheng, X. and van Breemen, R.B. (2005). Mass spectrometry-based screening for inhibitors of beta-amyloid protein aggregation. *Anal. Chem.* 77 (21): 7012–7015.

32 Kamchonwongpaisan, S., Vanichtanankul, J., Tarnchompoo, B. et al. (2005). Stoichiometric selection of tight-binding inhibitors by wild-type and mutant forms of malarial (*Plasmodium falciparum*) dihydrofolate reductase. *Anal. Chem.* 77 (5): 1222–1227.

33 Musson, D.G., Birk, K.L., Kitchen, C.J. et al. (2003). Assay methodology for the quantitation of unbound ertapenem, a new carbapenem antibiotic, in human plasma. *J. Chromatogr. B Anal. Technol. Biomed. Life Sci.* 783 (1): 1–9.

34 Siegel, M.M. (2007). Drug screening using gel permeation chromatography spin columns coupled with ESI-MS. In: *Mass Spectrometry in Medicinal Chemistry* (ed. K.T. Wanner and G. Höfner), 63–120. Wiley.

35 Zehender, H., Le Goff, F., Lehmann, N. et al. (2004). SpeedScreen: the "missing link" between genomics and lead discovery. *J. Biomol. Screening* 9 (6): 498–505.

36 Zehender, H. and Mayr, L.M. (2007). Application of high-throughput affinity-selection mass spectrometry for screening of chemical compound libraries in lead discovery. *Expert Opin. Drug Discovery* 2 (2): 285–294.

37 Muckenschnabel, I., Falchetto, R., Mayr, L.M., and Filipuzzi, I. (2004). SpeedScreen: label-free liquid chromatography-mass spectrometry-based high-throughput screening for the discovery of orphan protein ligands. *Anal. Biochem.* 324 (2): 241–249.

38 Brown, N., Zehender, H., Azzaoui, K. et al. (2006). A chemoinformatics analysis of hit lists obtained from high-throughput affinity-selection screening. *J. Biomol. Screening* 11 (2): 123–130.

39 Musetti, C., Bean, M.F., Quinque, G.T. et al. (2018). High-throughput assessment of structural continuity in biologics. *Anal. Chem.* 90 (4): 2970–2975.

40 Salcius, M., Bauer, A.J., Hao, Q. et al. (2014). SEC-TID: a label-free method for small-molecule target identification. *J. Biomol. Screening* 19 (6): 917–927.

41 O'Connell, T.N., Ramsay, J., Rieth, S.F. et al. (2014). Solution-based indirect affinity selection mass spectrometry – a general tool for high-throughput screening of pharmaceutical compound libraries. *Anal. Chem.* 86 (15): 7413–7420.

42 Choi, Y. and van Breemen, R.B. (2008). Development of a screening assay for ligands to the estrogen receptor based on magnetic microparticles and LC-MS. *Comb. Chem. High-Throughput Screening* 11 (1): 1–6.

43 Imaduwage, K.P., Go, E.P., Zhu, Z., and Desaire, H. (2016). HAMS: high-affinity mass spectrometry screening. A high-throughput screening method for identifying the tightest-binding lead compounds for target proteins with no false positive identifications. *J. Am. Soc. Mass Spectrom.* 27 (11): 1870–1877.

44 Lu, Y., Liu, H., Yang, D. et al. (2021). Affinity mass spectrometry-based fragment screening identified a new negative allosteric modulator of the adenosine A(2A) receptor targeting the sodium ion pocket. *ACS Chem. Biol.* 16 (6): 991–1002.

45 Quartararo, A.J., Gates, Z.P., Somsen, B.A. et al. (2020). Ultra-large chemical libraries for the discovery of high-affinity peptide binders. *Nat. Commun.* 11 (1): 3183.

46 Rush, M.D., Walker, E.M., Prehna, G. et al. (2017). Development of a magnetic microbead affinity selection screen (MagMASS) using mass spectrometry for ligands to the retinoid X receptor-α. *J. Am. Soc. Mass Spectrom.* 28 (3): 479–485.

47 Ratnayake, A.S., Flanagan, M.E., Foley, T.L. et al. (2021). Toward the assembly and characterization of an encoded library hit confirmation platform: bead-assisted ligand isolation mass spectrometry (BALI-MS). *Bioorg. Med. Chem.* 41: 116205.

48 Allenby, G., Bocquel, M.T., Saunders, M. et al. (1993). Retinoic acid receptors and retinoid X receptors: interactions with endogenous retinoic acids. *Proc. Natl. Acad. Sci. U.S.A.* 90 (1): 30–34.

49 Boehm, M.F., Zhang, L., Zhi, L. et al. (1995). Design and synthesis of potent retinoid X receptor selective ligands that induce apoptosis in leukemia cells. *J. Med. Chem.* 38 (16): 3146–3155.

50 Steen Redeker, E., Ta, D.T., Cortens, D. et al. (2013). Protein engineering for directed immobilization. *Bioconjug. Chem.* 24 (11): 1761–1777.

51 Mrksich, M. (2008). Mass spectrometry of self-assembled monolayers: a new tool for molecular surface science. *ACS Nano* 2 (1): 7–18.

52 Gurard-Levin, Z.A., Scholle, M.D., Eisenberg, A.H., and Mrksich, M. (2011). High-throughput screening of small molecule libraries using SAMDI mass spectrometry. *ACS Comb. Sci.* 13 (4): 347–350.

53 Gurard-Levin, Z.A., Kim, J., and Mrksich, M. (2009). Combining mass spectrometry and peptide arrays to profile the specificities of histone deacetylases. *ChemBioChem* 10 (13): 2159–2161.

54 Gurard-Levin, Z.A. and Mrksich, M. (2008). Combining self-assembled monolayers and mass spectrometry for applications in biochips. *Annu. Rev. Anal. Chem.* 1: 767–800.

55 Kim, J. and Mrksich, M. (2010). Profiling the selectivity of DNA ligases in an array format with mass spectrometry. *Nucleic Acids Res.* 38 (1): e2.

56 VanderPorten, E.C., Scholle, M.D., Sherrill, J. et al. (2017). Identification of small-molecule noncovalent binders utilizing SAMDI technology. *SLAS Discovery* 22 (10): 1211–1217.

57 Atwal, J.K., Chen, Y., Chiu, C. et al. (2011). A therapeutic antibody targeting BACE1 inhibits amyloid-β production in vivo. *Sci. Transl. Med.* 3 (84): 84ra43.

58 Pusch, W. and Kostrzewa, M. (2005). Application of MALDI-TOF mass spectrometry in screening and diagnostic research. *Curr. Pharm. Des.* 11 (20): 2577–2591.

59 Zhang, N., Doucette, A., and Li, L. (2001). Two-layer sample preparation method for MALDI mass spectrometric analysis of protein and peptide samples containing sodium dodecyl sulfate. *Anal. Chem.* 73 (13): 2968–2975.

60 Harlan, J.E., Egan, D.A., Ladror, U.S. et al. (2003). Driving affinity selection by centrifugal force. *Assay Drug Dev. Technol.* 1 (4): 507–519.

61 Jin, Y., Cheng, X., Yang, F., and Fu, L. (2015). Ultracentrifugation-based multi-target affinity selection mass spectrometry. *RSC Adv.* 5 (130): 107616–107622.

62 Liu, J., Jiang, F., Jin, Y. et al. (2012). Design, synthesis, and evaluation of 2-substituted ethenesulfonic acid ester derivatives as protein tyrosine phosphatase 1B inhibitors. *Eur. J. Med. Chem.* 57: 10–20.

9

Online Affinity Selection Mass Spectrometry

Hui Zhang[1] *and Juncai Meng*[2]

[1] *Entos Inc., Department of Analytical Technologies, Entos, San Diego, CA, USA*
[2] *Discovery Technologies and Molecular Pharmacology (DTMP), Janssen Research & Development, LLC, Spring House, PA, USA*

9.1 Introduction of Online Affinity Selection-Mass Spectrometry

Affinity selection-mass spectrometry (ASMS) techniques can directly detect ligands that bind to specific biomolecular targets by virtue of their molecular weights. Since the initial demonstration in the early 1990s, label-free, direct MS detection of binding events was recognized as a significant advantage over other approaches such as traditional high-throughput screening (HTS), where chemical modifications (radioisotopes, fluorophores, or other types of tags) are needed for either targets or ligands [1]. Modern MS technology, especially time-of-flight (TOF) or Orbitrap-MS, renders exquisitely high mass accuracy, selectivity, and detection sensitivity [2, 3]. These great features enable ASMS experiments to be performed using very low (ng) amounts of purified proteins or other biomolecular targets and detect specific binders from very complex mixtures. Enabled by MS technological advancement, ASMS application has seen fast growth in the recent decade and has been established as an essential technique for hits generation and optimization [4]. Indeed, ASMS has become a key high-throughput drug discovery screening tool, complementary to traditional HTS and DNA-encoded library (DEL). Several pharmaceutical companies have developed and implemented their in-house ASMS technology [5–8]. More and more CROs are establishing this service to meet the increasing demands from pharmaceutical and biotech customers.

High-Throughput Mass Spectrometry in Drug Discovery, First Edition. Edited by Chang Liu and Hui Zhang.
© 2023 John Wiley & Sons, Inc. Published 2023 by John Wiley & Sons, Inc.

One key attribute of ASMS technology is that it simply finds binders to the target of interest. No deep knowledge of structure or functional activity of the target of interest is required. Tool compounds are preferred, but it is not a requirement for ASMS. With minimal assay development, ASMS can be utilized to screen potentially any soluble target of interest. As an "unbiased" binding assay, ASMS finds not only specific "active" site binders but also allosteric binders against the target of interest regardless of its functional activity [9]. In a traditional HTS, depending on how the assay is designed, the HTS screen may be biased for the discovery of a specific type of ligands. The "unbiased" binding in ASMS enables the discovery of interesting ligands exhibiting multiple mechanisms of action, including agonists/activators, antagonists/inhibitors, silent binders, etc. This could potentially lead to the discovery of hits class of interest with novel mechanisms of action [10]. Different constructs or states of targets can be effectively screened with ASMS, such as full-length proteins or truncated versions, proteins in either active or inactive forms. ASMS experiments are carried out in solution, which means necessary cofactors, detergents, and other buffer components can be maintained to enable proper protein folding and ensure good stability [11]. This is an advantage compared with other surface-based binding assays (such as SPR), where protein reagents have to be affinity captured or artificially fixed via special tags on a surface.

Another advantage of ASMS is its ability to detect ligands from complex matrixes, including combinatorial libraries, natural product extracts, and chemical reaction mixtures. Most big pharmaceutical companies have their in-house compound collection library [12], and ASMS is a very attractive approach for screening millions of compounds in a high-throughput, cost-effective manner. Because mass is a universal and selective detection for small molecules, compression at a high level significantly enhances the throughput of ASMS. With the advancement of compound handling capabilities, especially the automation and Echo dispensing technology [13], compound libraries are readily available to be compressed into hundreds to thousands compounds per well, enabling ASMS screening of large libraries in a few 384-well assay plates. ASMS can also be applied to screen very complex mixtures of natural products in extracts of botanicals or microbial cultures. The very diverse chemical nature of those mixtures with various fluorophores and chromophores could be preventative for other HTS screens because of the concerns of potential assay interferences. ASMS has also been successfully coupled with nanoscale synthesis for screening crude chemical library reactions, requiring no purification [14].

All these combined, ASMS assays have less assay complexity and can be developed and optimized at a much faster pace than conventional HTS assays. For example, according to a recent Pfizer report [15] the overall turnaround time of a regular ASMS screen with 1 million library compounds usually takes a few weeks, including assay development time, in contrast to six to nine months for a traditional HTS campaign. Besides these remarkable time savings, significant cost-saving was also achieved primarily due to less FTE time, greatly simplified reagents demand, and no hits resynthesis requirement for confirmation since ASMS links binding directly with molecular identity.

The technique and general applications of ASMS have been reviewed extensively in the past decade. Therefore, this book chapter will only cover the principles of online ASMS or 2D LC/MS-based ASMS and discuss key considerations and recent developments of this technology. In addition, online ASMS applications in drug discovery will be highlighted, and some new application trends in supporting new modalities of drug discovery will also be discussed. Offline ASMS has been addressed in Chapter 8 in this book by Stratton et al.

9.2 Online ASMS Fundamental

There are varieties of ASMS technologies reported in literatures. However, all ASMS screen includes four major steps: (i) affinity selection: ligand(s) are incubated and equilibrated with the target of interest to form a target-ligand complex in solution. Protein targets are usually in the low μM concentration range while ligands can present as either singleton or high compression format; (ii) separation stage where the target-ligand complex is separated from the unbound, free ligands. The key differentiator between online and offline ASMS is the separation technique employed. Offline ASMS utilizes ultrafiltration, centrifugation, gel permeation, or other immobilization-centered techniques to separate large molecules from small molecules. In contrast, online ASMS exclusively uses fast size-exclusion chromatography (SEC) to perform the separation. During SEC, small, unbound test ligands are retained in the column, while larger molecules such as free target protein or protein–ligand complex pass through at void volume. An inline valve switch system is precisely controlled to trap the protein–ligand complex fraction and immediately transfer the complex fraction to a reverse-phase chromatography; (iii) dissociation stage: The protein–ligand complex will dissociate and release the bound ligands under high-temperature acidic conditions on a regular C18 reverse-phase column; (iv) detection stage where the released ligands are desalted, enriched on the C18 column, and then eluted into a high-resolution MS for detection. MS is set up at different scanning modes enabling the detection of different binders based on their intrinsic mass. In addition to SEC and LC/MS, usually, UV detectors are employed to monitor the effective peak cutting and transfer of the protein–ligand complex peak. A typical online ASMS instrument setup is shown in Figure 9.1 [8].

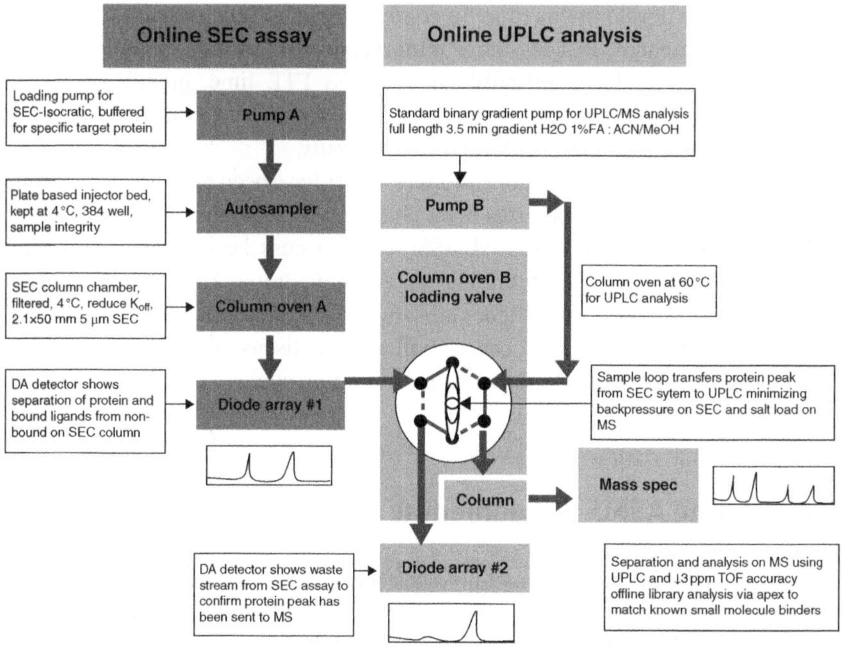

Figure 9.1 Typical setup of the online ASMS with 2-D LC/MS configurations. *Source:* Reproduced with permission from O'Connell et al. [8]/American Chemical Society.

9.3 Instrument Hardware and Software Consideration

9.3.1 SEC Selection, Fast Separation, and Temperature

SEC selection is the crucial step of the ASMS experiment, where noncovalently bound protein–ligand complex should be separated from the unbound free ligands and captured with enough recovery to be successfully detected by two dimensional LC/MS analysis. The three most critical parameters that determine the final ligand responses are the overall binding affinity of the ligand to the target (K_d), off-rate of the complex (K_{off}), and SEC separation time. It is easy to understand that the stronger the binding affinity K_d, the higher likelihood that the ligand can be detected as a binder. As a general rule of thumb, compounds with a dissociation constant (K_d) value with low double-digit μM or less can be readily identified by ASMS. What is frequently overlooked is that K_{off} and resident (SEC separation) time are equally, if not more critical, for ASMS detection. That is because ligands and protein targets may not reach complete equilibrium during SEC separation, and isolation of the bound complex from the free unbound ligand favors target/ligand complex dissociation. As a result, fast off-rate will decrease the recovery efficiency (thus the final ASMS signal of the ligands), and a similar effect is seen for a longer SEC isolation time. Such effect has been demonstrated earlier with both simulation (Figure 9.2) and experimental results [16]. The author used 10% recovery as an arbitrary surrogate of the ASMS detectability and demonstrated a weak compound with K_d ~30 μM could be detected due to a very slow off-rate of $0.005\,s^{-1}$ ($t_{1/2} = 138\,s$). On the other hand, a very potent compound with K_d ~1 nM but with a faster off-rate than $0.06\,s^{-1}$ ($t_{1/2} = 11.5\,s$) could still be missed in ASMS. Minimizing dissociation of the protein–ligand complex is an effective way to improve the ASMS detectability. The most frequently used approach is to maintain the temperature of the SEC column compartment at a low temperature of 4 °C. Similarly, a longer SEC separation time will have a negative impact on detecting ligands: a much shorter SEC separation time is required for detection of a ligand with faster K_{off} (Figure 9.2). For this reason, SEC separation should be carried out as fast as possible to capture ligands with a fast off-rate (such as $0.1\,s^{-1}$) while being sufficient to separate large complexes from small ligands. In general, a separation time of less than 20 seconds is ideal. Such requirement imposes a high challenge for the SEC separation in which special SEC columns need to be designed for ASMS applications to provide fast separation of the bound

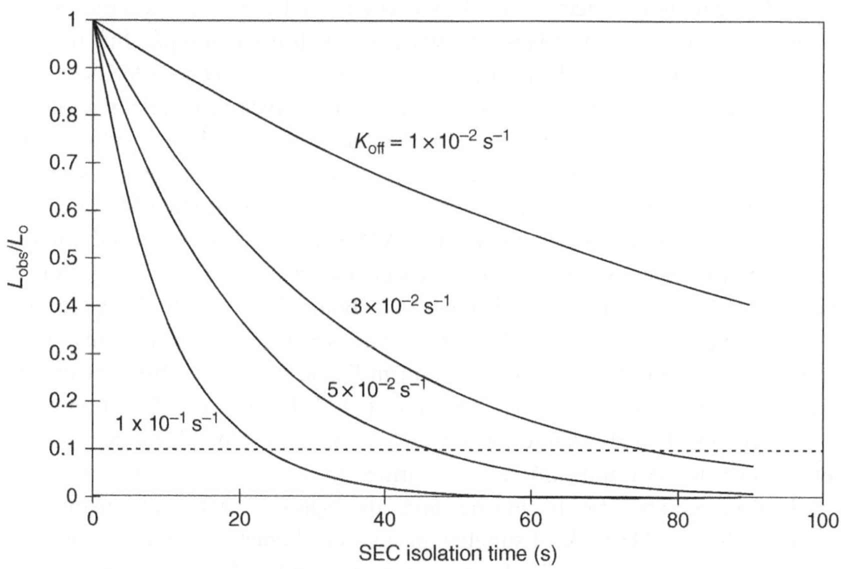

Figure 9.2 The time-dependent dissociation profiles simulated for a potent ligand (K_d = 1 nM) at differing K_{off} rates. Using a 10% recovery cut off, a fast-off compound with $K_{off} = 0.1\,s^{-1}$ ($t_{1/2} = 7\,s$) needs to have a much shorter (<25 s) isolation time than a slow off compound with $K_{off} = 0.01\,s^{-1}$ ($t_{1/2} = 69\,s$). *Source:* Reproduced with permission from Blom et al. [16]/American Chemical Society.

protein–ligand complex from the unbound ligands to capture the binders with weak K_d or fast K_{off}. To achieve a 20 seconds or shorter separation time, protein targets would have to be eluted from the void volume with minimum retention to the SEC resin. Traditional SEC columns usually operate at minimal separation within such a short separation time and will not be suitable for ASMS. In addition, SEC stationary phase should be compatible with the target proteins, or the protein could crush on the column resulting in variable protein signal and inconsistent affinity selection results. Unfortunately, there are not that many commercially available options. To our knowledge, the polyhydroxylethyl series from PolyLC (Columbia, MD) and Zenix series from Sepax (Newark, DE) are the most commonly used SEC columns in ASMS. Waters BEH SEC column has also been explored in ASMS experiment with a longer lifetime to approximately a couple of thousands of injections. Beyond that, researchers in Neogenesis (later became part of Merck) have developed their own SEC column with proprietary methods [9].

9.3.2 MS: Low Resolution and High Resolution

When bound ligands are successfully separated and captured by the first-dimensional SEC, the remaining detection work is carried out by the second-dimensional reverse-phase LC/MS. The reverse-phase LC is usually used for separating the salts and biological target from the ligand to minimize the interference in MS detection. There are many different types of reverse-phase columns commercially available. Users usually do not have difficulties finding the right option based on the nature of the ligand types to be analyzed. Instead, MS is an interesting option as several distinct types of MS detectors are reported for ASMS, including QqQ, MALDI-TOF, ESI-TOF, and Orbitrap. Early competitive ASMS binding assays utilized a native ligand (non-labeled) as a marker, and the signal alternation of ASMS detection of such marker, when incubated with a potential competitor, would indicate the competitive binding interactions. Low-resolution MS works well for this probe approach, and QqQ MS was used to confirm ligand binding to different targets, including dopamine D1 and D2 receptors [17, 18]. Early ultrafiltration-based ASMS also relied on low-resolution MS such as LCQ ion-trap MS, where noncovalent binders to HSA and BSA were confirmed from neat and complex matrices (such as fermentation broth). Furthermore, the high selectivity and sensitivity of triple quadrupole MS make it suitable for binding characterization (K_{on}, and K_{off} measurement, etc.), especially for very potent ligands where both protein and ligand concentrations are kept low (at nM instead of the regular μM range for hit identification) [7]. On the other hand, low-resolution MS becomes obsolete and has difficulties in identifying the true binder when a high compression level of ligands is used, as many ligands could share similar *m/z* within 1 Da unit or if there are co-eluting matrix ions in the similar *m/z* window. For those situations, high-resolution MS such as different types of TOF and orbitrap MS are much more suitable options because they can provide high mass accuracy (<5 ppm) and high mass resolution (>10k). Such high mass accuracy and mass resolution enable true ligands to be identified without ambiguity, even from highly compressed compound libraries or very complex biological matrices such as herbal extracts. For example, Annis et al. have demonstrated that approximately 75% or more of the library compounds could be interfered with each other because of multiple nominally isobaric components while simply limiting the mass accuracy to 20 ppm or lower level could reduce all these potential ambiguities [19]. It is not surprising that, as reported, for most known ASMS operations in recent years, TOF and orbitrap high-resolution instruments are the dominant MS platforms.

9.3.3 Software: Key Features, False Positives, and False Negatives

Like many other MS applications, software is very crucial for ASMS operation. Binding confirmation for ligands heavily relies on the software capabilities after a successful ASMS wet experiment, especially when high compression libraries are screened. Ideally, the software should be able to process the raw LC/MS data and compare the observed *m/z* values to identify primary hits from the compound library that are screened. Also, the software should have intelligent bioinformatic features to triage and confirm the primary hit lists from the potential false-positive signals. Many parameters must be evaluated before a tentative ligand can be called a binder, including mass accuracy, isotope pattern match, mass and chromatography peak intensity, background noise level (both mass spec level and extracted ion chromatogram level), retention time, etc. After that, all the tentative binders must be further evaluated by looking at other samples to eliminate potential false positives from the background or carryover. The searching algorithm must be robust and fast to facilitate fast data processing. Software provided by major MS instrument vendors has some basic functionality but is far short of the capability to support large-scale ASMS screens in general. Many research groups have developed their in-house software tools. For example, the Merck team (through their legacy Neogenesis team) has built their informatics tools for binder identification from up to 2500-fold compression [4]. Similarly, Abbvie had their in-house ASMS software for similar functions [20]. On the other hand, there are not many commercial solutions in the market, even at the current stage, for ASMS data analysis. To the authors' best knowledge, there are only two commercially available software that ASMS users have well adopted: APEX by Sierra Analytics [8] and Analytical Studio by Virscidian [21].

False positives are one of the major concerns for ASMS screens, especially for large library screening at the primary screen stage, where high hit rates would cause an extra burden for hits confirmation. False positives are usually resulted from nonspecific binding. Such nonspecific binding events can be monitored by running the hits against a decoy/irrelevant target and comparing the binder signals with the real target of interest. For the real bona fide binders, there should be no ASMS signal observed with a decoy target or at least much lower than with the real target of interest. Analytical Studio software from Virscidian allows users to compare the mass signal of potential hits in the target well with that of other wells on the plates. If a compound is identified in both target well and other wells, it is considered a false positive. By careful design of hits selection criteria, such as LC peak width, mass accuracy, isotope pattern similarity, and S/N ratio. Analytical Studio can filter out potential false positives from MS spikes or broad peaks from the background. When multiple compounds sharing the same molecular formula are present in the same well, by comparing the retention time of each compound between direct LC-MS and ASMS, potential false positives have the same m/z but different retention times to real hits can also be removed. Another way to tease out the nonspecific binding is to run those binders at a saturation experiment (such as those used for K_d determination) with the target of interest: the binding signal should reach saturation against fixed target concentration with increasing concentration of ligands. Otherwise, a continuingly increasing ligand signal would suggest the binding event is not specific. However, saturation experiments in ASMS can be time-consuming. Another primary reason for high false positives is SEC chromatographic breakthrough, where the unbound ligand is not effectively trapped by the SEC stationary phase and eluted with the protein band like the real binding ligands. For that case, elevated hit rates than normal/expected would be a strong indication. An ASMS counter screen against decoy protein can remove potential false positives caused by compounds breakthrough. Compound aggregation can also produce false positives, and the main indication is the high ligand signal even in the absence of protein. Different detergents, including Dodecyl-β-D-maltoside (DDM), Triton, and Chaps, have been used in ASMS assay buffer to reduce compound aggregation [22].

False negatives are a more complex problem in ASMS as it often represents an unknown issue unless targeted troubleshooting is performed with known binders. This becomes more difficult when a novel target is pursued without any previous knowledge or known positive controls in hand. Several reasons could cause false negative ASMS signal, including (i) undesirable binding kinetics of binders, such as a fast K_{off} rate which leads to quick dissociation of protein–ligand complex and eventually loss of binders during SEC separation; (ii) ligands are not compatible for reverse-phase liquid chromatography (too hydrophilic) or not ionize well in MS; (iii) software settings may be too restrictive for the hits ID criteria. In addition, poor SEC column compatibility to the target protein is another leading cause of false negatives. The main indication for this situation is an unstable (or even diminishing) protein signal from the one-dimensional SEC separation. SEC buffer also plays an essential role in ASMS compatibility. Phosphate buffer and ammonium acetate buffer (pH 7–8) are commonly used SEC buffers in literature, but we have noticed in an ASMS screen of a large size polymerase protein (~260 kDa), where the target protein is only stable in SEC buffer containing 300–500 mM salt (unpublished data). Therefore, a brief SEC buffer scouting may be necessary during ASMS assay development. Further investigation and troubleshooting can be carried out for the protein compatibility, ligand characterization, and software settings correspondingly.

9.3.4 Compound Libraries and Compression Level

One key advantage of ASMS for hits discovery is its high-throughput capability to rapidly screen a large library of compounds for the targets of interest. Because the detection is based on the unique exact mass of small molecules, compounds can be pooled and screened as mixtures in ASMS, as long as they do not share the exact same chemical formula. In fact, library compression is the most straightforward but most effective approach to increase the screening throughput for ASMS, as the average time to sample one test compound is reversely proportional to the compression level. Many organizations have demonstrated high compression levels of hundreds to thousands of compounds per well. The maximum compression ratio is limited by MS detection sensitivity. Not surprisingly, over 2000 : 1 compression became a reality with the advancement of modern MS instruments, as demonstrated by both Merck and Abbott teams [4, 20].

Besides MS detection sensitivity, another major limitation would be DMSO tolerance for the testing protein target. Predominantly library compounds are stored as DMSO solutions in all major pharmaceutical companies and CROs because of DMSO's low volatility and capacity to dissolve compounds with a wide range of chemical properties [23]. In general, 4% DMSO is the upper limit in the final incubation solution for many protein targets, and higher than that level, many protein targets could become unstable. With that final DMSO concentration limit, higher compression of testing compounds would require a higher initial stock solution of test molecules if the final individual test compound concentration is maintained (due to MS detection sensitivity requirements). Assuming the initial concentration of the individual test compound stock solutions is 30 mM and the targeted final concentration of each ligand at 0.4 μM, 3000 : 1 compression would ensure that the final DMSO concentration is not exceeding 4%. Practically, Pfizer team has chosen the 2600 : 1 compression ratio for the convenience of library preparation: at such compression, one 384-well plate would make up about just one million ($2600 \times 384 = 998{,}400$) test compounds in total [21].

As the compression level increases, a general concern about ASMS is whether there would come more nonspecific/promiscuous binders. Pfizer had reported in the recent conferences that they had compared the hits identified by ASMS from the same chemical library running at 400 : 1, 1300 : 1, or 2600 : 1 compression ratio. Overall the reproducibility was very promising (70% overlap between different screens), indicating high compression of library compounds is compatible for ASMS screen as long as the total DMSO concentration was not exceeding 4%, and there was still enough detection sensitivity [21].

Recently, a group from Human Institute at ShanghaiTech University demonstrated a very high compression ratio of up to 20k compound/pool [24]. The critical step to enable the identification of binders from such as large pool seems to be the iterative affinity selection steps – the first round of compound extracts was dried down, reconstituted, and re-incubated and such process was repeated three times before the final detection. Besides combinatorial libraries generated from the pooling of purified individual compounds, ASMS has also proven useful in screening unpurified mixtures such as natural product extracts [25] or unpurified crude reaction solutions [14], because the one-dimensional SEC separation can effectively separate the hits from the majority of other irrelevant small-molecule components.

9.4 Type of Assays Using ASMS

ASMS has been utilized for several drug discovery applications, from target validation to hits generation and lead optimization (Figure 9.3). Here we briefly summarize the general principle and most recent updates for each of the key applications, as several review papers have been published with great details for interested readers [11, 27–30].

9.4.1 Target Identification and Validation

ASMS can be a valuable tool for target evaluation. When a pathway is well understood and known ligand(s) are available, ASMS can be employed to assay a panel of protein targets to confirm the selective target-ligand interactions of

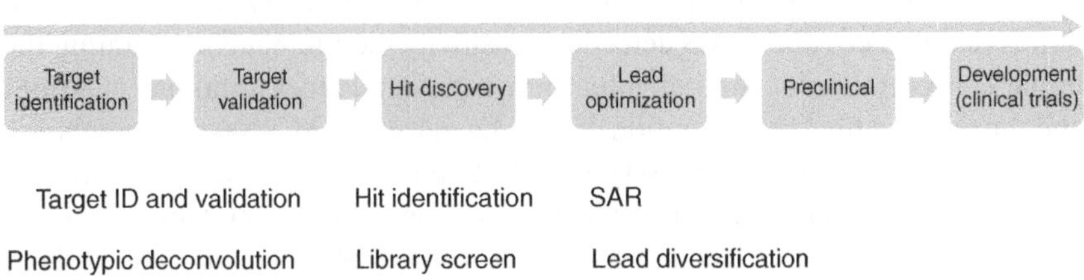

Figure 9.3 Potential ASMS applications throughout the drug discovery and development stages. *Source:* Yang [26]/The Pittsburgh Conference on Analytical Chemistry and Applied Spectroscopy, Inc. (with modification)

well-characterized clinical compounds and to discover potential off-target interactions [31].

For first-in-class targets with no pre-knowledge about what type of chemistry might interact with that target, ASMS can directly probe the ligandability of a given target with preexisting compound libraries by quickly generating hit rates analysis. This could serve as one of the key criteria for prioritizing different target options against a particular type of disease or a specific signaling pathway. For example, Kutilek et al. screened a 50k member chemically diverse compound library against 15 targets with known function within NF-κB pathway and identified 382 target-selective hits binding to 14 of the total 15 proteins tested [6]. Combining with functional cell-based assays, two chemical series were further confirmed for the target-driven cellular activity of two proteins (TRAF5 and NF-κB1/P105). These results would serve as novel drug discovery opportunities and prove the ASMS utility in finding the most tractable drug target within a pathway. The same group also applied a similar approach for a more extensive study with 311 antibacterial target proteins against a 57k antimicrobially active compound library [32]. The authors confirmed the known antibacterial compounds that selectively bind to their corresponding targets (trimethoprim-dihydrofolate reductase, actinonin-peptide deformylase, erythromycin-ribosome, etc.). In addition, and more excitingly, they identified several hits bound to 24 novel proteins with no marketed drugs, making this ASMS approach very appealing to discovering new targets for antibacterial therapies.

It is noticeable that one major limitation of applying ASMS for large-scale target identification is the availability of purified protein reagents. Even though the ASMS screen is very sample sparing compared with conventional HTS, generating enough protein (mg level in general) to screen big libraries is not trivial, especially when hundreds or more different protein targets are involved. Two approaches have been developed to solve this challenge. One uses a mini-array expression system based on a cDNA open reading frame bank. McMillan et al. from UT Southwestern expressed a genome-scale protein panel (14,000 proteins arrayed in 96-well plate format with one protein in each well) and identified glucose transporter GLUT8 as the lone hit to one of their active compounds SW157765 [33]. Another exciting approach is called cytosolic proteome and affinity-based target identification (CPATI). Yang et al. from Merck combined quantitative proteomics analysis for cytosolic protein fractions and ASMS ligand library screening results against each cytosolic fraction. Potential interaction pairs can be assigned based on the distribution pattern match between protein level and bound ligand signal. This novel approach enables simultaneous evaluation of the whole proteome in cellular matrixes without taking the burden of protein expression/purification, and was successfully validated with known target pairs such as AUY922-HSP90A and TH588-NUDT1 [34].

9.4.2 Hits ID from Combinatorial Libraries or Compound Collections

ASMS effectively finds all different types of binders to a given target of interest (e.g. orthosteric agonists/activators, antagonists/inhibitors, or allosteric modulators regardless of functional activity). ASMS has demonstrated the greatest success for hits ID across a broad range of drug targets, including kinases, transferases, reductases, polymerases, and other enzymes. ASMS is not only suitable for screening of cytosolic soluble protein targets but also reported for successful hits ID with membrane-bound targets such as GPCRs and other receptors. A thorough list of various hits ID campaigns can be found in a recent review [30]. Many of those hits were cross-validated with other orthogonal biophysical and functional (biochemical or cellular) assays, validating the fidelity of hits identification via the ASMS approach. Also, it is worth highlighting that a few clinical assets emerged from various ASMS campaigns since the initial report of applying ASMS for hits generation in the early 2000s. Merck (including its predecessors Schering Plough and Neogenesis) had identified many lead molecules, including the ERK1/2 inhibitor MK8353, which is currently in phase II for advanced solid tumors [35]. Pfizer team identified the initial hit of RAC-4, a noncovalent modulator of hemoglobin for treating sickle cell disease, from an ASMS screen of 35k compound library based on virtue screen results. The initial hit was further developed to the final candidate PF-07059013 (Figure 9.4), which has been advanced to Phase I clinical trials [36]. Due to the proprietary nature of the early drug discovery programs, outcomes from most ASMS-based hits discovery campaigns were not publicly disclosed. However, it is reasonable to anticipate that more successful ASMS campaign results will become available when clinical assets are disclosed.

9.4.3 Hits Characterization and Leads Optimization

In addition to hits identification, ASMS is also a very powerful tool for characterizing ligand–protein interactions, generating structure–activity relationship (SAR), and supporting lead optimization. For example, primary hits can be

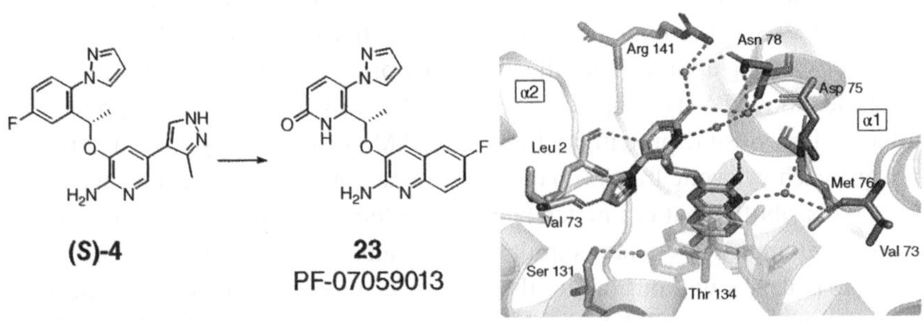

Figure 9.4 One successful example of identifying clinical drug candidates with the ASMS approach. *Source:* Reproduced with permission from Gopalsamy et al. [36]/American Chemical Society.

further evaluated with a ligand titration experiment to determine K_d simply by analyzing a series of experiments with a fixed target protein concentration and serially increased ligand concentration. If the initial hit is a specific binder, a saturation binding curve can be obtained from which a simple nonlinear regression can derive the K_d value. On the other hand, if the binding is non-saturable, the hits are more likely to be a nonspecific binder. Furthermore, initial hits can be rank-ordered directly by their K_d values. However, it is crucial to understand that those K_d values derived from ASMS could be much higher than those determined by other biophysical methods such as SPR. This is because the K_d derived from such ASMS experiments are "apparent" values and are heavily limited by the protein target concentration. No matter how potent the actual K_d is, the measured apparent K_d is usually floored at half the protein concentration. Also, ligand solubility could impact the determination of the inflection point of the saturation curve. Nevertheless, apparent K_d assay by ASMS would serve as a good tool for hits confirmation and optimization, generating quick rank order to compare initial hits and guiding selection and prioritization of different hit series.

ASMS can be used to characterize hits to determine binding mode through a competition experiment. For example, with the addition of known competitors at serially increasing concentration, the equilibrated test ligand could be gradually competed off the target. The changes in the MS signals of the test ligand versus competitor can reveal the binding mode of the test ligand: for competitive binding, the ratio of competitor-to-ligand signals should generate a straight line, as demonstrated for warfarin and heavy(D6)-labeled warfarin signals as they are the perfect orthosteric competitors to their binding target of human serum albumin, or between Staurosporine and NGD-6380, binders to protein ZAP-70 even though they do not share a structural similarity [9]. On the other hand, when an allosteric ligand is titrated by a noncompetitive titrant, the signal ratio of the two would form a parabolic instead of linear relationship, as demonstrated by the pair of staurosporine/NGD-28835, which was later confirmed to bind a different site of the same target. Also, based on the signal level of the testing ligand, negative or positive cooperativity can be easily assigned between sites.

ASMS competition experiment can also be carried out with a mixture of testing compounds instead of single ligand–competitor pair. In this setting, testing compounds would compete with each other, and weaker ones will be competed out at conditions when protein/binding pockets are not in excess. For example, Adam et al. [5] carried out a simple experiment with 10 μM dipeptidyl peptidase IV (DPP-4) protein and 20 of its inhibitors (potency ranges from 2 nM to 2 μM) at an equal concentration of each. Protein is in excess at lower concentrations of compounds, and all binders can be detected. However, when ligand mixture concentration increases, the condition changes from ligand limited to protein limited, creating a competitive environment where strong binders compete weak binders off the target protein. As a result, the signal of weaker binders gradually disappears in ASMS, with only the stronger ones remaining. The sequence of ligand disappearance correlates well with the potency ranking (DPP-4 inhibition IC_{50}). Another variation in the ASMS competition experiment is protein titration, where ligand concentrations are kept the same, but the target concentration is gradually diluted. Similarly, weaker binders will only be detected at higher concentrations of targets, while more potent binders in the same pool can remain being detected at lower concentrations of the target. This protein titration experiment is beneficial to save reagent consumption or for scenarios where testing ligand concentrations are not consistent, such as the crude chemical mixtures.

A more sophisticated competition experiment is called ACE50 [9]. ACE50 experiments resemble the traditional radioligand displacement assays but have significant advantages. First, it relies on label-free mass analysis with stable isotopes and does not require extra radiolabel steps. Second, MS detection also renders lower background than radiolabeled assays. Third, ionization discrepancy is a significant concern for quantitation in MS analysis without an internal standard. However, ACE50 determination relies on normalized MS responses to each compound itself, and no signal calibration with purified standards is needed. This unique feature enables ACE50 to be used directly for unpurified samples without knowing the exact concentration of the components of interest.

The ACE50 method is suitable for simultaneously rank-ordering compounds by affinity, particularly for structurally similar analogs synthesized in library chemistry approaches. The technique is especially valuable for SAR generation and lead optimization through affinity enhancement from a progenitor or weaker hits [9]. The ACE50 approach can be applied to studies with protein–ligand interactions with various affinities, covering the low nM to μM range. This is more advantageous than the K_d experiment, which is greatly limited by the assay floor. ACE50 experiment should be carried out with conditions that the target protein concentration is maintained at higher than the sum of the library components' concentrations. Otherwise, ACE50 curves would collapse, and affinity differences would become indistinguishable.

ASMS experiments with different variations can shed light on understanding binding interactions between ligands and targets with multi-domain or multi-subunits and provide valuable and complementary structural insights to other biophysical endpoints. For example, this approach is handy for the ligand binding study of large proteins or protein complexes. Traditional surface-based SPR assay or thermal shift assay may not be suitable due to small binding response in SPR assay and likely multiple transitions of protein complexes in thermal shift assay. In addition, ASMS methods are promising for discovering and developing allosteric ligands by quickly accessing target selectivity among similar protein family members that share conserved active binding sites and improve ligand binding specificity through SAR development.

9.5 Applications Examples and New Modalities of ASMS for Drug Discovery

ASMS has proved to be a very effective tool for both hits discovery and lead optimization for a broad range of protein targets. There have been a few interesting studies where ASMS showed great success with the very difficult or poorly ligandable targets by conventional wisdom. One example is the stimulator of interferon genes (STING) [37], an important target in inflammatory diseases. STING is well known for the difficulty to target because of the unique dimerized structure with the most putative ligand-binding site located at the interface between the two monomers. Such binging pocket has an abnormally large polar solvent accessible surface area (390 $Å^2$) and a large volume (952 $Å^3$). Taking advantage of the intrinsic symmetry of the dimer protein, the Merck team applied ASMS and successfully identified small-molecule ligands to occupy the ample binding space with 2 : 1 stoichiometry of ligand: STING homodimer. The special 2 : 1 ligand:protein dimer binding was confirmed by both SPR (yielding ~50 μM and ~2 nM of binding K_d of the first and second bound molecule, respectively, with their best lead molecule) and X-ray crystallography. Compared with one single, large, and polar ligand (like its endogenous ligand cGAMP), the smaller ligands identified by the ASMS approach have more favorable physicochemical properties that enable better oral bioavailability.

PCSK9 is another hot target for cardiovascular diseases with strong genetic and clinical evidence in cholesterol management. It is another poorly druggable target, and many organizations have tried but could not find small molecules that directly inhibit its function. Interestingly, a few years ago, the Pfizer team applied ASMS to confirm the unique mechanism of action – modulation of PCSK9 protein translation – of the single active hit identified through the phenotypic screen [38]. This was elegantly demonstrated by binding confirmation of the hit molecule to the human 80S ribosome directly. In addition, ASMS confirmed that one enantiomer (R-IMPP) binds much stronger with human ribosome than the other (S-IMPP), while neither of the enantiomers binds to *Escherichia coli* ribosome. All these ASMS findings matched well with functional results. It is worth noting that ribosomes are much bigger proteins (millions of Da) than regular protein targets for ASMS (usually in tens to hundreds of Da), suggesting good compatibility of ASMS to very large proteins or protein complexes. Indeed, more physiologically relevant complexes are compatible with ASMS, such as mitochondria [39] or even whole-cell membrane fractions [40, 41].

In recent years, targeted protein degradation (TPD) or proteolysis targeting chimeric (PROTAC) has become a very attractive approach to drug "undruggable" proteome, because they can degrade proteins of interest and do not rely on the modulation of their function. ASMS screening is a perfect tool to generate hits for the PROTAC or molecular glue approach because it can identify hits and drive hits optimization purely from binding, regardless of its functional activity. For the same example of PCSK9, the Merck team recently reported their success in applying ASMS to discover one functionally inactive allosteric binder through a campaign of a 200,000-compound collection [10]. This single hit showed robust and reproducible binding across orthogonal biophysical assays, including SPR, thermo shift, and differential scanning calorimetry assays. Based on this single hit, a high-affinity agent was developed as an effective degrader to reduce intracellular PCSK9 levels without displaying direct functional activity. This approach is an excellent example that ASMS could be used as a powerful strategy to develop degrader as the new modality for difficult/non-druggable targets.

Only a small fraction of the genome is encoded by proteins, while RNA constitutes the majority of the genome with diverse functions. With the significant advancement and understanding of RNA structural and functional information, RNA has become an exciting drug target. ASMS screening has recently proved very useful for screening RNA targets. Rizvi et al. first reported the successful confirmation of five different small-molecule ligands (such as FMN) to their respective riboswitch RNA [42], demonstrating the general compatibility of RNA to ASMS screening. Later this study was expanded to the screening of 42 different RNA targets from various classes, ranging from 75 to 2222 nucleotides in size, against a 50k drug-like compound collection plus another 5.5k functionally annotated tool compound library. Interestingly, binders were found for 40 out of 42 total RNA targets, with 30 targets having at least one selective binder. These results strongly suggest that RNA could be another class of targets for various diseases.

Compared with RNA, targeting DNA with small-molecule agents is even more challenging because there is a less well-defined 3-D structure formed by DNA double helix. One particular type of G-rich DNA sequence called G-Quadruplex (or G4) can modulate downstream gene products by inhibiting transcription when being stabilized in their folded forms. In addition, different G4s can fold differently to form unique tertiary structures that suggest a certain degree of specificity. A recent study [43] proved that small-molecule ligands could be identified for the important oncogene Myc G4 quadruplexes by ASMS screening, and selectivity was demonstrated by control experiments with other G4 quadruplexes and non-quadruplex DNA fragments. With these demonstrated successful screening results with both RNA and DNA targets, ASMS technology will enable a significant expansion of the drug target space.

9.6 Future Perspectives

Since its inception almost two decades ago, ASMS has seen tremendous growth in small molecule drug discovery and development applications through technological advancement, including high-resolution mass spectrometry, fast SEC chromatography, and software capability enhancement for million-compound-level parallel processing. Online SEC-based ASMS has been developed into a compelling analytical technology supporting both target identification and validation and hits identification and optimization. In addition to functionally active agents such as inhibitors or activators, as a label-free, unbiased screening assay, ASMS can also identify silent binders as a potential warhead for PROTAC or molecular glue. Besides protein targets, other modalities can be screened with ASMS, such as DNA or RNA. ASMS can also be applied as a complementary approach for target deconvolution in phenotypic screens. On the library side, ASMS is very effective in screening large libraries of purified compounds and has proven very useful for complex mixtures such as unpurified combinatorial chemical reactions or natural product extracts. Recent years also see ASMS techniques being developed as a tandem and complementary tool for very large libraries like DELs. Overall, it is estimated that hundreds of various targets have been screened, and many have yielded validated ligands with diverse mechanisms of action. Many of those hits are being developed in various drug discovery stages, including clinical trials that one day will lead to real therapy for some diseases.

References

1 Gribbon, P. and Sewing, A. (2003). Fluorescence readouts in HTS: no gain without pain? *Drug Discovery Today* 8 (22): 1035–1043.

2 Haslam, C., Hellicar, J., Dunn, A. et al. (2016). The evolution of MALDI-TOF mass spectrometry toward ultrahigh-throughput screening: 1536-well format and beyond. *J. Biomol. Screening* 21 (2): 176–186.

3 Hu, Q., Noll, R.J., Li, H. et al. (2005). The Orbitrap: a new mass spectrometer. *J. Mass Spectrom.* 40 (4): 430–443.

4 Annis, D.A., Athanasopoulos, J., Curran, P.J. et al. (2004). An affinity selection-mass spectrometry method for the identification of small molecule ligands from self-encoded combinatorial libraries: discovery of a novel antagonist of *E. coli* dihydrofolate reductase. *Int. J. Mass Spectrom.* 238 (2): 77–83.

5 Adam, G.C., Meng, J., Athanasopoulos, J. et al. (2007). Affinity-based ranking of ligands for DPP-4 from mixtures. *Bioorg. Med. Chem. Lett.* 17 (9): 2404–2407.

6 Kutilek, V.D., Andrews, C.L., Richards, M.P. et al. (2016). Integration of affinity selection-mass spectrometry and functional cell-based assays to rapidly triage druggable target space within the NF-κB pathway. *J. Biomol. Screening* 21 (6): 608–619.

7 Shimizu, Y., Ishii, T., Ogawa, K. et al. (2015). Biochemical characterization of smoothened receptor antagonists by binding kinetics against drug-resistant mutant. *Eur. J. Pharmacol.* 764: 220–227.

8 O'Connell, T.N., Ramsay, J., Rieth, S.F. et al. (2014). Solution-based indirect affinity selection mass spectrometry – a general tool for high-throughput screening of pharmaceutical compound libraries. *Anal. Chem.* 86 (15): 7413–7420.
9 Annis, D.A., Nazef, N., Chuang, C.C. et al. (2004). A general technique to rank protein-ligand binding affinities and determine allosteric versus direct binding site competition in compound mixtures. *J. Am. Chem. Soc.* 126 (47): 15495–15503.
10 Petrilli, W.L., Adam, G.C., Erdmann, R.S. et al. (2020). From screening to targeted degradation: strategies for the discovery and optimization of small molecule ligands for PCSK9. *Cell Chem. Biol.* 27 (1): 32–40.e3.
11 Annis, D.A., Nickbarg, E., Yang, X. et al. (2007). Affinity selection-mass spectrometry screening techniques for small molecule drug discovery. *Curr. Opin. Chem. Biol.* 11 (5): 518–526.
12 Gong, Z., Hu, G., Li, Q. et al. (2017). Compound libraries: recent advances and their applications in drug discovery. *Curr. Drug Discovery Technol.* 14 (4): 216–228.
13 Griffith, D., Northwood, R., Owen, P. et al. (2012). Implementation and development of an automated, ultrahigh-capacity, acoustic, flexible dispensing platform for assay-ready plate delivery. *J. Lab. Autom.* 17 (5): 348–358.
14 Gesmundo, N.J., Sauvagnat, B., Curran, P.J. et al. (2018). Nanoscale synthesis and affinity ranking. *Nature* 557 (7704): 228–232.
15 Troutman, M.D. ed. (2021). Lessons learned from Pfizer's experience in utilizing affinity selection by mass spectrometry for hit identification. *1st Edelris Symposium on Affinity Selection-Mass Spectrometry in Drug Discovery*, Lyon, France (2–3 June).
16 Blom, K.F., Larsen, B.S., and McEwen, C.N. (1999). Determining affinity-selected ligands and estimating binding affinities by online size exclusion chromatography/liquid chromatography-mass spectrometry. *J. Comb. Chem.* 1 (1): 82–90.
17 Höfner, G. and Wanner, K.T. (2003). Competitive binding assays made easy with a native marker and mass spectrometric quantification. *Angew. Chem. Int. Ed.* 42 (42): 5235–5237.
18 Zepperitz, C., Hofner, G., and Wanner, K.T. (2006). MS-binding assays: kinetic, saturation, and competitive experiments based on quantitation of bound marker as exemplified by the GABA transporter mGAT1. *ChemMedChem* 1 (2): 208–217.
19 Annis, D.A., Cheng, C.C., Chuang, C.C. et al. (2009). Inhibitors of the lipid phosphatase SHIP2 discovered by high-throughput affinity selection-mass spectrometry screening of combinatorial libraries. *Comb. Chem. High-Throughput Screen.* 12 (8): 760–771.
20 Comess, K.M., Schurdak, M.E., Voorbach, M.J. et al. (2006). An ultraefficient affinity-based high-throughout screening process: application to bacterial cell wall biosynthesis enzyme MurF. *J. Biomol. Screening* 11 (7): 743–754.
21 Thomas O'Connell, W.H. and Zhang, H. (2019). Affinity selection mass spectrometry for hit ID and early hit expansion. *Pittcon Conference and Exposition*, Philadelphia, PA (17–21 March).

22 Lu, Y., Liu, H., Yang, D. et al. (2021). Affinity mass spectrometry-based fragment screening identified a new negative allosteric modulator of the adenosine A2A receptor targeting the sodium ion pocket. *ACS Chem. Biol.* 16 (6): 991–1002.

23 Balakin, K.V. (2003). DMSO solubility and bioscreening. *Curr. Drug Discovery* 8: 27–30.

24 Lu, Y., Qin, S., Zhang, B. et al. (2019). Accelerating the throughput of affinity mass spectrometry-based ligand screening toward a G protein-coupled receptor. *Anal. Chem.* 91 (13): 8162–8169.

25 van Breemen, R. (2020). Affinity selection-mass spectrometry: defining the bioactive compounds in complex mixtures of natural products and combinatorial libraries. *LCGC Suppl., Special Issues-03-01-2020* 18 (1): 18–23.

26 Yang, X. ed. (2019). ALIS affinity selection in pharmaceutical discovery. *Pittcon Conference and Exposition 2019*, Philadelphia, PA (17–21 March).

27 Zehender, H. and Mayr, L.M. (2007). Application of high-throughput affinity-selection mass spectrometry for screening of chemical compound libraries in lead discovery. *Expert Opin. Drug Discovery* 2 (2): 285–294.

28 Zhu, Z. and Cuozzo, J. (2009). Review article: high-throughput affinity-based technologies for small-molecule drug discovery. *J. Biomol. Screening* 14 (10): 1157–1164.

29 Motoyaji, T. (2020). Revolution of small molecule drug discovery by affinity selection-mass spectrometry technology. *Chem. Pharm. Bull. (Tokyo)* 68 (3): 191–193.

30 Prudent, R., Annis, D.A., Dandliker, P.J. et al. (2021). Exploring new targets and chemical space with affinity selection-mass spectrometry. *Nat. Rev. Chem.* 5 (1): 62–71.

31 Salcius, M., Bauer, A.J., Hao, Q. et al. (2014). SEC-TID: a label-free method for small-molecule target identification. *J. Biomol. Screening* 19 (6): 917–927.

32 Nickbarg, E. ed. (2017). Protein array-ALIS: an affinity selection-mass spectrometry/phenotypic screening approach for identification of protein drug targets. *Proceedings of ASMS Conference*, Indianapolis, IN (4–8 June).

33 McMillan, E.A., Ryu, M.J., Diep, C.H. et al. (2018). Chemistry-first approach for nomination of personalized treatment in lung cancer. *Cell* 173 (4): 864–78 e29.

34 Yang, X. ed. (2020). Cytosolic proteome and affinity based target identification (CPATI). *SLAS*, San Diego, CA (25–29 January).

35 Moschos, S.J., Sullivan, R.J., Hwu, W.J. et al. (2018). Development of MK-8353, an orally administered ERK1/2 inhibitor, in patients with advanced solid tumors. *JCL Insight* 3 (4): e92352.

36 Gopalsamy, A., Aulabaugh, A.E., Barakat, A. et al. (2021). PF-07059013: a noncovalent modulator of hemoglobin for treatment of sickle cell disease. *J. Med. Chem.* 64 (1): 326–342.

37 Siu, T., Altman, M.D., Baltus, G.A. et al. (2019). Discovery of a novel cGAMP competitive ligand of the inactive form of sting. *ACS Med. Chem. Lett.* 10 (1): 92–97.

38 Petersen, D.N., Hawkins, J., Ruangsiriluk, W. et al. (2016). A small-molecule anti-secretagogue of PCSK9 targets the 80S ribosome to inhibit PCSK9 protein translation. *Cell Chem. Biol.* 23 (11): 1362–1371.

39 Yang, X.X., Xu, F., Wang, D. et al. (2015). Development of a mitochondria-based centrifugal ultrafiltration/liquid chromatography/mass spectrometry method for screening mitochondria-targeted bioactive constituents from complex matrixes: herbal medicines as a case study. *J. Chromatogr. A* 1413: 33–46.

40 Tao, Y., Yan, J., and Cai, B. (2021). Label-free bio-affinity mass spectrometry for screening and locating bioactive molecules. *Mass Spectrom. Rev.* 40 (1): 53–71.

41 Qin, S., Meng, M., Yang, D. et al. (2018). High-throughput identification of G protein-coupled receptor modulators through affinity mass spectrometry screening. *Chem. Sci.* 9 (12): 3192–3199.

42 Rizvi, N.F., Howe, J.A., Nahvi, A. et al. (2018). Discovery of selective RNA-binding small molecules by affinity-selection mass spectrometry. *ACS Chem. Biol.* 13 (3): 820–831.

43 Flusberg, D.A., Rizvi, N.F., Kutilek, V. et al. (2019). Identification of G-quadruplex-binding inhibitors of Myc expression through affinity selection-mass spectrometry. *SLAS Discovery* 24 (2): 142–157.

10

Native Mass Spectrometry in Drug Discovery and Development

Mengxuan Jia[1], *Jianzhong Wen*[1], *Olivier Mozziconacci*[2], *and Elizabeth Pierson*[3]

[1] *Preclinical Development ADMET/BA, Merck & Co., Inc, South San Francisco, CA, USA*
[2] *Discovery Pharmaceutical Sciences, Merck & Co., Inc, South San Francisco, CA, USA*
[3] *Analytical R&D, Merck & Co., Inc., Rahway, NJ, USA*

10.1 Introduction

The functioning of biological systems involves molecular recognition and assembly mechanisms through non-covalent interactions. These reversible interactions are fundamental to ensure the stability role in cellular processes such as cell division, gene expression and transcription, and the regulation of major biological processes such as cell apoptosis.

In this chapter, we will discuss the importance of the study of non-covalent biological complexes and how supramolecular mass spectrometry has advanced this research field. This chapter is intended to be practical by providing technical advice to tune specific instruments and presenting the contribution of such techniques to the study of these biological complexes. This chapter will conclude with the provision of alternative approaches to provide additional structural information on non-covalent complexes.

10.1.1 The Significance of Non-Covalent Protein Complexes in Biology

The all-genetic approach to comprehend cellular mechanisms creates a paradigm. Indeed, genes code for proteins, which in turn enables them to perform a specific cellular function. However, knowledge of the genetic code alone is not sufficient to

High-Throughput Mass Spectrometry in Drug Discovery, First Edition. Edited by Chang Liu and Hui Zhang.
© 2023 John Wiley & Sons, Inc. Published 2023 by John Wiley & Sons, Inc.

understand the protein network and the nature of the organisms. In the 1990s, with efforts to map interaction networks at the proteome level [1, 2], network biology or interactive biology has emerged [3, 4]. In a network biology paradigm, proteins are no longer independent biological objects but rather biomolecules involved in the formation of intermittent complexes, or networks that mediate the genotype and phenotype [5]. The structure and topology of these networks, modulated by internal (genomic changes) or external (environmental) influences, determine the phenotype. In this sense, any perturbation in the functioning of complexes, or even networks of complexes, can lead to perturbations in the phenotype.

Studies of biological complexes and molecular networks have led to a better understanding of several diseases such as Alzheimer [6], cystic fibrosis [7], or viral infections [8, 9], including hepatitis C viruses [10], H1N1 influenza viruses [11], or HIV [12, 13]. The understanding of molecular networks involved in a different type of cancer [14] permitted the prediction of the response of cancer cells to certain drugs and, therefore, enabled the development of pharmaceutical drugs [15].

Most often, subunits of biological complexes interact with each other through reversible non-covalent interactions, allowing the regulation of biological processes. These interactions may involve electrostatic bonds, van der Waals bonds, hydrogen bonds, and the hydrophobicity group effect. They ensure not only the tertiary and quaternary structure of large biomolecules such as DNA, RNA, proteins, and carbohydrates but also the structural integrity of protein–protein, protein–ligand, protein–DNA, DNA–DNA, and DNA or RNA–ligand complexes [16]. Because of the importance of these non-covalent biological constructs, considerable efforts have been made to develop methodologies and analytical tools to elaborate their structural and functional relationships.

10.1.2 Advantages and Disadvantages of Conventional Structural Analytical Techniques

Numerous approaches are currently used to study biological complexes. Among those, we can mention Spectroscopic methods (absorption, fluorescence), isothermal titration calorimetry (ITC), X-ray diffraction (XRD), nuclear magnetic resonance (NMR), circular dichroism (CD), surface plasmon resonance (SPR), electron microscopy (EM), dynamic light scattering (DLS), static light scattering (SLS), analytical Ultracentrifugation (AUC). Below are brief summaries of these applications.

Spectroscopic methods highlight molecular interactions leading to a modification of the absorption/emission of the molecule of interest [17]. They require the presence of a chromophore in at least one of the subunits of the complex, and whose optical properties must be affected because of the interaction.

ITC provides thermodynamic parameters of the interactions, such as changes in free enthalpy of binding (ΔH), free energy (ΔG), entropy (ΔS); dissociation constant K_d, number of binding sites, and stoichiometry [18]. However, ITC experiments require a non-negligible number of samples from a few milligrams to tens of milligrams of protein. Moreover, they require that the heat variation linked to the interaction is large enough to be detected. Finally, ITC analysis becomes difficult in the case of complex interactions involving cooperative binding.

XRD gives information on the three-dimensional structure of biological compounds without size limitations and can accurately locate atoms. The main difficulties are the high material consumption and the challenges of obtaining high-quality crystals [19]. In addition, XRD requires ordering (atomic, molecular, or macromolecular), and it is possible that the stacking of crystals does not reflect the major conformation in solution.

NMR provides structural information close to the atomic scale of protein complexes. However, as the size of biological macromolecules increases, two major difficulties are encountered. On one hand, the signals broaden due to a greater dipolar relaxation; on the other hand, they can overlap, leading to more complexity in the spectra to be interpreted [20]. Therefore, NMR studies of complexes larger than 100 kDa becomes difficult.

CD allows to characterize the structures of asymmetric compounds, whether they are small molecules or supramolecular edifices, and allows to follow changes in the conformation of biomolecules as a function of the environment (pH, temperature) [21]. However, unlike XRD and NMR, which provide structural information at the atomic scale, the information obtained by CD is of lower resolution, even though CD analysis is faster and consumes less sample [22]. In addition, commonly used aqueous buffers such as acetate, sulfate, or Tris buffers can absorb light and interfere with sample analyses.

SPR allows the detection of molecular interactions and gives access to the dissociation constant K_d as well as the association kinetic constants k_{on} and dissociation k_{off}. Nevertheless, the immobilization of the sample on the metal-coated surface during the analysis can alter the properties and conformational flexibility and, thus, the function of biological compounds [23]. The main difficulties are both the implementation of a reproducible experimental protocol, such as the creation of a homogeneous layer of the analyte, and the reproducibility of nonspecific adsorption phenomena of the molecules on the surface.

EM is widely used in biology, often in parallel with XRD, for the structural study of complexes with a resolution of about 6 Å [24]. The difficulty consists in optimizing the intensity of the electron beam to avoid the degradation of the biological sample while limiting the noise to have good precision and enough contrast [25]. In addition, heterogeneous or dynamic complexes can limit resolution.

DLS and SLS are commonly used to determine the size, size distribution, diffusion, and interactions of the complexes. DLS provides the hydrodynamic size of particles in solution, and SLS provides the molecular weight of particles in solution [26]. The methods require high sample concentration compared to mass spectrometry [27]. Due to the nature of the method, DLS and SLS are unable to resolve small difference in diameter of the nanoparticles [28].

AUC allows the quantitative measurement of molecular weight, homogeneity, and aggregation of macromolecules in solution. It is compatible with a wide range of solvents over a wide analyte concentration range; therefore, AUC allows the analysis of biomolecules in their native states [29, 30].

All these methods can be used in parallel and will be chosen according to criteria such as the amount of material available, the homogeneity, affinity, and size of the complex of interest, and the biological questions asked.

The development of electrospray ionization (ESI), matrix-assisted laser desorption ionization (MALDI), and mass spectrometry (MS) offers new possibilities to study biology [31, 32]. These techniques are complementary methods to classical structural analyses and are now techniques of choice for the structural characterization of non-covalent biological complexes.

10.2 Fundamentals of Native MS

10.2.1 Principles of Native Electrospray Ionization

The base of native mass spectrometry is the development of electrospray ionization (ESI), where the analytes can be ionized with the structure being kinetically trapped in droplets. When sprayed from a nondenaturing solvent, the "native-like" structure can be preserved [33]. The noncovalent interactions of the analytes, e.g. electrostatic interactions, hydrogen bonding, van der Waals interactions, and hydrophobic interactions, can be retained when the analyte ions are flying through the mass spectrometer. Therefore, the proteins or nucleic acid ions preserve their folded structures in the gas phase, allowing the measurements and quantification of their higher order complex structures.

The simplified process of ESI in positive mode is shown in Figure 10.1. First, a high voltage, typically around 4 kV for ESI and 1 kV for nano ESI, is applied between the ESI capillary and the sample cone or heated capillary, inducing the formation of a Taylor cone at the capillary tip. The Taylor cone emits a mist of charged droplets in the micrometer size range. Next, the droplets undergo solvent evaporation, causing increased Coulombic repulsion on the shrinking droplets. When the repulsion force is balanced by the surface tension of the droplets, at the Rayleigh limit, the droplets become smaller droplets by jet fission [34]. With

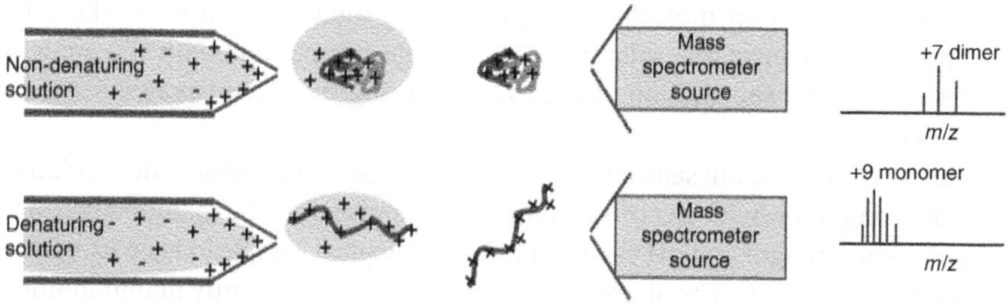

Figure 10.1 Illustration of the electrospray ionization process of protein analytes in a non-denaturing condition and in a denaturing condition.

repeating evaporation and fission processes, charged droplets in the nanometer size range are formed. The structure of the analyte retains during the process in the droplets. Finally, naked analyte ions are formed and introduced into the mass spectrometer for analysis [35]. Thus, the gas-phase structure of the analyte ion reflects the solution structure of the analyte.

The difference between denaturing mass spectrometry and native mass spectrometry is presented in Figure 10.1. When in denaturing solution, which typically contains organic solvent and acid, e.g. 50% acetonitrile and 0.1% formic acid, the protein is denatured. The denatured protein with a chain conformation has an extended surface area, allowing more charges on the surface. The ionization model was described as the chain ejection model (CEM). In contrast, when the analyte is presented in a non-denaturing solution, e.g. ammonium acetate solution, the protein retains its folded structure. The droplets with smaller sizes contain fewer charges. The charge nanodroplets containing a single analyte evaporate to dryness while the charges are retained on the analyte. This model is called charged residue model (CRM) [36–38]. As shown in Figure 10.1, native MS generates ions with low charge states and preserves the complexes formation, while in the denaturing conditions, only highly charged monomers are observed.

10.2.2 Specific Sample Preparation to Preserve Non-Covalent Interactions and Be Compatible with ESI-MS Analysis

To analyze non-covalent complexes in MS, it is necessary to be in conditions preserving these complexes while considering the intolerance of ESI-MS to the presence of non-volatile salts. To mimic in vivo conditions or pharmaceutical formulations in drug products, biological samples are stabilized in buffers containing a non-volatile salt (NaCl, KCl), which is not compatible with ESI-MS analysis. Indeed, these inorganic salts interact in a non-specific way with proteins leading to the formation of adducts in the mass spectrum. The latter results in a large heterogeneity in molecular weight making interpretation impossible. In addition, these nonvolatile salts introduce charge competition with the analyte and lead to signal suppression of the analyte during the electrospray ionization process [39–42].

The use of ammonium salts has proven to be a good compromise. These volatile salts are compatible with electrospray ionization and ensure the maintenance of a certain ionic strength and pH essential for the biological sample to keep its native conformation [32, 43]. The ammonium salts appear to sufficiently maintain non-covalent constructs in the gas phase in a mass spectrometer while helping minimize peak broadening in the mass spectrum. The most used ammonium salts are ammonium acetate (pH 6.8) and ammonium bicarbonate (pH 7.5), at concentrations ranging from a few mM to hundreds of mM, and even up to 1 M in some cases [32, 43–46]. Lars Konermann and his collaborators demonstrated that ammonium bicarbonate buffers lead to the formation of CO_2 gas bubbles that can denature proteins during ESI-MS analysis under native conditions [47]. Therefore, ammonium acetate remains the buffer of choice despite its low buffering capacity at neutral pH.

A buffer exchange step is mandatory prior to ESI-MS analysis. It allows for the exchange of the buffer at the end of the purification or conservation of biological samples against the ammonium buffer. The most used techniques are ultracentrifugation (such as Millipore's Centricon and Vivaspin columns) and gel filtration (such as GE Healthcare's NAP 5 or Biorad's Biospin columns). Buffer exchange sometimes can present certain challenges, so it is necessary to adopt the right technique for each scenario. A sample can be degraded during ultracentrifugation, just as it is possible to have a material loss during column filtration. The compromise always lies in a suitable buffer. When the presence of a nonvolatile salt is necessary for complex stabilization, analysis by supramolecular ESI-MS proves to be challenging if not impossible. We present in Section 10.3.2 the example of an inline buffer exchange protocol. Inline desalting presents both advantages to maintain the proteins in their original formulations, and to increase protein recovery reproducibility.

Depending on protein concentration in the formulations, some precautions must be taken. The formation of complexes depends on the concentration of the partners in relation to the affinity constant. For an AB complex, this results in the following thermodynamic equilibrium:

$$AB \leftrightarrows A + B, K_d = [A][B]/[AB]$$

Thus, the concentration of the partners must be higher than K_d to favor the non-covalent association of species A and B. In addition, the concentration of A and B must be sufficient for the species present to be detectable by MS. Sometimes high concentrations of analytes can lead to non-specific aggregates in the gas phase [48]. It is therefore important to check a few different concentrations to work at the optimal concentration for complex formation and MS detection. Finally, instrumental parameters of the mass spectrometer at the interface and the analyzer also influence the species detected.

10.3 Instrumentation

10.3.1 Nano-ESI and ESI

Native mass spectrometry generally performs better with nano electrospray ionization (nESI). Nano ESI typically applies gentle source conditions compared with ESI, which preserves the native conformations of the analyte. Nano ESI requires a smaller amount of sample with better sensitivity. However, obtaining a stable nESI is challenging. The reproducibility of quantification and even qualitative measurement may vary upon the sprays. The major parameter that influence the spray flow rate and stability is the aperture size [49, 50]. A well-shaped and small opening on the spray capillary leads spectra with less salt, even directly spray from the nonvolatile solution without buffer exchange to the volatile solution [51, 52]. Obtaining reproducible and long-lasting tips for nESI is challenging. Williams group published tips on making these tiny tips to help broaden the application of submicron nESI emitters [53].

Due to the small opening of the emitter, there is a higher possibility of clogging of the tip in nESI. In industry, due to the workload, efficiency, and quality control requirements, a more reliable system is desired. As a result, ESI has a wider application in industry compared with nESI.

The use of ESI and nESI will depend on the experimental conditions such as the amount of protein samples and sensitivity required to detect low concentrations (<1 μM) of protein complexes. The choice between ESI and nESI will also depend on the chromatographic conditions and the availability of column chemistries in nano-capillaries. We detail below some chromatography conditions that are amenable for ESI and nESI.

10.3.2 Inline Desalting and Separations Coupled to Native Mass Spectrometry

The combination of liquid chromatography (LC) and mass spectrometry has become a common analytical technique from small molecules to intact protein analyses. With the development of native MS in the past decade, increasingly inline LC-nMS methods are developed to achieve orthogonal information from the analytes.

There are a few challenges for inline LC – native MS method development: the first is converting the traditional nonvolatile mobile phases to MS-friendly volatile mobile phases; the second is stabilizing the spray under high salt aqueous flow; the third is choosing MS-compatible columns. For the mobile phases, ammonium acetate is commonly used to replace the nonvolatile salt. To stabilize the ESI spray, an optimized desolvation settings allow the evaporation of the aqueous solvent. Currently, due to the increasing needs for native mass spectrometry in industry, most mass spectrometer vendors provide suggested native mass spectrometry settings for different flow rate ranges as a start point for tuning. Another cause of unstable spray or sparking is the high electro current caused by high conductivity of the salt solution. To reduce the electro current, one solution is avoiding use high ionic strength. If high salt is not avoidable, reducing the inner diameter (e.g. PEEK Tubing 1/16″ OD×0.0025″ ID) and increasing the length of the tubing between the grounded holder and the spraying needle will help reduce the spray current [54]. Additionally, using a flow splitter to reduce the flow rate is another way to reduce the current, especially when system pressure is too high with the longer thinner tubing.

The most used LC-nMS methods are size-exclusion chromatography (SEC), ion exchange chromatography (IEX), and hydrophobic interaction chromatography (HIC). These separation methods are compatible with the requirement of nMS.

10.3.2.1 Inline SEC and Desalting

SEC is currently applied to characterize protein aggregates, protein fragments, impurities, and antibody–drug conjugates [55–57]. The process of inline-SEC/desalting process is shown in Figure 10.2. Nonvolatile separated from the large molecules through the SEC column using volatile mobile phase. The proteins and

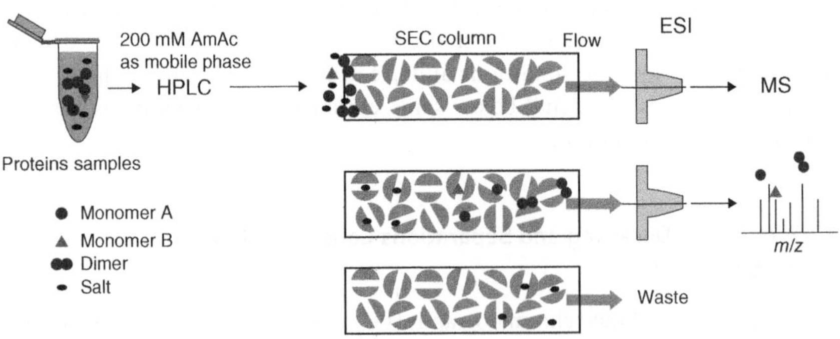

Figure 10.2 Illustration of the inline-desalting process. Non-volatile salt is separated from the large molecules through the SEC column using a volatile mobile phase. The nonvolatile salt is diverted to waste after the elution of protein peak.

protein complexes are separated based on their sizes. After the elution of large molecules, the non-volatile salt is diverted to waste in the end.

Short-length SEC columns, i.e. below 5 cm, are currently used as a rapid inline buffer exchange (BE) method for sample screening, allowing fast inline buffer-exchange into ammonium acetate buffers, reducing time and sample variability often observed when such steps are performed separately and offline [54, 58, 59]. VanAernum et al. introduced a rapid online buffer exchange for screening of protein, protein complexes, and cell lysate by nMS in their protocol [54], the in-house made column is now commercially available through ThermoFisher (PN 43803-052130).

In addition to separation purely based on size, Yan et al. took the advantage of the unwanted secondary interactions between the protein analytes and the column matrix and developed the Mixed-Mode SEC. Chromatographic separation of four bispecific antibody mixtures was accomplished successfully with either electrostatic interaction or hydrophobic interaction. Moreover, this method is believed to be highly sensitive with an improved sensitivity in detecting low abundance homodimer impurities (LOD from 0.01% to 0.1%) [60].

10.3.2.2 Inline IEX

IEX is used to separate charge variants [61–64] as well as complex protein mixtures [65, 66]. The separation based on the net charge of the analytes in solution. This allows for a detailed characterization of biotherapeutics such as monoclonal antibodies (mAbs) [67–70]. There are a few mobile phases developed for IEX-nMS. The first is the ammonium acetate salt gradient, simply from low to high ionic strength. The second is combination of salt and pH gradient, limiting the high spray current due to high ionic strength. Due to the variety of analyte isoelectric points, typically a broader gradient is used first to identify the range of elution condition. A further detailed tuning of gradient will help separate charge variants of the targeted protein. Both strong and weak ion exchange columns can be coupled to native mass spectrometry, as well as both anion and cation exchange columns. The column should be chosen based on the property of the analytes.

10.3.2.3 Inline HIC

HIC has been applied to characterize mAb variants as well as drug to antibody ratio and payload distribution [71, 72]. In recent years, HIC was successfully coupled to native mass for separation and characterization of mAb variants and ADC [73–75]. HIC-MS is able to separate the oxidation and free thiol variants of the mAbs with less than 250 mM ammonium acetate concentration [74]. In another case, high salt up to 3 M ammonium acetate was used, a water makeup plus splitting flow system was applied to overcome the challenge caused by high salt mobile phase concentration [75].

10.3.2.4 Inline 2D LC

Two-dimensional (2D) LC allows for the combination of two solution phase separation methods on the analytes. The protein samples can be separated based on different properties (e.g. mass, charge, hydrophobicity). Thus, inline 2D LC-MS with complementary separations has the advantage of higher peak capacity, more straightforward sample preparation, and reduces sample consumption with fewer runs [64, 76]. Among the potential combinations of 2D LC methods, HIC×SEC, SEC×SEC, IEX×SEC, SEC×IEX have been coupled to mass spectrometry [64, 77–79].

An example shown in Figure 10.3 is the application of SEC×IEX coupled to native MS. In the first dimension of separation, i.e. SEC, the mixture of photoproducts from the mixture of mAbs were separated based on size, allowing separation of dimer, monomers, and fragments. The three fractions from SEC dimension were further run through IEX dimension, respectively. In dimer fraction, two homodimer and one heterodimer were successfully separated and characterized. In the monomer fraction, each monomer and the charge variants were separated. In the last fraction, the figments, e.g. light chain and heavy chains of the mAbs were identified [64]. The extension of solution phase separation allows characterization of the complicated mixtures.

10.3.2.5 Compatibility with nESI

As introduced in prior paragraphs, nESI has advantages in sensitivity and gentle conditions. However, usage of the fragile spray tips limited its usage in industry. Recently, a microflow-nano spray electrospray ionization (MnESI) source and a microfabricated monolithic multinozzle (M3) emitter were developed for native mass spectrometry applications [80–82].

Monolithic multinozzle (M3) emitter work flow rate is lower than 20 μL/min. With capillary columns, e.g. 0.3 mm inner diameter SEC from POLYLC, the MnESI system can be directly used. However, capillary columns are more fragile and sometimes not available in desired separation material. A flow splitter can be used to make normal analytical column and flow rate compatible with the system. As shown in Figure 10.4, flow was split to UV, allowing an additional dimension of information at the same time. A bottle containing modifier, e.g. imidazole, triethylamine, IPA, ACN, or 5% (v/v) ammonia in ACN is used for charge reduction [80]. This nLC-MS platform is suitable for industrial laboratory application. This system is capable of handling 0.1–0.8 mL/min wide flow rate range and up to 600 mM mobile phase salt concentration.

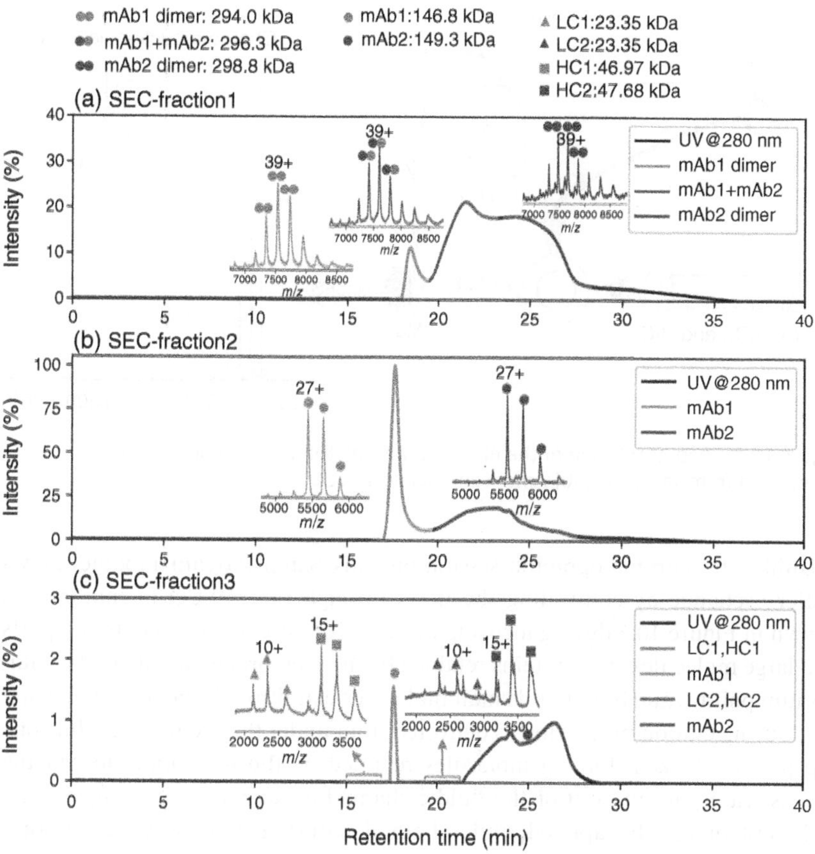

Figure 10.3 SEC-IEX-nMS chromatogram and spectra of reslizumab and bevacizumab mixture. (a) The IEX chromatogram of the first inline SEC fraction; inserts are the dimer mass spectra extracted from the peaks; mAb1 homodimer, mAb1+mAb2 heterodimer, and mAb2 heterodimer were separated and identified. (b) The IEX chromatogram of the second inline SEC fraction, inserts are the monomer mass spectra extracted from the peaks; mAb1 and mAb2 were separated and identified. (c) The IEX chromatogram of the third inline SEC fraction; inserts are the fragment mass spectra extracted from the peaks. The scale of the *y*-axis is different for each plot. *Source:* Reproduced with permission from Jia et al. [64]/Elsevier.

10.3.3 High-Throughput Native Mass Spectrometry

It is crucial to convert the analytical workflows into automated high-throughput workflows in biopharmaceuticals. In addition to the rapid inline buffer exchange method introduced in Section 10.3.2.1, another fluidic device called SampleStream is introduced in this section [83, 84].

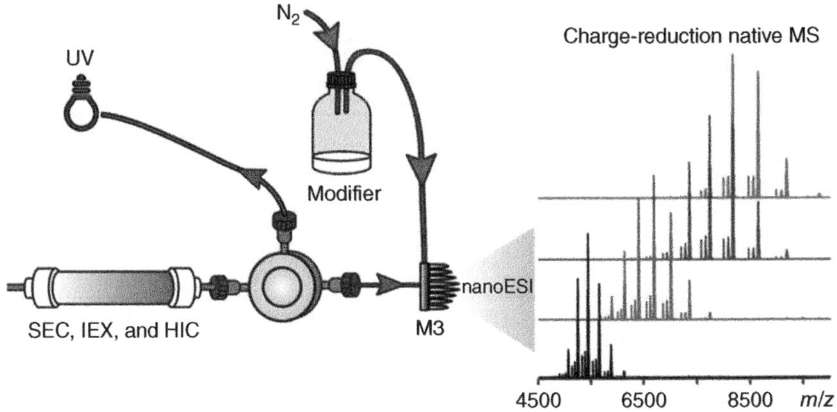

Figure 10.4 Representation of the nLC-MS platform. *Source:* Reproduced with permission from Yan et al. [80]/American Chemical Society.

Unlike the chromatographic separation, this SampleStream module allows buffer exchange by controlling the flows through the MWCO membrane. As shown in Figure 10.5 during focusing mode, a flow stream enters both top ports and large molecules, e.g. proteins retained by the membrane migrate to the equilibrium point while buffer and small molecules exit through the membrane and frit. During elution mode, flow is delivered from under the membrane while one top port is blocked, forcing molecules retained by the membrane to exit the downstream port and out of the fluidic channel for downstream analysis [83]. This system can be applied under both denaturing and native conditions.

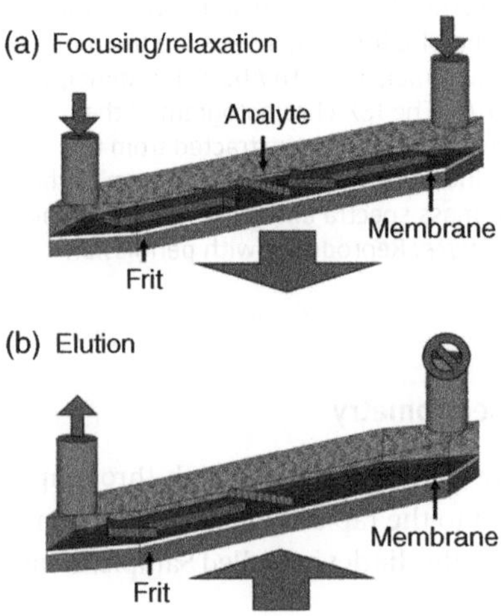

Figure 10.5 Depiction of the two modes of operation for the fluidic channel where sample concentration and vigorous buffer exchange occur above a membrane. (a) During focusing, a flow stream enters both top ports and molecules retained by the membrane migrate to the equilibrium point while buffer and small molecules exit through the membrane and frit. (b) During elution, flow is delivered from under the membrane while one top port is blocked, forcing molecules retained by the membrane to exit the downstream port and out of the fluidic channel for downstream analysis. *Source:* Reproduced with permission from Hae-Min Park et al. [83]/ American Chemical Society.

The system theoretically could generate a throughput of 15 seconds per sample with two channels. This high-throughput platform is attractive for industrial applications.

10.3.4 Mass Analyzers

Time of flight (TOF) is typically used for native MS due to its theoretically unlimited *m*/*z* detection range. In the past decade, the extension of Orbitrap *m*/*z* detection range allows the application of native mass spectrometry on the Orbitrap analyzers [85]. Orbitrap provides data with higher resolution than the TOF. For samples that need ultrahigh resolution, e.g. metal ligands, Fourier transform ion cyclotron resonance (FTICR) can be used [86]. These analyzers can measure native protein/complexes from a few kDa to hundreds kDa. For the complexes in MDa range, charge detection mass analyzers can be used.

Native analysis by charge detection mass analyzers involves single-particle measurements (i.e. one ion measured at a time) whereby the individual measurements of the ion mass-to-charge ratio and the charge are used to calculate ion mass directly. Though charge detection mass spectrometry (CDMS) has a rich, decades long history [87], recent technical developments involving improved resolution and sensitivity have led to broader applicability of the technique in academia and industry alike.

CDMS measurements, in general, are driven by the practical upper mass limit of conventional mass spectrometry techniques. Briefly, challenges for large, native MS measurements by conventional mass spectrometry involve low-detection efficiencies for large ions [88–90], resolution of adjacent charge states in the electrospray mass spectrum, and the intrinsic and extrinsic mass heterogeneity associated with large proteins and native complexes for the former, and salt/solvent adducts for the latter. Complex systems are being studied that include larger analytes (>3–4 MDa) and heterogeneous mixtures of species. These more challenging measurements can be addressed by CDMS.

Single particle and CDMS instrumentation and approaches have been reviewed in the literature [91].

10.3.5 Data Processing

10.3.5.1 Contrasts Between Non-Native and Native MS Data Processing and Interpretation

The major difference of the non-native and native MS data is the number of charges. For denatured proteins, charge states are much higher because the extended surface allows more charges during the ionization. The deconvolution principle is basically the same for the non-native and native MS data. There are two practical challenges for native MS data handling. The first comes from the salt adducts. Due to the gentle ionization process, non-covalent interactions are preserved, including some nonspecific metal ion–protein interactions, e.g. Na+ and K+. By tuning the source to "hotter" condition, which can be achieved by increasing cone voltage source collision energy, or source temperature, the formation of nonspecific adducts is reduced. The latter may break native interactions, so the optimal result is a balance in the fine tunings of the mass spectrometer. Adducts will broaden the peak, spread the signal, and build challenges for deconvolution. It is normal to see higher than theoretically mass for the larger molecules after deconvolution due to the adducts. Another challenge comes from the overlapping mass-to-charge ratio (m/z) of different charge states among the homo-oligomers. For example, monomer of charge state 5+ will have the same m/z as the dimer of charge state 10+. The intensity of monomer and dimer may not be accurate after deconvolution if only MS dimension is used for measurement. Additionally, separation e.g. SEC or ion mobility can be applied to ensure accurate quantification of homo-oligomers.

10.3.5.2 Software for Native MS

With the development of native MS, more software and packages were developed for native LC-MS data processing. There are three main categories: commercial software, open-source software, and open-source scripts. Typically, vendors offer commercial software for their raw data processing. Other platforms, e.g. Protein Metrics – Intact Mass, Genedata Expressionist, are compatible with various raw data types across different mass spectrometry vendors. UniDec is an open-source software targeted for native MS data process; it is provided by Dr. Marty [92].

For additional needs, raw data can also be extracted through scripts written from C, java, Python using the vendors provided open-source scripts, e.g. MSFileReader from Thermo.

10.4 Application Highlights

When looking for application of native MS, we should ask ourselves when can native MS *not* be used? Native MS is well suited to study intact proteins including assessing protein isoforms within a mixture and defining post-translational modifications that are fundamental information in the context of drug product development. As protein formulation are becoming more complex (i.e. co-formulation of multiple proteins in a single drug product), the access to intact protein/subunit folding dynamics; intact protein conformations and conformational changes; protein complexes and stoichiometry; and protein complex assembly states are mandatory. In early discovery and development, the knowledge of biologically active complexes and determining protein binding affinity are essential. At the frontier of small and large molecules, other key interests include the observation and understanding of conformational changes that occur in proteins (or DNA/RNA) when a complex is formed with a small molecule, such as a drug or metabolite and/or nucleic acids to develop more effective drug compounds. It is only recently that pharmaceutical industry has started to implement native MS in its toolbox for quality control assessment of impurities and/or aggregates of proteins in drug products (e.g. antibody-based immunoconjugates, biosimilars, and antibody–drug conjugates) since such data are critical to obtain approval by the governing authorities. The analysis of a biotherapeutic drug product under non-denaturing conditions not only confirms the molecular mass of the product but also provides invaluable information on homogeneity, conformation, and oligomerization state; all of which are key factors in generating a quality product. In addition, analysis under native conditions may also reveal the existence of natural or unexpected ligands that may have co-purified with the protein sample. Undoubtedly, native MS will gain a permanent and central position in the routine analysis of biological pharmaceuticals. Below, we present a few examples of applications of native MS or research fields in which native MS could lead to new discoveries.

Protein–ligand interactions modulate various biological processes such as the interaction of small molecules as protein cofactors. For example, the inhibition of pathogenic proteins by specific inhibitor binding to the active site of the target protein is necessary. Experimentally, to observe in vitro an interaction between a protein and small molecules, the target protein, which is overexpressed in host cells, is thoroughly purified. Then, libraries of small molecules are used to screen any interaction. Such protocol is time consuming and relies on highly purified molecules. That is why systems to screen the interactions between proteins and small molecules in a biologically crude environment were developed to determine potential inhibitors of target proteins. In this quest to isolate any potential interaction between small molecules and target proteins, nMS presents several advantages [93]. The combination of nMS and nanoESI permit the observation of multimeric proteins in crude cell lysate without chromatographic purification of the analyte proteins [94]. Secreted proteins from cultivated eukaryotic cells could also be measured without the need of any purification step of the proteins [95]. Protein–ligand interactions modulate various biological processes such as the interaction of small molecules as protein cofactors. For example, the inhibition of pathogenic proteins by specific inhibitor binding to the active site of the target protein is necessary [93–95]. The core of native MS is to preserve non-covalent interactions and acquire accurate mass measurements. Coupling with different separation techniques, native MS has found increasingly diverse applications in drug discovery such as high-throughput drug screening, the characterization of therapeutic molecules, and elucidating the interactions between the drug and a target.

10.4.1 Using Native MS to Develop Stable Protein Formulations

Native MS is gaining momentum toward monoclonal antibody (mAb) analysis during developability assessment and forced degradation studies, which aimed at assessing potential critical quality attributes in antibody drug molecules during drug discovery and development [96]. It offers a rapid screening technique to characterize mAb charge heterogeneity [70], glycan identification [97], as well as disulfide bonding among IgG subclasses [98]. As part of pre-developability assessment for antibody-based drugs, Ma et al. [70] directly coupled online strong cation exchange chromatography to native mass spectrometry to characterize the charge heterogeneity of therapeutic monoclonal antibodies. Combining pH and salt gradient, the method provided a universal separation of mAbs from a wide range of pI values (6.4–9.2). The characterization of antibodies in the native conditions shifts the charge envelope to higher m/z with lower noise level. The presence of fewer charge states for the deconvolution not only simplifies the deconvolution algorithm but also reduces the possibility of overlapping charge envelopes from different ion species. The integrated strategy allows unbiased characterization of numerous charge variant species and low-molecular weight fragments.

MAb could form aggregates due to environmental changes and stresses during production, shipping, and storage. Commonly, the level of aggregates in formulated mAbs are a critical quality attribute, as it may affect the pharmacokinetics and pharmacodynamics of the biopharmaceutical product or lead to adverse side effects and immunogenic responses. Thus, it is important to monitor and quantify protein aggregates during drug manufacturing. Kükrer et al. [99] applied native MS coupled with SEC to monitor pH-stressed mAb aggregation. The monomeric and aggregated IgG fractions in the stressed IgG formulation can be separated by SEC. Native mass enabled the authors to determine the molecular weight of the monomeric IgG as well as the aggregates, including dimers, trimers, and tetramers.

Rouby et al. investigated the nature and structure of the high-molecular weight (HMW) species present at less than 1% in an unstressed formulated roledumab biopharmaceutical [100]. The authors purified the HMW species through preparative SEC and then analyzed them by a combination of chromatographic methods (ion-exchange chromatography [IEX] and SEC) coupled to native mass spectrometry. Both covalently and non-covalently bound dimers were identified at a proportion of 50/50. Native MS revealed the presence of three mAb dimer formats having the same mass but differing by their charge and size. They were attributed to different compact and elongated dimers. Finally, high-resolution middle-up approaches using different enzymes (IdeS and IgdE) were performed to determine the mAb domains implicated in the dimerization and revealed that the roledumab dimers were associated mainly by a single Fab-to-Fab arm-bound association.

In addition to characterizing traditional mAbs, native MS has found applications in new therapeutic protein scaffolds, such as understanding the structure of bispecific antibodies (bsAb) and antibody-drug conjugates (ADCs). Assembly of the bsAb construct normally requires correct pairing of cognate light and heavy chains, as well as heterodimerization of two different half-molecules. Unfortunately, this process sometimes results in the formation of misassembled molecular constructs, such as monospecific molecules (e.g. homodimer species). Detection and quantitation of homodimer impurities in therapeutic bispecific antibody drug products is essential to support development and quality control release. Yan et al. [60] developed a mixed-mode size exclusion chromatography coupled with native mass spectrometry method that allow them to sensitively detect and quantitate homodimer impurities in bispecific IgG.

ADCs are an emerging drug family that are gaining momentum in advancing cancer treatment. Their ability to effectively target disease-associated cells while minimizing off-target effects on healthy or non-diseased cells has drawn significant interest. Eleven ADCs are currently on the market, and seven were approved in the past three years. ADCs are complex, multicomponent drug substances designed to deliver highly potent pharmaceutically active agents to specific cells known to play crucial roles in various diseases. Due to their complex structure and heterogeneity, the biophysical characterization of ADCs presents a significant challenge. The main advantage of using native MS for ADC measurements lies in its ability to directly detect the intact molecules, which simplifies the drug-antibody-ratio (DAR) evaluation, and frequently preserving the noncovalent associations of light and heavy chains for interchain cysteine conjugation are necessary for DAR measurement.

SEC coupled to native mass spectrometry has also been reported to quantitatively determine the DAR on cysteine-linked ADCs by Jones et al. [101]. They applied a microbore SEC column to exchange the ADC into ammonium acetate, thus facilitating online native electrospray ionization analysis, and established that the method for DAR evaluation was comparable to hydrophobic interaction chromatography (HIC), the gold standard for ADC quality control. Debaene et al. [102] described the combination of native MS to ion mobility (IM) MS for ADC characterization. The authors highlighted the benefits of high-resolution native MS and native IM-MS for the determination of ADCs' drug load profile, naked antibody content, and average DAR determination. The innovative coupling of IM with native MS allowed the authors to directly measure the collisional cross sections of each payload species and witnessed slight conformational changes for ADC with multiple drug loading. The same group also demonstrated the use of native IM-MS to characterize trastuzumab emtansine (T-DM1, Kadcyla), an ADC with non-cleavable linker and random lysine conjugation. They showed that native MS combined with high-resolution measurements and/or charge reduction was beneficial in terms of DAR determination and the drug load profiles [103].

10.4.2 Native MS to Understand Drug/Target Interaction

A key step in the development of new pharmaceutical drugs is the identification of the molecular target. Target identification remains a crucial process and a current bottleneck for advancing hits through the discovery pipeline. Liu et al. [104] reported a method that takes advantage of the specific detection of protein–ligand complexes by native mass spectrometry to probe the protein partner of a ligand in an untargeted method. As proof of concept, the authors demonstrated the use of native MS to identify known ligand–protein interactions in a protein mixture under various experimental conditions. The benefit of native MS allows the authors to use unmodified small molecules for binding and, thereby, it does not require labeled ligands and is not limited by the chemistry required to tag the molecule.

Proteolysis-targeting chimeras (PROTACs) and related molecules that induce targeted protein degradation by the ubiquitin–proteasome system represent a new therapeutic modality and are the focus of great interest, owing to potential advantages over traditional occupancy-based inhibitors with respect to dosing, side effects, drug resistance, and modulating "undruggable" targets. These heterobifunctional small-molecule degraders allow simultaneous binding of both the target protein and an E3-Ubiquitin ligase, bringing the two proteins into close spatial proximity to allow ubiquitinoylation and degradation of the target protein via the cell's endogenous protein degradation pathway. Sternicki et al. [105] utilized native MS to study the ternary complexes promoted by the PROTAC molecule between Brd4 bromodomains 1 and 2, and Von Hippel Lindeau E3-Ubiquitin Ligase. Native MS at high resolution allowed the authors to measure ternary complex formation as a function of PROTAC concentration to provide a high-throughput measure of complex affinity and stability while measuring other intermediate protein species.

A crucial aspect of antibody drug discovery is to map the epitope. Mehaffey et al. [106] demonstrated a workflow utilizing native MS and ultraviolet photodissociation (UVPD) to map the antigenic determinants of a model antibody–antigen complex involving hemagglutinin (HA), the primary immunogenic antigen of the influenza virus, and the D1 H1-17/H3-14 antibody, which has been shown to confer potent protection to lethal infection in mice despite lacking neutralization activity. Comparing the sequence coverage upon UV photoactivation of HA and of the HA–antibody complex indicated the elimination of some sequence ions that originated from backbone cleavages exclusively along the putative epitope regions of HA in the presence of the antibody. UVPD coupled to native MS shows promise as a method to rapidly map epitope regions along antibody–antigen complexes as novel antibodies are discovered or developed.

10.4.3 Native Mass Spectrometry and Tractable Protein–Protein Interactions for Drug Discovery

Native mass spectrometry has proven to be a powerful tool to screen the change of conformation and surface of proteins. nMS has thereof been implemented in several structural biology studies. In drug discovery, protein–protein interaction (PPI) modulators are an emerging class of molecular targets. As mentioned in the introduction of this chapter, many functions in cells, if not all, are regulated by multiprotein complexes (e.g. cell growth, DNA replication, transcriptional activation, transmembrane transduction). It is well established that human diseases can be caused by aberrant PPIs, which are the results of either the loss of an essential interaction or through the formation and/or stabilization of a protein complex at an inappropriate time or location. Some examples include cervical cancer, leukemia, and neurodegenerative diseases. That is why PPIs have developed into attractive molecular targets for novel therapies. Small organic molecules can selectively interfere with PPIs involved in the regulation of transcriptional machinery and therefore serve as molecular tools to investigate the chemical biology of cell signaling.

Several successful examples have demonstrated the possibility to develop PPIs modulators as drugs. The general assumption for PPIs as targets in drug discovery were described as large surfaces, almost flat and often lacking suitable pockets for small molecules to bind. If those assumption happened to be correct, clearly, these features would not support PPIs as targets for drug modulators. Nussinov and Keslin provided a better description of the requirements to address the principles of PPIs [107]. Small grooves on the exposed surface of an unbound protein may accommodate water molecules, in the bound state, a few of them may also accommodate fragments of a partner protein, thus defining the contact surfaces of a PP complex. In addition, only a subset of few amino acids available of the PP-contact area are required to maintain a robust modulation of a PPI. In oncology, the protein regulators of apoptotic cell death, i.e. the Bcl-2 and Bcl-XL protein family, and HDM2, an oncoprotein with both p53-dependent and independent functions, are good examples of signaling proteins where PPIs have been targeted and small-molecule inhibitors have entered clinical trials. A detailed review of the state of art of the design of PPI modulators was published by Zinzalla and Thurston [108].

The diversity of new PPIs being discovered will ensure that both PPI inhibitors and stabilizers remain attractive opportunities over the next decades. The number of targets that will become available and will make their way to a drug depends essentially on the inventiveness of biologists, chemists, and in the development of tools to discover and validate the novel PPIs as suitable for therapeutic intervention. The potential drug ability of new PPI targets tingly depends on advanced technologies relating to structure determination such as X-ray crystallography,

NMR, and molecular modeling. Parallel developments are required to improve assays to assess PPI modulators. There is a need to rapidly identify and eliminate false positives in assays. High-mass MALDI, surface plasmon resonance, and native mass spectrometry techniques appear to be efficient and simple tools for directly monitoring ligand–protein interactions and are likely to become crucial techniques in the future.

10.4.4 Structural Stability Using Collision-Induced Unfolding

Ion mobility (IM) is a mass spectrometry compatible gas-phase separation method that separates ions based on their mass, charge, and shape [109]. IM-MS has been applied to proteins and protein oligomers to characterize their conformation and stability [110, 111]. IM-MS, together with collision-induced unfolding (CIU), was used to characterize the conformational change and stability of mAbs [112–114]. The collision cross section itself usually does not have enough resolution to distinguish the minor conformational changes in antibodies, e.g. antibody-drug conjugates (ADC). By introducing the collisional activation of the analyte ions in the gas phase before IM separation, studies demonstrated the ability of such technique to differentiate IgG subclasses and to reveal the shifts in the stabilities of the site-specific ADCs [98, 115]. A software called CIUsuit2 was developed and provided by the Ruotolo lab to automate the CIU data acquisition and processing [116].

10.4.5 Vaccines and Virus Proteins Using CDMS

CDMS has been used for analysis of large molecules and complexes, such as virus capsids [117–121]. It is crucial to be able to characterize the structure of viruses. As mentioned in Section 10.3.4, conventional mass spectrometry techniques typically were limited by the analyte size. Therefore, CDMS has its unique applications for such large molecules in mega Daltons range.

An example for application of CDMS is the characterization of the heterogeneity associated with glycosylation of the 66 *N*-glycan sites on the protein trimer making up the spike (S) region of the SARS-CoV-2 virus [122]. Miller et al. measured the mass distribution for the S protein with a trimerization domain and calculated the glycan mass distribution. Their results showed that the average glycan mass calculated from the direct “top-down” CDMS measurements is much larger than the average obtained from “bottom-up” MS-based glycoproteomic studies. This finding indicates the heterogeneity in the extent of glycan processing of the S protein trimers. It is interpreted as some trimers being much more heavily processed than others. The observation of heterogeneity may have evolved as a way of further confounding the host’s immune system [122].

In another study, Miller et al. characterized three classical multivalent vaccines, inactivated poliomyelitis vaccine (IPOL), RotaTeq, and Gardasil-9, using CDMS. CDMS could provide mass distribution for the antigens well as the relative abundances of impurities (i.e. empty and defective particles). The extent of formaldehyde cross-linking was measured from IPOL. A slightly lower mass was observed for RotaTeq, indicating a potential missing of 10 trimers. The analysis time for such an informative CDMS assay is less than one hour and the assay only requires 10–20 μL of samples. Additionally, CDMS is sensitive enough to detect low-abundance intermediates and misassembled particles. This method is expected to accelerate vaccine development [123].

10.5 Conclusions and Future Directions

Native MS is becoming an unavoidable orthogonal technique in structural biology. With small quantities of protein material (<500 ng), this technique rapidly provides fundamental information to a structural biologist. Native MS is increasingly implemented as a routine characterization method in biopharmaceutical laboratories. Indeed, nMS provides a high-throughput technique to screen the combinations of excipients and evaluate their effects on protein stability.

Integration of native MS with data generated from other sources, e.g. cross-linking MS, hydrogen-deuterium exchange MS, surface labeling and more recently, cryo-EM, provides data-rich information that led to better understand the protein network in biological systems as well as in complex biopharmaceutical formulations.

According to Heck's, "There will be a reduction in sample preparation, and it will be possible to analyze protein structures in situ – direct from cell to MS." In addition, the author believes that: "There is no limitation on how big you can go," [124] for example, mass cytometry is a method that already uses ESI on whole cells. Then, the questions we should ask ourselves are "What useful biological information can we get? And with what precision?"

References

1 Finley, R.L. and Brent, R. (1994). Interaction mating reveals binary and ternary connections between Drosophila cell cycle regulators. *Proc. Natl. Acad. Sci. U.S.A.* 91 (26): 12980–12984.

2 Bartel, P.L., Roecklein, J.A., SenGupta, D., and Fields, S. (1996). A protein linkage map of *Escherichia coli* bacteriophage T7. *Nat. Genet.* 12 (1): 72–77.

3 Strohman, R.C. (1997). The coming Kuhnian revolution in biology. *Nat. Biotechnol.* 15 (3): 194–200.

4 Arkin, A.P. and Schaffer, D.V. (2011). Network news: innovations in 21st century systems biology. *Cell* 144 (6): 844–849.

5 Bensimon, A., Heck, A.J.R., and Aebersold, R. (2012). Mass spectrometry-based proteomics and network biology. *Annu. Rev. Biochem.* 81: 379–405.

6 Rembach, A., Stingo, F.C., Peterson, C. et al. (2015). Bayesian graphical network analyses reveal complex biological interactions specific to Alzheimer's disease. *J. Alzheimers Dis.* 44 (3): 917–925.

7 Guggino, W.B. and Stanton, B.A. (2006). New insights into cystic fibrosis: molecular switches that regulate CFTR. *Nat. Rev. Mol. Cell Biol.* 7 (6): 426–436.

8 Calderwood, M.A., Venkatesan, K., Xing, L. et al. (2007). Epstein-Barr virus and virus human protein interaction maps. *Proc. Natl. Acad. Sci. U.S.A.* 104 (18): 7606–7611.

9 Mendez-Rios, J. and Uetz, P. (2010). Global approaches to study protein–protein interactions among viruses and hosts. *Future Microbiol.* 5 (2): 289–301.

10 de Chassey, B., Navratil, V., Tafforeau, L. et al. (2008). Hepatitis C virus infection protein network. *Mol. Syst. Biol.* 4: 230.

11 Shapira, S.D., Gat-Viks, I., Shum, B.O.V. et al. (2009). A physical and regulatory map of host-influenza interactions reveals pathways in H1N1 infection. *Cell* 139 (7): 1255–1267.

12 Ptak, R.G., Fu, W., Sanders-Beer, B.E. et al. (2008). Short communication: Cataloguing the HIV type 1 human protein interaction network. *AIDS Res. Hum. Retroviruses* 24 (12): 1497–1502.

13 Jäger, S., Gulbahce, N., Cimermancic, P. et al. (2011). Purification and characterization of HIV-human protein complexes. *Methods* 53 (1): 13–19.

14 Kreeger, P.K. and Lauffenburger, D.A. (2010). Cancer systems biology: a network modeling perspective. *Carcinogenesis* 31 (1): 2–8.

15 Pe'er, D. and Hacohen, N. (2011). Principles and strategies for developing network models in cancer. *Cell* 144 (6): 864–873.

16 Fan, C.; Basri, G.; Simon, W.; Stephan, R.F.; Alfredo, J.I and Renato, Z. 2016. Applying Mass Spectrometry to Study Non-Covalent Biomolecule Complexes. *Mass Spectrom. Rev.* 35 (1), 48–70. https://doi.org/10.1002/mas.21462.

17 Sirajuddin, M., Ali, S., and Badshah, A. (2013). Drug-DNA interactions and their study by UV-visible, fluorescence spectroscopies and cyclic voltametry. *J. Photochem. Photobiol. B* 124: 1–19.

18 Ghai, R., Falconer, R.J., and Collins, B.M. (2012). Applications of isothermal titration calorimetry in pure and applied research – survey of the literature from 2010. *J. Mol. Recognit.* 25 (1): 32–52.

19 Giegé, R. (2013). A historical perspective on protein crystallization from 1840 to the present day. *FEBS J.* 280 (24): 6456–6497.

20 Göbl, C., Madl, T., Simon, B., and Sattler, M. (2014). NMR approaches for structural analysis of multidomain proteins and complexes in solution. *Prog. Nucl. Magn. Reson. Spectrosc.* 80: 26–63.

21 Ranjbar, B. and Gill, P. (2009). Circular dichroism techniques: biomolecular and nanostructural analyses – a review. *Chem. Biol. Drug Des.* 74 (2): 101–120.

22 Kelly, S.M., Jess, T.J., and Price, N.C. (2005). How to study proteins by circular dichroism. *Biochim. Biophys. Acta* 1751 (2): 119–139.

23 Homola, J. (2008). Surface plasmon resonance sensors for detection of chemical and biological species. *Chem. Rev.* 108 (2): 462–493.

24 Peplow, M. (2020). Cryo-electron microscopy reaches resolution milestone. *ACS Cent. Sci.* 6 (8): 1274–1277.

25 Milne, J.L.S., Borgnia, M.J., Bartesaghi, A. et al. (2013). Cryo-electron microscopy – a primer for the non-microscopist. *FEBS J.* 280 (1): 28–45.

26 Wishard, A. and Gibb, B.C. (2019). Dynamic light scattering – an all-purpose guide for the supramolecular chemist. *Supramol. Chem.* 31 (9): 608–615.

27 Engelke, J., Brandt, J., Barner-Kowollik, C., and Lederer, A. (2019). Strengths and limitations of size exclusion chromatography for investigating single chain folding – status and future perspectives. *Polym. Chem.* 10 (25): 3410–3425.

28 Fissan, H., Ristig, S., Kaminski, H. et al. (2014). Comparison of different characterization methods for nanoparticle dispersions before and after aerosolization. *Anal. Methods* 6 (18): 7324–7334.

29 Edwards, G.B., Muthurajan, U.M., Bowerman, S., and Luger, K. (2020). Analytical ultracentrifugation (AUC): an overview of the application of fluorescence and absorbance AUC to the study of biological macromolecules. *Curr. Protoc. Mol. Biol.* 133 (1): e131.

30 Berkowitz, S.A. and Philo, J.S. (2020). Characterizing biopharmaceuticals using analytical ultracentrifugation, Chapter 9. In: *Biophysical Characterization of Proteins in Developing Biopharmaceuticals*, 2e (ed. D.J. Houde and S.A. Berkowitz), 225–283. Elsevier https://www.sciencedirect.com/science/article/pii/B9780444641731000093 (accessed 18 August 2021).

31 Katta, V. and Chait, B.T. (1991). Observation of the heme-globin complex in native myoglobin by electrospray-ionization mass spectrometry. *J. Am. Chem. Soc.* 113 (22): 8534–8535.

32 Ganem, B., Li, Y.T., and Henion, J.D. (1991). Detection of noncovalent receptor-ligand complexes by mass spectrometry. *J. Am. Chem. Soc.* 113 (16): 6294–6296.

33 van den Heuvel, R.H.H. and Heck, A.J.R. (2004). Native protein mass spectrometry: from intact oligomers to functional machineries. *Curr. Opin. Chem. Biol.* 8 (5): 519–526.

34 Rayleigh, L. (1882). XX. On the equilibrium of liquid conducting masses charged with electricity. *Lond. Edinb. Dublin Philos. Mag. J. Sci.* 14 (87): 184–186.

35 Jia, M. (2019). Native mass spectrometry for characterization of protein: protein and protein: RNA complexes. The Ohio State University. https://etd.ohiolink.edu/apexprod/rws_olink/r/1501/10?clear=10&p10_accession_num=osu156597018800034 (accessed 15 May 2022).

36 Konermann, L., Ahadi, E., Rodriguez, A.D., and Vahidi, S. (2013). Unraveling the mechanism of electrospray ionization. *Anal. Chem.* 85 (1): 2–9.

37 Iavarone, A.T. and Williams, E.R. (2003). Mechanism of charging and supercharging molecules in electrospray ionization. *J. Am. Chem. Soc.* 125 (8): 2319–2327.

38 Kebarle, P. and Verkerk, U.H. (2009). Electrospray: from ions in solution to ions in the gas phase, what we know now. *Mass Spectrom. Rev.* 28 (6): 898–917.

39 Tang, L. and Kebarle, P. (1993). Dependence of ion intensity in electrospray mass spectrometry on the concentration of the analytes in the electrosprayed solution. *Anal. Chem.* 65 (24): 3654–3668.

40 Enke, C.G. (1997). A predictive model for matrix and analyte effects in electrospray ionization of singly-charged ionic analytes. *Anal. Chem.* 69 (23): 4885–4893.

41 Constantopoulos, T.L., Jackson, G.S., and Enke, C.G. (1999). Effects of salt concentration on analyte response using electrospray ionization mass spectrometry. *J. Am. Soc. Mass Spectrom.* 10 (7): 625–634.

42 Tang, K., Page, J.S., and Smith, R.D. (2004). Charge competition and the linear dynamic range of detection in electrospray ionization mass spectrometry. *J. Am. Soc. Mass Spectrom.* 15 (10): 1416–1423.

43 Ganem, B., Li, Y.T., and Henion, J.D. (1991). Observation of noncovalent enzyme-substrate and enzyme-product complexes by ion-spray mass spectrometry. *J. Am. Chem. Soc.* 113 (20): 7818–7819.

44 Heck, A.J.R. (2008). Native mass spectrometry: a bridge between interactomics and structural biology. *Nat. Methods* 5 (11): 927–933.

45 Sterling, H.J., Batchelor, J.D., Wemmer, D.E., and Williams, E.R. (2010). Effects of buffer loading for electrospray ionization mass spectrometry of a noncovalent protein complex that requires high concentrations of essential salts. *J. Am. Soc. Mass Spectrom.* 21 (6): 1045–1049.

46 Banerjee, S. and Mazumdar, S. (2012). Electrospray ionization mass spectrometry: a technique to access the information beyond the molecular weight of the analyte. *Int. J. Anal. Chem.* 12: e282574.

47 Hedges, J.B., Vahidi, S., Yue, X., and Konermann, L. (2013). Effects of ammonium bicarbonate on the electrospray mass spectra of proteins: evidence for bubble-induced unfolding. *Anal. Chem.* 85 (13): 6469–6476.

48 Terrier, P., Tortajada, J., and Buchmann, W. (2007). A study of noncovalent complexes involving single-stranded DNA and polybasic compounds using nanospray mass spectrometry. *J. Am. Soc. Mass Spectrom.* 18 (2): 346–358.

49 Wilm, M. and Mann, M. (1996). Analytical properties of the nanoelectrospray ion source. *Anal. Chem.* 68 (1): 1–8.

50 Schmidt, A., Karas, M., and Dülcks, T. (2003). Effect of different solution flow rates on analyte ion signals in nano-ESI MS, or: when does ESI turn into nano-ESI? *J. Am. Soc. Mass Spectrom.* 14 (5): 492–500.

51 Susa, A.C., Lippens, J.L., Xia, Z. et al. (2018). Submicrometer emitter ESI tips for native mass spectrometry of membrane proteins in ionic and nonionic detergents. *J. Am. Soc. Mass Spectrom.* 29 (1): 203–206.

52 Panczyk, E.M., Gilbert, J.D., Jagdale, G.S. et al. (2020). Ion mobility and surface collisions: submicrometer capillaries can produce native-like protein complexes. *Anal. Chem.* 92 (3): 2460–2467.

53 Jordan, J.S., Xia, Z., and Williams, E.R. (2022). Tips on making tiny tips: secrets to submicron nanoelectrospray emitters. *J. Am. Soc. Mass Spectrom.* 33 (3): 607–611.

54 VanAernum, Z.L., Busch, F., Jones, B.J. et al. (2020). Rapid online buffer exchange for screening of proteins, protein complexes and cell lysates by native mass spectrometry. *Nat. Protoc.* 15 (3): 1132–1157.

55 den Engelsman, J., Garidel, P., Smulders, R. et al. (2011). Strategies for the assessment of protein aggregates in pharmaceutical biotech product development. *Pharm. Res.* 28 (4): 920–933.

56 Hong, P., Koza, S., and Bouvier, E.S.P. (2012). A review size-exclusion chromatography for the analysis of protein biotherapeutics and their aggregates. *J. Liq. Chromatogr. Relat. Technol.* 35 (20): 2923–2950.

57 Fekete, S., Beck, A., Veuthey, J.L., and Guillarme, D. (2014). Theory and practice of size exclusion chromatography for the analysis of protein aggregates. *J. Pharm. Biomed. Anal.* 101: 161–173.

58 Cavanagh, J., Benson, L.M., Thompson, R., and Naylor, S. (2003). In-line desalting mass spectrometry for the study of noncovalent biological complexes. *Anal. Chem.* 75 (14): 3281–3286.

59 Jia, M., Sen, S., Wachnowsky, C. et al. (2020). Characterization of [2Fe–2S]-cluster-bridged protein complexes and reaction intermediates by use of native mass spectrometric methods. *Angew. Chem. Int. Ed.* 59 (17): 6724–6728.

60 Yan, Y., Xing, T., Wang, S. et al. (2019). Coupling mixed-mode size exclusion chromatography with native mass spectrometry for sensitive detection and quantitation of homodimer impurities in bispecific IgG. *Anal. Chem.* 91 (17): 11417–11424.

61 Leblanc, Y., Ramon, C., Bihoreau, N., and Chevreux, G. (2017). Charge variants characterization of a monoclonal antibody by ion exchange chromatography coupled on-line to native mass spectrometry: case study after a long-term storage at +5 °C. *J. Chromatogr. B* 1048: 130–139.

62 Muneeruddin, K., Nazzaro, M., and Kaltashov, I.A. (2015). Characterization of intact protein conjugates and biopharmaceuticals using ion-exchange chromatography with online detection by native electrospray ionization mass spectrometry and top-down tandem mass spectrometry. *Anal. Chem.* 87 (19): 10138–10145.

63 Sankaran, P.K., Kabadi, P.G., Honnappa, C.G. et al. (2018). Identification and quantification of product-related quality attributes in bio-therapeutic monoclonal

antibody via a simple, and robust cation-exchange HPLC method compatible with direct online detection of UV and native ESI-QTOF-MS analysis. *J. Chromatogr. B* 1102, 1103: 83–95.

64 Jia, M., Mozziconacci, O., Abend, A. et al. (2022). Multi-dimensional plug-and-play liquid chromatography-native ion mobility mass spectrometry method for the analysis of biotherapeutics. *Int. J. Mass Spectrom.* 471: 116725.

65 Leblanc, Y., Bihoreau, N., and Chevreux, G. (2018). Characterization of human serum albumin isoforms by ion exchange chromatography coupled on-line to native mass spectrometry. *J. Chromatogr. B* 1095: 87–93.

66 Chen, Z., Boyken, S.E., Jia, M. et al. (2019). Programmable design of orthogonal protein heterodimers. *Nature* 565 (7737): 106–111.

67 Chen, S., Lau, H., Brodsky, Y. et al. (2010). The use of native cation-exchange chromatography to study aggregation and phase separation of monoclonal antibodies. *Protein Sci.* 19 (6): 1191–1204.

68 Yan, Y., Liu, A.P., Wang, S. et al. (2018). Ultrasensitive characterization of charge heterogeneity of therapeutic monoclonal antibodies using strong cation exchange chromatography coupled to native mass spectrometry. *Anal. Chem.* 90 (21): 13013–13020.

69 Füssl, F., Cook, K., Scheffler, K. et al. (2018). Charge variant analysis of monoclonal antibodies using direct coupled ph gradient cation exchange chromatography to high-resolution native mass spectrometry. *Anal. Chem.* 90 (7): 4669–4676.

70 Ma, F., Raoufi, F., Bailly, M.A. et al. (2020). Hyphenation of strong cation exchange chromatography to native mass spectrometry for high-throughput online characterization of charge heterogeneity of therapeutic monoclonal antibodies. *MAbs* 12 (1): 1763762.

71 Haverick, M.A., Mengisen, S., Shameem, M., and Ambrogelly, A. (2014). Separation of mAbs molecular variants by analytical hydrophobic interaction chromatography HPLC: overview and applications. *MAbs* 6 (4): 852–858.

72 Bobály, B., Fleury-Souverain, S., Beck, A. et al. (2018). Current possibilities of liquid chromatography for the characterization of antibody-drug conjugates. *J. Pharm. Biomed. Anal.* 147: 493–505.

73 Chen, B., Lin, Z., Alpert, A.J. et al. (2018). Online hydrophobic interaction chromatography–mass spectrometry for the analysis of intact monoclonal antibodies. *Anal. Chem.* 90 (12): 7135–7138.

74 Wei, B., Han, G., Tang, J. et al. (2019). Native hydrophobic interaction chromatography hyphenated to mass spectrometry for characterization of monoclonal antibody minor variants. *Anal. Chem.* 91 (24): 15360–15364.

75 Yan, Y., Xing, T., Wang, S. et al. (2020). Online coupling of analytical hydrophobic interaction chromatography with native mass spectrometry for the characterization of monoclonal antibodies and related products. *J. Pharm. Biomed. Anal.* 186: 113313.

76 Chapel, S., Rouvière, F., Sarrut, M., and Heinisch, S. (2020). Two-dimensional liquid chromatography coupled to high-resolution mass spectrometry for the analysis of ADCs. In: *Antibody-Drug Conjugates: Methods and Protocols*, Methods in Molecular Biology (ed. L.N. Tumey), 163–185. New York, NY: Springer https://doi.org/10.1007/978-1-4939-9929-3_11 (accessed 13 January 2020).

77 Ehkirch, A., D'Atri, V., Rouviere, F. et al. (2018). An online four-dimensional HIC×SEC-IM×MS methodology for proof-of-concept characterization of antibody drug conjugates. *Anal. Chem.* 90 (3): 1578–1586.

78 Ehkirch, A., Goyon, A., Hernandez-Alba, O. et al. (2018). A novel online four-dimensional SEC×SEC-IM×MS methodology for characterization of monoclonal antibody size variants. *Anal. Chem.* 90 (23): 13929–13937.

79 Yan, A., Shikha, V., Chen, Y. et al. (2017). Forced degradation study of monoclonal antibody using two-dimensional liquid chromatography. *J. Chromatogr. Sep. Tech.* 08 (03): https://www.omicsonline.org/open-access/forced-degradation-study-of-monoclonal-antibody-using-twodimensional-liquid-chromatography-2157-7064-1000365.php?aid=90825 (accessed 19 August 2020).

80 Yan, Y., Xing, T., Wang, S., and Li, N. (2020). Versatile, sensitive, and robust native LC-MS platform for intact mass analysis of protein drugs. *J. Am. Soc. Mass Spectrom.* 31 (10): 2171–2179.

81 Liu, A.P., Yan, Y., Wang, S., and Li, N. (2022). Coupling anion exchange chromatography with native mass spectrometry for charge heterogeneity characterization of monoclonal antibodies. *Anal. Chem.* 94 (16): 6355–6362.

82 Yan, Y., Xing, T., Liu, A.P. et al. (2021). Post-column denaturation-assisted native size-exclusion chromatography–mass spectrometry for rapid and in-depth characterization of high molecular weight variants in therapeutic monoclonal antibodies. *J. Am. Soc. Mass Spectrom.* 32 (12): 2885–2894.

83 Park, H.M., Winton, V.J., Drader, J.J. et al. (2020). Novel interface for high-throughput analysis of biotherapeutics by electrospray mass spectrometry. *Anal. Chem.* 92 (2): 2186–2193.

84 Santos Seckler, H.d., Park, H.M., Lloyd-Jones, C.M. et al. (2021). New interface for faster proteoform analysis: immunoprecipitation coupled with samplestream-mass spectrometry. *J. Am. Soc. Mass Spectrom.* 32 (7): 1659–1670.

85 Eliuk, S. and Makarov, A. (2015). Evolution of orbitrap mass spectrometry instrumentation. *Annu. Rev. Anal. Chem.* 8 (1): 61–80.

86 Zhou, M., Yan, J., Romano, C.A. et al. (2018). Surface induced dissociation coupled with high resolution mass spectrometry unveils heterogeneity of a 211 kDa multicopper oxidase protein complex. *J. Am. Soc. Mass Spectrom.* 29 (4): 723–733.

87 Shelton, H., Hendricks, C.D., and Wuerker, R.F. (1960). Electrostatic acceleration of microparticles to hypervelocities. *J. Appl. Phys.* 31 (7): 1243–1246.

88 Gilmore, I.S. and Seah, M.P. (2000). Ion detection efficiency in SIMS: dependencies on energy, mass and composition for microchannel plates used in mass spectrometry. *Int. J. Mass Spectrom.* 202 (1): 217–229.

89 Westmacott, G., Frank, M., Labov, S.E., and Benner, W.H. (2000). Using a superconducting tunnel junction detector to measure the secondary electron emission efficiency for a microchannel plate detector bombarded by large molecular ions. *Rapid Commun. Mass Spectrom.* 14 (19): 1854–1861.

90 Fraser, G.W. (2002). The ion detection efficiency of microchannel plates (MCPs). *Int. J. Mass Spectrom.* 215 (1): 13–30.

91 Keifer, D.Z., Pierson, E.E., and Jarrold, M.F. (2017). Charge detection mass spectrometry: weighing heavier things. *Analyst* 142 (10): 1654–1671.

92 Marty, M.T., Baldwin, A.J., Marklund, E.G. et al. (2015). Bayesian deconvolution of mass and ion mobility spectra: from binary interactions to polydisperse ensembles. *Anal. Chem.* 87 (8): 4370–4376.

93 Takano, K., Arai, S., Sakamoto, S. et al. (2020). Screening of protein–ligand interactions under crude conditions by native mass spectrometry. *Anal. Bioanal. Chem.* 412 (17): 4037–4043.

94 Gan, J., Ben-Nissan, G., Arkind, G. et al. (2017). Native mass spectrometry of recombinant proteins from crude cell lysates. *Anal. Chem.* 89 (8): 4398–4404.

95 Ben-Nissan, G., Vimer, S., Warszawski, S. et al. (2018). Rapid characterization of secreted recombinant proteins by native mass spectrometry. *Commun. Biol.* 1: 213.

96 Campuzano, I.D.G. and Sandoval, W. (2021). Denaturing and native mass spectrometric analytics for biotherapeutic drug discovery research: historical, current, and future personal perspectives. *J. Am. Soc. Mass Spectrom.* 32 (8): 1861–1885.

97 Lalonde, M.E. and Durocher, Y. (2017). Therapeutic glycoprotein production in mammalian cells. *J. Biotechnol.* 251: 128–140.

98 Tian, Y. and Ruotolo, B.T. (2018). Collision induced unfolding detects subtle differences in intact antibody glycoforms and associated fragments. *Int. J. Mass Spectrom.* 425: 1–9.

99 Kükrer, B., Filipe, V., van Duijn, E. et al. (2010). Mass spectrometric analysis of intact human monoclonal antibody aggregates fractionated by size-exclusion chromatography. *Pharm. Res.* 27 (10): 2197–2204.

100 Rouby, G., Tran, N.T., Leblanc, Y. et al. (2020). Investigation of monoclonal antibody dimers in a final formulated drug by separation techniques coupled to native mass spectrometry. *MAbs* 12 (1): https://www.ncbi.nlm.nih.gov/pmc/articles/PMC7531515 (accessed 2 November 2020).

101 Jones, J., Pack, L., Hunter, J.H., and Valliere-Douglass, J.F. (2020). Native size-exclusion chromatography-mass spectrometry: suitability for antibody–drug conjugate drug-to-antibody ratio quantitation across a range of chemotypes and drug-loading levels. *MAbs* 12 (1): 1682895.

102 Debaene, F., Bœuf, A., Wagner-Rousset, E. et al. (2014). Innovative native MS methodologies for antibody drug conjugate characterization: high resolution native MS and IM-MS for average DAR and DAR distribution assessment. *Anal. Chem.* 86 (21): 10674–10683.

103 Marcoux, J., Champion, T., Colas, O. et al. (2015). Native mass spectrometry and ion mobility characterization of trastuzumab emtansine, a lysine-linked antibody drug conjugate. *Protein Sci.* 24 (8): 1210–1223.

104 Liu, M., Van Voorhis, W.C., and Quinn, R.J. (2021). Development of a target identification approach using native mass spectrometry. *Sci. Rep.* 11 (1): 2387.

105 Sternicki, L.M., Nonomiya, J., Liu, M. et al. (2021). Native mass spectrometry for the study of PROTAC GNE-987-containing ternary complexes. *ChemMedChem* 16 (14): 2206–2210.

106 Mehaffey, M.R., Lee, J., Jung, J. et al. (2020). Mapping a conformational epitope of hemagglutinin A using native mass spectrometry and ultraviolet photodissociation. *Anal. Chem.* 92 (17): 11869–11878. https://pubs.acs.org/doi/10.1021/acs.analchem.0c02237 (accessed 15 May 2022).

107 Keskin, O., Gursoy, A., Ma, B., and Nussinov, R. (2008). Principles of protein-protein interactions: what are the preferred ways for proteins to interact? *Chem. Rev.* 108 (4): 1225–1244.

108 Zinzalla, G. and Thurston, D.E. (2009). Targeting protein-protein interactions for therapeutic intervention: a challenge for the future. *Future Med. Chem.* 1 (1): 65–93.

109 Clemmer, D.E., Hudgins, R.R., and Jarrold, M.F. (1995). Naked protein conformations: cytochrome c in the gas phase. *J. Am. Chem. Soc.* 117 (40): 10141–10142.

110 Collings, B.A. and Douglas, D.J. (1996). Conformation of gas-phase myoglobin ions. *J. Am. Chem. Soc.* 118 (18): 4488–4489.

111 Hunter, C.L., Mauk, A.G., and Douglas, D.J. (1997). Dissociation of heme from myoglobin and cytochrome b5: comparison of behavior in solution and the gas phase. *Biochemistry* 36 (5): 1018–1025. https://pubs.acs.org/doi/pdf/10.1021/bi961993%2B (accessed 15 May 2022).

112 Robinson, C.V., Chung, E.W., Kragelund, B.B. et al. (1996). Probing the nature of noncovalent interactions by mass spectrometry. A study of protein—CoA ligand binding and assembly. *J. Am. Chem. Soc.* 118 (36): 8646–8653.

113 Wu, Q., Gao, J., Joseph-McCarthy, D. et al. (1997). Carbonic anhydrase-inhibitor binding: from solution to the gas phase. *J. Am. Chem. Soc.* 119 (5): 1157–1158.

114 Gabelica, V. and Pauw, E.D. (2005). Internal energy and fragmentation of ions produced in electrospray sources. *Mass Spectrom. Rev.* 24 (4): 566–587.

115 Deslignière, E., Ehkirch, A., Duivelshof, B.L. et al. (2021). State-of-the-art native mass spectrometry and ion mobility methods to monitor homogeneous site-specific antibody-drug conjugates synthesis. *Pharmaceuticals* 14 (6): 498.

116 Polasky, D.A., Dixit, S.M., Fantin, S.M., and Ruotolo, B.T. (2019). CIUSuite 2: next-generation software for the analysis of gas-phase protein unfolding data. *Anal. Chem.* 91 (4): 3147–3155.

117 Pierson, E.E., Keifer, D.Z., Selzer, L. et al. (2014). Detection of late intermediates in virus capsid assembly by charge detection mass spectrometry. *J. Am. Chem. Soc.* 136 (9): 3536–3541.

118 Pierson, E.E., Keifer, D.Z., Kukreja, A.A. et al. (2016). Charge detection mass spectrometry identifies preferred non-icosahedral polymorphs in the self-assembly of woodchuck hepatitis virus capsids. *J. Mol. Biol.* 428 (2, Part A): 292–300.

119 Keifer, D.Z., Motwani, T., Teschke, C.M., and Jarrold, M.F. (2016). Measurement of the accurate mass of a 50 MDa infectious virus. *Rapid Commun. Mass Spectrom.* 30 (17): 1957–1962.

120 Lutomski, C.A., Lyktey, N.A., Zhao, Z. et al. (2017). Hepatitis B virus capsid completion occurs through error correction. *J. Am. Chem. Soc.* 139 (46): 16932–16938.

121 Bond, K., Tsvetkova, I.B., Wang, J.C.Y. et al. (2020). Virus assembly pathways: straying away but not too far. *Small* 16 (51): 2004475.

122 Miller, L.M., Barnes, L.F., Raab, S.A. et al. (2021). Heterogeneity of glycan processing on trimeric SARS-CoV-2 spike protein revealed by charge detection mass spectrometry. *J. Am. Chem. Soc.* 143 (10): 3959–3966.

123 Miller, L.M., Bond, K.M., Draper, B.E., and Jarrold, M.F. (2021). Characterization of classical vaccines by charge detection mass spectrometry. *Anal. Chem.* 93 (35): 11965–11972.

124 Keiryn, L.B. (2019). Native mass spectrometry: a glimpse into the machinations of biology. Biopharma from Technology Networks. https://www.technologynetworks.com/biopharma/articles/native-mass-spectrometry-a-glimpse-into-the-machinations-of-biology-319755 (accessed 2 May 2021).

Section 5

Other Ambient Ionization Other than ESI

11

Laser Diode Thermal Desorption-Mass Spectrometry (LDTD-MS): Fundamentals and Applications of Sub-Second Analysis in Drug Discovery Environment

Pierre Picard[1], Sylvain Letarte[1], Jonathan Rochon[1], and Réal E. Paquin[2]

[1] *R&D Department, Phytronix Technologies Inc., Québec, QC, Canada*
[2] *Département de physique, Université Laval, Québec, QC, Canada*

11.1 A Historical Perspective of the LDTD

Thermal desorption as a sample introduction method for nonvolatile compounds goes back to the early days of mass spectrometry, where the main ionization technique was then by electron impact (EI). During the 1950s, techniques were developed [1, 2] to introduce a solid sample directly into the ionization chamber of a mass spectrometer via a probe that vaporizes the sample within a few centimeters of the ionization region. These techniques expanded mass analysis by adding a large number of compounds that could be analyzed, which were not very volatile but had a relatively good thermal stability. Work continued over the next decade to achieve the analysis of nonvolatile and thermally unstable compounds. Beckey [3] reported the use of field desorption for the analysis of monosaccharide glucose to exemplify the capability of the technique. Meanwhile, the work on thermal desorption continued and helped to set strategies to obtain the mass spectrum of such compounds. Most of the research was conducted on the transfer of energy to compounds that were involatile and easily altered by thermal processes. The authors [4] showed that there is a competition of kinetic nature between the vaporization (i.e. dissociation of the bonds between molecules) and the decomposition (i.e. dissociation of the bonds inside molecules) of the sample when heat is applied. In most cases, the decomposition of the samples was the result. Friedman and his coworkers [5] suggested two concepts to increase the volatility of the samples being analyzed: reduce the energy that is necessary to

High-Throughput Mass Spectrometry in Drug Discovery, First Edition. Edited by Chang Liu and Hui Zhang.
© 2023 John Wiley & Sons, Inc. Published 2023 by John Wiley & Sons, Inc.

vaporize the molecules and choose conditions that favor vaporization over decomposition. From this work emerged the rapid heating desorption [5, 6] and flash desorption [7] techniques. The thermal probes were also used with the atmospheric pressure inlet in the triple quadrupole during the 1980s [8]. At the same time, other techniques such as electrospray and atmospheric pressure chemical ionization (APCI) coupled to high-performance liquid chromatography (HPLC) progressed to a level of performance for sensitivity and reproducibility that outmatched the thermal probe. Since then, HPLC-MS has become the method of choice in analytical chemistry. Even with the great efforts put into improving the thermal techniques with flash heating, material coatings, and others, the overall performance did not reach the usability needed for such analyses.

Over the years, the analytical performance of mass spectrometers increased the demand from pharmaceutical research laboratories for a higher throughput, considering that this change in technology would broaden the application fields available for drug discovery. Since a liquid chromatography analysis is required, and still requires, minutes to be performed, efforts were made to optimize the instrument throughput down to the physical processes involved in the ionization process itself. Ionic suppression in electrospray and ionization efficiency in APCI were and are still part of the limitations. Thermal desorption characteristics, such as speed and versatility, justified revisiting this physical process, which resulted in the creation of a new ionization source, the LDTD® (Laser Diode Thermal Desorption). The uses of the laser diode thermal desorption technique (LDTD) were first presented in 2005 at the ASMS conference in San Antonio, USA [9]. In this chapter, we present a fundamental characterization of the source in terms of optimal heating, sample deposition, gas dynamics, and atmospheric pressure chemical ionization. Knowledge of the physical processes involved is essential for an efficient analytical method development. Applications to several domains such as pharmacokinetics, adsorption distribution metabolism, and excretion (ADME) and high-throughput screening (HTS) are described.

The highest achievable throughput using a mass spectrometer is reached when the ionized molecules of interest spend minimal time at the entrance of the instrument, which directly influences the smallest acquisition period allowed. Samples in liquid form require sprays and imply the presence of solvent in the ionization region. The analytes constitute only parts per trillion or parts per billion of the total quantity of ions produced and are subject to suppression upon sample content [10–12]. Removing the liquid from the ionization zone is considered as the best option to limit suppression compared to a sample dilution or a cleanup. However, when working with a dried material, it is necessary to get it in the gaseous phase before it enters the mass spectrometer. To fulfill this requirement, a thermal process was selected for its efficiency and practicality. As mentioned earlier, the thermal probes of the 1980s had limitations in terms of sensitivity and reproducibility; therefore, they were mostly used in methods where qualitative results were needed. Thermal stability of the compounds was also necessary to successfully attain mass detection. By studying the thermal probe, key factors were determined, which led to working on several physical aspects of the desorption process that were the crucial aspects for the development of the LDTD ion source.

11.2 Instrumentation

11.2.1 LDTD Process

The laser diode thermal desorption (LDTD) ion source (Figure 11.1) is an ultrafast technique that desorbs samples in solid phase and creates ions in a matter of seconds. The technology uses a sample holder well plate in which samples are deposited and dried prior to analysis. The sample holder is placed between a laser beam and an extraction nozzle. The laser beam then heats up the back of the well and desorbs neutral molecules into the gas phase, meaning that the optical beam does not come into direct contact with the sample. The ramp, the duration, and the maximum power of the laser beam throughout the desorption is customizable depending on the sample analyzed. The vaporized molecules are carried away by a constant flow of compressed air through a transfer tube toward the ionization region. The neutral molecules are then ionized by a corona discharge at the tip of a needle (APCI process). The ions created enter the mass spectrometer inlet guided by the electric field and the gas flow. This whole process, from desorption

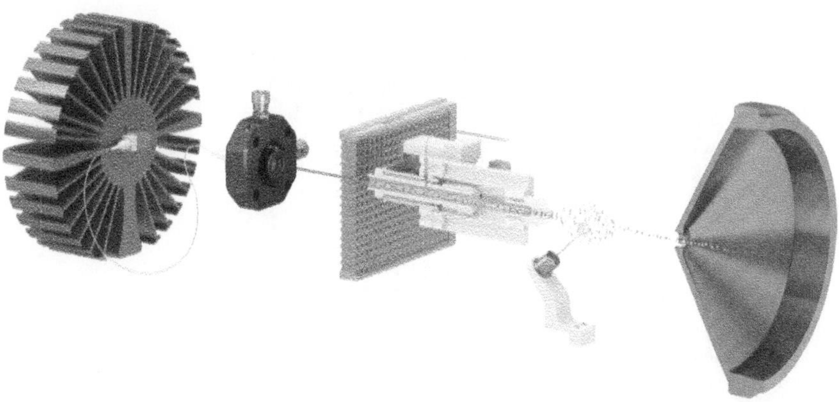

Figure 11.1 LDTD process. The laser path is shown heating the back of the sample holder where the molecules were deposited. They are then desorbed and directed through an extraction nozzle by a carrier gas to the APCI ionization region. The molecules are ionized and drift accordingly with the electric field and the dynamic flow.

to the end of the analysis, can reach the sub-second range. At the end of a sample analysis, the sample holder moves to a new position in hundreds of milliseconds. As can be deduced from the explanation of this process, once the sample preparation is completed, the LDTD process does not involve solvents (as for LC-MS) or ionization enhancement matrices (as for the matrix-assisted laser desorption/ionization [MALDI] technique) in the MS analysis. The measurement of femtograms on the sample holder has been reported [13] with accurate quantification. As it will be demonstrated, it performs advantageously compared to other high-throughput MS techniques when the density and the time dedicated to the analysis must be reduced.

11.2.2 Sample Holder Design

The LDTD ion source requires a specific sample holder. It is designed to be compatible with liquid handling instrumentation used in high-throughput environments for an efficient integration in the workflow. To achieve this integration, the sample holder has the standardized microplate format of 96, 384, and 1536 (Figure 11.2) [14]. The well in which the sample is deposited has a hexagonal imprint (see Figure 11.3) to concentrate the liquid in the center of the well during the drying process. The well is made of a 25.4 μm thick metal alloy foil with a high thermal conductivity and low heat capacity. This foil thickness was selected to maximize the heat transfer to the dried deposit. The metal foil absorbs the continuous infrared laser beam, which has a wavelength of 980 nm. The volume that can be dispensed in a well can reach up to 10 μL for the 96 format, 2 μL for the 384 format, and 0.5 μL for the 1536 format.

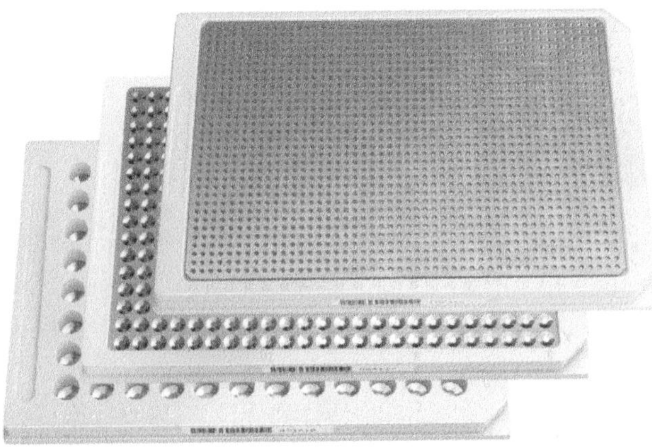

Figure 11.2 The sample holder with standard dimensions (ANSI footprint) for the 96, 384, and 1536 formats.

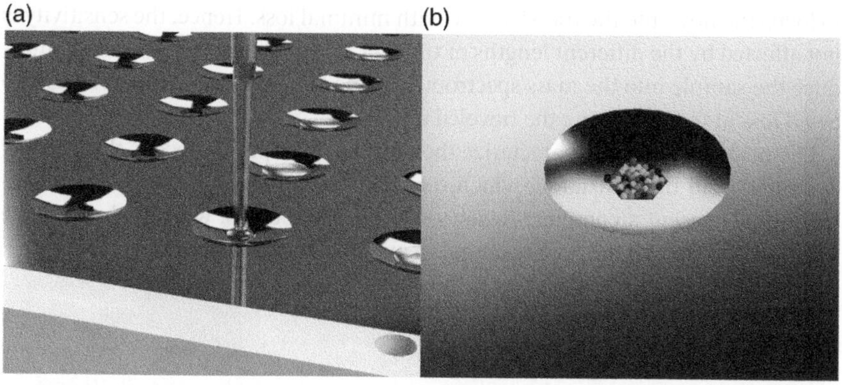

Figure 11.3 Schematic view of (a) the sample preparation deposited on the well of the sample holder (384 format) and (b) the molecules left on the well after the evaporation of the solvents.

11.2.3 Vapor Extraction Nozzle

An extraction nozzle has been designed to collect the desorbed samples near the surface of the well and guide them toward the ionization region. The extraction nozzle is made of two concentric tubes. A carrier gas flow originating from the gap between the concentric tubes is directed toward the well surface. The carrier gas flow follows the well surface and turns 90° into the inner tube (Figure 11.9). The inner tube is called the transfer tube and its function is to guide the desorbed compound to the ionization region. A schematic sectional view of the extraction nozzle shows its concentric design in Figure 11.4. The dimensions are optimized

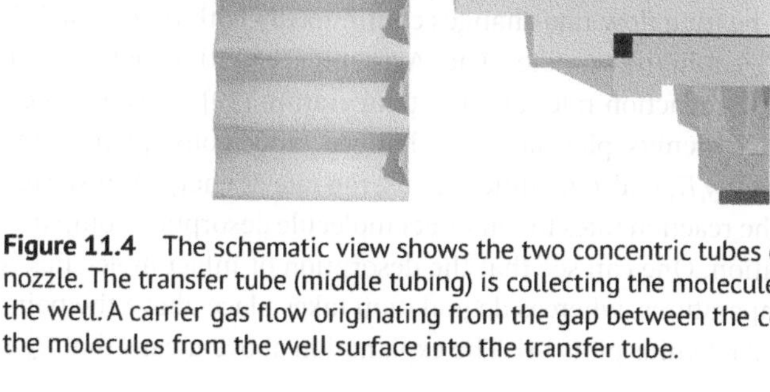

Figure 11.4 The schematic view shows the two concentric tubes of the extraction nozzle. The transfer tube (middle tubing) is collecting the molecules near the surface of the well. A carrier gas flow originating from the gap between the concentric tubes drives the molecules from the well surface into the transfer tube.

to focus the flow into the transfer tube with minimal loss. Hence, the sensitivity is not affected by the different lengths of transfer tubes (100–180 mm) used to introduce the sample into the mass spectrometer. The flow is in the laminar regime to avoid loss at the wall along the traveled distance.

The next sections will characterize the thermal desorption process, the gas flow dynamics, and the ionization, which represent the core of the LDTD technique. We do not present a complete review of APCI ion sources, which is beyond the scope of this chapter and can be found in any book on mass spectrometry [15].

11.3 Theoretical Background

11.3.1 Thermal Process

Thermal desorption is a phenomenon where a substance is released from a surface by the application of heat. It is in equilibrium with sorption, which is the opposite phenomenon following Arrhenius' equation [16, 17].

$$k = Ae^{-E_a/RT}$$

where k is the rate constant, T the absolute temperature (kelvin), A the pre-exponential factor constant for each desorption reaction, E_a is the activation energy of desorption, and R the universal gas constant. The key factor influencing the desorption is the activation energy, which is needed to liberate the molecule from its bound state. Thermal degradation will take place if the bound dissociation energy of the molecules, E_d, is reached before E_a. The reduction of the activation energy of desorption needed is crucial to maximize the number of molecules analyzed. The sorption process is subdivided in physisorption and chemisorption for different levels of enthalpy. Physisorption represents the lowest level governed by Van der Waals forces. On the other hand, the chemical bounds of chemisorption lead to energy levels in the order of E_d [18, 19].

Hence, in thermal desorption, some compounds are desorbed at an apparent E_a higher than the dissociation energy E_d. The process of thermal desorption is dynamic because the heating flow rate changes continuously with time, which is in contrast with the equilibrium state. The Arrhenius equation defines the dependency between the reaction rates and the temperature [20]. During a thermal desorption, the Arrhenius plot slope is changed, and, consequently, the effective activation energy E_a is also modified. Thus, the rate of energy transferred (Figure 11.5) shows the reaction rates for an intact molecule desorption compared to a result of degradation. One can see that the desorption of intact molecules is favored at higher temperatures, whereas degradation takes place above the point of intersection in the Arrhenius plots [4]. The explanation above means that E_a is

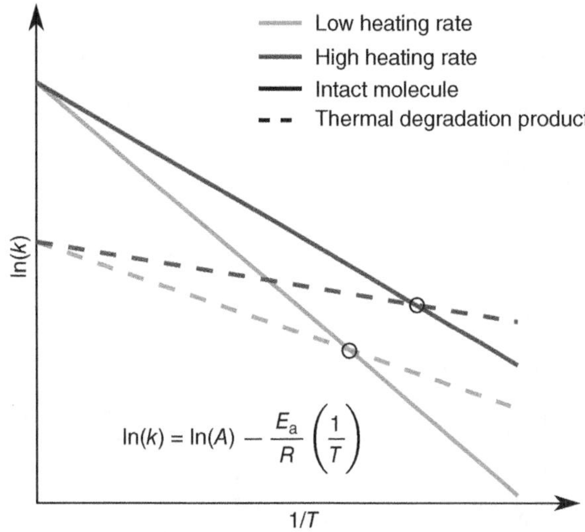

Figure 11.5 Logarithmic relationship between the desorption rate and the heating rate for an intact molecule and a molecule which undergoes thermal degradation.

dependent on dynamic parameters and that it can be reached at lower temperatures than what is expected in materials in bulk form [21]. These observations led to the development of flash heating in mass spectrometry, which attempts to limit the breakdown of the molecules [22]. The changes observed in the effective activation energy cannot be explained by the desorption enthalpy, which is linked to the binding forces present inside the dried residue of the sample. The apparent variation of E_a is connected to the energy transfer process.

Experimental evidence, as shown in Figure 11.6, with the LDTD ion source has shown that thermal desorption occurs at a temperature lower than the melting point of the molecule. This is verified using in situ temperature measurements with a microscopic thermocouple placed at the surface of the well, which allows the instrument to be operated normally while the temperature is being monitored. As an example, for sulfadiazine, desorption begins at 60 °C as shown in Figure 11.6. The maximum intensity area is obtained at 95 °C and remains stable until it reaches 140 °C even though the sulfadiazine has a reported bulk melting point between 252 and 256 °C [23]. This result seems to contradict the statement made earlier about the high yield desorption rate at high temperatures but can be explained by the fact that the temperature monitored is an average value over the whole surface and takes into account heat diffusion throughout the bulk material.

Exploring the dynamic behavior of heat transfer between the well and the layer of residual deposit may bring to light the parameters allowing the high yield of desorption at lower temperatures. The classical approach using the relaxation time failed to explain the two conflicting values of E_a for one temperature. This brings us to explore the solid-state physics associated with the heat transfer

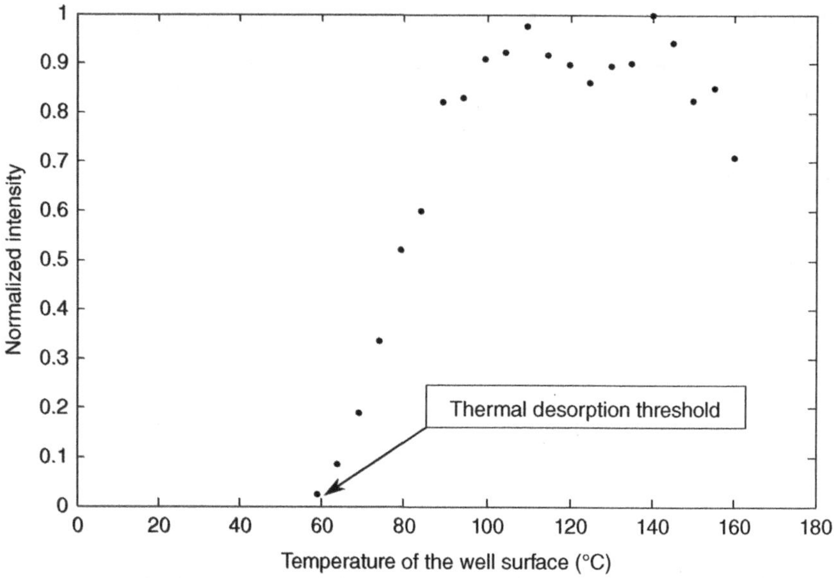

Figure 11.6 Normalized intensity of sulfadiazine compared to the temperature of the well measured with a microscopic thermocouple. This graph shows that, for this molecule, the intensity area is maximized when the well surface reached 95 °C and starts dropping around 140 °C.

process. We will present an overview of the associated theory to highlight the key parameters of the thermal process. Vibrational energy (heat) is represented by phonon, a collective excitation in a periodic, elastic arrangement of atoms or molecules in condensed matter, specifically in solids and some liquids. Most theories of energy transport in solids employ the Boltzmann transport equation (BTE). For both electron and phonon transport, the form of the equation and the form of the various scattering mechanisms are very well known [24, 25]. This theory can explain, in bulk homogeneous materials, the dependence of the electrical conductivity and the thermal conductivity [24]. Changes from bulk physical properties arise when the phonon wavelength has the same order of magnitude than the size of the lattice structure [26]. An example of these physical differences is used in the semiconductor industry with the vaporization of nanocrystals of silicon at a temperature well below the reported melting point [21]. The following question arises: How can we desorb at a temperature lower than the melting point? The energy at a specific temperature follows a Maxwell–Boltzmann distribution [27]. Using the phonon representation, the energy associated with the vibration wavelength is $E_\lambda = h2\pi/\lambda$, where h is the Plank constant and λ the phonon wavelength. The desorption is possible when $E_\lambda > E_a$ and will occur at a sufficiently short phonon wavelength as can be seen in Figure 11.7. This quantum representation

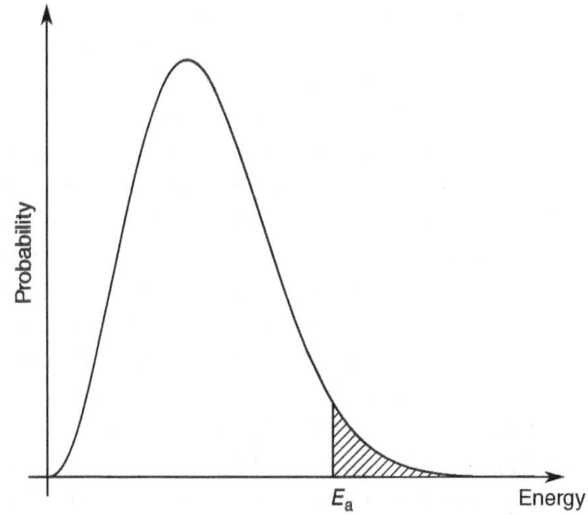

Figure 11.7 Maxwell–Boltzmann distribution.

shows the possibility of reaching E_a at a lower average temperature. Even if the probability of occurrence of high-energy phonons is small compared to low-energy ones, practical observations, as seen using the LDTD, show a high desorption yield.

The phonon wavelengths are defined by the structure and dimensions of the lattice [26]. It has been reported that lattice thickness affects the thermal conductivity [24]. As the thickness decreases, the thermal conductivity tends to rely on the physical dimension of the lattice and moves away from the conductivity of bulk materials. Consequently, for structures in the nanometer size range, only phonons with wavelengths in the order magnitude of the structure dimensions will propagate into the solid in the axial direction. In this context, the lattice can be visualized as acting as a high-pass filter that can only transmit high-energy phonons. The resulting temperature distribution in the lattice will not follow a Maxwell–Boltzmann distribution, explaining the observed differences in the physical characteristics of the material. The structural properties that were presented so far do not include the dynamics of the energy flow into the thin layer which is, in this case, the sample deposited on the surface of the well. One may state that the high-energy phonons will diffuse in a steady state by scattering into the lattice at lower frequencies in the axis perpendicular to the propagation and eventually reach the shape of a Maxwell–Boltzmann distribution [24, 28].

The LDTD heating system configuration (Figure 11.1) involves a flow of energy that starts at the back of the well, goes through the metal alloy foil, and is transferred to the nanometer deposit before exiting on the other side. The source of energy comes from a continuously emitting laser diode at a wavelength of 980 nm, which delivers a high rate of energy, up to 10 optical Watts. The metallic surface absorbs the energy through the photon–electron interaction. The energy diffusion follows a classical approach through the metallic foil, which, as mentioned previously, has a thickness of 25.4 μm [24]. There is a complex boundary effect [29] at the interface between the metal and the deposit that depends on the rugosity, the oxide layer of the stainless-steel, and the lattice composition. Phonons that have a wavelength smaller than the layer size will be favored to propagate into the layer. The flow of energy is continuously carried away by the carrier gas hitting the surface of the well transporting the desorbed molecules at the same time.

The assessment of the physical aspects of the sample deposited on a hexagonal well surface from a sample holder plate called LazWell™ is presented in Figure 11.8 to confirm the adequation between the LDTD heating system and the theory. Drying the liquid sample produces a residual deposit solidified in an amorphous form on the surface. Figure 11.8 presents images of deposits taken by a scanning electron microscope (SEM). Figure 11.8a shows the dried deposit from a solution of 500 pg of prednisone in ethyl acetate. The presence of a ringed pattern indicates that the deposit covers the whole surface. Figure 11.8b is a magnified image of Figure 11.8a and highlights the lattice structure of the sample on the surface. A dried blank plasma crashed from a 2 μL protein precipitate (1 : 3 volume ratio in MeCN) shows a similar structure in Figure 11.8c measured from a few nanometers to hundreds of nanometers. Similar lattice structures have been reported by Segura et al. [30, 31] for the residual deposit of ethylenediaminetetraacetic acid (EDTA) and bovine serum albumin (BSA) coatings, which enhance the thermal desorption of specific compound families (more details about the coatings in Section 11.4.2.1).

11.3.2 Gas Dynamics

The main function of the gas flow is to carry away the desorbed molecules and dissipate the energy coming from the heating at the back of the well to obtain an optimal E_a. In static (no flow) conditions, desorption–adsorption would be in equilibrium with a partial pressure relative to the molecule of interest. Removing

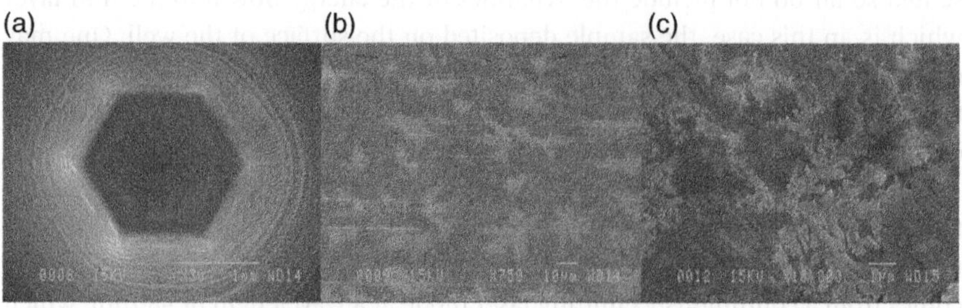

Figure 11.8 Dried deposit of prednisone on a surface well magnified (a) 30 X and (b) 750 X (c) shows dried blank plasma crash sample.

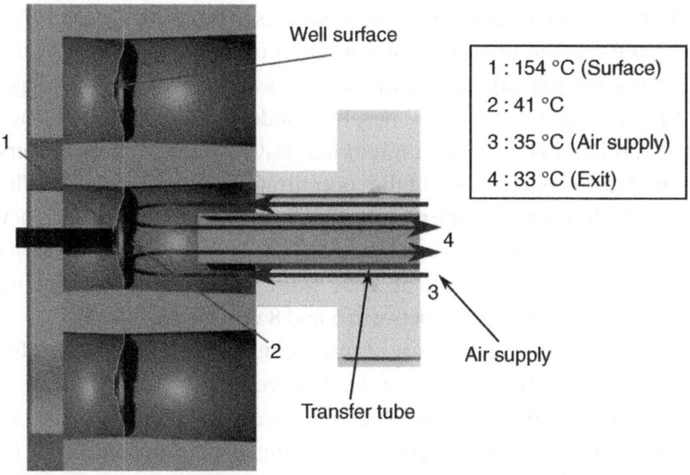

Figure 11.9 Carrier gas flow and temperature profile. The carrier gas enters the nozzle at room temperature and is heated by the contact with the well to 41 °C and then cools back down to room temperature at the end of the transfer tube.

the freshly desorbed molecules in a dynamic system changes the equilibrium state to favor the desorption [32]. The flow in LDTD is designed to sweep the surface in a parallel motion from the outer edge toward the center of the well. The concentric design of the nozzle forces the gas to turn 180° from its initial direction (Figure 11.9). The transfer tube position is optimized to focus the flow into the center of it with minimal loss on the entrance wall. Each well size has its specific transfer tube diameter to obtain a focusing flow between 2 and 3 L/min. To minimize loss during the transfer process from the desorption region to the entrance of the mass spectrometer, the size of the tube and its length are chosen to keep the flow laminar and are parametrized to respect both the flow rate and the focusing of the gas.

Figure 11.9 shows the gas temperature recorded in the system at various positions in the flow path. The surface of the well is heated by ramping the power from 0 to 4.5 W in six seconds. As one can see, the gas temperature decreases along the flow path to reach room temperature at the end of the transfer tube. The difference between the temperature at the back of the well ($t = 154$ °C) and the one recorded 1 mm above the surface (position 2) ($t = 41$ °C) illustrates the rapid cooling of the desorbed molecules.

11.3.3 Ionization

Thermal desorption is the obvious choice to eliminate ionic suppression coming from the electrospray and liquid APCI techniques – since there is no solvent involved in the desorption process, ionic suppression is not present, and it is possible to analyze, in a high-throughput manner, multiple molecules in less than a second. The main ionization process is the proton transfer reaction of the water cluster to the molecules [33]. The carrier gas is dried air with traces of water in the low ppm range. Air is used, and not nitrogen, because the negative mode deprotonation sequence requires oxygen to initiate [34]. The hydronium ion, H_3O^+, tends to form clusters with other water molecules as a function of the concentration of water in the air. The size of the water cluster ions changes the proton affinity, and therefore, the efficiency of the reaction. The proton affinities are as follows: H_3O^+ (691 kJ/mol) < $H_3O^+H_2O$ (820 kJ/mol) < $H_3O^+(H_2O)_2$ (835 kJ/mol) at 25 °C with a 21% relative humidity (6600 ppm). Most water clusters contain between 5 and 8 water molecules [35].

Figure 11.10 reports data on the intensity of water clusters with n molecules for different moisture levels in the carrier gas in positive APCI [13]. The data is obtained with a gas management system which includes a cryogenic trap used to remove water vapor from the carrier gas and a syringe pump dosing water or solvent for quantitative addition prior to the entry in the LDTD ion source. The inlet of a mass spectrometer usually has a declustering mechanism that may disturb the distribution; therefore, the declustering potential (DP) was set to zero. Similar results have been reported in negative mode [34]. In optimal operational conditions, the recommended level of water in the air ranges from 10 to 2000 ppm in which the Gaussian type distribution is at its maximum intensity at $n = 4$.

Drops in the ionization efficiency have been observed when using air saturated with water vapor in the analysis of fully saturated fatty acids in positive mode [36]. During this experience, it was discovered that the infusion of a few ppm of

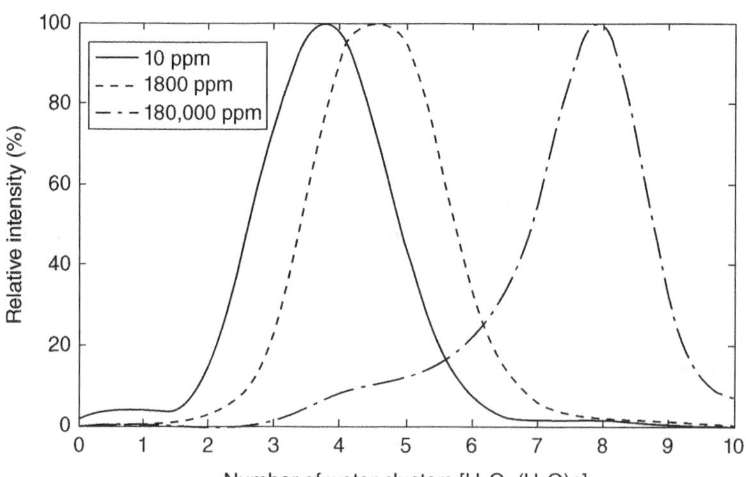

Figure 11.10 Relative abundance versus number of water cluster for different concentrations of water.

methanol or acetonitrile into the carrier gas led to a quench of the proton transfer reaction observed as an important drop of signal intensity. This is explained by the fact that the proton affinities of the solvent were higher than the analytes, which shows the major difference between LDTD and LC APCI. Experiments have shown that using air with traces of water is optimal to cover a broad range of molecule types.

The other characteristics of the ionization process are related to the mechanical configuration relative to the mass spectrometer inlet. The corona discharge is generated by the electric field when a high difference of potential is applied between the needle and the inlet of the mass spectrometer. The discharge plume forms an envelope that will reach the maximum of space-charge in the air at atmospheric pressure [37]. For a current of 3 μA, the inlet area of the mass spectrometer is covered by the plume.

The atmospheric pressure chemical ionization is limited by the number of charges created at the tip of the needle. The number of charges depends on the strength of the electric field, which relies on the voltage at the tip and its sharpness. Data reported [13] the APCI capacity of ionization at 2×10^{-10} M/s of charges for a corona discharge current of 3 μA, which assumes a 100% yield proton transfer [38]. This is demonstrated using a binary mixture of clomiphene and dextrorphan. These compounds are used since they desorb and ionize well in the LDTD ion source. To highlight the capacity limit, one analyte (probe) is kept constant at 100 ng/mL in a MeOH:H_2O solution (75 : 25), while the other is spiked at 50, 100, 200, 300, 400, and 500 μg/mL. Figure 11.11 presents the signal level as a function of desorbed molecules in M/s. The effect of APCI saturation for both compounds

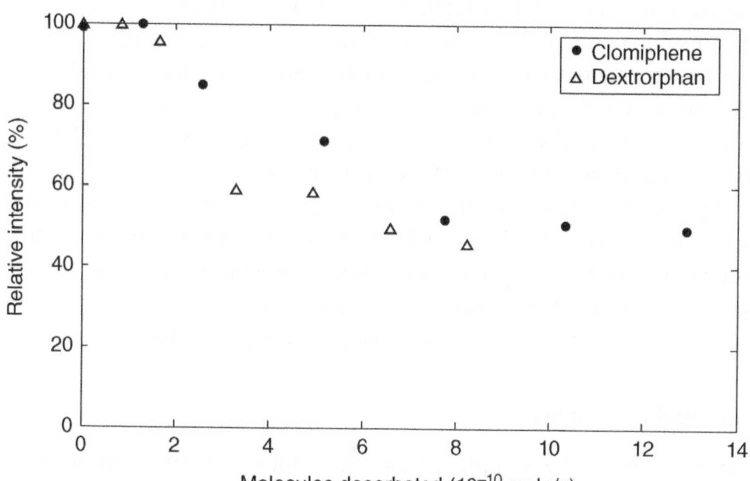

Figure 11.11 Relative signal intensity versus the rate of molecules desorbed.

begins at 2×10^{-10} with an increased signal suppression as the concentration increases.

For liquid chromatography, ion suppression is typically observed in sample extracts from biological samples and not likely to be caused by reactions occurring in the gas phase [10, 39]. It is most likely that ionization suppression is the result of high concentrations of nonvolatile materials present in the spray with the analyte. However, there is no liquid process involved in LDTD and ionic suppression occurs only in the gas phase. The term ionic saturation will be used to avoid any confusion with other techniques that use a liquid spray. Ionic saturation occurs when there are more molecules to ionize than the maximum number of charges available. Even though the ionic saturation is limited [10], one should always be aware that overloading (high concentration) the surface of the well may result in a decrease in sensitivity.

11.4 Sample Preparation

11.4.1 Motivations

Drug discovery and development is a high risk–high reward business. The development cost for new drugs draws substantial resources from pharmaceutical companies [40]. This industry is constantly searching for ways to optimize time and materials dedicated to this research, which has brought them to the following solution: increase the high-throughput analysis capacity by using mass spectrometry. The LDTD technology has been developed with this precise goal, i.e. reducing the sample analysis time. The first work reported by Bateman et al. [41] in drug discovery has demonstrated a throughput capacity of less than 10 seconds per sample in comparison to more than six minutes per sample with LC-MS for the analysis of cytochrome P450 (CYP), phospholipids, and pharmacokinetics (PK) studies. Bateman stated that it "requires minimal sample preparation, consumes small quantity of samples and has no carryover." The LDTD–tandem mass spectrometry (MS/MS) method yields the same sensitivity, linearity, and reproducibility as the LC-MS/MS [42], which makes the technology suitable for high-throughput analysis in a drug discovery environment since its speed of analysis is exceptionally high. The next sections present the guidelines for the sample preparation, the analytical characteristics of the LDTD process, and the applications in high-throughput analysis.

11.4.2 General Guidelines

Despite being simple, the sample preparation does not differ from other analytical techniques, as it needs to separate the compound of interest from the biological matrix to enhance the signal with minimal interferences.

An equilibrium in the sample preparation must be found to maximize the deposit on the surface with the compound of interest without overloading it. This approach requires some compromises and may at first seem counterintuitive compared to classical sample preparations in liquid chromatography, as the compound of interest must not be overconcentrated on the surface of the sample holder. As stated in the previous section, the activation energy, E_a, is dependent on the molecular binding properties and the sample preparation development must be undertaken with these considerations. Even with a high purity solvent, there will always remain some impurities from unidentified sources. The sample preparation and residual material content will influence the level of energy needed to thermally desorb the deposit. Dried residues with a strong binding must be avoided. As an example, many organic compounds may form salt and become nonvolatile. This makes the activation energy greater than the dissociation energy ($E_a > E_d$) and results in thermal degradation of the compounds on the surface. The dried deposited sample will have its own unique lattice properties, and as a result, the energy transferred through the heating process with the laser will need to be adjusted and characterized. The addition of compounds, like potassium phosphate, will form salt residues, which are more appropriate to phonon transfer, without creating an unbreakable bond with the molecules of interest. The final sample quantity deposited onto the well surface must respect the nanoscale residue thickness for the thermal desorption process to occur.

11.4.2.1 Compound Detection Background

Active pharmaceutical ingredients (APIs) under their salt forms present different biological and physicochemical properties. Formulations are designed upon several parameters such as stability, toxicity, and absorption to name a few. Often, the selection of the counterion is based on the pK_a rule to ensure a strong binding energy between the opposite ionic species that form the complex [43]. On the other hand, thermal desorption will perform better with a low binding energy. Dissolving compounds from their salt forms in a pure solvent will recreate the strong binding complex on the sample holder once dried. The highly polar compounds will be difficult to desorb and will result in a limited sensitivity and reproducibility. Mixing additives into the solution allows the formation of lower energy binding complexes for these polar compounds. The following agents can be added to prepare samples containing most of the small molecules: ethylenediaminetetraacetic acid (EDTA) and bovine serum albumin (BSA) for acidic and basic functional groups, respectively. Pre-coated sample plates are available with an optimized quantity of reagent to achieve adequate desorption with unknown compound structures, which is particularly useful when screening a bank of a hundred thousands of compounds [44].

Sample preparation methods used for the LDTD process remain simple and applicable for a large range of small molecules. Nevertheless, some molecular functional groups will not desorb even in a heterogeneous deposit that favors a low E_a, such as the carboxylic acid and amino radical (NH_2) groups. They are the most common types of groups that will require certain additives to be mixed into the final solution to desorb efficiently.

Therefore, to reduce the formation of strong bonds in the dried residue, the addition of ethylenediaminetetraacetic acid (EDTA), which is 4-carboxylic acid groups linked together, chelates the counterion and the sites where bonds are formed [30, 31]. A few hundred micrograms per milliliter are sufficient to allow the desorption of the molecules containing a carboxylic acid group. The EDTA will not vaporize and will remain on the plate. This addition does not influence the APCI process. In the case where the molecule contains an amino radical (NH_2), it will benefit from the addition of bovine serum albumin (BSA) to produce a similar effect. The combination of both BSA and EDTA is possible and a good compromise for methods where both groups are present in the sample.

The EDTA and BSA may be applied directly to the sample holder as a coating and these plates are commercially available to simplify the automation of the method to attain a high-throughput. Where other difficult compounds are concerned, their desorption is occasionally enhanced with the use of a phosphate salt and/or a HEPES buffer.

The site chelating approach may be applied to more challenging assays with strong binding compounds – amino acids are an excellent example of such difficult compounds since they have both acidic and basic sites on a small-sized molecule. Most of them can be desorbed when diluted into a solution of KH_2PO_4 10 mM and may be measured quantitatively without derivatization.

Other compounds, like phospholipids, will desorb partially and additives like EDTA, phosphate, or BSA do not help. Recent results tend to show that the addition of another phospholipid into the solution in the 100 μg/mL range produces a similar effect and allows a quantitation in the low ng/mL range. Current research on peptides shows a promising behavior with the addition of phosphate and amino acids. However, the multiplicity of the amino acid sequence arrangements offers an additional challenge.

11.4.2.2 Details on Ionic Saturation

11.4.2.2.1 Effects of Ionic Saturation on the Signal Intensity The quantity of sample in the well greatly impacts the ionic saturation in gas phase APCI. The maximum intensity is achieved when the desorbed material equals the charge produced in the ionization. Whenever this maximum is exceeded, a competition takes place for the charges. This behavior in the LDTD process is called plate overloading. The dilution associated with a deposited volume on the sample

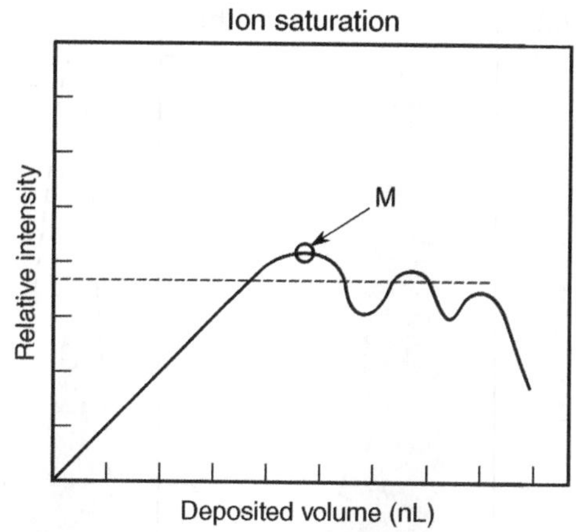

Figure 11.12 Schematic of the signal intensity versus the volume deposited on the well for the same dilution ratio. Ion saturation occurs when the number of molecules exceeds the total number of charges produced by the APCI process. The dashed line shows that it is possible to have a similar signal intensity for different volumes when the saturation point (M) is exceeded.

holder must be determined for each unique sample preparation. Figure 11.12 presents the schematic plot of the relative intensity versus the volume deposited on a plate with a fixed dilution ratio. The signal intensity increases proportionally with the deposited volume until it reaches a maximum where there are more molecules than charges present in the APCI process. The relative intensity comes to a plateau before decreasing when a heavier competition for the charges occurs. This maximum quantity is reached at point M marked on the curve of Figure 11.12 just before the plateau. At this point, the LDTD process will produce the highest data quality it can achieve for the specific preparation method chosen for a particular biological matrix. Caution must be exercised during a method development, as slightly overloading the sample holder gives a similar signal intensity plot. If this situation arises, the reproducibility of the analysis will degrade rapidly.

11.4.2.2.2 Effects of Ionic Saturation on the Signal Peak Shape The peak shape obtained from the desorption of an extract gives a great deal of information and may indicate an overload of the sample holder. Figure 11.13 illustrates the different peak shapes obtained by thermal desorption by LDTD. The normal desorption profile presented at (a) is almost symmetric. The laser beam covers the entire part of the concentric well to transfer the heat as uniformly as possible. The thermal conductivity of the outer edge of the metallic foil causes a gradient of temperature from the center to the external wall. The center of the well reaches the desorption temperature earlier than the circumference of the well, making the peak asymmetric by adding a tail. Figure 11.13b presents a partially saturated desorption with a substantial shoulder and a 50% signal loss. Figure 11.13c shows a completely attenuated peak due to a heavy overloading characterized by the presence of a split peak.

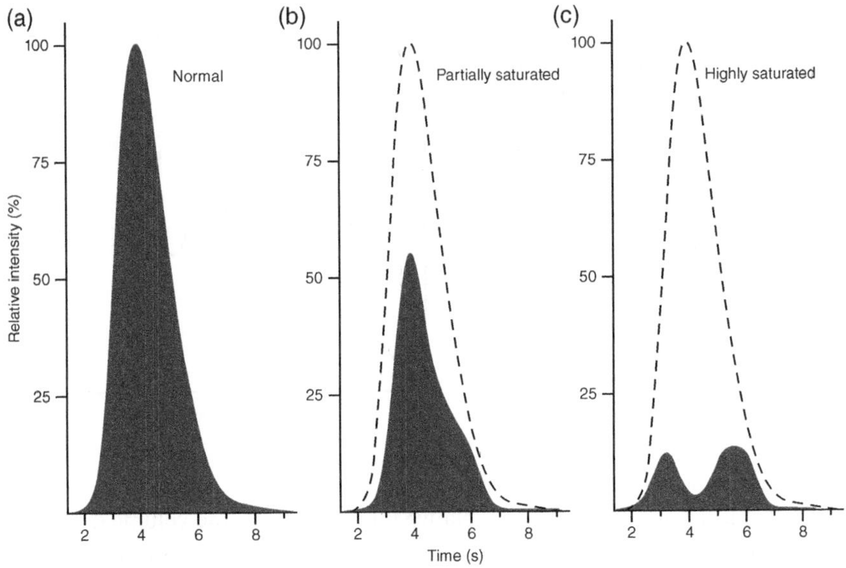

Figure 11.13 Peak shape characterization when ionic saturation occurs, with (a) being normal peak shape with a small tailing, (b) a shoulder peak caused by a partially saturated desorption, and (c) a split peak due to a heavy overloading of the well.

11.4.2.2.3 Effects of LDTD Parameters on Ionic Saturation The laser pattern (see Figure 11.14) determines the laser power needed in time to achieve the desorption. The speed of the rise in power is linked to the quantity and the nature of the sample deposited on the well, as it will generate a flow of neutral molecules toward the ionization region that have a maximum rate of charge available. Heating too quickly may produce a saturation of the ionization as it will release desorbed molecules at a higher rate than the charges generated. It is possible to heat the well very rapidly only when using simple solutions since the limited quantity of compounds will not be saturated by the limited number of charges available. In more complex biological matrix extracts, the typical laser pattern has a ramping time that ranges between three and six seconds with a plateau of two seconds. To maximize the throughput when using simpler extracts, the ramp is minimized at less than 0.1 seconds with a plateau of 0.3 seconds, but, in this case, the mass spectrometer scan rate capability may be a limiting factor to attain such a speed. Since, the laser pattern can be customized, the user can adjust it to the mass spectrometer in use to achieve the best data quality. To start developing a new method, a laser pattern with a three-second ramp that increases to a maximum of 65% with a two-second plateau at the top is the initial setting to begin with. Minor adjustments may be implemented depending on the compounds and the extraction type of the biological matrix. The LDTD instrument is calibrated to uniform the surface temperature for 96, 384, and 1536 well plate formats.

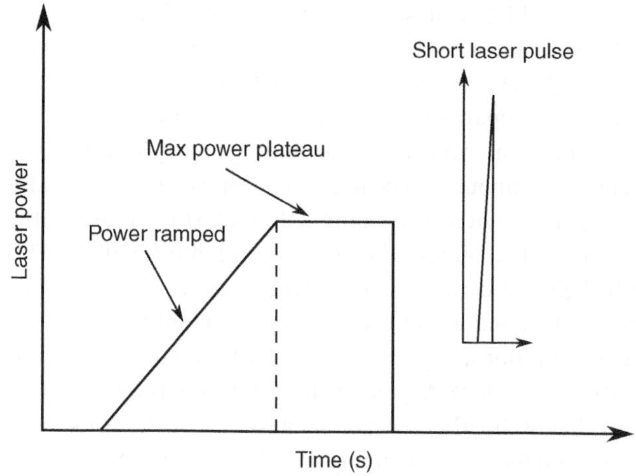

Figure 11.14 Typical laser pattern shape. The length of the power ramp and the plateau can be adjusted depending on the sample extraction. The smaller graph shows a short laser pulse used in the case where the deposited sample used a simple solution to extract the molecule of interest.

The gas flow parameter and the quantity of sample deposited on the well are also linked, as they change the concentration of vaporized neutral molecules heading to the ionization region. The main function of the flow is to remove the desorbed molecules from the heated surface and focus them into the transfer tube. The system is designed to operate at a 3 L/min flow rate as an optimal value. It matches most of the mass spectrometer atmospheric pressure inlet flow rates. The flow will remain laminar up to the maximum of 10 L/min. Each increase of the flow has an effect of dilution in the gas phase for the desorbed sample. This behavior compensates for an overloading of the sample holder and may be useful when the sample preparation has limitations on the volumes that can be used.

11.4.2.3 Consideration for Biological Matrices

Original or synthetic biological matrices are generally too concentrated to be applied to the sample holder directly. Minimal preparation is required to obtain optimal signal intensity. The crucial parameter is to determine the optimal dilution to reach the maximum loading of the sample holder in accordance with the laser pattern and carrier gas flow rate.

11.4.2.3.1 In Vitro Applications An ideal in vitro sample preparation would be to dilute with the appropriate solvent. Simpler samples, mainly constituted of buffered solutions and small quantities of proteins, require a simple dilution to achieve an adequate measurement. Nevertheless, the buffer type, its concentration, and the presence of sugar, which can strongly disturb the desorption and ionization process, are determinant. The latest developments [45, 46] have shown that a phosphate buffer (K_2HPO_4 and KH_2PO_4), which is an inorganic salt, improves the overall sample preparation. It seems to equilibrate the lattice structure of the deposit, form counterions with low binding energy and enhance the desorption of challenging compounds. The use of a phosphate buffer can also help with method development when using pure solutions since most of the constituents composing inorganic buffers do not vaporize during the LDTD process having very high E_a [47]. A certain improvement of the desorption can be seen when adding these buffers and are linked to the final quantity left on the sample holder. At high concentrations, large crystals will form, and the compounds of interest will be trapped on the surface. A typical value for the final concentration of the KH_2PO_4 solution is below 10 mM to avoid any adverse effects. These salts influence the desorption behavior by changing the binding properties and reducing the activation energy needed as well as the thermal conductivity of the residue [48].

Organic buffers (TRIS, HEPES, ammonium formate) behave differently and desorb; therefore, they are part of the ionization process. Experiments using these buffers are designed and validated for other detection techniques such as optical diffraction or electro-chemiluminescence are often transferred to mass spectrometry for a label-free analysis. However, these organic buffering solutions are optimized for optical or radioactive detection and may cause difficulties during the LDTD process. They are often used at high concentrations and large dilutions are necessary to avoid an ionic saturation during the APCI process. Different options are available to alleviate this problem: change the organic buffer for an inorganic one (when possible), reduce the quantity used, or perform an additional sample treatment step.

11.4.2.3.2 In Vivo Applications The sample preparation for in vivo sample is very similar to ones for LC-MS. It is usually simpler and requires small changes to the volume and the reagents. LDTD can be used with a wide variety of biological matrices like urine, whole blood, serum, plasma, saliva, nail and hair. The preparation method depends upon the biological matrices, the specificity of the mass spectrometer analysis, and ultimately, on the required limit of quantitation.

The preparation of a blood plasma sample is an excellent example. The first step is to precipitate the proteins with an organic solvent and centrifuge the solution to separate the organic phase from the precipitated material. The upper layer is deposited on the well surface and dried prior to the analysis. Acetonitrile is usually added to the sample [49] for the precipitation, with a dilution ratio of 1 : 5. A standard LC-MS method suggests a 1 : 2 or 1 : 3 ratio to limit the dilution of the sample. The electrospray may require pushing the preparation further to remove any co-extracts causing suppression in this ionization technique [50]. With dilution ratios equal or greater than 1 : 3, LDTD does not exhibit any ionic saturation in the gas phase and remains suitable for any compound that undergoes thermal desorption. Other organic solvents can be used for the precipitation but may require a ratio optimization of the dilution to maximize the sensitivity. For example, methanol needs a higher dilution ratio as more biological matrix (especially cholesterol) is extracted into the organic phase. Solid phase extraction (SPE) approach is not required and is used only in the case where there is an interference in the MS/MS transition from the biological matrix extracts. Without a chromatographic separation, the selectivity relies mainly on the mass spectrometer and on the thermal properties to a lesser extent. Nonvolatile material remains on the sample holder, making de facto a separation with desorbed molecules.

11.4.2.3.3 Focused Targeted Compound Applications with targeted compounds as probes can benefit from a focused sample preparation to optimize the yield of the assay, using the generic approach as a foundation. The addition of an organic solvent in an appropriate ratio is the starting point of these preparations. Once mixed, the organic layer is deposited onto the sample holder and dried prior to the analysis. This technique refers to a liquid–liquid extraction (LLE) method and is analogue to the protein precipitation technique discussed in the previous section. Cleaner extracts can be obtained by applying simple modifications to this approach without increasing its complexity or cost. [51] If a user is facing an undesirable background signal for a monitored transition or if they cannot attain the wanted sensitivity, they have the possibility of improving the results by tuning the extraction.

To adapt the sample preparation to a particular molecule of interest, the first approach to use relies on determining its partition coefficient [52] in the solution. Compounds having ionizable groups exist in solution as a mixture of different ionic forms. This behavior may be predicted using the distribution coefficient (log D), the appropriate descriptor for ionizable compounds, since it is a measure of the pH-dependent differential solubility of all species in the octanol/water system. The choice of solvent and pH adjustment should be based on the value of log D. Figure 11.15 shows examples of log D curves as a function of pH for two molecules. The extraction will be maximized at the value where the log D is at its highest point. The solvent polarity is used as a differentiator. Less polar solvents dissolve compounds with high positive log D values and generate cleaner extracts. On the other hand, highly polar compounds require polar solvents for negative log D (low lipophilicity) values to extract a large part of the chemical space. Acetonitrile used in combination with a salt at high concentration generates one of the most polar partitions. The salt-assisted liquid–liquid

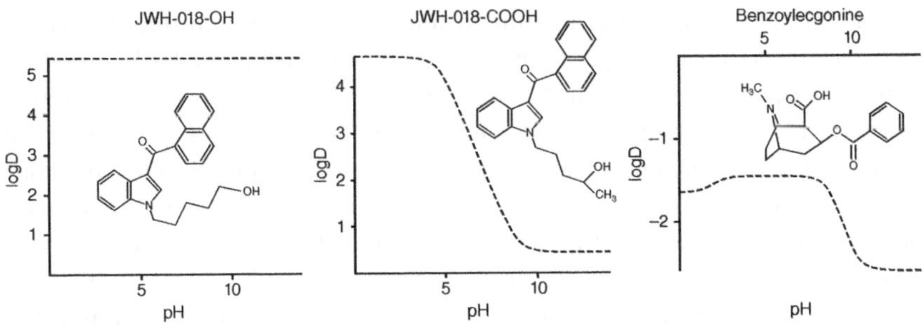

Figure 11.15 LogD plot for three compounds with distinct behavior.

extraction (SALLE), also named QuEChERS (quick, easy, cheap, effective, rugged, and safe), is the simplest technique from a practical point of view. For example, the addition of a salt solution in an acetonitrile protein precipitation in blood plasma creates a phase separation between the aqueous and the organic solvent. The upper layer is cleaner and may lower the possible interference in the MS signal [42].

The presence of sodium causes no issue to the LDTD technique, as there is no adduct formation in the desorption process. When using the SALLE technique, if the organic phase is still not clean enough, changing the solvent may eliminate the interferences. Using hexane, ethyl acetate and methyl tert-butyl ether allows to use the same preparation steps for the protein precipitate and the LLE. For many, liquid–liquid extractions are a synonym of complexity with drying and reconstitution steps, but it is not the case with the LDTD. Other complex biological matrices such as tissues are prepared using a similar approach. Homogenates mixed with acetonitrile at an optimum solid-solvent ratio would be the initial approach followed, and, if necessary, adding SALLE or liquid–liquid extraction with a pH adjustment.

11.5 Applications

The LDTD process was designed with speed and ruggedness in mind. It can be used for HTS purposes without being limited to these types of applications. It allows the automation and a complete integrated workflow from the sample preparation to the data analysis. In the following sections, applications which can be regarded as typical for the domain are presented. The first published application, which is a cytochrome inhibition assay, is presented, followed by permeability, protein binding, and pharmacokinetics. These assays include parameters that may be easily applied to other HTS applications for targeted or nontargeted analysis types.

11.5.1 CYP Inhibition Analysis

It is well known that most drugs are metabolized by cytochrome P450 enzymes [53]. In 2007, Wu et al. [54] were the first to report a CYP inhibition analysis with the LDTD technology. A method was developed to analyze different cytochrome families: 1A2, 2C9, 2D6, and 3A4. Testosterone, dextromethorphan, diclofenac, phenacetin, and their corresponding metabolites were used as probes for the CYP. The CYPs were incubated in a solution composed of a phosphate buffer, probes stock, human liver microsomes (HLM) protein, and NADPH. With the LDTD-MS, the specificity of detection relies strictly on the primary ion/daughter ion pairs (MS/MS). The CYP solution is complex and can be subject to isobaric interferences since there is no separation (LC) prior to the entry in the mass spectrometer. The incubation solution was quenched with a different solution (water/acetonitrile) containing the internal standards (stable label isotope) for each metabolite. The solution was directly used with LC-MS to cross validate the results. For the LDTD, the solution was further diluted in acetonitrile and 2 μL was deposited on the sample holder.

A good linearity and sensitivity were reported for the metabolites monitored. The signal intensity and the S/N were higher compared to LC-MS (electrospray ionization [ESI] or APCI ionization) for three of the four compounds. The S/N ratio went through a two-fold improvement for acetaminophen and 4-hydroxydiclofenac and by 90-fold for the 6β-hydroxy-testosterone. This improvement in the S/N ratio between the two techniques is the result of a low background in the transition monitored due to the absence of solvents during the thermal desorption. Competitive inhibition assays with several drugs were analyzed by both LC-MS and LDTD-MS and the half maximal inhibitory concentration (IC_{50}) values were compared. Figure 11.16 shows the comparative plots for 87 assays and a correlation factor of 0.94 is reported. The LDTD-MS gives similar results as the LC-MS for a throughput that has a cycle time of 10 seconds per sample.

In 2016, Haarhoff et al. [55] reported an application using acoustic deposition and the multiplexing of the assays in one well to further enhance the throughput in CYP inhibition assays. The acoustic sample deposition system deposits nanoliter volumes rapidly and reproducibly. In this work, the 384-sample holder format was used. Figure 11.17 shows the linearity response for the acoustic deposition of volumes ranging from 5 to 200 nL for a constant concentration of Dextrorphan. The metabolites were monitored for the CYP2D6 – dextrorphan, the CYP2C9 – 4-OH-diclofenac, and the CYP3A4 – 1-OH-midazolam. The method used a 50 nL volume for all the assays. The response has shown linearity for all three metabolites at this volume for limits of quantification (LOQ) of 3.6, 15.6, and 62.5 nM for CYP2D6, CYP2C9, and CYP3A4, respectively.

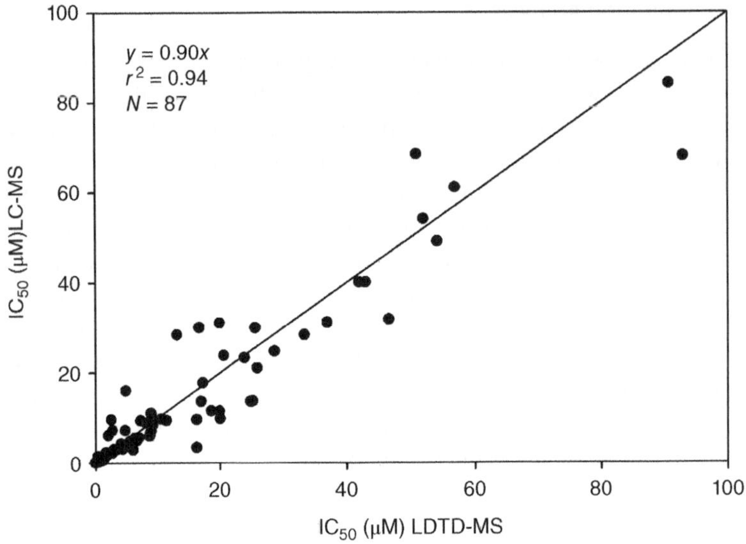

Figure 11.16 Correlation plot between the IC_{50} values obtained form LDTD-APCI-MS and LC-MS methods (N = 87). *Source:* Wu et al. [54]/With permission of American Chemical Society.

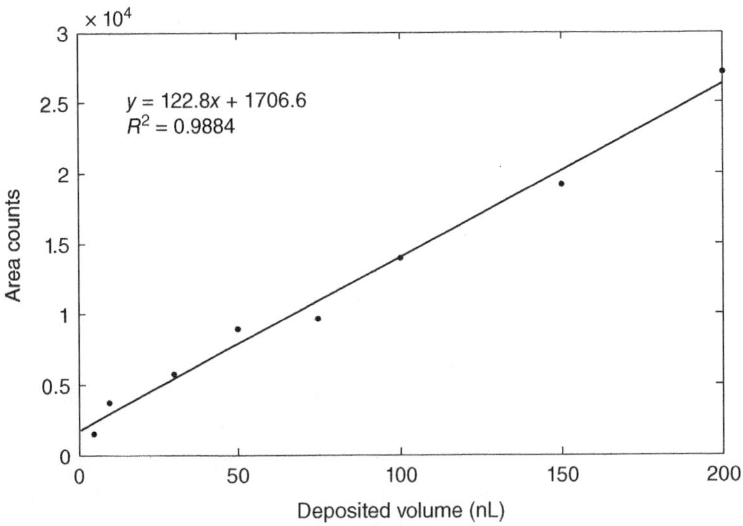

Figure 11.17 Area counts for different volumes deposited in the well for the same concentration of dextrorphan (7.9 ng/mL).

Normal laser pattern settings have a slow ramp to attain the maximum laser power to allow time for an efficient heat transfer to the residual deposit. The low volume used in these assays allows the use of a short laser pulse (1.2 seconds) to rapidly desorb the compounds. The cycle time for the three CYPs was 2.14 seconds per sample. The runtime for the entire sample holder plate was less than 14 minutes. Hence, by multiplexing [55] the three postreaction assays in the same well, the throughput by analyte does a threefold decrease to 0.7 seconds per sample.

Based on the same strategy and with the new generation of LDTD ion source, we reported the possibility to acquire 384 samples of OH-midazolam at a cycle time of 0.91 seconds per well (Figure 11.18a,b). The laser pattern used is in the form of a square pulse of a length of 0.5 seconds at maximum power. Thus, the runtime for the entire sample holder plate is reduced to less than six minutes. A curve of dextrorphan was acquired at the sub-second throughput and it was validated that the linearity and the sensitivity were not impacted by the short pulse method. Figure 11.19 shows a correlation coefficient over 0.99 for the calibration curve.

LDTD-MS shows similar IC50 values (Figure 11.20) compared to other technologies such as LC-MS (classic LC or Rapidfire™). As part of the method cross-validation, authors explicitly cite that one of the main advantages is the inexistence of carryover between the samples [54, 55].

11.5.2 Permeability

Adsorption, Distribution, Metabolism, and Excretion (ADME) studies include the investigation of the permeability (P_e) through membranes for candidate compounds. The permeability model for the gastrointestinal absorption uses the Caco-2 cells. To measure P_e, the use of a mass spectrometer as a detector compared to other methods of detection (fluorescence or radiolabeling) has enhanced the sensitivity and the selectivity of the analysis [56]. Nevertheless, the throughput achieved with the LC-MS/MS technique can be measured in minutes per sample [56]. This section presents a comparative study between LC-MS/MS and LDTD-MS/MS that demonstrates the equivalency of both methods, but with a critical reduction of the analytical time when the latter is used as the sample introduction technology [57].

In this ADME study, the transport of compounds through the intestines was measured using Caco-2/TC-7 cell monolayers with an apical compartment (HBSS; pH 6.5; 0.5% BSA) and a basal compartment (HBSS; pH 7.4; 5% BSA). The final concentration of the test compound added to the apical compartment is 20 μM and was incubated for two hours [57] as shown in Figure 11.21. The sample treatment used is designed for UPLC-MS. The treatment was used without making any functional adjustments to the sample preparation and implemented directly

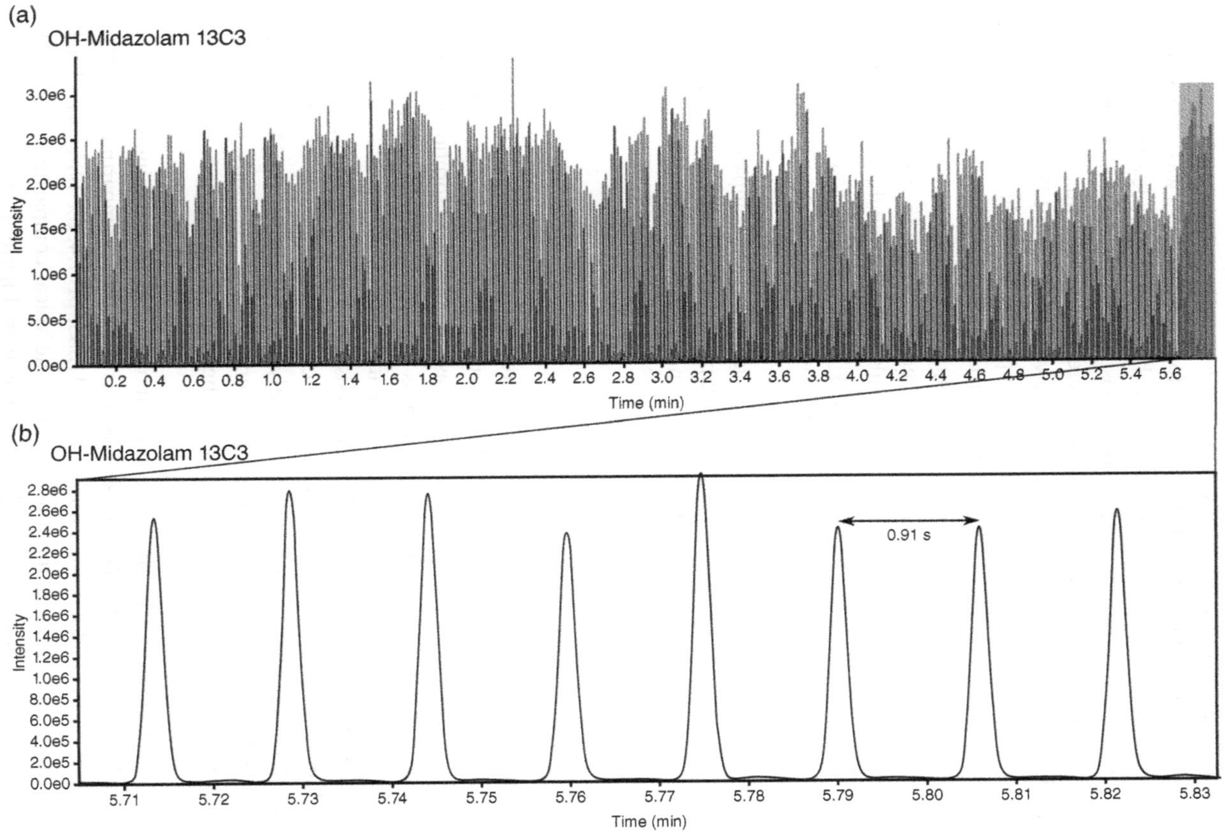

Figure 11.18 (a) 384 samples of the same concentration of OH-Midazolam acquired with a throughput of 0.91 seconds in a single file. (b) Zoom on 8 sample peaks showing the 0.91 second separation between peaks.

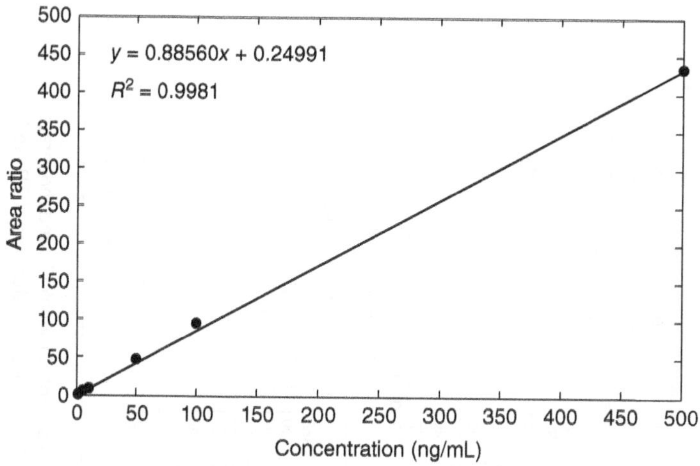

Figure 11.19 Calibration curve of dextrorphan acquired with a throughput of 0.91 seconds per sample.

with LDTD-MS/MS. For the LDTD parameters, the laser pattern used had a three-second ramp going up to 35% of the maximum laser power with a plateau of one second and the gas flow was set to 3 L/min. A volume of 2 μL of the apical and basal solution was deposited in the well.

The bottleneck when using a mass spectrometer as a detector is attributable to the optimization of each selected reaction monitoring (SRM) or multiple reaction monitoring (MRM) transitions (sample prep and mass spectrometer [MS] tuning). In these studies, a large number of components are analyzed and the tuning of each of them is time-consuming even when an automation software is used. The compound optimization for the UPLC method includes a chromatographic separation, which raises the time per run to four minutes for a compound. As the LDTD does not require any chromatographic separation, the tuning time is reduced to 25 seconds per run, which includes laser pattern time and the time for the mass spectrometer to transition between samples in a batch. As a result, this gave a success rate of 85% compared to 97% in LC-MS. Figure 11.22 shows the correlation obtained for the P_e measurement between LDTD-MS/MS and LC-MS/MS for 107 compounds.

Moreover, the dilution of the buffer solution (HBSS) was not optimized during this assay. Since HBSS includes glucose at a concentration of 1 mg/mL, it is enough to create ionic saturation in the gas phase and irregularities during the desorption. A suggested preparation would be to dilute the buffer solution 10 times with a water-acetonitrile 50 : 50 v/v mix. With this dilution, the centrifugation step is not useful or required for the LDTD process to be efficient. Also, a large range of molecules would be covered if an analysis with both polarities is used – in large

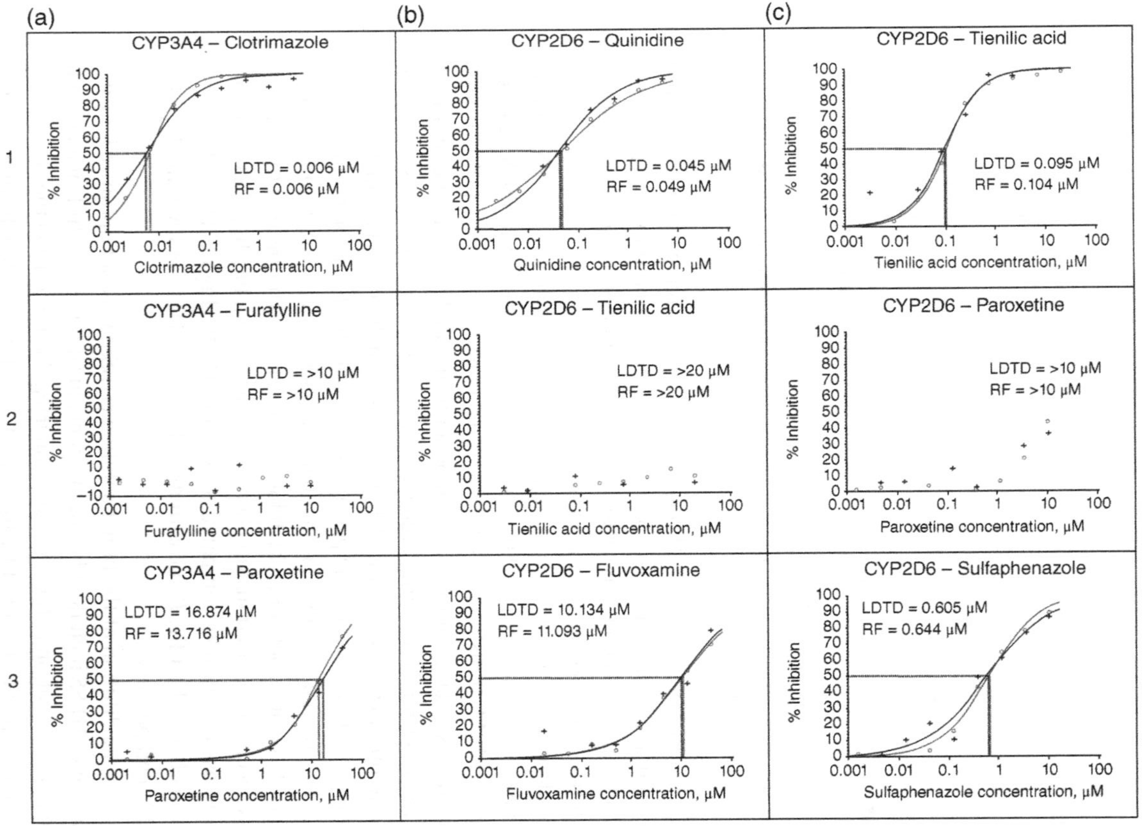

Figure 11.20 Comparison of IC_{50} curves obtained by LDTD and RapidFire, showing potent (1), weak (2), and moderate (3) inhibitors for each cytochrome P450 enzyme: CYP3A4 (a), CYP2D6 (b), and CYP2C9 (c). The IC_{50} plots for LDTD (solid crosses) and RapidFire (open circles) are shown. Percent CYP inhibition and inhibitor concentrations are shown on the *y* and *x* axes, respectively. *Source:* Haarhoff et al. [55]/With permission of SAGE Publications.

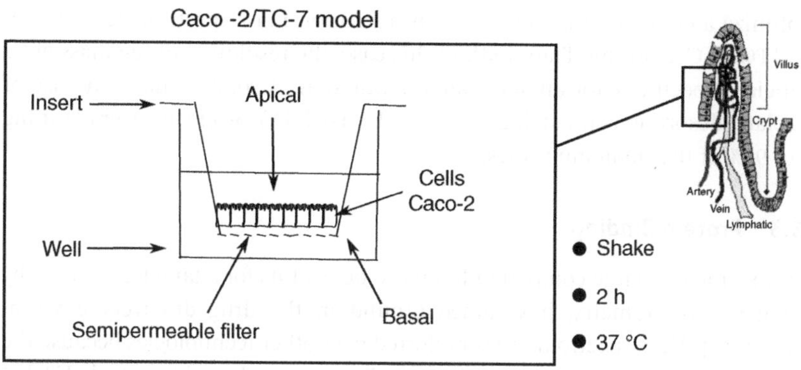

Figure 11.21 Caco-2 model.

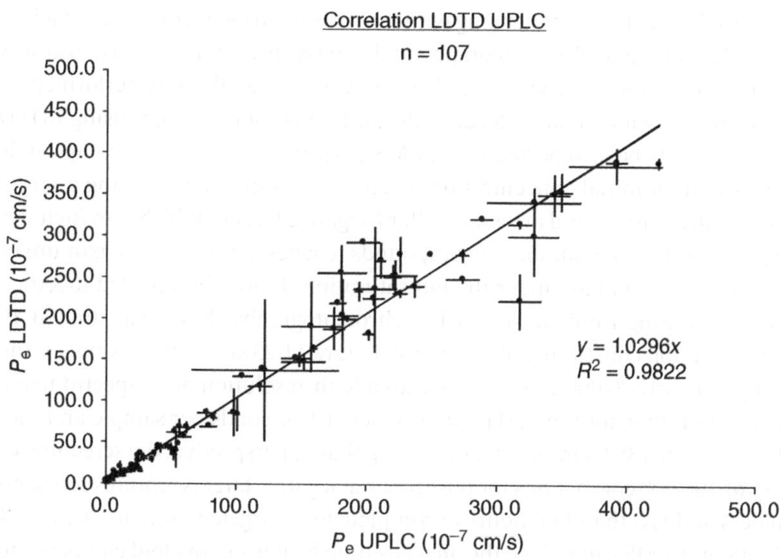

Figure 11.22 Cross validation of permeability analysis of Caco-2/TCY between UPLC and LDTD systems.

databanks of compounds, 5–7% of molecules are only sensitive enough to be quantified in negative mode.

The high-throughput in an application is achieved by cutting any time-consuming steps along the way. The permeability comparative studies have shown that the total time, which includes the compound optimization, the incubation of Caco-2, the sample preparation, and the sample analysis, was divided by three by using the LDTD-MS/MS system [57]. One way to further reduce the time needed to optimize a compound is to use a high-resolution mass spectrometer (time of flight [TOF], Cyclotronic Trap) [58]. In this case, the resolution of the mass gives a sufficient specificity for an adequate quantitation. Also, nowadays, the use of DEC plate (desorption enhancing coating of EDTA-BSA) lowers the overall tuning rate of 95% of the small molecules.

11.5.3 Protein Binding

The screening of a large compound library to detect hits for a targeted protein, by using mass spectrometry, has gained ground in the drug discovery environment [59–62]. Mass spectrometry is preferred over other technologies because the detection is made without any labels or tags. This approach enables a truly HTS of a library of thousands of compounds in hours. The affinity selection mass spectrometry (AS-MS) technique enables the detection of various receptors, including the ones that are more challenging with the traditional biochemical approaches.

The tandem of liquid chromatography and a mass spectrometer as a detector is a technique traditionally used by end-users when an analysis is performed in mass spectrometry for small molecules. Recently, a proof of concept using LDTD instead of LC has been reported for AS-MS [44]. In this study, the compounds were pooled at an initial concentration of 1 μM in a solution containing a target protein (positive target) and an assay buffer (negative target), which also includes a phosphate buffer. The number of compounds reaches a functional maximum of 200 compounds to remain under the ion saturation limits when pooled together. The elution coming from size-exclusion chromatography (SEC) was diluted 10 times in a solution constituted of methanol. Figure 11.23 shows the results for the screening of nearly 70,000 compounds with a high-resolution mass spectrometer. The acquisition time for the LDTD-MS was set at 12 seconds per sample and 2 μL was deposited on a 384-sample holder. Taking that in perspective, the screening of a library of 100,000 compounds for positive and negative targets requires 384 sample holders and less than four hours to complete for a targeted protein. The use of LDTD-MS in AS-MS can reduce the analysis time by four- to fivefold compared to other fast MS techniques [44].

11.5.4 Pharmacokinetic

The development of new drugs requires in vivo studies in animals and humans. Samples from biological matrices are typically analyzed by liquid chromatography tandem mass spectrometry (LC-MS/MS) because of its selectivity, sensitivity, and high precision. Samples that require a high-throughput analysis, because of the sheer number sent to laboratories are typically in plasma or serum rather than tissue. The preparation steps must be amenable to automation with an equivalent yield to be consistent throughout the entire workflow.

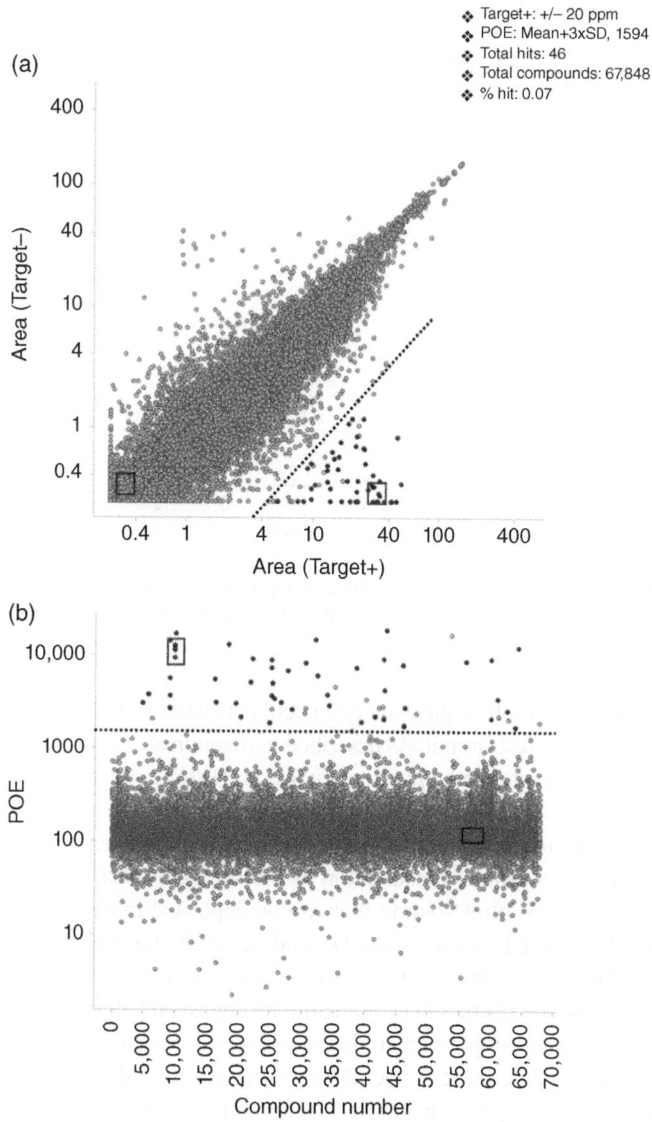

Figure 11.23 LDTD-AMS SWATH method data for a full 384 well plate. (a) shows a plot of "filtered" mass spectral area of all library compounds with a target protein (*x*-axis) and without a target protein (*y*-axis). (b) shows POEs of all library compounds. Dotted line represents compound filtering criteria; compounds with a target protein (Target+): ±20 ppm, POE > Mean filtered mass spectral area for all compounds +3X standard deviation of filtered mass spectral area for all compounds. Dark gray and light gray dots represent compounds that passed and failed the compound identification filter, respectively. The boxes around dark gray dots indicate positive controls and the boxes around light gray dots indicate negative controls. LDTD-AMS SWATH method identified 46 unique compounds out of 67,848 total compounds, i.e. 0.07% positive hits.
Source: Sahasrabuddhe et al. [44]/With permission of SAGE Publications.

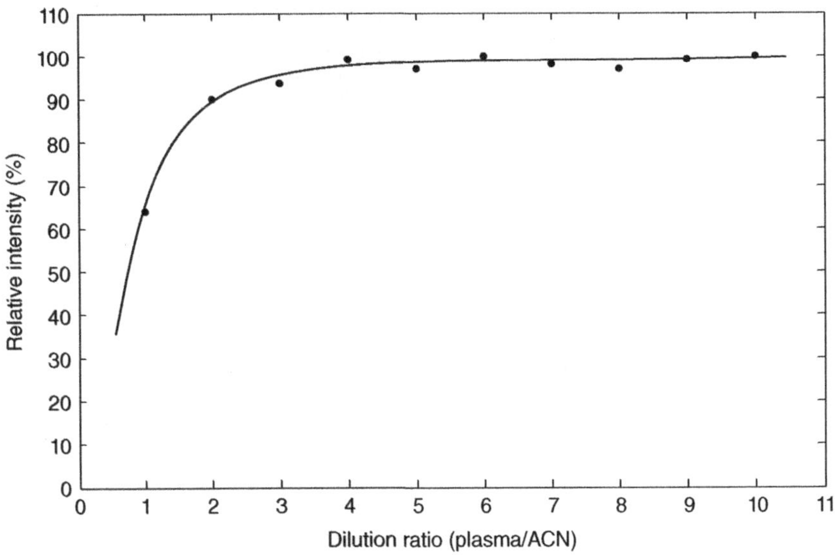

Figure 11.24 Dilution ratio optimization for a protein precipitate in plasma with acetonitrile. The maximal intensity is reached when the dilution ratio is 1:3. *Source:* Picard et al. [13].

Generally, the first step with a biological sample such as plasma is to clean and separate the compound of interest from it. As described in detail in Section 11.4.2.3, one can proceed by a simple protein precipitation. Figure 11.24 shows the optimization of the dilution ratio (sample/solvent) [13] for plasma. The data shows that a minimum of 1/3 is necessary to get an optimal analysis process. The use of acetonitrile is preferred to methanol for a protein precipitation as the sample is cleaner when using a smaller ratio. The maximum volume that can be deposited in the sample holder wells must be determined for each ratio. Depending on the precision of the liquid handler used, the ratio of the dilution may be adapted to transfer larger volumes: for example, a ratio of 1/10 could be used and the deposited volume on the plate would be 6 μL. A typical sample preparation starts with desorbing the compounds and ionizing them to determine the best MS/MS transitions and polarity for them. As there is no chromatographic separation, the chosen transition must be specific. This is the first criteria to work on before optimizing the intensity of the signal.

The analysis time of eight seconds per sample using a 96-well plate format is considered long for LDTD, but it may be reduced to attain a higher throughput by using 384- and 1536-well plates. The maximum volume deposited in these plates is smaller, but the APCI capacity remains the same. The possibility of depositing a smaller volume at a higher concentration gives an equivalent sensitivity. This is demonstrated by doing a sample protein precipitation in human plasma focused toward a 1536-well plate. The overall signal remains at the same level as the 384-well format. As the 1536-well plate allows a maximum volume of 500 nL per well, we assume that increasing the sample concentration will overcome the lower plate spotting volume. This assumption is based on the maximum loading capacity of the LDTD technique, which is characterized by the number of charges produced in the APCI. The usual dilution used is 6× for the 384 when spotting 2 μL. It is compared to 3× dilution and spotting 250 nL on the 1536. Good accuracy (90.4–110.3%), precision (3.7–9.7%), and excellent linearity ($R^2 > 0.9995$) are obtained for both methods and they exhibit the same LOQ of 1 ng/mL for the analysis of Quetiapine as shown in Figure 11.25.

The sample volume required for LDTD analysis is small and allows the implementation of micro sampling strategies. The challenge focuses more on the development of the automation development of the preparation step as the LDTD-MS/MS remains the same. As an example, a pharmacokinetic (PK) study analysis of dextrorphan was performed using 1 μL of blood from a finger prick. The protein precipitate had a 1 : 5 dilution ratio in acetonitrile with an internal standard, and 2 μL were deposited in the well. The studies show a R^2 of 0.9945 for

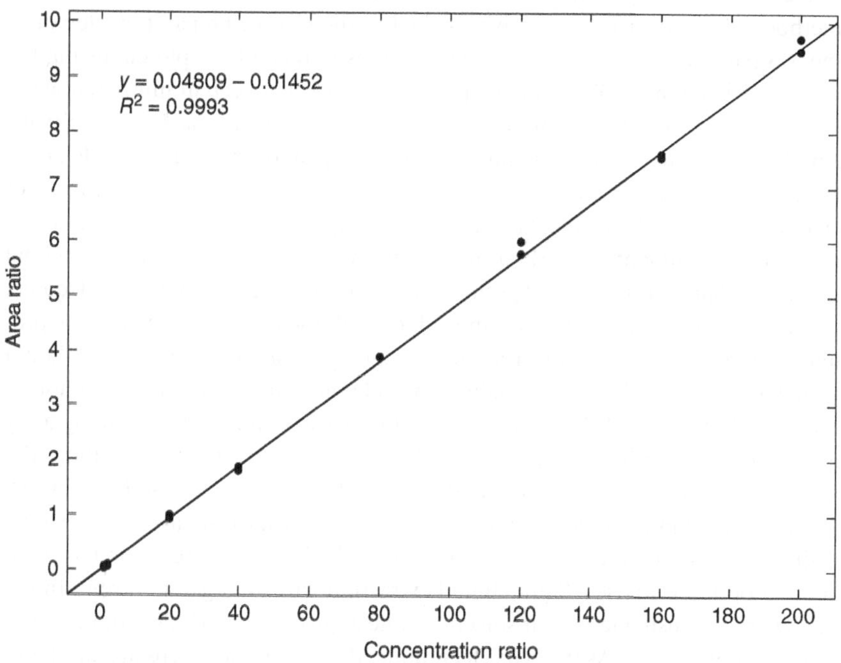

Figure 11.25 Results for the Quetiapine with a 250 nL volume deposited on the sample holder for a range of 1–200 ng/mL.

a concentration range of 0.78–50 ng/mL. The PK study results were supported by literature values [63]. Sampling in the μL range allows for a wide range of possibilities: performing PK studies from a single mouse/rat, running early drug discovery tests with a learner approach to synthesize compounds and reagents as well as providing fast micro sampling from fingertips.

Heudi et al. [64] and Lanshoeft et al. [42] have reported comparative studies between the LC-MS/MS and the LDTD-MS/MS in the field of pharmacokinetics in a regulated environment. To alleviate the burden of analysis coming from the large number of samples in clinical studies, methods were developed for inhouse syntheses of molecules [64] and Buparlisib (BKM120) [42]. Figure 11.26 shows the reported data for 27 subjects that were administrated a single dose of 50 mg of BKM120. One can see the correlation in between LC-MS/MS and LDTD-MS/MS for the release of the drug in the subjects. The results show an equivalent result in terms of sensibility, recovery, precision, and accuracy with LC-MS/MS. The use of the LDTD-MS/MS instead of LC-MS/MS can increase the throughput by 27-fold.

11.5.5 Preparation Tips

A lack of selectivity and sensitivity can cause difficulties in some analytical methods. These difficulties can be resolved in the sample preparation step or in the MS parameters. Here are some possibilities that can be explored during the method development. To avoid any problems, one must keep in mind that water loss transitions should not be used with LDTD due to the lack of specificity. Another step is to evaluate the intensity of the signal of the blank sample at the transitions of interest. The selectivity may be enhanced using mass spectrometer characteristics such as increasing the quadrupole resolution, using a high-resolution instrument, or working at higher/lower collision energies than optimal to differentiate it from the background signal. If the selectivity is still inadequate, the sample preparation can be optimized to reach the required level. A basic protein precipitation with acetonitrile can be upgraded to a salt assisted liquid–liquid extraction by the addition of a sodium chloride saturated solution. Production of sodium adducts in LDTD is unlikely, by contrast with a liquid chromatography ion source. The NaCl will not desorb and will not take part in the ionization. For a typical preparation with a sample/MeCN 1/6 dilution ratio, the addition of two parts of a sodium chloride saturated water solution will cause layers to separate, giving a cleaner extract. If the signal of the blank solution remains too high, another option is to use a liquid–liquid extraction with an organic solvent. In the LDTD process, using less polar solvents while adjusting the pH gives the cleanest extraction procedure. As it is a dry technique, the preparation steps are similar to a simple protein precipitation, which means the upper layer is directly transferred to the analyzing plate. Pure or mixed hexane, ethyl acetate and MTBE are the

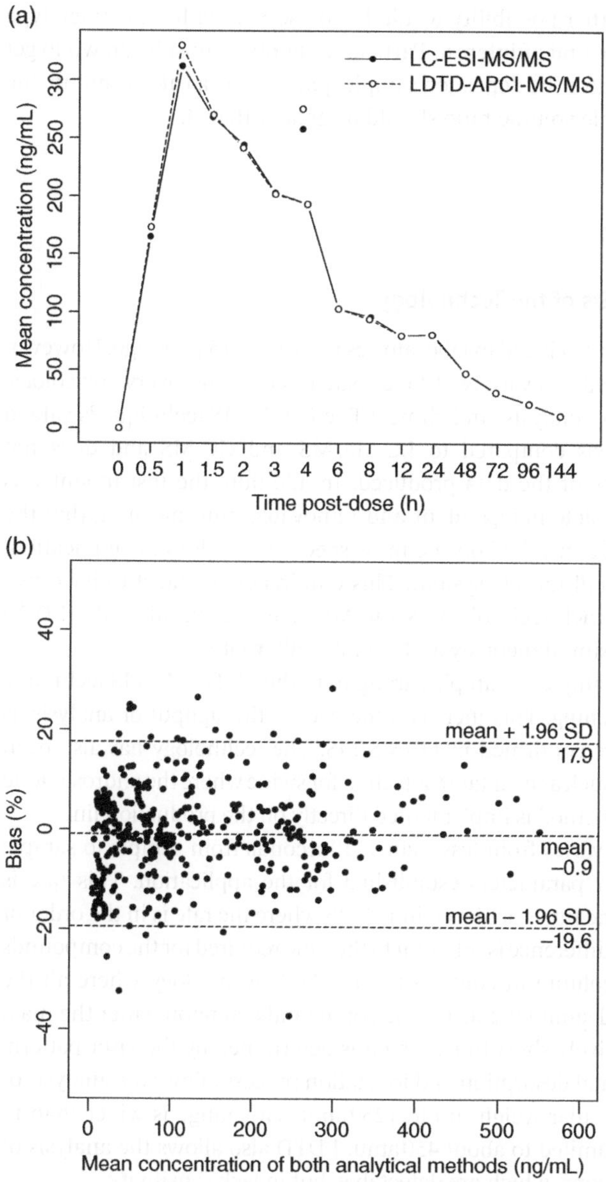

Figure 11.26 (a) Mean plasma concentration–time profiles of 27 subjects administrated a single dose of 50 mg BKM120 once daily in clinical study 1 measured either with LC-ESI-MS/MS or LDTD-APCI-MS/MS (n = 405) and (b) Bland–Altman plot with 95% limits of agreement (*dashed lines*) for assessment of agreement between both analytical assays (n = 368). *Source:* Lanshoeft et al. [42]/With permission of Springer Nature.

preferred choices. Another possibility would be to use SPE with a cartridge type optimized for the compound of interest. Particular attention must be drawn to get the best elution volume for the deposited sample quantity. As a rule of thumb, the elution/deposited sample volume ratio should be greater than 4.

11.6 Conclusion

11.6.1 Use and Merits of the Technology

LDTD technology was introduced in laboratories more than 15 years ago. However, its use has rapidly spread to a variety of fields, such as drug discovery, toxicology, police laboratories, food analysis, and clinical The LDTD-MS technique is valued for its speed of analysis compared to LC-MS/MS and GC-MS and does not compromise the quality of the data produced. In addition, the instrument was developed to be a complete independent and removable unit, meaning that the LDTD ion source can be installed on the mass spectrometer for one application, then removed to reinstall the LC system. This eliminates the need to have two dedicated systems for each technology. Some MS system even allow the LDTD and LC to be installed simultaneously and operated alternately.

Due to its very high sample-to-sample throughput, the LDTD-MS/MS technique is widely used for screening. This increases the overall throughput of analysis as only positive results are confirmed in LC-MS/MS. The technology has also been adopted in other areas, such as boar taint detection for swine where the androstenone and skatole screening method is implemented directly on the production line.

The sampling speed varies from less than 1–10 seconds from sample to sample depending on the laser parameters established for the application. This rate is clearly different from the times achieved in LC-MS where the rate is in the order of a minute or more. This difference is inherent to the time required for the compounds to elute from the LC column in contrast to the LDTD technology where all the molecules are desorbed simultaneously. The compounds therefore enter the mass spectrometer in a relatively short time, which is determined by the laser pattern. Furthermore, the thermal desorption and ionization process allows the analysis of molecules with a molecular weight up to 1250 amu. This range is wider than in GC-MS where we are limited to about 450 amu. LDTD also allows the analysis of fatty acids or hydrocarbons, which are detectable, but in lack sensitivity.

Reproducibility and sensitivity are comparable to the performances obtained in LC-MS. The stability of the ionization at the mass spectrometer inlet is critical to performance and is dependent on certain source parameters such as the corona needle voltage or the discharge current and the curtain or sweep gases. The stability is also related to the composition of the gas in the source housing, which is composed of air with traces of water with a concentration between 10 and 2000 ppm For a negative ionization to be successful, air is required to achieve an efficient deprotonation process, and, in the case of a positive ionization, traces of water is the only source of protons when the samples are vaporized from a dry state. These parameters can vary from instrument to instrument, so it is important to optimize them.

The LDTD technology becomes interesting when there is a need for speed and when the targeted and non-targeted compounds are small molecules. The instrument itself does not require heavy and expensive maintenance. Only the corona needle and the transfer tube need to be cleaned at regular intervals. In addition, the absence of solvents and the small number of molecules desorbed for each sample minimizes the overall cleaning and maintenance of the mass spectrometer dedicated to the LDTD source.

11.6.2 Limitations

Although the technology has a marked advantage in terms of sampling rate, it is not universal and has certain limitations. First, the range of molecular weight (up to 1250 amu) is advantageous compared to GC-MS, but limited compared to what can be done in LC-MS. This limitation is inherent to the desorption process, i.e. in LDTD we must vaporize the molecules in the carrier gas and the more massive molecules tend to dissociate under the effect of heat before being desorbed. Molecules such as proteins and peptides generally dissociate before getting into the gas phase making the intact structure difficult to detect.

The amount of material, in terms of concentration, deposited on the analytical plate is limited to avoid ion saturation in the APCI. Samples are usually diluted, and concentration during the preparation step is rarely possible. Sample processing should be oriented to increase the ratio of molecules of interest to the matrix. The ion saturation effect results in a limit of quantitation generally in the ng/mL range and rarely in the pg/mL range.

The lack of chromatographic separation is a problem when the panel of compounds includes isobars or isomers. Although not all isomeric pairs are separable in a chromatographic column, most are. Finding an MS/MS transition specific to each isomer is one avenue. Another avenue is to couple the LDTD source with a differential ion mobility instrument that allows some filtration of the molecules entering the mass spectrometer inlet. It is obvious that even with these solutions some molecules remain problematic.

Finally, for more polar molecules the desorption efficiency is limited as they are not volatile or require temperatures outside of the working range. To increase the desorption efficiency, additives (phosphate buffer, BSA, etc.) must be added during the sample preparation or deposited as a coating solution in the well beforehand, but it does not work in every situation.

11.6.3 Perspectives

The LDTD ion source is an ultrafast technique that desorbs samples in solid phase and creates ions in a manner of seconds. It is designed to work efficiently in a high-throughput environment using robust technologies and standardized analyzing plates running hundreds of thousands of samples without requiring maintenance. The thermal desorption and the APCI ionization makes the process sensitive and highly reproducible. The common trend of this chapter is the importance of the quantity deposited on the sample holder and the composition of the residue to achieve a successful LDTD-MS/MS analysis. The high resistance to ionic saturation originating from the liquid-free APCI allows the simultaneous analysis of compounds with a sensitivity in the low nM range. Most of the sample treatments feature simple dilutions with the appropriate solvents to get an adequate dilution and buffer concentration when used. The molecular coverage is maximized by the addition of enhancers or with the use of DEC plates, which are acquired pre-coated. The plate formats available are 96, 384, and 1536, which match the method requirements or the laboratory liquid handler capabilities.

The applications have shown a brief overview of throughput capacity of the LDTD ion source in the drug discovery environment. The throughput reaches the maximum speed of acquisition of the mass spectrometer with a sample-to-sample analyzing time of less than a second. To attain such a throughput in targeted screening, only a few MS/MS transitions are monitored to get enough points per peak. The minimal analysis time of the LDTD system of 0.67 seconds is reached using the 1536 model [65]. Untargeted assays should be done with the use of high-resolution instruments with lower scan speeds. Even though they are slower, they allow the monitoring of an unlimited number of compounds simultaneously. Screening libraries of hundreds of thousands of compounds with the pooled approach is accomplished within a few hours. The LDTD process can be applied to most of HTS applications for small-molecule analysis with a well-oriented sample treatment to respect its functional characteristics.

LDTD technology is now recognized and accepted in a wide range of fields. The rapidly expanding use of the technology is elevating the understanding of the technology and enabling the development of new applications in new areas. LDTD technology has undergone instrumental developments over the years that are reflected in increased speed of analysis. As previously mentioned, the recent advent of coating solutions or the addition of additives (phosphate, EDTA, or BSA) has increased the molecular coverage. It is undeniable that the growing number of application fields will bring new ideas and solutions to the field of LDTD knowledge.

References

1 Beynon, J.H., Baitinger, W.E., Amy, J.W., and Komatsu, T. (1969). The study of meta-stable positive ions with a double-focussing mass spectrometer: Part I. Description of the instrument used and its performance. *Int. J. Mass Spectrom.* 3: 47–54. https://doi.org/10.1016/0020-7381(69)80058-6.

2 Beynon, J.H., Baitinger, W.E., and Amy, J.W. (1969). The study of meta-stable positive ions with a double-focussing mass spectrometer: Part II. Methods used to study "meta-stable peaks" and the sensitivity achieved with a toluene sample. *Int. J. Mass Spectrom.* 3: 55–62. https://doi.org/10.1016/0020-7381(69)80059-8.

3 Beckey, H.D. (1969). Field desorption mass spectrometry: a technique for the study of thermally unstable substances of low volatility. *Int. J. Mass Spectrom.* 2: 500–502. https://doi.org/10.1016/0020-7381(69)80047-1.

4 Daves, G.D. (1979). Mass-spectrometry of involatile and thermally unstable molecules. *Acc. Chem. Res.* 12: 359–365.

5 Beuhler, R.J., Flanigan, E., Greene, L.J., and Friedman, L. (1972). Volatility enhancement of thyrotropin releasing hormone for mass-spectrometric studies. *Biochem. Biophys. Res. Commun.* 46: 1082–1088.

6 Soltmann, B., Sweely, C.C., and Holland, J.F. (1977). Electron impact ionization mass spectrometry using field desorption activated emitters as solid sample probes. *Anal. Chem.* 49: 1164–1166.

7 Anderson, W.R., Frick, W., Daves, G.D. et al. (1977). Mass-spectra of underivatized peptide amides related to substance-P. *Biochem. Biophys. Res. Commun.* 78: 372–376.

8 Hunt, D.F., Shabanowitz, J., and Giordani, A.B. (1980). Collision activated decompositions in mixture analysis with a triple quadrupole mass spectrometer. *Anal. Chem.* 52: 386–390. https://doi.org/10.1021/ac50053a004.

9 Picard, P., Carrier, V., Lessard, D., and Paquin, R. (2005). Atmospheric pressure laser diode thermal desorption mass spectrometry: a new high performance ionization source. In: *53rd ASMS Conference on Mass Spectrometry and Allied Topics*, (ed. ASMS) (6 June 2005), p. 52. San Antonio, TX: ASMS.

10 King, R., Bonfiglio, R., Fernandez-Metzler, C. et al. (2000). Mechanistic investigation of ionization suppression in electrospray ionization. *J. Am. Soc. Mass Spectrom.* 11: 942–950. https://doi.org/10.1016/S1044-0305(00)00163-X.

11 Annesley, T.M. (2003). Ion suppression in mass spectrometry. *Clin. Chem.* 49: 1041–1044.

12 Larger, P.J., Breda, M., Fraier, D. et al. (2005). Ion-suppression effects in liquid chromatography-tandem mass spectrometry due to a formulation agent, a case study in drug discovery bioanalysis. *J. Pharm. Biomed. Anal.* 39: 206–216.

13 Picard, P., Auger, S., Blachon, G., and Paquin, R. (2012). Theoretical and experimental characterization of solvent free atmospheric pressure chemical

ionization applied to LDTD-MS/MS analysis. *60th ASMS Conference on Mass Spectrometry and Allied Topics* (23 May 2012), p. 153, Vancouver, BC: ASMS.

14 ANSI SLAS 1-2004 (2004). *Microplates – Footprint Dimensions (9 January 2004)*. American National Standards Institute and Society for Laboratory Automation and Screening.

15 Gross, J.H. (2017). *Mass Spectrometry: A Textbook*, 3e. Cham, Switzerland: Springer https://doi.org/10.1007/978-3-319-54398-7.

16 Ionov, N.I. (1972). Surface ionization and its applications. *Prog. Surf. Sci.* 1: 237–354.

17 Bénard, J., Berthier, Y., Delarnare, F. et al. (ed.) (1983). Adsorption-desorption kinetics and catalysis on single crystals, Chapter 7. In: *Adsorption on Metal Surfaces*, Studies in Surface Science and Catalysis, vol. 13, 177–210. Elsevier https://doi.org/10.1016/S0167-2991(08)64968-X.

18 Lafferty, J.M. (1998). *Foundations of Vacuum Science and Technology*, vol. 1. New York: Wiley.

19 Sims, R.A., Harmer, S.L., and Quinton, J.S. (2019). The role of physisorption and chemisorption in the oscillatory adsorption of organosilanes on aluminium oxide. *Polymers* 11: 410. https://doi.org/10.3390/polym11030410.

20 Lord, F.M. and Kittelberger, J.S. (1974). On the determination of activation energies in thermal desorption experiments. *Surf. Sci.* 43: 173–182. https://doi.org/10.1016/0039-6028(74)90226-X.

21 Riegler, H. and Kohler, R. (2007). How pre-melting on surrounding interfaces broadens solid-liquid phase transitions. *Nat. Phys.* 3: 890–894.

22 Beuhler, R.J., Flanigan, E., Greene, L.J., and Friedman, L. (1974). Proton-transfer mass-spectrometry of underivatized peptides. *Biochemistry* 13: 5060–5068.

23 O'Neil, M.J. (2001). *The Merck Index: An Encyclopedia of Chemicals, Drugs, and Biologicals*. Whitehouse Station, NJ: Merck.

24 Cahill, D., Ford, W., Goodson, K. et al. (2003). Nanoscale thermal transport. *J. Appl. Phys.* 93. https://doi.org/10.1063/1.1524305.

25 Sääskilahti, K., Oksanen, J., Tulkki, J. et al. (2016). Vibrational mean free paths and thermal conductivity of amorphous silicon from non-equilibrium molecular dynamics simulations. *AIP Adv.* 6: 121904. https://doi.org/10.1063/1.4968617.

26 Choi, S.-H., Maruyama, S., Kim, K.-K., and Lee, J.-H. (2003). Evaluation of the phonon mean free path in thin films by using classical molecular dynamics. *J. Korean Phys. Soc.* 43. https://doi.org/10.3938/jkps.43.747.

27 Laurendeau, N.M. (2005). *Statistical Thermodynamics: Fundamentals and Applications*. Cambridge: Cambridge University Press https://doi.org/10.1017/CBO9780511815928.

28 Balandin, A., Pokatilov, E., and Nika, D. (2007). Phonon engineering in hetero- and nanostructures. *J. Nanoelectron. Optoelectron.* 2: 140–170. https://doi.org/10.1166/jno.2007.201.

29 Anufriev, R., Ramiere, A., Maire, J., and Nomura, M. (2017). Heat guiding and focusing using ballistic phonon transport in phononic nanostructures. *Nat. Commun.* 8: 15505. https://doi.org/10.1038/ncomms15505.

30 Dion-Fortier, A., Gravel, A., Guérette, C. et al. (2019). Signal enhancement in laser diode thermal desorption-triple quadrupole mass spectrometry analysis using microwell surface coatings. *J. Mass Spectrom.* 54 (2): 167–177. https://doi.org/10.1002/jms.4328.

31 Gravel, A., Guérette, C., Fortin, D. et al. (2019). Further studies on the signal enhancement effect in laser diode thermal desorption-triple quadrupole mass spectrometry using microwell surface coatings. *J. Mass Spectrom.* 54. https://doi.org/10.1002/jms.4455.

32 Eduard, H. (1967). *Vapour-Liquid Equilibrium*, 2e. New York: Pergamon Press.

33 Shahin, M.M. (1966). Mass-spectrometric studies of corona discharges in air at atmospheric pressures. *J. Chem. Phys.* 45: 2600–2605.

34 Sekimoto, K., Sakai, M., and Takayama, M. (2012). Specific interaction between negative atmospheric ions and organic compounds in atmospheric pressure corona discharge ionization mass spectrometry. *J. Am. Soc. Mass Spectrom.* 23: 1109–1119. https://doi.org/10.1021/jasms.8b04300.

35 Sunner, J., Nicol, G., and Kebarle, P. (1988). Factors determining relative sensitivity of analytes in positive mode atmospheric-pressure ionization mass-spectrometry. *Anal. Chem.* 60: 1300–1307.

36 Picard, P., Tremblay, P., and Paquin, R. (2009). Mechanisms involved in positive atmospheric pressure chemical ionization (APCI) of an LDTD source. In: *57th ASMS Conference on Mass Spectrometry & Allied Topics* (ed. ASMS) (2 June 2009), p. 101. Philadelphia, PA: American Society for Mass Spectrometry.

37 Adamiak, K. and Atten, P. (2004). Simulation of corona discharge in point-plane configuration. *J. Electrost.* 61: 85–98.

38 Wilson, R.G. and Brewer, G.R. (1973). *Ion Beams with Application to Ion Implantation*. New-York: Wiley-Interscience Publication.

39 Antignac, J.-P., de Wasch, K., Monteau, F. et al. (2005). The ion suppression phenomenon in liquid chromatography-mass spectrometry and its consequences in the field of residue analysis. *Anal. Chim. Acta* 529: 129–136. https://doi.org/10.1016/j.aca.2004.08.055.

40 Kumar, R.A. and Clark, D.S. (2006). High-throughput screening of biocatalytic activity: applications in drug discovery. *Bioinorg. Chem. Biocatal. Biotransformation* 10: 162–168. https://doi.org/10.1016/j.cbpa.2006.02.033.

41 Bateman, K.P., Wu, J., Gagne, S. et al. (2007). Application of a laser diode thermal (LDTD) ion source for mass spectrometry in a drug discovery environment. In: *55th ASMS Conference on Mass Spectrometry and Allied Topics*, (ed. ASMS) (7 June 2007), p. 122. Indianapolis, IN: ASMS.

42 Lanshoeft, C., Heudi, O., Leuthold, L.A. et al. (2014). Laser diode thermal desorption atmospheric pressure chemical ionization tandem mass spectrometry applied for the ultra-fast quantitative analysis of BKM120 in human plasma. *Anal. Bioanal. Chem.* 406: 5413–5423. https://doi.org/10.1007/s00216-014-7966-6.

43 Gupta, D., Bhatia, D., Dave, V. et al. (2018). Salts of therapeutic agents: chemical, physicochemical, and biological considerations. *Molecules* 23: 1719. https://doi.org/10.3390/molecules23071719.

44 Sahasrabuddhe, A., Oakley, D., Chen, K., and McCarter, J.D. (2020). Development of a high-throughput affinity mass spectrometry (ASM) platform using laser diode thermal desorption ionization coupled to mass spectrometry (LDTD-MS). In: *68th ASMS Conference on Mass Spectrometry and Allied Topics*, (ed. ASMS) (3 June 2020), p. 122. ASMS.

45 Imrazene, S., Auger, S., Plante, P.L. et al. (2019). Development of a screening method for illicit drugs in hair using LDTD-MS/MS at 8 seconds per sample. In: *67th ASMS Conference on Mass Spectrometry and Allied Topics*, (ed. ASMS) (4 June 2019), p. 118. Atlanta, GA: ASMS.

46 Rochon, J., Lacoursière, J., Auger, S. et al. (2020). Urine LDTD-MS/MS drugs of abuse screening in urine at 9 seconds per sample using dry-and-dissolve preparation. In: *68th ASMS Conference on Mass Spectrometry and Allied Topics*, (ed. ASMS) (1 June 2020), p. 44. ASMS.

47 Guérette, C., Dion-Fortier, A., Auger, S. et al. (2019). Ultra-fast screening of glyphosate, glufosinate and AMPA in surface water by LDTD-QqQMS. In: *67th ASMS Conference on Mass Spectrometry and Allied Topics*, (ed. ASMS) (4 June 2019), p. 116. Atlanta, GA: ASMS.

48 Guerrieri, P., Jarring, K., and Taylor, L.S. (2010). Impact of counterion on the chemical stability of crystalline salts of procaine. *J. Pharm. Sci.* 99: 3719–3730. https://doi.org/10.1002/jps.22009.

49 Stone, J. (2017). Sample preparation techniques for mass spectrometry in the clinical laboratory, Chapter 3. In: *Mass Spectrometry for the Clinical Laboratory* (ed. H. Nair and W. Clarke), 37–62. San Diego: Academic Press https://doi.org/10.1016/B978-0-12-800871-3.00003-1.

50 Ismaiel, O.A., Halquist, M.S., Elmamly, M.Y. et al. (2008). Monitoring phospholipids for assessment of ion enhancement and ion suppression in ESI and APCI LC/MS/MS for chlorpheniramine in human plasma and the importance of multiple source matrix effect evaluations. *J. Chromatogr. B* 875: 333–343.

51 Berendsen, B.J.A., Stolker, L.A.M., and Nielen, M.W.F. (2013). The (Un)certainty of selectivity in liquid chromatography tandem mass spectrometry. *J. Am. Soc. Mass Spectrom.* 24: 154–163. https://doi.org/10.1021/jasms.8b04369.

52 Younggil, K. (2002). *Handbook of Essential Pharmacokinetics, Pharmacodynamics and Drug Metabolism for Industrial Scientists*. New York: Kluwer Academic Publishers.

53 White, R.E. (2000). High-throughput screening in drug metabolism and pharmacokinetic support of drug discovery. *Annu. Rev. Pharmacol. Toxicol.* 40: 133–157. https://doi.org/10.1146/annurev.pharmtox.40.1.133.

54 Wu, J., Hughes, C.S., Picard, P. et al. (2007). High-throughput cytochrome P450 inhibition assays using laser diode thermal desorption-atmospheric pressure chemical ionization-tandem mass spectrometry. *Anal. Chem.* 79: 4657–4665.

55 Haarhoff, Z., Wagner, A., Picard, P. et al. (2015). Coupling laser diode thermal desorption with acoustic sample deposition to improve throughput of mass spectrometry-based screening. *J. Biomol. Screening* 21. https://doi.org/10.1177/1087057115607184.

56 van Breemen, R.B. and Li, Y. (2005). Caco-2 cell permeability assays to measure drug absorption. *Expert Opin. Drug Metab. Toxicol.* 1: 175–185. https://doi.org/10.1517/17425255.1.2.175.

57 Tremblay, P., Picard, P., Auger, S., and Blachon, G. (2010). LDTD384-MS/MS for in vitro assays: different buffer environment. In: *58th ASMS Conference on Mass Spectrometry and Allied Topics*, (ed. ASMS) (27 May 2010), p. 42. Salt Lake City, UT: ASMS.

58 Julian, B., Bertrand, C., Fedeli, O. et al. (2011). High-throughput caco-2/TC7 human intestinal absorption studies using laser diode thermal desorption-atmospheric pressure chemical ionization coupled to exactive benchtop orbitrap. In: *59th ASMS Conference on Mass Spectrometry and Allied Topics*, (ed. ASMS) (8 June 2011), p. 125. Denver, CO: ASMS.

59 Motoyaji, T. (2020). Revolution of small molecule drug discovery by affinity selection-mass spectrometry technology. *Chem. Pharm. Bull. (Tokyo)* 68: 191–193. https://doi.org/10.1248/cpb.c19-00832.

60 Gabriel, J., Höfner, G., and Wanner, K.T. (2019). A library screening strategy combining the concepts of MS binding assays and affinity selection mass spectrometry. *Front. Chem.* 7: 665. https://doi.org/10.3389/fchem.2019.00665.

61 Annis, D.A., Nickbarg, E., Yang, X. et al. (2007). Affinity selection-mass spectrometry screening techniques for small molecule drug discovery. *Anal. Tech. Mech.* 11: 518–526. https://doi.org/10.1016/j.cbpa.2007.07.011.

62 Zhu, Z. and Cuozzo, J. (2009). High-throughput affinity-based technologies for small-molecule drug discovery. *J. Biomol. Screening* 14: 1157–1164. https://doi.org/10.1177/1087057109350114.

63 Lacoursière, J., Auger S, Dion-Fortier, A., Picard, P. (2013). Microsampling method using 1 μL of human blood to determine pharmacokinetic parameters of dextrorphan using LDTD-MS/MS. In: *61st ASMS Conference on Mass Spectrometry and Allied Topics*, (ed. ASMS) (10 June 2013), p. 54. Minneapolis, MN: ASMS.

64 Heudi, O., Barteau, S., Picard, P. et al. (2011). Laser diode thermal desorption–positive mode atmospheric pressure chemical ionization tandem mass spectrometry for the ultra-fast quantification of a pharmaceutical compound in human plasma. *J. Pharm. Biomed. Anal.* 54: 1088–1095. https://doi.org/10.1016/j.jpba.2010.11.025.

65 Picard, P., Plante, P.-L., Demers, S. et al. 2019. Effect of increased plate density on sensitivity in high-throughput LDTD-MS. In: *67th ASMS Conference on Mass Spectrometry and Allied Topics*, (ed. ASMS) (5 June 2019), p. 155. Atlanta, GA: ASMS.

12

Accelerating Drug Discovery with Ultrahigh-Throughput MALDI-TOF MS

Sergei Dikler

Life Science Mass Spectrometry Division, Bruker Scientific, LLC, Billerica, MA, USA

12.1 Introduction

Matrix-assisted laser desorption/ionization (MALDI) is a soft ionization technique discovered more than 30 years ago by Karas and Hillenkamp and by Tanaka et al. [1, 2]. MALDI ionization sources are most commonly coupled to time-of-flight (TOF) mass analyzers. Since its development, MALDI-TOF mass spectrometry has been successfully applied to the analysis of peptides, proteins, DNA, RNA, lipids, carbohydrates, small organic compounds, and synthetic polymers. MALDI applications have gradually become more complex, and now include bacterial and fungi identification based on protein profiling as well as mass spectrometry imaging (MSI) [3, 4]. MALDI-TOF MS was readily amenable to automation from the early days of commercially available MALDI instrumentation, as the automation of laser firing and movement of a MALDI target plate by an X, Y stage was relatively straightforward compared to other mass spectrometric techniques.

One of the early implementations of automation on MALDI-TOF instruments was achieved with the use of fuzzy logic feedback control, also known as a fuzzy logic engine [5]. The key aspect of this control algorithm was automated adjustment of laser power based on signal intensity and mass resolution of the most abundant signal in a spectrum, thus allowing unattended operation with optimized quality data collection. This algorithm was realized in a commercially available instrument control software and is still currently in use.

Analysis of single-nucleotide polymorphisms (SNPs) was one of the first successful applications of automated MALDI-TOF MS [6, 7]. In 2001, Buetow et al. described an approach that allowed rapid identification, development, and validation of reagents

0000-0003-2915-0451

High-Throughput Mass Spectrometry in Drug Discovery, First Edition. Edited by Chang Liu and Hui Zhang.
© 2023 John Wiley & Sons, Inc. Published 2023 by John Wiley & Sons, Inc.

for more than 9100 SNPs [8]. Nanoliter aliquots of primer extension products were loaded onto silicon chips preloaded with 3-hydroxypicolinic acid matrix, followed by MALDI-TOF analysis in fully automated mode [8]. Another approach was to identify DNA sequence variations by using four base-specific cleavage reactions with subsequent MALDI-TOF analysis of each mixture of cleavage products. In this approach, tens of nanoliter aliquots of the cleavage reaction mixtures were robotically dispensed onto silicon chips preloaded with matrix, followed by MALDI analysis [9].

A cornerstone of the early proteomic research was the combination of 2D gel electrophoresis and automated peptide mass fingerprinting (PMF) by MALDI-TOF [10, 11]. One of these early studies focused on creating a prototype of an integrated, automated high-throughput system for gel-based proteomics. The prototype system included a device for scanning and automated excision of protein spots from polyvinylidene difluoride (PVDF) membranes, a 4-channel liquid handling robot for tryptic digestion and spotting of MALDI plates, and a MALDI-TOF instrument for automated acquisition of PMF spectra [10]. This was an important development in MALDI automation, even though multiple steps in the protocol, including digest plate and MALDI plate transfers, were done manually. Automated PMF was later expanded to include automated MS/MS measurements of peptide parent ions when MALDI-TOF/TOF instruments were introduced in the early 2000s [12, 13]. Hyphenated LC-MALDI approaches were introduced in the mid-2000s and involved nanoflow or microflow LC separation of complex enzymatic digest mixtures with automated spotting of the eluate onto MALDI plates, followed by automated MALDI-TOF and MALDI-TOF/TOF analyses and database searching for protein identification [14–17].

The next frontier conquered by automated MALDI-TOF was high-throughput screening (HTS). Application of MALDI-TOF to HTS and ultrahigh-throughput screening (uHTS) is often described as high-throughput MALDI (HT-MALDI) and ultrahigh-throughput MALDI (uHT-MALDI). uHT-MALDI is defined here as automated MALDI-TOF with analysis times less than 0.5 seconds per sample.

In one of the earliest examples of HT-MALDI for multiple target screening of molecular libraries, a high-density spotting format was applied. More than 4000 samples were spotted on a square MALDI plate (4.5×4.5 cm) using a custom robot equipped with spotting pins with a diameter of 211.6 μm [18]. Another early application of MALDI-TOF to screening described quantitation using percent conversion and percent maximal activity calculated from peak area ratios to produce IC_{50} curves [19]. In the mid-2000s, MALDI-TOF analysis time in screening applications was limited to 10 seconds per sample [19, 20], a rate more than 30 times slower in comparison to the present analysis speeds due to substantially lower laser repetition rates. This limitation was significant and prevented a wide-scale adoption at that time.

Current MALDI-TOF systems are ideally suited for HTS, as analysis time requirement of less than one second per sample, which remain elusive for many other mass spectrometric techniques including multiplexed LC-MS and multiplexed solid-phase extraction MS (SPE-MS), can be easily met. Further, HT-MALDI is a label-free technique, which gives it an advantage over traditional HTS detection via time-resolved fluorescence energy transfer (TR-FRET), fluorescence intensity, luminescence, absorbance, and/or radioactivity-based methods [21].

The integration of HT-MALDI with liquid handling robots, robotic arms, and other front-end robotic devices is one of the key requirements for HTS. One of the first successful examples was the integration of an automated MALDI-TOF system with a robotic arm, liquid handling robot, random-access plate storage system, and lidding/delidding device achieved by us in collaboration with Novartis Institutes for Biomedical Research and HighRes Biosolutions teams [22]. The system used disposable steel HTS MALDI plates (1 mm thickness) mounted onto adapters and spotted in 1536 format by a positive displacement 16-channel liquid handling robot (mosquito HTS, SPT Labtech, Melbourn, Hertfordshire, UK). This system later became known as MALDI PharmaPulse (Bruker Daltonics, Billerica, MA) [23]. The introduction of a rapifleX MALDI-TOF/TOF mass spectrometer with a 10 kHz scanning beam laser substantially improved the analysis time of the MALDI PharmaPulse system to 0.3 seconds per sample or 8–11 minutes per MALDI plate in 1536 format [24]. An increase of the laser repetition rate by a factor of 5 was one of the key developments in the reduction of the analysis time per sample, as the amount of time needed to fire the same number of laser shots per sample spot was dramatically reduced. The improved laser repetition rate allowed increasing the number of laser shots for sample spots of lower concentration without significant penalties to the analysis time. A twofold increase in the movement speed of the X, Y stage also contributed to the reduction of the analysis time per sample. These developments made uHT-MALDI a more appealing alternative to traditional HTS detection methods.

Leveridge and coworkers increased the spotting density to 6144, the equivalent of four 1536-well assay plates spotted onto one MALDI plate. The spotting volume was 25 nL [25]. The speed of MALDI plate spotting in 1536 format was further improved with cycle times of only 0.6 seconds per sample, bringing it closer to the analysis times per sample of uHT-MALDI. This was achieved by using an advanced liquid handling robot with a 1536-channel pipetting head equipped with multiuse ceramic tips (CyBio Well vario, Analytik Jena, Jena, Germany). A sophisticated integrated sample preparation system that included a robotic arm, two liquid handling robots, three plate storage devices, plate peeler, and plate sealer was positioned separately from the MALDI system that was also integrated with a simpler robotic system tasked with automated handling of the prepared MALDI plates [26].

The application of HT-MALDI and uHT-MALDI to HTS and uHTS is one of the most important applications of automated MALDI in drug discovery and is the main focus of this chapter. The application of HT-MALDI to enzymatic (biochemical) assays where substrates and products are small molecules, peptides, and proteins will be reviewed, followed by the applications to screening of chemical reactions, screening of cell-based assays, screening of other types of assays and libraries, bead-based workflows, and the use of functionalized and modified MALDI plate surfaces.

12.2 uHT-MALDI MS of Assays and Chemical Reactions

12.2.1 HT-MALDI of Enzymatic Assays

Enzymatic (biochemical) assays represent the majority of HT-MALDI screening applications. Table 12.1 contains 12 examples of these assays. Generally, the adaptation of an enzymatic assay to HT- or uHT-MALDI involves optimizing both assay conditions and automated MALDI sample preparation. Optimizing assay conditions included optimizing or modifying assay buffers and detergents to minimize ionization suppression and maximize data quality and sensitivity. Chandler et al. analyzed the ionization suppression effects of 43 buffer components, detergents, and cell culture media using acetylcholinesterase assay as a model and identified polymer-based detergents Triton X100 and Pluronic F127, cell culture media Opti-MEM and CD-CHO, and sodium bicarbonate as the most detrimental components for screening by HT-MALDI. Sixteen of the studied buffer components were found to be MALDI friendly and did not require a dilution or other protocol changes [42]. The most commonly used MALDI matrix for assays where substrates and products were small molecules or peptides was α-cyano-4-hyroxycinnamic acid (CHCA), while 2,5-dihydroxyacetophenone (DHAP) was used for assays where substrates and products were intact proteins. Most of the automated MALDI sample preparation was done using previously mentioned mosquito HTS [25, 27, 29] or CyBio Well vario with 1536-channel pipetting head [28, 31].

Examples of the assays adapted to HT-MALDI where substrate and product were small molecules included acetylcholinesterase and butyrylcholinesterase [25, 27]. The HT-MALDI acetylcholinesterase assay was validated using concentration–response curves for 23 known inhibitors, and then scaled up to a single-concentration screening of 10,000 library compounds. The coefficient of variation (CV) was 2%, remarkably low for HT-MALDI assays [25].

A more recent example of the enzymatic assay screened by uHT-MALDI with small molecules as substrates and a product was cyclic GMP-AMP synthase (cGAS). This enzyme converted ATP and $^{13}C/^{15}N$-labeled GTP to isotopically labeled cyclic GMP-ATP (cGAMP) (Table 12.1). The unlabeled cGAMP was added as an internal standard. The uHT-MALDI screening of the cGAS assay

Table 12.1 Examples of enzymatic assays, chemical reactions, and cell-based assays screened using HT-MALDI MS.

Type	Analyte type	Assay or chemical reaction	Substrate or starting material	Product	References
Enzymatic	Small molecules	Acetylcholinesterase	Acetylcholine	Choline	[25]
Enzymatic	Small molecules	Butyrylcholinesterase	Butyrylcholine	Choline	[27]
Enzymatic	Small molecules	Cyclic GMP-AMP synthase	ATP and GTP	Cyclic GMP-ATP (cGAMP)	[28]
Enzymatic	Peptides	cAMP-dependent protein kinase	Kemptide	Phosphorylated kemptide	[19]
Enzymatic	Peptides	Protein kinase C-α (PKCα)	PKCα peptide	Phosphorylated PKCα peptide	[19]
Enzymatic	Peptides	Histone demethylase	Histone trimethylated peptide	Histone dimethylated peptide, histone monomethylated peptide	[25]
Enzymatic	Peptides	Salt-inducible kinase	CHKtide peptide	Phosphorylated CHKtide peptide	[29]
Enzymatic	Peptides	c-MET tyrosine kinase	SRCtide peptide	Phosphorylated SRCtide peptide	[30]
Enzymatic	Peptides	Protein tyrosine phosphatase 1B	ETDpYYRKG-NH_2	ETDYYRKG-NH_2	[31]
Enzymatic	Peptides	β-Secretase	[Asn670, Leu671]-Amyloid β/A4 protein precursor (667–676)	SEVNL and DAEFR	[32]
Enzymatic	Proteins	Deubiquitilase	Diubiquitin	Ubiquitin	[33, 34]

(*Continued*)

Table 12.1 (Continued)

Type	Analyte type	Assay or chemical reaction	Substrate or starting material	Product	References
Enzymatic	Proteins	E2/E3 ligase	Ubiquitin	Polyubiquitin, ubiquitilated E3 ligase	[35]
Chemical	Small molecules	Buchwald–Hartwig	Aryl bromides and secondary amines	Tertiary aryl amines	[24]
Chemical	Small molecules	Lipidoid synthesis	Pyridyl disulfide, thiolactone, and amine	Lipidoid	[36, 37]
Cell-based	Small molecules	BCR-ABL tyrosine kinase	Heme B (inhibitor response marker)		[38]
Cell-based	Small molecules	Bacterial TMA-lyase	Choline	Trimethylamine (TMA), iodoacetamide derivative of TMA	[39]
Cell-based	Small molecules	Fatty acid synthase	Malonyl-coenzyme A (inhibitor response marker)		[40]
Cell-based	Small molecules	Organic anion transporting polypeptide 2B1	Estrone-3-sulfate (inhibitor response marker)		[41]

represented a large-scale full-diversity uHTS campaign that included analysis of more than 1.13 million samples (1,080,879 library compounds and 51,486 controls) [28]. The library compounds were in 384-well plates, which had been reformatted into 1536-well assay plates, and the MALDI plates were spotted in 1536 format. Batches of up to 45 assay plates were processed per day, resulting in the analysis of an average of 60,000 library compounds per day. The inhibition of cGAS was tracked using percentage of control (PoC) calculated using signal ratios of the sample, the low control (no library compound, with assay buffer replacing the enzyme solution) and the high control (no library compound). The assay had an excellent average Z′ value of 0.90% and 3.8% CV of Z′ values (Figure 12.1a). The CV of PoC values for the entire campaign was 3.5% (Figure 12.1b). The screen had a hit rate of 0.67% with more than 7200 compounds identified as inhibitors [28]. The dose–response experiments allowed the confirmation of 5229 hits (hit confirmation rate of 71%), and there was a good correlation between IC_{50} values determined by MALDI-TOF and by multiplexed SPE-MS (RapidFire-MS) for 40 representative hit compounds (Figure 12.1d, f). This work was one of the largest uHT-MALDI screening campaigns published to date.

Peptides were substrates and products for the largest subclass of enzymatic assays screened by HT-MALDI, with seven examples shown in Table 12.1. Two of the earliest examples in this assay category included cAMP-dependent protein kinase, which converted kemptide into phosphorylated kemptide, and protein kinase C-α (PKCα), which phosphorylated PKCα peptide. CV values less than 7% were calculated from percent product for controls in the former example [19].

Kinase assays were the most common enzymatic assays adapted to HT-MALDI. In another example CHKtide peptide was converted to phosphorylated CHKtide peptide by salt-inducible kinase 2 (SIK2). A library of over 2600 compounds was screened and 45 hits were identified [29].

c-MET tyrosine kinase assay adaption to uHT-MALDI was an interesting example since most of the other kinase assays were based on serine/threonine kinase family. An acoustic liquid handling robot was used to deposit 75 nL of the assay and MALDI matrix mixture to HTS MALDI plates positioned in custom 3D-printed plastic adapters, in contrast to the contact deposition liquid handling robots used in other described assay examples. The HTS MALDI plates were spotted in only 430 seconds; however, there was an upper volume limit per well for the assay (source) plate. The IC_{50} values obtained by uHT-MALDI correlated well with the values obtained using fluorescence-based detection on a capillary electrophoresis system [30].

In another uHT-MALDI-adapted assay, protein tyrosine phosphatase 1B was used to dephosphorylate an insulin receptor substrate peptide containing phosphotyrosine (ETDpYYRKG-NH_2). The IC_{50} values were determined for 103 compounds, and these correlated well to the values obtained using AlphaScreen

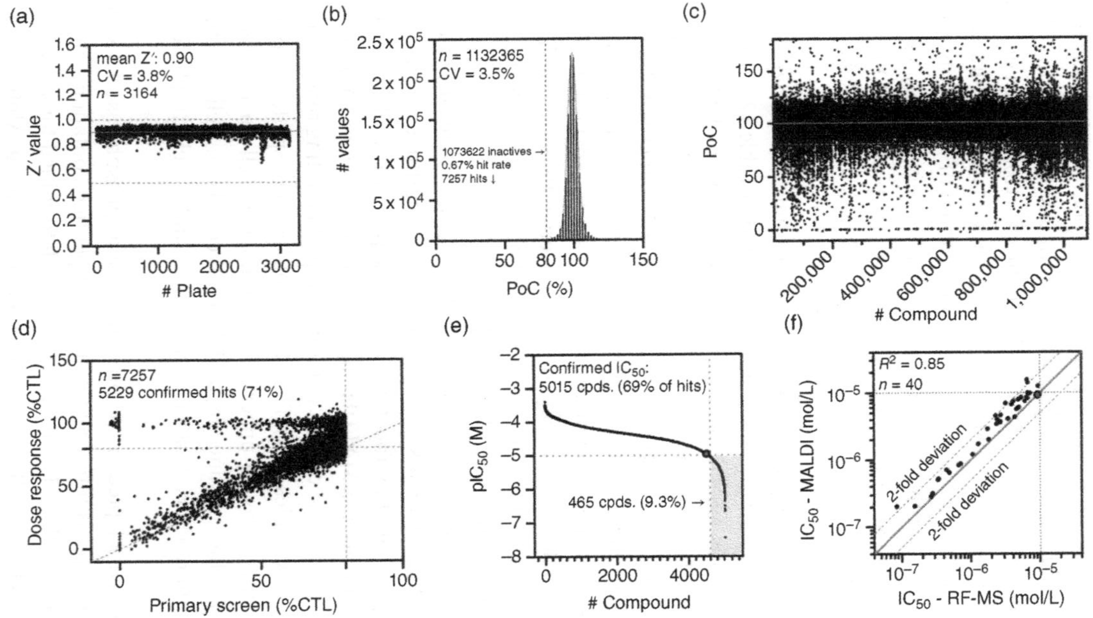

Figure 12.1 Results from the cGAS HTS assay campaign and hit triaging. (a) Z′ values of deconvoluted 384-well plates throughout the screening campaign. Each plate contained 16 high and 16 low controls. All values stayed within the predefined quality threshold of Z′ ≥ 0.5. (b) Histogram of PoC values for samples from the primary screen of 1,080,879 compounds and 51,486 controls. Compounds with PoC values <80% were considered hits. (c) Scatterplot of PoC values of library compounds. The hit threshold of 80% PoC is indicated as a dashed line. (d) Confirmation of primary screening hits. The average PoC values (at 5 μg/mL compound concentration) of duplicate determinations from two individual dose–response experiments are plotted against PoC values from the primary screen. The hit threshold of 80% PoC is indicated as dashed lines. (e) Distribution of IC50 values for confirmed hits. The average pIC50 values from two occasions are plotted in decreasing order. A total of 465 compounds with pIC50 < −5 (10 μM) are found in the shaded area. (f) Correlation of IC50 values for 40 representative compounds from the hit pool (and Ref. Inh. 1) generated by either MALDI-TOF or RF-MS analysis. In panels (c–f), reference inhibitor 1 was tracked throughout the assay campaign and is highlighted. *Source:* Reproduced with permission from Simon et al. [28]/SAGE Publications.

technology with luminescence-based detection. The Z′ value was 0.87 averaged from two replicates [31]. Other examples of enzymatic assays adapted to HT-MALDI with peptide substrates and products include histone demethylase and β-secretase [25, 32].

Most of the HT-MALDI assays with intact proteins as substrates and products focused on targets in the ubiquitylation cascade pathway. Changes in this pathway have been associated with the onset of neurodegenerative diseases, cancer, autoimmunity, and other diseases [34]. As such, multiple enzymes in this pathway have been identified as targets in drug discovery efforts. A deubiquitilase (DUB) assay was the first assay in this group to be developed for automated MALDI-TOF detection, based on monitoring the conversion of diubiquitin into ubiquitin. ^{15}N-labeled ubiquitin was added as an internal standard for quantitation purposes. Ritorto et al. characterized activity and specificity of 42 DUBs and tested 11 potential inhibitors against a panel of 32 DUBs [33]. Very recently, the DUB assay was further developed for automated MALDI plate preparation and uHT-MALDI [34]. It was reported that small molecule inhibitors of USP15 deubiquitilase were licensed to develop potential new treatments of neurodegenerative diseases, including multiple sclerosis [34].

Another example of an assay for a target in ubiquitylation cascade pathway is the E2 conjugating enzyme, E3 ligase (E2/E3 ligase) assay (Table 12.1). This uHT-MALDI assay monitored disappearance of ubiquitin substrate, which was measured using ^{15}N-labeled ubiquitin internal standard. This assay was used to screen a library of 1430 compounds and nine compounds were identified as positive hits (0.63% hit rate) [35].

12.2.2 Screening Chemical Reactions Using uHT-MALDI

Very recently, applications of uHT-MALDI have been extended to screening nanomole-scale, synthetic chemical reactions. Merck & Co. team and us created the workflow for this approach starting with the design of reaction conditions followed by extended nanomole synthesis in glass 1536-well plates and then uHT-MALDI with analysis time 0.3 seconds per sample (Figure 12.2). We explored four catalytic methods including Cu, Pd, Ir/Ni photoredox, and Ru/Ni photoredox applied to Buchwald–Hartwig reaction [24]. This reaction is a C-N coupling between a secondary amine and aryl bromide, and it plays an important role in the synthesis methodology for lead compounds in drug discovery efforts (Table 12.1).

In the first group of experiments, a simple secondary amine reacted with a simple aryl bromide in the presence of 383 single and polyfunctional molecules using four catalytic methods, resulting in 1536 reactions on one glass assay plate in order to determine catalytic poisons (Figure 12.3a). The correlation between uHT-MALDI results and LC-MS results was very high (R^2 = 0.85) following the

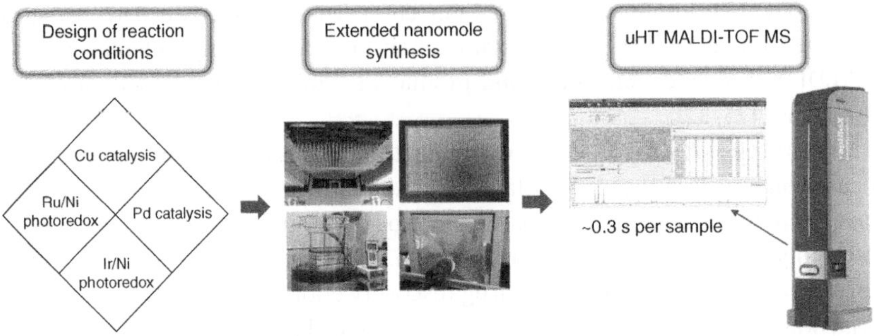

Figure 12.2 Workflow for uHT-MALDI screening of chemical reactions. The first step is the design of reaction conditions, followed by nanomole-scale synthesis in glass 1536-well plates using a multichannel liquid handling robot to dispense reagents, LabRam resonant acoustic mixing, heating in aluminum blocks, and a photochemistry setup for plate-based experiments. The last step is uHT-MALDI with analysis time of 0.3 seconds per sample.

normalization of the MALDI signal against the deuterated reaction product (internal standard) (Figure 12.3b). In the second group of experiments, a simple secondary amine was reacted with 192 aryl bromides of increasing complexity and a simple aryl bromide was reacted with 192 secondary amines of increasing complexity using four catalytic methods, totaling 1536 reactions. These uHT-MALDI experiments not only helped to determine catalytic poisons but also allowed the establishment of key structural features in starting materials that led to Buchwald–Hartwig reaction failures, including functional groups, number of hydrogen bond acceptors in aryl bromides, and number of hydrogen bond donors in secondary amines [24].

A novel approach to screening chemical reactions was developed by creating a microarray platform that allowed solution-based organic synthesis, library characterization by automated MALDI-TOF, and on-chip cell-based screening. This platform was called chemBIOS [36]. A proof-of-concept study was conducted using lipidoid synthesis from an amine, thiolactone, and pyridyl disulfide. Twenty-five lipidoids were synthesized by sandwiching a specially patterned slide containing an array of amines in 2.83 mm diameter spots with another slide containing an array of thiolactone and pyridyl disulfide mixtures [36].

The chemBIOS platform was substantially improved by miniaturizing the array spots to a diameter of 500 μm and by using glass slides with indium tin oxide (ITO) coating for special patterning to create an omniphilic–omniphobic surface. The use of ITO slides allowed direct uHT-MALDI analysis, IR analysis, and reaction monitoring using UV-Vis spectroscopy. Regular stainless steel MALDI plates were patterned with a 280 × 180 array (50,400 sample spots) [37].

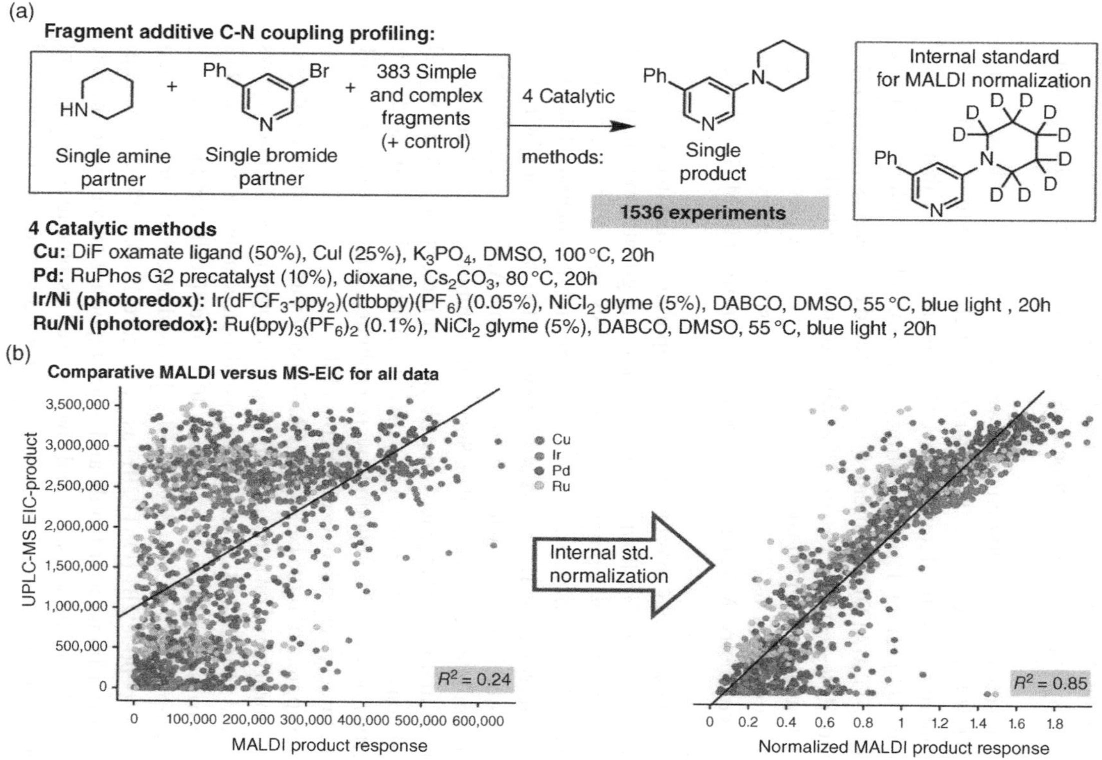

Figure 12.3 uHT-MALDI MS approach to screening potential reaction poisons. (a) 1536 reactions were run with four leading C–N coupling methods to evaluate the impact of 383 single and polyfunctional fragments on one simple coupling reaction, requiring 11 minutes of MALDI acquisition time. Ph, phenyl; DiF, difluoro; RuPhos, 2-dicyclohexylphosphino-2′,6′-diisopropoxybiphenyl; ppy, 2-phenyl pyridinyl; dtbbpy, di-*tert*-butyl di-pyridyl; DABCO, 1,4-diazabicyclo[2.2.2]octane; bpy, bipyridinyl. (b) Raw MALDI product response is not well correlated with UPLC-MS or UPLC-UV measurements, but normalization using an internal standard reveals very good correlation. *Source:* Reproduced with permission from Lin et al. [24]/AAAS.

12.2.3 uHT-MALDI of Cell-Based Assays

Screening of cell-based assays using HT- or uHT-MALDI has been considered substantially more challenging than enzymatic assay screens. Among the impediments to successful screening were ionization suppression effects in very complex mixtures, particularly when monitored analytes were small molecules, the presence of cell culture media components that suppressed MALDI signals, and a reduction in dynamic range. Very recently, cell-based assays were adapted to uHT-MALDI by monitoring inhibitor response markers rather than enzymatic reaction products [38, 40] or by chemical derivatization of the diluted assay mixtures [39].

BCR-ABL tyrosine kinase inhibition assay was one of the first cell-based assays adapted to HT-MALDI detection. The workflow included treating cells with defined concentrations of inhibitors followed by spotting assay mixture onto a MALDI plate in 384-position format, coating the MALDI plate with matrix using a matrix sprayer, HT-MALDI measurements, automated data processing and plotting of inhibitor dose–response curves. Two inhibitor response markers, heme B and the $[M+K]^+$ adduct of phosphatidylcholine PC(36 : 1), were identified by high resolution measurements using MALDI Fourier-transform ion cyclotron resonance mass spectrometry (MALDI-FTICR MS) and confirmed by MALDI-FTICR MS/MS [38].

Fatty acid synthase (FASN) has emerged as an important target in drug discovery due to its importance in lipid metabolism and the growth and survival of tumors. The very recent development of cell-based FASN assay with uHT-MALDI detection was a significant step in filling a gap in high-throughput, label-free assays for this target (Table 12.1). The workflow was the same as for BCR-ABL tyrosine kinase inhibition assay, with malonyl-coenzyme A, a substrate of FASN, selected as the inhibitor response marker. Successful inhibition of FASN led to the accumulation of malonyl-coenzyme A in cells and an increase in MALDI signal abundance. Malonyl-$^{13}C_3$-coenzyme A was used as an internal standard and inhibition properties of several compounds were evaluated. The Z′ value was 0.61 averaged from three independent experiments [40].

An alternate approach to the cell-based FASN assay was realized by creating a microarray platform based on superhydrophobic–hydrophilic droplet microarrays on ITO slides and MALDI-MSI. Concentration of droplets on hydrophilic patches allowed miniaturization of the assay down to 40 nL and 10 cells per spot [43].

Another example of a recently developed cell-based assay was organic anion transporting polypeptide 2B1 (OATP2B1) assay (Figure 12.4). OATP2B1 belongs to the solute carrier transporter family and is an important target due to its role in the drug uptake in the intestine and liver as well as potential involvement in mechanisms of drug–drug interactions. Estrone-3-sulfate, a physiological

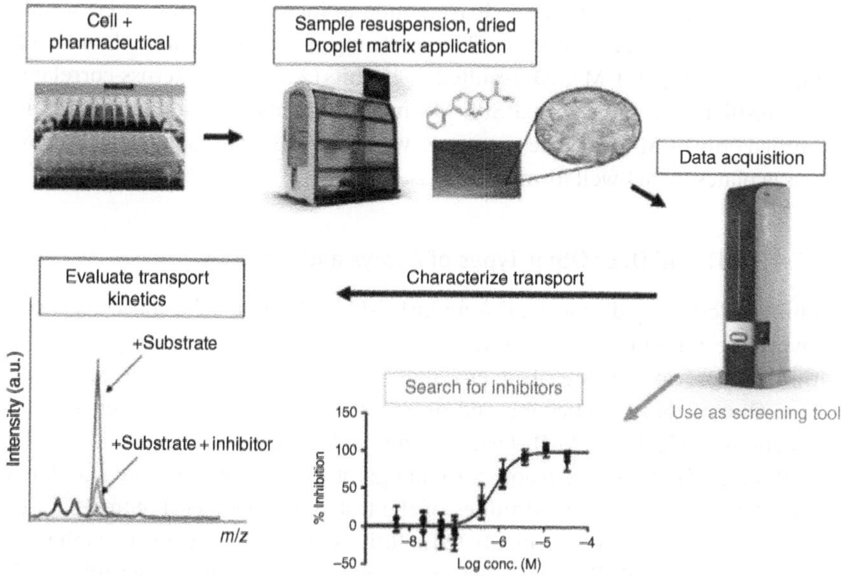

Figure 12.4 Automated workflow of the cellular MALDI MS uptake assay: Adherent HEK293 cells overexpressing OATP2B1 are grown and treated with compound in a 96-well plate (automated application of dilution series). After washing, the plate containing intact cell layers is snap-frozen in liquid nitrogen. Cell resuspension, mixing with internal standard, and dried droplet sample preparation application on MALDI targets are all automated on a pipetting platform. Data is acquired on a rapifleX MALDI-TOF MS, and ion intensities normalized on the internal standard can be used either as a method for transporter (kinetics) characterization or as an inhibitor screening tool. *Source:* Unger et al. [41]/with permission form American Chemical Society.

substrate of OATP2B1, was chosen as the inhibitor response marker, and the deuterated form of estrone-3-sulfate was used as an internal standard. The assay was optimized for automated sample preparation using 4-phenyl-α-cyanocinnamic acid amide as a matrix followed by HT-MALDI detection in negative ion mode. A library of nearly 300 frequently prescribed small-molecule drugs was screened in this inhibition assay and 67 hits were confirmed. The determined pIC_{50} values had low variability with an average CV 5.2% based on six measurements [41].

For the uHT-MALDI detection of microbial trimethylamine-lyase (TMA-lyase) activity, a different strategy was developed based on the dilution of the assay mixture followed by chemical derivatization. TMA-lyase converts choline into trimethylamine and is a target in drug development due to its role in cardiovascular and kidney diseases. Deuterated substrate and product were used, and ^{13}C, ^{15}N-labeled trimethylamine was added as an internal standard. The 10× dilution of the assay mixture was used to reduce the complexity followed by the chemical derivatization with substituted iodoacetamide to increase the molecular weight and reduce volatility of the product. The screening of a library of more than 10,000 compounds using uHT-MALDI resulted in 252 hits (2.5% hit rate) cross-correlated to the results from multiplexed SPE-MS measurements, and additional 61 hits identified only by MALDI. The average Z′ values of 0.73 were calculated based on 32 assay plates in 384-well format [39].

12.2.4 uHT-MALDI of Other Types of Assays and Libraries

Within modern drug discovery, HT- and uHT-MALDI approaches have also been shown to be relevant in other types of assays and screens, not easily classified within the previously described enzymatic, chemical reactions, or cell-based assays. One of these studies focused on screening non-covalent protein–ligand interactions using HT-MALDI-based affinity selection mass spectrometry. The workflows utilized size-exclusion chromatography (SEC) and internal standards to calculate relative binding affinities of the tested compounds (Figure 12.5). In the main screening workflow, incubated solutions of protein tyrosine phosphatase 1B and small-molecule ligands were transferred to 384-well filter plates filled with SEC resin. The protein–ligand complexes and unbound protein passed through the resin, while small-molecule ligands and SEC quality control standard were retained (Figure 12.5a). The bound small-molecule ligands were eluted, then detected by HT-MALDI and classified as primary hits. The primary hits were validated using a hit confirmation workflow. This secondary workflow included repeated incubated solutions of the protein-primary hit compound complexes measured against low controls that included a mixture of an unrelated protein with the primary hit compound and high controls that included pure DMSO with the protein, followed by spiking the eluate with the primary hit compound (Figure 12.5b). A validation study was performed with more than 23,000 compounds and resulted in 240 primary hits (1.02% hit rate) and five confirmed hits (0.021% validated hit rate) [44].

Several studies were devoted to compound library analysis by HT- or uHT-MALDI. Acoustic sample deposition onto ITO-coated glass slides in 384 or 1536 format was accomplished by first depositing nanoliter volumes of matrix solution, followed by vacuum drying, then depositing tens of nanoliters of library compound solutions in DMSO and a second vacuum dry. HT-MALDI analysis of an array containing 384 compounds resulted in 75% positive identification rate in the first pass, which was improved to 85% in the second pass [45].

Very recently, Bennett et al. developed uHT SPE-MALDI workflow for analysis of peptide libraries. Automated SPE was performed in 96-well plates using four types of ion exchange resin including strong-cation, weak-anion, strong-anion, and weak-cation exchange resins. These plates were reformatted into 384-well

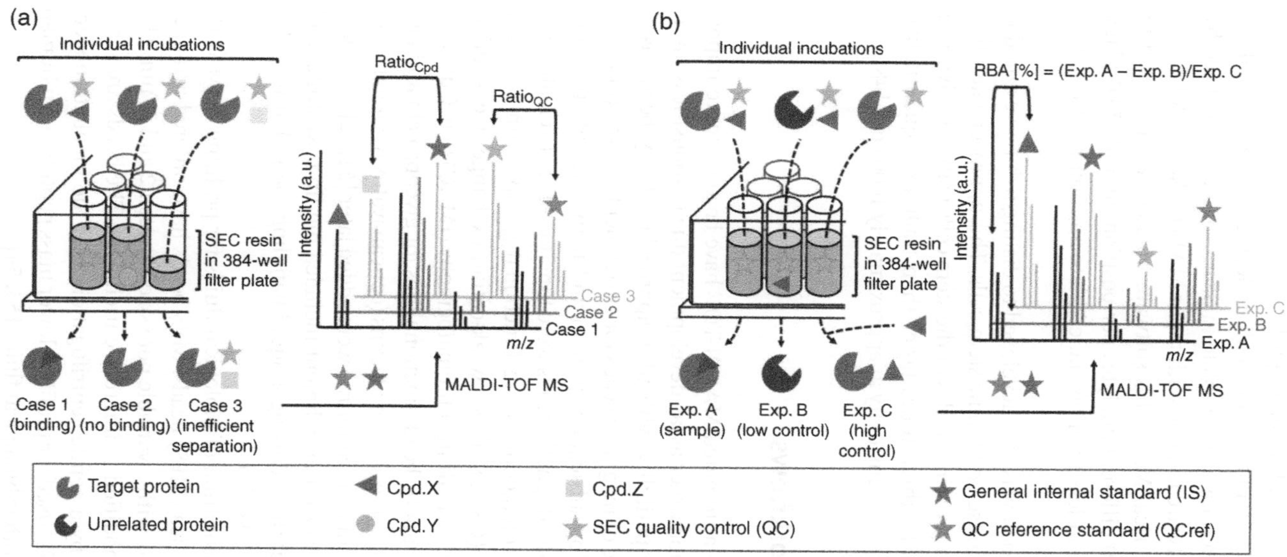

Figure 12.5 Scheme of MALDI-TOF affinity selection mass spectrometry (AS-MS) assay workflows. (a) Screening workflow: Individual incubation solutions are transferred from 384-well assay plates onto size-exclusion chromatography (SEC) resin-loaded 384-well filter plates. Protein and protein–ligand complexes are separated from unbound components via centrifugation, and eluates are captured. After addition of standards, samples are spotted onto MALDI target plates and analyzed by MALDI-TOF MS. Compounds binding to the protein target (Case 1) produce specific signals in MS analysis compared to samples in which the compound is retained by the SEC resin (Case 2). Intactness of the SEC-based separation procedure is monitored by quantifying a quality control (QC) peptide to detect false positives caused by inefficient separation (Case 3). (b) Hit confirmation workflow: In addition to regular sample incubation (Exp. A), controls containing either an unrelated target (Exp. B) or pure DMSO instead of test compound (Exp. C) are prepared. In the case of high controls, compound is spiked to eluted high controls to simulate full target binding. The amount of protein binding for samples (Exp. A) is then quantified in relation to low and high controls, respectively. *Source:* Reproduced with permission from Simon et al. [44]/SAGE Publications.

format followed by automated spotting of MALDI plates and uHT-MALDI analysis. The relative quantitation was accomplished by calculating percent recovery and using neurotensin as an internal standard [46].

In our laboratory, we prepared a library of small-molecule pharmaceuticals in the mass range of 155–760 Da on a MALDI plate in 384 format and analyzed by uHT-MALDI using a new software module within our MALDI PharmaPulse software (Bruker Daltonics, Billerica, MA). This software module was designed to enable detection of up to 100 different analytes per position on a MALDI plate. The automated runs resulted in less than one failed plate position per run (averaged from three runs) [47].

A HT-MALDI-based approach was also developed to distinguish naïve and differentiating populations of mouse embryonic stem cells. This approach classified signal features extracted from protein profiles of the stem cells using multivariate analysis. It was estimated to be three times faster than traditional quantitative PCR (qPCR) while having similar sensitivity at approximately 1000 cells [48].

12.3 Bead-Based Workflows

Bead-based workflows using automated MALDI-TOF have become more important in drug discovery and biomarker assay development due to robustness, miniaturization ability, and compatibility with uHT approaches. Early bead-based workflows focused on the use of magnetic beads for enrichment and fractionation of biological fluid samples for subsequent protein and native peptide profiling using automated MALDI-TOF. Magnetic beads were functionalized to create bead surfaces with hydrophobic interaction using functional groups with different length of the carbon chain (C3, C8, C18), weak-cation exchange, weak-anion exchange, immobilized metal ion affinity chromatography (IMAC) (e.g. Cu^{2+} and Fe^{3+}), and others [49, 50]. Magnetic bead functionalities for glycoprotein and glycopeptide capture based on boronic acid and lectins (concanavalin A, wheat germ agglutinin, and agglutinin from *Lens culinaris*) were added later [51, 52]. A typical magnetic bead workflow for serum enrichment included binding, washing, and elution steps, which were automated using a liquid handling robot that was also used for spotting MALDI targets [50].

Another approach to fractionation of biological fluids for profiling by MALDI was based on mesoporous silica (MPS) beads. MPS beads have ordered arrangement of cavities and channels with a variable pore size between 2 and 50 nm. The cavities can be chemically modified to improve enrichment. The beads are separated from the supernatant by a gentle centrifugation after the incubation step. The MPS beads were primarily used for profiling lower mass peptidome in human plasma, urine, sputum, and other biological fluids [53, 54].

Several immuno-affinity enrichment workflows based on magnetic beads were coupled to detection by MALDI-TOF to develop targeted biomarker assays. SISCAPA (stable isotope standards and capture by anti-peptide antibodies) workflow was originally developed for LC-MS [55] and later adapted to MALDI-TOF [56]. This workflow is based on binding anti-peptide antibodies to protein G-coated magnetic beads for immuno-affinity capture of target peptides and includes automated sample preparation. The SISCAPA approach provides very accurate quantitation, with average CVs of 2.2% as calculated from light/heavy peptide peak area ratios [56]. This workflow was applied to develop quantitative assays for a protein C inhibitor relevant to the recurrence of prostate cancer and for apolipoproteins A-I and B-100 important in the development of cardiovascular disease [57, 58].

Immuno-MALDI (iMALDI) is very similar to the SISCAPA workflow, also using anti-peptide antibodies bound to protein G-coated magnetic beads for immuno-affinity capture. iMALDI was optimized for automated sample preparation in 96-well plates using a liquid handling robot with a 96-channel pipetting head equipped for working with magnetic beads. The automated iMALDI workflow was used to develop assays for plasma renin activity based on the detection of angiotensin I peptide, AKT1 (RAC-alpha serine/threonine-protein kinase) and AKT2 (RAC-beta serine/threonine-protein kinase) in breast and colorectal cancer cell lines and tumors, and PI3-kinase signaling pathway proteins [59–62].

Bead-assisted mass spectrometry (BAMS) is a microarray assay platform developed by Bergo and coworkers based on immuno-affinity capture using magnetic beads and detection by HT-MALDI [63]. BAMS beads are agarose beads with a diameter of approximately 375 μm with magnetic particle cores that contain covalently attached antibodies with specificity for a proteolytic peptide generated from a protein of interest (bottom-up approach) or specificity for an intact protein (top-down approach). In the bottom-up workflow, biological samples of interest (cells, tissues, etc.) are lysed and digested with a protease, followed by incubation with BAMS beads to capture analytes of interest and then several washing steps (Figure 12.6). The beads are transferred to a "sandwich" assembly that includes a gold-coated glass slide (25 × 75 mm) tightly covered with a multi-well silicone gasket with a multi-chamber frame on top, held together with metal clips (Figure 12.6). The wells in the silicone gasket have a diameter 500 μm, such that only one BAMS bead can fit into the well [63]. After the BAMS beads settle into their respective wells, the "sandwich" is disassembled and the BAMS slide (still covered with the multi-well gasket) is sprayed with a MALDI matrix solution to simultaneously elute the analyte and prepare the slide for HT-MALDI analysis on an autoflex maX mass spectrometer (Bruker Daltonics, Billerica, MA). The completed BAMS slides have 88 × 26 microarrays containing 2288 sample positions.

A wide array of BAMS assays with applications in neurology, oncology, epigenetics, and virology have been developed. These BAMS assays include multiplexed

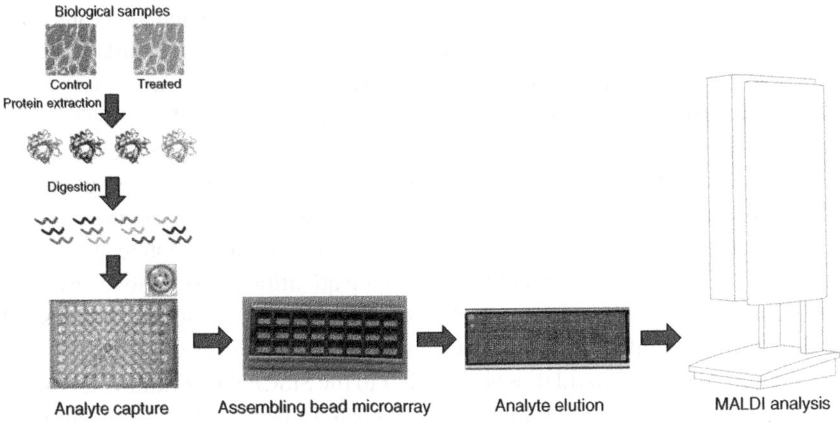

Figure 12.6 Bead-assisted mass spectrometry (BAMS) workflow. Lysed and digested biological samples (cells, tissues, etc.) are incubated with BAMS magnetic beads to capture analytes of interest. The beads are washed several times followed by a transfer into a "sandwich" assembly that includes a gold-coated glass slide, multi-well silicone gasket, and a multi-chamber frame held together with metal clips. Once beads settle into wells, the assembly is disassembled and the slide (still covered with the multi-well gasket) is sprayed with MALDI matrix. The prepared BAMS slide is then placed in an adapter and loaded into a mass spectrometer for HT-MALDI analysis.

amyloid beta (Aβ) peptide 1-x (from Aβ1-10 to Aβ1-42) assay, six tau protein assays, two ERK1/ERK2 (extracellular signal-regulated kinase 1 and 2) assays, three 4EBP1 (eukaryotic translation initiation factor 4E-binding protein 1) assays, three histone H3 assays, histone H4 acetyl-lysine assay, SARS-CoV-2 S1 glycoprotein assay, and many others [63–65].

A different microarray approach called bead-based global proteomic screening (Bead-GPS) was developed for proteome-wide drug screening using MALDI-MSI. Bead-GPS is based on photocleavable mass-tags (PC-Mass-Tags) for coding and arrayed proteins attached to 30–34 μm diameter streptavidin-coated beads. Following an initial study, where agarose, glass, and polymer (poly[styrene-oxyethylene] copolymer) beads were tested, agarose bead microarrays were formed on ITO-coated custom fiber optic microwell plates (25 × 75 mm) with 45 μm well diameter containing approximately one million wells [66, 67]. The binding of two previously known kinase-targeted drugs was studied against an immobilized library containing 50 kinases. MALDI-MSI was used to simultaneously detect the PC-Mass-Tags after photocleavage and the drug molecules which had bound to the kinases on the beads [67].

12.4 Using Functionalized, Modified, and Microarrayed MALDI Plates for HT-MALDI

Substantial efforts have been dedicated to the development of functionalized, modified, or microarrayed MALDI plates and chips since the early days of MALDI MS. Some of these MALDI plates facilitated sample concentration while others allowed matrix-free analysis, increased the density of sample spots for higher throughput, or facilitated selective immobilization, capture, or enrichment of substrates and products. One of the early examples was a prestructured MALDI plate with hydrophobic surface and hydrophilic anchors called the AnchorChip plate. The anchors had diameter range 400–2000 μm and allowed concentrating samples into spots of smaller surface area, thus improving sensitivity of the measurements [68]. Currently, AnchorChip plates in 1536 format are used to improve sensitivity in assay development and screening by HT-MALDI [29, 33–35].

Several different types of MALDI plates were developed for matrix-free analysis of small molecules and other analytes with suggested applications in drug discovery and pharmaceutical analysis. These matrix-free plates included DIOS (desorption ionization on silicon) chips based on etched silicon wafers [69–71], NALDI (nanostructure-assisted laser desorption/ionization) plates based on nanowire technology [72], NIMS (nanostructure-initiator mass spectrometry) chips based on silicon chips with fluorous liquid-coated surface [73], and NAPA-LDI (nanopost array-laser desorption/ionization) chips based on silicon nanopost array technology [74, 75], among others.

An example of a microarray plate designed to improve throughput was a high-density micropatterned array (HDMA). The HDMA plate was a 25 × 75 mm stainless steel plate with thickness 0.3 mm that contained 5076 sample spots in 108 × 47 array, a barcode, and teaching spots. The hydrophilic spots had 300 μm diameter and were separated from each other by 200 μm of a hydrophobic silane surface. The HDMA plates were used to analyze 20,304 identical spots of a mixture containing a MALDI matrix and four small peptides, as well as a library of 47 primary amines [76].

Self-assembled monolayers for MALDI (SAMDI) mass spectrometry have been one of the more developed approaches combining modified MALDI plate surfaces and HT-MALDI. SAMDI is based on monolayers of alkanethiolates attached to gold surfaces with oligo(ethylene glycol) groups extending out. The glycol groups are important in preventing nonspecific binding, and some are functionalized with groups such as maleimide that react with substrates of interest and form covalent conjugates. These conjugates are desorbed in MALDI process and detected as alkanethiolates or dialkyl disulfides [77, 78]. The monolayers with covalently attached substrates and products can be washed to remove buffer components that are detrimental to MALDI analysis. SAMDI plates were made by coating MALDI stainless steel plates with 4–10 nm layers of titanium followed by a 22–35 nm layer of gold to form standard 384 position arrays with gold circles surrounded by a hydrophobic surface. The plates were then soaked in a solution containing a mixture of a symmetric disulfide and an asymmetric disulfide with a specific functional group to form a monolayer presenting the functional group of choice [79, 80].

SAMDI approach was successfully used for a range of enzymatic assays where substrates and products were peptides including lysine deacetylase [79, 81], kinase [78], phosphatase [82, 83], lysine demethylase [84], arginase [85], and others. In most cases, the peptide substrates had a C-terminal cysteine residue which was covalently attached to the monolayer by the reaction between the thiol group of the cysteine and the maleimide group presented by the functionalized monolayer. While some of the published studies were proof-of-concept experiments, several others focused on larger scale screens with 100,000–175,000 library compounds tested. The compounds were screened in pools of eight to compensate for limited throughput, which necessitated a second round of individual compound screening of the pools with inhibitory properties, followed by hit validation [79, 81, 85].

Recently, the SAMDI approach was applied to RNA methyltransferase METTL3/METTL14 assay, where the substrate and product were RNA oligonucleotides containing 11 residues. The oligonucleotide substrate had a biotin moiety at the 3′ terminus and was captured on a self-assembled monolayer presenting neutravidin. The SAMDI assay was validated using radiometric flashplate-based assay with one of the substrates labeled with tritium [86]. Very recently, the SAMDI workflow was applied to develop inhibition assays for SARS-CoV-2 3-chymotrypsin-like cysteine protease (3CLpro) and nonstructural protein 14 (NSP14) that exhibited exonuclease activity. These protein targets are considered to be important in antiviral drug discovery efforts. The 3CLpro assay used a biotinylated peptide as a substrate and the NSP14 exonuclease assay used an RNA biotinylated at the 5′ terminus. The substrates were captured by the monolayers presenting neutravidin in both assays [87, 88]. More than 10,000 library compounds were screened in the NSP14 exonuclease assay.

The SAMDI workflow is not only applicable to standard enzymatic assays, as it has also been used in cell-based assays to probe the activity of lysine deacetylase and phosphatases. In one approach, the substrates captured on monolayers were treated with cell lysates [89, 90]. In a second approach, known as tandem culture and lysis SAMDI (TCAL-SAMDI), cells were cultured on monolayers presenting both cell adhesion ligands and a phosphotyrosine peptide, which was a substrate for protein tyrosine phosphatase. The cells attached to the monolayer were incubated with lysis buffer, washed, and then the monolayer was covered with a MALDI matrix. The dephosphorylated peptide product was detected from gold-coated spots containing as few as tens of cells [80]. Another application of the SAMDI workflow was SAMDI-based affinity selection MS. In this proof-of-concept study, biotinylated proteins (β-secretase or dual leucine zipper kinase) were incubated with pools of 3–20 small-molecule compounds. The noncovalent compound–protein complexes and proteins with no compounds attached were captured by the monolayer presenting neutravidin. MALDI analysis of the washed monolayers detected and quantified only the bound compounds [91].

A relatively recent example of a different approach to microarray slide design for automated MALDI-TOF involved depositing MoS_2 nanoflakes in 16×6 array (96 positions with diameter 1.9 mm) on a standard size, disposable ITO slide. MoS_2 acted as a matrix for MALDI analysis in negative ion mode. The slides were used for identification and quantitation of sulfonamide antibiotics in food samples. The sulfonamides were enriched using immuno-affinity capture on magnetic beads prior to deposition onto the slides [92].

12.5 Summary and Future Trends

The last eight years have been a particularly exciting period in the development of HT- and uHT-MALDI MS. The paradigm shifted from standalone MALDI-TOF instruments used for screening and other applications to fully integrated systems with front-end robotics for uHTS. The analysis time per sample (cycle time) dropped from seconds to as little as 0.3 seconds as laser repetition rates increased by nearly an order of magnitude to 10 kHz. The variety of assays suitable for uHT-MALDI has increased, from relatively simple enzymatic assays to more complex enzymatic assays involving wider array of substrates and products, to screening nanomole-scale chemical reactions, and to cell-based assays of increasing complexity. The scale of screening projects increased dramatically, from proof-of-concept studies involving thousands of library compounds to screens of hundreds of thousands and even more than a million compounds.

The trend for further miniaturization of assays will likely lead to further increases in sample array density on regular size MALDI plates and on 25×75 mm slides with conductive coatings. Increases in sample array density will lead in turn to higher throughput, as fewer MALDI plates will be needed to complete the screen and less time will be required for plate loading and unloading.

The general mass spectrometry trend of improving resolution and mass accuracy will undoubtedly apply to uHT-MALDI. Orthogonally designed instruments with ultrahigh resolution QTOF analyzers and dual MALDI/ESI ionization sources have a lot of application possibilities. Ion mobility spectrometry, particularly TIMS (trapped ion mobility spectrometry) [93], will be beneficial to uHT-MALDI in separating isobaric compounds and compounds with overlapping isotopic patterns. Very recently, this technology had been implemented in a timsTOF MALDI PharmaPulse mass spectrometer (Bruker Daltonics, Billerica, MA). MALDI-TIMS is expected to provide the most significant advantage to assays where substrates and products are small molecules or peptides. Laser-induced postionization (MALDI-2) [94] also holds potential to improve measurements of assays with small molecules as substrates and products.

The rapid technological developments in uHT-MALDI described in this chapter led to the quickly growing range of assays and chemical reactions amenable to this detection method. HT- and uHT-MALDI MS have been successfully applied to some of the most relevant target classes for drug discovery [95], which will insure further growth and development of these techniques in the future.

Acknowledgment

The author thanks Dr. Amanda Bulman Monteiro Rodrigues for reviewing the manuscript and providing helpful suggestions.

References

1 Karas, M. and Hillenkamp, F. (1988). Laser desorption ionization of proteins with molecular masses exceeding 10,000 daltons. *Anal. Chem.* 60 (20): 2299–2301.

2 Tanaka, K., Waki, H., Ido, Y. et al. (1988). Protein and polymer analyses up to *m/z* 100 000 by laser ionization time-of-flight mass spectrometry. *Rapid Commun. Mass Spectrom.* 2 (8): 151–153.

3 Kostrzewa, M. (2018). Application of the MALDI Biotyper to clinical microbiology: progress and potential. *Expert Rev. Proteomics* 15 (3): 193–202.

4 Spraggins, J.M., Rizzo, D.G., Moore, J.L. et al. (2016). Next-generation technologies for spatial proteomics: integrating ultrahigh speed MALDI-TOF and high mass resolution MALDI FTICR imaging mass spectrometry for protein analysis. *Proteomics* 16 (11, 12): 1678–1689.

5 Jensen, O.N., Mortensen, P., Vorm, O. et al. (1997). Automation of matrix-assisted laser desorption/ionization mass spectrometry using fuzzy logic feedback control. *Anal. Chem.* 69 (9): 1706–1714.

6 Griffin, T.J. and Smith, L.M. (2000). Single-nucleotide polymorphism analysis by MALDI-TOF mass spectrometry. *Trends Biotechnol.* 18 (2): 77–84.

7 Pusch, W., Wurmbach, J.H., Thiele, H. et al. (2002). MALDI-TOF mass spectrometry-based SNP genotyping. *Pharmacogenomics* 3 (4): 537–548.

8 Buetow, K.H., Edmonson, M., MacDonald, R. et al. (2001). High-throughput development and characterization of a genomewide collection of gene-based single nucleotide polymorphism markers by chip-based matrix-assisted laser desorption/ionization time-of-flight mass spectrometry. *Proc. Natl. Acad. Sci. U.S.A.* 98 (2): 581–584.

9 Stanssens, P., Zabeau, M., Meersseman, G. et al. (2004). High-throughput MALDI-TOF discovery of genomic sequence polymorphisms. *Genome Res.* 14 (1): 126–133.

10 Traini, M., Gooley, A.A., Ou, K. et al. (1998). Towards an automated approach for protein identification in proteome projects. *Electrophoresis* 19 (11): 1941–1949.

11 Yan, J.X., Sanchez, J.C., Binz, P.A. et al. (1999). Method for identification and quantitative analysis of protein lysine methylation using matrix-assisted laser desorption/ionization--time-of-flight mass spectrometry and amino acid analysis. *Electrophoresis* 20 (4, 5): 749–754.

12 Medzihradszky, K.F., Campbell, J.M., Baldwin, M.A. et al. (2000). The characteristics of peptide collision-induced dissociation using a high-performance MALDI-TOF/TOF tandem mass spectrometer. *Anal. Chem.* 72 (3): 552–558.

13 Suckau, D., Resemann, A., Schuerenberg, M. et al. (2003). A novel MALDI LIFT-TOF/TOF mass spectrometer for proteomics. *Anal. Bioanal. Chem.* 376 (7): 952–965.

14 Mirgorodskaya, E., Braeuer, C., Fucini, P. et al. (2005). Nanoflow liquid chromatography coupled to matrix-assisted laser desorption/ionization mass spectrometry: sample preparation, data analysis, and application to the analysis of complex peptide mixtures. *Proteomics* 5 (2): 399–408.

15 Chen, H.S., Rejtar, T., Andreev, V. et al. (2005). Enhanced characterization of complex proteomic samples using LC-MALDI MS/MS: exclusion of redundant peptides from MS/MS analysis in replicate runs. *Anal. Chem.* 77 (23): 7816–7825.

16 Hofmann, S., Glückmann, M., Kausche, S. et al. (2005). Rapid and sensitive identification of major histocompatibility complex class I-associated tumor peptides by nano-LC MALDI MS/MS. *Mol. Cell. Proteomics* 4 (12): 1888–1897.

17 Sparbier, K., Asperger, A., Resemann, A. et al. (2007). Analysis of glycoproteins in human serum by means of glycospecific magnetic bead separation and LC-MALDI-TOF/TOF analysis with automated glycopeptide detection. *J. Biomol. Tech.* 18 (4): 252–258.

18 Hsieh, F., Keshishian, H., and Muir, C. (1998). Automated high-throughput multiple target screening of molecular libraries by microfluidic MALDI-TOF MS. *J. Biomol. Screening* 3 (3): 189–198.

19 Greis, K.D., Zhou, S., Burt, T.M. et al. (2006). MALDI-TOF MS as a label-free approach to rapid inhibitor screening. *J. Am. Soc. Mass Spectrom.* 17 (6): 815–822.

20 Greis, K.D. (2007). Mass spectrometry for enzyme assays and inhibitor screening: an emerging application in pharmaceutical research. *Mass Spectrom. Rev.* 26 (3): 324–339.

21 Ma, H., Deacon, S., and Horiuchi, K. (2008). The challenge of selecting protein kinase assays for lead discovery optimization. *Expert Opin. Drug Discovery* 3 (6): 607–621.

22 Staab, D., Busby, S., Peltier, J.M. et al. (2013). Implementation of a high-throughput screening platform using MALDI-MS as a readout. *2nd Annual Society for Laboratory Automation & Screening Conference*, Orlando, FL (12–16 January 2013).

23 Dikler, S., Busby, S., Hill, W.A.G. et al. (2015). High-throughput screening MALDI-TOF assays beyond 100,000 samples per day. *Proceedings of the 63rd ASMS Conference on Mass Spectrometry and Allied Topics*, St. Louis, MO (31 May–4 June 2015).

24 Lin, S., Dikler, S., Blincoe, W.D. et al. (2018). Mapping the dark space of chemical reactions with extended nanomole synthesis and MALDI-TOF MS. *Science* 361 (6402): eaar6236.

25 Haslam, C., Hellicar, J., Dunn, A. et al. (2016). The evolution of MALDI-TOF mass spectrometry toward ultrahigh-throughput screening: 1536-well format and beyond. *J. Biomol. Screening* 21 (2): 176–186.

26 Winter, M., Ries, R., Kleiner, C. et al. (2019). Automated MALDI target preparation concept: providing ultrahigh-throughput mass spectrometry-based screening for drug discovery. *SLAS Technol.* 24 (2): 209–221.

27 Dikler, S. and Kowalski, P. (2016). Biochemical assay development using novel MALDI high-throughput screening system. *Proceedings of the 64th ASMS Conference on Mass Spectrometry and Allied Topics*, San Antonio, TX (5–9 June 2016).

28 Simon, R.P., Winter, M., Kleiner, C. et al. (2020). MALDI-TOF mass spectrometry-based high-throughput screening for inhibitors of the cytosolic DNA sensor cGAS. *SLAS Discovery* 25 (4): 372–383.

29 Heap, R.E., Hope, A.G., Pearson, L.-A. et al. (2017). Identifying inhibitors of inflammation: a novel high-throughput MALDI-TOF screening assay for salt-inducible kinases (SIKs). *SLAS Discovery* 22 (10): 1193–1202.

30 Beeman, K., Baumgartner, J., Laubenheimer, M. et al. (2017). Integration of an in situ MALDI-based high-throughput screening process: a case study with receptor tyrosine kinase c-MET. *SLAS Discovery* 22 (10): 1203–1210.

31 Winter, M., Bretschneider, T., Kleiner, C. et al. (2018). Establishing MALDI-TOF as versatile drug discovery readout to dissect the PTP1B enzymatic reaction. *SLAS Discovery* 23 (6): 561–573.

32 Machalkova, M., Schejbal, J., Glatz, Z. et al. (2018). A label-free MALDI TOF MS-based method for studying the kinetics and inhibitor screening of the

Alzheimer's disease drug target β-secretase. *Anal. Bioanal. Chem.* 410 (28): 7441–7448.

33 Ritorto, M.S., Ewan, R., Perez-Oliva, A.B. et al. (2014). Screening of DUB activity and specificity by MALDI-TOF mass spectrometry. *Nat. Commun.* 5: 4763.

34 De Cesare, V., Moran, J., Traynor, R. et al. (2020). High-throughput matrix-assisted laser desorption/ionization time-of-flight (MALDI-TOF) mass spectrometry–based deubiquitylating enzyme assay for drug discovery. *Nat. Protoc.* 15 (12): 4034–4057.

35 De Cesare, V., Johnson, C., Barlow, V. et al. (2018). The MALDI-TOF E2/E3 ligase assay as universal tool for drug discovery in the ubiquitin pathway. *Cell Chem. Biol.* 25 (9): 1117–1127.

36 Benz, M., Molla, M.R., Böser, A. et al. (2019). Marrying chemistry with biology by combining on-chip solution-based combinatorial synthesis and cellular screening. *Nat. Commun.* 10 (1): 2879.

37 Benz, M., Asperger, A., Hamester, M. et al. (2020). A combined high-throughput and high-content platform for unified on-chip synthesis, characterization and biological screening. *Nat. Commun.* 11 (1): 5391.

38 Weigt, D., Sammour, D.A., Ulrich, T. et al. (2018). Automated analysis of lipid drug-response markers by combined fast and high-resolution whole cell MALDI mass spectrometry biotyping. *Sci. Rep.* 8 (1): 11260.

39 Winter, M., Bretschneider, T., Thamm, S. et al. (2019). Chemical derivatization enables MALDI-TOF-based high-throughput screening for microbial trimethylamine (TMA)-lyase inhibitors. *SLAS Discovery* 24 (7): 766–777.

40 Weigt, D., Parrish, C.A., Krueger, J.A. et al. (2019). Mechanistic MALDI-TOF cell-based assay for the discovery of potent and specific fatty acid synthase inhibitors. *Cell Chem. Biol.* 26 (9): 1322–1331.

41 Unger, M.S., Schumacher, L., Enzlein, T. et al. (2020). Direct automated MALDI mass spectrometry analysis of cellular transporter function: inhibition of OATP2B1 uptake by 294 drugs. *Anal. Chem.* 92 (17): 11851–11859.

42 Chandler, J., Haslam, C., Hardy, N. et al. (2017). A systematic investigation of the best buffers for use in screening by MALDI-mass spectrometry. *SLAS Discovery* 22 (10): 1262–1269.

43 RamalloGuevara, C., Paulssen, D., Popova, A.A. et al. (2021). Fast nanoliter-scale cell assays using droplet microarray-mass spectrometry imaging. *Adv. Biol.* 5 (3): e2000279.

44 Simon, R.P., Winter, M., Kleiner, C. et al. (2021). MALDI-TOF-based affinity selection mass spectrometry for automated screening of protein-ligand interactions at high-throughput. *SLAS Discovery* 26 (1): 44–57.

45 Chin, J., Wood, E., Peters, G.S. et al. (2016). Acoustic sample deposition MALDI-MS (ASD-MALDI-MS): a novel process flow for quality control screening of compound libraries. *J. Lab. Autom.* 21 (1): 204–207.

46 Bennett, R., Pirrone, G.F., Nowak, T. et al. (2021). Ultrahigh-throughput SPE-MALDI workflow: blueprint for efficient purification and screening of peptide libraries. *Anal. Chim. Acta* 1142: 10–18.

47 Dikler, S. (2020). Novel MALDI-TOF MS workflow for ultrahigh-throughput screening of different analytes at each position on a plate. *9th Annual Society for Laboratory Automation & Screening Conference*, San Diego, CA (25–29 January 2020).

48 Heap, R.E., Segarra-Fas, A., Blain, A.P. et al. (2019). Profiling embryonic stem cell differentiation by MALDI TOF mass spectrometry: development of a reproducible and robust sample preparation workflow. *Analyst* 144 (21): 6371–6381.

49 Pusch, W., Flocco, M.T., Leung, S.M. et al. (2003). Mass spectrometry-based clinical proteomics. *Pharmacogenomics* 4 (4): 463–476.

50 Pusch, W. and Kostrzewa, M. (2005). Application of MALDI-TOF mass spectrometry in screening and diagnostic research. *Curr. Pharm. Des.* 11 (20): 2577–2591.

51 Sparbier, K., Koch, S., Kessler, I. et al. (2005). Selective isolation of glycoproteins and glycopeptides for MALDI-TOF MS detection supported by magnetic particles. *J. Biomol. Tech.* 16 (4): 407–413.

52 Sparbier, K., Wenzel, T., and Kostrzewa, M. (2006). Exploring the binding profiles of ConA, boronic acid and WGA by MALDI-TOF/TOF MS and magnetic particles. *J. Chromatogr. B Analyt. Technol. Biomed. Life Sci.* 840 (1): 29–36.

53 Savino, R. and Terracciano, R. (2012). Mesopore-assisted profiling strategies in clinical proteomics for drug/target discovery. *Drug Discovery Today* 17 (3, 4): 143–152.

54 Hajduk, J., Matysiak, J., and Kokot, Z.J. (2016). Challenges in biomarker discovery with MALDI-TOF MS. *Clin. Chim. Acta* 458: 84–98.

55 Anderson, N.L., Anderson, N.G., Haines, L.R. et al. (2004). Mass spectrometric quantitation of peptides and proteins using stable isotope standards and capture by anti-peptide antibodies (SISCAPA). *J. Proteome Res.* 3 (2): 235–244.

56 Anderson, N.L., Razavi, M., Pearson, T.W. et al. (2012). Precision of heavy-light peptide ratios measured by MALDI-TOF mass spectrometry. *J. Proteome Res.* 11 (3): 1868–1878.

57 Razavi, M., Johnson, L.D., Lum, J.J. et al. (2013). Quantification of a proteotypic peptide from protein C inhibitor by liquid chromatography-free SISCAPA-MALDI mass spectrometry: application to identification of recurrence of prostate cancer. *Clin. Chem.* 59 (10): 1514–1522.

58 van den Broek, I., Nouta, J., Razavi, M. et al. (2015). Quantification of serum apolipoproteins A-I and B-100 in clinical samples using an automated SISCAPA-MALDI-TOF-MS workflow. *Methods* 81: 74–85.

59 Popp, R., Malmström, D., Chambers, A.G. et al. (2015). An automated assay for the clinical measurement of plasma renin activity by immuno-MALDI (iMALDI). *Biochim. Biophys. Acta* 1854 (6): 547–558.

60 Li, H., Popp, R., Frohlich, B. et al. (2017). Peptide and protein quantification using automated immuno-MALDI (iMALDI). *J. Vis. Exp.* 126: e55933.

61 Popp, R., Li, H., LeBlanc, A. et al. (2017). Immuno-matrix-assisted laser desorption/ionization assays for quantifying AKT1 and AKT2 in breast and colorectal cancer cell lines and tumors. *Anal. Chem.* 89 (19): 10592–10600.

62 Froehlich, B.C., Popp, R., Sobsey, C.A. et al. (2020). Systematic optimization of the iMALDI workflow for the robust and straightforward quantification of signaling proteins in cancer cells. *Proteomics Clin. Appl.* 14 (5): e2000034.

63 Hamza, G.M., Bergo, V.B., Mamaev, S. et al. (2020). Affinity-bead assisted mass spectrometry (Affi-BAMS): a multiplexed microarray platform for targeted proteomics. *Int. J. Mol. Sci.* 21 (6): 2016.

64 Silva, J.C., Dikler, S., Mamaev, S. et al. (2020). A targeted multiplexed MALDI MS assay platform using affinity-bead assisted mass spectrometry (Affi-BAMS) for monitoring brain and CSF biomarkers. *Association of Biomolecular Resource Facilities Annual Conference 2020*, Palm Springs, CA (29 February-3 March 2020).

65 Bergo, V.B., Hamza, G.M., Dikler, S. et al. (2020). Multiplexed analytical platform using affinity capture and MALDI MS enables novel assay development for screening Biomarkers in neurological diseases. *Proceedings of the 68th ASMS Conference on Mass Spectrometry and Allied Topics*, Online Meeting (1–12 June 2020).

66 Lim, M.J., Liu, Z., Braunschweiger, K.I. et al. (2014). Correlated matrix-assisted laser desorption/ionization mass spectrometry and fluorescent imaging of photocleavable peptide-coded random bead-arrays. *Rapid Commun. Mass Spectrom.* 28 (1): 49–62.

67 Zhou, Y., Liu, Z., Rothschild, K.J. et al. (2016). Proteome-wide drug screening using mass spectrometric imaging of bead-arrays. *Sci. Rep.* 6: 26125.

68 Schuerenberg, M., Luebbert, C., Eickhoff, H. et al. (2000). Prestructured MALDI-MS sample supports. *Anal. Chem.* 72 (15): 3436–3442.

69 Wei, J., Buriak, J.M., and Siuzdak, G. (1999). Desorption-ionization mass spectrometry on porous silicon. *Nature* 399 (6733): 243–246.

70 Thomas, J.J., Shen, Z., Crowell, J.E. et al. (2001). Desorption/ionization on silicon (DIOS): a diverse mass spectrometry platform for protein characterization. *Proc. Natl. Acad. Sci. U.S.A.* 98 (9): 4932–4937.

71 Shen, Z., Thomas, J.J., Averbuj, C. et al. (2001). Porous silicon as a versatile platform for laser desorption/ionization mass spectrometry. *Anal. Chem.* 73 (3): 612–619.

72 Daniels, R.H., Dikler, S., Li, E. et al. (2008). Break free of the matrix: sensitive and rapid analysis of small molecules using nanostructured surfaces and LDI-TOF mass spectrometry. *J. Assoc. Lab. Autom.* 13 (6): 314–321.

73 Northen, T.R., Lee, J.-C., Hoang, L. et al. (2008). A nanostructure-initiator mass spectrometry-based enzyme activity assay. *Proc. Natl. Acad. Sci. U.S.A.* 105 (10): 3678–3683.

74 Walker, B.N., Stolee, J.A., and Vertes, A. (2012). Nanophotonic ionization for ultratrace and single-cell analysis by mass spectrometry. *Anal. Chem.* 84 (18): 7756–7762.

75 Korte, A.R., Morris, N.J., and Vertes, A. (2019). High-throughput complementary analysis and quantitation of metabolites by MALDI- and silicon nanopost array-laser desorption/ionization-mass spectrometry. *Anal. Chem.* 91 (6): 3951–3958.

76 Lee, J.H., Choi, H.S., Nasr, K.A. et al. (2011). High-throughput small molecule identification using MALDI-TOF and a nanolayered substrate. *Anal. Chem.* 83 (13): 5283–5289.

77 Su, J. and Mrksich, M. (2002). Using mass spectrometry to characterize self-assembled monolayers presenting peptides, proteins, and carbohydrates. *Angew. Chem. Int. Ed.* 41 (24): 4715–4718.

78 Mrksich, M. (2008). Mass spectrometry of self-assembled monolayers: a new tool for molecular surface science. *ACS Nano* 2 (1): 7–18.

79 Gurard-Levin, Z.A., Scholle, M.D., Eisenberg, A.H. et al. (2011). High-throughput screening of small molecule libraries using SAMDI mass spectrometry. *ACS Comb. Sci.* 13 (4): 347–350.

80 Berns, E.J., Cabezas, M.D., and Mrksich, M. (2016). Cellular assays with a molecular endpoint measured by SAMDI mass spectrometry. *Small* 12 (28): 3811–3818.

81 Patel, K., Sherrill, J., Mrksich, M. et al. (2015). Discovery of SIRT3 inhibitors using SAMDI mass spectrometry. *J. Biomol. Screen.* 20 (7): 842–848.

82 Szymczak, L.C., Huang, C.F., Berns, E.J. et al. (2018). Combining SAMDI mass spectrometry and peptide arrays to profile phosphatase activities. *Methods Enzymol.* 607: 389–403.

83 Huang, C.F. and Mrksich, M. (2019). Profiling protein tyrosine phosphatase specificity with self-assembled monolayers for matrix-assisted laser desorption/ionization mass spectrometry and peptide arrays. *ACS Comb. Sci.* 21 (11): 760–769.

84 Wigle, T.J., Swinger, K.K., Campbell, J.E. et al. (2015). A high-throughput mass spectrometry assay coupled with redox activity testing reduces artifacts and false positives in lysine demethylase screening. *J. Biomol. Screen.* 20 (6): 810–820.

85 Scholle, M.D. and Gurard-Levin, Z.A. (2021). Development of a novel label-free and high-throughput arginase-1 assay using self-assembled monolayer desorption ionization mass spectrometry. *SLAS Discovery* 26 (6): 775–782.

86 Buker, S.M., Gurard-Levin, Z.A., Wheeler, B.D. et al. (2020). A mass spectrometric assay of METTL3/METTL14 methyltransferase activity. *SLAS Discovery* 25 (4): 361–371.

87 Gurard-Levin, Z.A., Liu, C., Jekle, A. et al. (2020). Evaluation of SARS-CoV-2 3C-like protease inhibitors using self-assembled monolayer desorption ionization mass spectrometry. *Antiviral Res.* 182: 104924.

88 Scholle, M.D., Liu, C., Deval, J. et al. (2021). Label-free screening of SARS-CoV-2 NSP14 exonuclease activity using SAMDI mass spectrometry. *SLAS Discovery* 26 (6): 766–774.

89 Kuo, H.Y., DeLuca, T.A., Miller, W.M. et al. (2013). Profiling deacetylase activities in cell lysates with peptide arrays and SAMDI mass spectrometry. *Anal. Chem.* 85 (22): 10635–10642.

90 Szymczak, L.C., Sykora, D.J., and Mrksich, M. (2020). Using peptide arrays to profile phosphatase activity in cell lysates. *Chemistry* 26 (1): 165–170.

91 VanderPorten, E.C., Scholle, M.D., Sherrill, J. et al. (2017). Identification of small-molecule noncovalent binders utilizing SAMDI technology. *SLAS Discovery* 22 (10): 1211–1217.

92 Zhao, Y., Tang, M., Liao, Q. et al. (2018). Disposable MoS(2)-arrayed MALDI MS chip for high-throughput and rapid quantification of sulfonamides in multiple real samples. *ACS Sens.* 3 (4): 806–814.

93 Fernandez-Lima, F., Kaplan, D.A., Suetering, J. et al. (2011). Gas-phase separation using a trapped ion mobility spectrometer. *Int. J. Ion Mobil. Spectrom.* 14 (2, 3): 93–98.

94 Soltwisch, J., Heijs, B., Koch, A. et al. (2020). MALDI-2 on a trapped ion mobility quadrupole time-of-flight instrument for rapid mass spectrometry imaging and ion mobility separation of complex lipid profiles. *Anal. Chem.* 92 (13): 8697–8703.

95 McLaren, D.G., Shah, V., Wisniewski, T. et al. (2021). High-throughput mass spectrometry for hit identification: current landscape and future perspectives. *SLAS Discovery* 26 (2): 168–191.

13

Development and Applications of DESI-MS in Drug Discovery

Wenpeng Zhang

State Key Laboratory of Precision Measurement Technology and Instruments, Department of Precision Instrument, Tsinghua University, Beijing, China

13.1 Introduction

Drug discovery involves screening of new drug targets, identification of drug candidates, evaluation of therapeutic efficacy, testing of the quality of pharmaceuticals, and monitoring of therapeutic drugs and their metabolism in vivo [1, 2]. It is also associated with the biological processes such as protein interactions and metabolism. Mass spectrometry (MS) is a powerful tool in the field of drug discovery due to its capability in the analysis of complex samples and powerful structure identification ability [3–5]. MS analysis can provide highly specific molecular information of a wide range of compounds based on their molecular weights and chemical structures. In comparison with other analytical technologies, the most outstanding advantage of MS is its wide applicability and excellent quantitative capability even at low concentrations, which is challenging for technologies such as immunoassay. Besides, MS is essential for pharmacokinetic studies and identification of new drug metabolites [6, 7], and MS-based proteomics, metabolomics, and lipidomics are important platforms for drug development. With the development of instrumentation and methods, MS has also been widely employed for investigation of interactions between biological molecules, which can be used for studying drug–protein interactions in drug discovery [8–10].

Liquid chromatography (LC)-MS with carefully optimized sample pretreatment procedures is the primary MS method in most drug analysis applications [3]. Sample pretreatment and chromatographic separation allow for removal of matrices and isolation of targeted compounds with other endogenous or exogenous compounds before MS analysis, which can provide high reproducibility for

High-Throughput Mass Spectrometry in Drug Discovery, First Edition. Edited by Chang Liu and Hui Zhang.
© 2023 John Wiley & Sons, Inc. Published 2023 by John Wiley & Sons, Inc.

quantitation of different types of compounds. LC-MS is also the main method of proteomics, metabolomics, and lipidomics. The combination of retention behavior and native MS is useful for characterization of molecular interactions. However, chromatography-based MS methods required rigorous sample preparation or purification are labor- and time-consuming. The complicated procedures, evolvement of multiple expensive equipment pieces, as well as overall long turnaround time make LC-MS limited in some drug discovery applications.

The concept of ambient ionization MS was introduced in order to minimize sample preparation and chromatographic separation in MS analysis [11]. With ambient ionization, direct ionization of compounds from raw samples can be achieved in their ambient states [12–14]. It should be noted that ambient ionization is different from the atmospheric pressure ionization methods such as ESI and APCI, which usually need purified samples and cannot use raw samples. Since the invention of desorption electrospray ionization (DESI) [11], more than 40 ambient ionization methods have been developed during past years [14, 15], which have successfully contributed to environmental, biological, medical, and food analysis.

Ambient ionization MS also found its role in drug discovery [16–18]. As the first ambient ionization method, DESI has been well developed in methodologies, instrumentation, and application manners. DESI-MS is now able to identify a wide range of molecules with different masses, quantify drugs and metabolites, and provide spatial distribution of drugs and metabolites via MS imaging. Several reviews have been published by focusing on the general method development of DESI-MS [19, 20], and its applications in forensic analysis [21] and food analysis [22–24]. This chapter will focus on the applications of DESI in drug discovery, including pharmaceutical analysis and therapeutic drug monitoring, natural product analysis, MS imaging, detection of drug–protein interactions, and high-throughput screening. Finally, the future development of DESI-MS in the improvement of drug discovery will also be prospected.

13.2 Development of DESI and Related Ambient Ionization Methods

The invention of ambient ionization is to ionize analytes directly in their native status in raw samples. Since ambient ionization is free of sample preparation, the MS analysis can be performed in real time. The first ambient ionization technology, desorption electrospray ionization (DESI, Figure 13.1a) was presented in 2004 [11]. In general, a DESI experiment includes the formation of charged droplets by electrospray ionization (ESI), transfer of the charged droplet toward the sample surface, collision of the sample surface with charged droplets or free gas

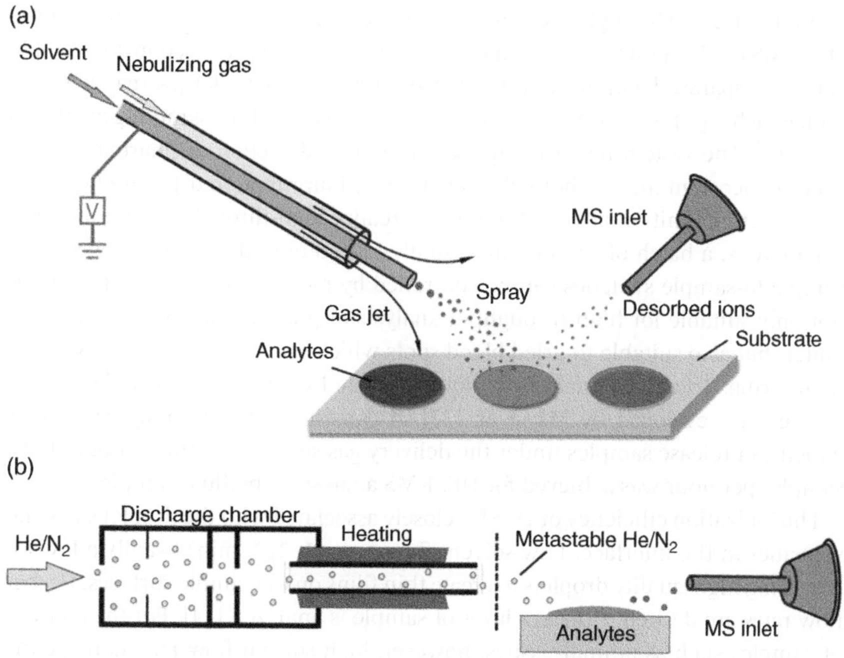

Figure 13.1 Schematics of (a) desorption electrospray ionization (DESI), (b) direct analysis in real time (DART).

phase ions, desorption of droplets, surface materials, and free gas ions from the sample surface, and transfer of these droplets and ions under atmospheric pressure into the interface of the mass spectrometer. During the journey the ionization will be realized through droplet evaporation and electrostatic fission that are similar with in ESI.

In comparison with many other types of ionization methods such as direct analysis in real time (DART, Figure 13.1b) [25], DESI only requires simple setups to realize effective MS analysis. A typical setup consists of a metal capillary with a flow of solvent, which is applied with high voltages for electrospray ionization [11]. A sheath gas (e.g. nitrogen gas) is applied to facilitate the formation of small droplets. The sample is usually placed on a 3D moving stage, which can move samples as demanded for MS analysis; the stage can also adjust the angle between the secondary droplet flow and the inlet of the MS system. It is worth noting that the DESI setup should be optimized to improve the efficiency of ion transport.

In a typical DESI-MS experiment, samples are placed in an insulative polymer substrate, sample-to-MS inlet distance is set to about 30 cm [11]. Relatively short injection distance is considered to be able to decrease ion transfer loss during DESI. However, electrospray ionization would be formed instead of the surface analysis mode. Although the ion transport distance is relatively short in most DESI-MS applications, long-distance detection would be preferred in some applications. Soparawalla et al. [26] developed a non-proximate large-area DESI system, which equipped with a large-area sprayer and capable of sampling an area of ~2.5 cm^2. The system has been applied into rapid detection of pharmaceuticals such as neostigmine, acebutolol, amiodarone, ibuprofen, and β-estradiol, with low detection limits of about 10 ng/cm^2. To realize high-throughput pharmaceutical analysis, a batch of samples are usually loaded onto the sample stage, while sample-to-sample switches can be controlled by moving the stage. This setup is not only suitable for high-throughput analysis of pharmaceuticals such as drug tablets but also suitable to dried blood spots which are usually used in vivo drug monitoring. To enable analysis of liquid sample by DESI-MS, a capillary-based device was developed by Ma et al. [27]. It consists of 16 sampling capillaries, which can release samples under the delivery gas stepwise. A throughput of 600 samples per hour was achieved for DESI-MS analysis of biofluid samples.

The ionization efficiency of DESI is closely associated with the droplet sizes and velocities in the interface. Low solvent flow rates (1–2 μL/min) usually allow for obtaining high-quality droplets to create thin films on the sample surfaces. Higher flow rates are desired if thicker layer of sample is analyzed [14]. For the analysis of samples such as animal tissues, however, high solvent flow may damage the morphology of the tissue, leading to loss of part of chemical information or spatial distribution information for MS imaging [28]. The nebulizing capillary angle can influence the shape of the solvent sprayed. When the capillary-surface angle is set to about 90°, obvious secondary droplets can be observed, and analyte ions are usually difficult to be injected into the mass spectrometer. Using smaller capillary-surface angles are important to obtain good signal for DESI-MS analysis; however, when very small angles (e.g. <45°) are used, ions produced by ESI can be directly injected into the mass spectrometer, which should be taken into consideration in the optimization of the angle for DESI-MS analysis.

The pressure of the gas and the distance between emitter and sample surface can affect the velocities of the droplets directly. Other conditions, such as the electrical potential applied onto the spray emitter and the flow rate of the nebulizing gas, are usually considered to be less important to the variation of the droplet velocities. The nature of using droplets for analyte pick-up, however, can limit the spatial resolution of DESI in comparison with laser-based methods such as matrix-assisted laser desorption ionization (MALDI). For example, in many cases, the spatial resolution of DESI is expected to be 100–200 μm, while resolution as good as 1.4 μm can be obtained by MALDI [29]. During past few years, a lot of efforts have been made to improve the resolution of DESI [30–33]. For example, a synchronization of DESI with the ion injection period was developed to improve spatial resolution by four to six times [30]. This strategy can also improve sensitivity and reduce the amount of sample depleted during the acquisition of each spectrum. The mechanism is associated with decreasing distribution and dispersion of the analytes influenced by the washing effect. Nanospray desorption electrospray ionization (nano-DESI) [34] is a modified DESI method with higher spatial resolution. Instead of using flying droplets for sample pick-up, nano-DESI uses localized flowing droplets for desorption of analytes from sample surfaces into a liquid bridge between two capillaries. Nanospray ionization is performed at the end of the capillary, which can give a higher ionization efficiency and spatial resolution as good as 10 μm [35]. To improve the analytical abilities of DESI-MS, some techniques have also been incorporated. For example, Dixon et al. [36] coupled DESI with a linear ion trap quadrupole mass spectrometer via air ejector, establishing a system for long-distance detection of samples. To improve the extraction performance of DESI, traditional sample preparation techniques such as solid-phase extraction (SPE) have also been introduced to develop integrated systems such as SPE-DESI-MS [37].

13.3 Applications in Drug Discovery

13.3.1 Pharmaceutical Analysis and Therapeutic Drug Monitoring

In the field of pharmaceutical analysis, LC- or GC-based methods are still the gold standard in most of official guidance [38, 39]. Pharmaceutical products are usually not as complicated as biological matrices, but their analysis by chromatography-based MS methods still requires standard, extensive sample preparation and chromatographic separation. This can be contradictory for the analysis of solid pharmaceuticals such as drug tablets which contain relatively simple matrices. Tablet analysis needs additional steps for sample preparation, including milling, meshing, weighing, and dissolution, resulting in longer analysis time [40]. However, solid pharmaceuticals such as tablets often require fast and high-throughput analysis for the purpose of quality inspection [41]. DESI has the advantage of direct analyzing solid pharmaceuticals without additional steps and enabling point-by-point ionization of a batch of samples. Such approach is potentially to be used in high-throughput impurity profiling and quality control in the pharmaceutical industry. In an early work, Chen et al. [42] built a variable-speed moving belt for high-throughput sampling of tablets. Sampling rate of three samples per second was achieved for rapid semiquantitative of drug constituents, and relative standard deviations of the relative ion abundances for major components less than 8% were obtained.

DESI was also found to be useful for in vivo drug analysis in combination with sample sampling methods such as DBS. In an example, Siebenhaar et al. [43]

employed DESI in quantitation of aspirin in DBS collected at different time points after dosing a healthy volunteer with the drug. The limits of detection (LOD) and limits of quantitation (LOQ) were determined to be 8 and 10 μg/mL, respectively, with a wide linear range from 10 to 2000 μg/mL. To improve the reproducibility of DESI-MS for quantitation, it is usually necessary to incorporate internal standards (IS) for calibration. For simultaneous quantitation of sitamaquine, terfenadine, and prazosin in DBS by DESI-MS, Wiseman et al. [44] used verapamil as the IS. LODs were found to be 10 ng/mL, with accuracy as good as 2.7% for some quality control samples. Coupling of DESI with microextraction methods has also been explored for therapeutic drug analysis. Silica-C18 stationary phase-coated fibers were used for solid-phase microextraction (SPME) of drugs from raw urine samples, followed by direct analysis of the SPME fiber by DESI-MS, by which LODs below 50 ng/mL in urine was obtained for therapeutic drugs including cocaine, benzoylecgonine, coca-ethylene, methadone, and norfentanyl [45].

It should be noted that ambient ionization approaches such as DESI-MS have not been included into official guidance for drug testing or therapeutic drug monitoring; therefore, it is worthy to validating the results obtained from DESI-MS with standard methods such as GC-MS and LC-MS [14, 46, 47]. In a previous work [48], urine samples from drug abusers were tested for drugs of abuse, and the urine metabolites were analyzed by DESI-MS. Compounds including codeine, morphine, oxymorphone, Δ^9-tetrahydrocannabinol, alprazolam, oxazepam, and hydroxytemazepam were tested as the metabolites of drugs such as amphetamines, opiates, cannabinoids, and benzodiazepines. The results obtained by DESI-MS was found to correlate well with that obtained by GC-MS for real urine samples and the sensitivity of DESI-MS was adequate for testing of these drugs in real samples.

13.3.2 Analysis of Drugs in Natural Products

Accurate characterization of chemical components in natural products is important of quality control of tradition medicines and discovery of new drugs [49, 50]. The chemical composition of natural drugs is usually very complicated and contains abundant active ingredients, which require separation and extensive sample preparation before analysis. The development of a direct, in situ, real-time analysis method is useful to obtain the most authentic information about the presence of drugs in natural products [51]. In recent years, the development of DESI-MS and other ambient ionization-based MS methods has provided new opportunities for direct and rapid analysis of active compounds in natural products [52–54].

For the analysis of natural produce by DESI-MS, the pretreatment of the samples is important to obtain good analytical results. There are three basic methods for preparation of plant samples: simply pasting onto the sample plate, frozen sections, and imprinting [55]. Samples with smooth and regular surfaces can be directly analyzed, while non-flat surfaces such as plant roots need to be prepared by frozen sections. For samples with soft, irregular surfaces and strong absorbency, it is usually difficult to ionize directly. Imprinting methods have been employed for these types of samples. However, metabolites will rapidly dissolve on the imprinting plate, resulting in loss of signal intensity, and incomplete imprinting will lead to reduced spatial resolution [56]. In order to retain the metabolites from natural drugs for MS analysis, a number of new surface materials have been introduced, such as functionalized porous silicon, functionalized glass, and sol–gel based materials [57–59]. Because of the plant cell wall, the direct ionization of plant tissue samples can be difficult; however, after proper treatment, good signal of active compounds can still be obtained by DESI-MS [60]. The changes of metabolites in the production process can be found through the detection and comparison of active ingredients, and the differences of plant varieties and the production of toxic metabolites can be identified, which makes DESI-MS a useful tool for the quality control of medicinal materials. It also allows qualitative or quantitative analysis of active ingredients in natural medicines to further control sample quality.

At the initial stage of development, the application of DESI-MS mainly focused on the detection of alkaloid active components. Talaty et al. [61] applied DESI-MS into determination of alkaloid-rich plant samples such as poison hemlock, jimsonweed, and deadly nightshade. Nineteen known alkaloids and 15 major alkaloids of belladonna were detected in *Datura stramonium* and the results were confirmed to be stable and reliable. At the perspective of the properties of compounds, alkaloids are usually easy to be ionized by DESI, making it possible to screen active compounds of a wide range of natural medicines such as *Ocotea spixiana* [62], *Pyrenacantha volubilis* Hook [63], *Dysoxylum binectariferum* [64], *Madagascar periwinkle* [65], and other alkaloid-rich plant samples. The chemical composition of natural products is complex, containing alkaloids, flavonoids, organic acids, and other active ingredients. It is important to consider the ability of DESI-MS for the detection of a relatively wide range of active compounds. With the improvement of the device and the method, DESI-MS has also been applied to active compounds that are usually considered to be difficult to ionize, such as diterpene glycosides [60]. DESI-MS with ion trap as the mass analyzer was used for qualitative analysis of phenolic components in cinnamon by multiple stage MS. Although low-content components such as erucic acid was not detected in cinnamon by DESI, the qualitative analysis of phenolic compounds in cinnamon was reliable [66]. DESI-MS has also been applied for detection of other active compounds, such as ginsenosides [67], flavonoids [68], and hyperin [69] in various natural products.

For the samples with simple composition and less interference from impurity, DESI-MS analysis can be carried out directly. However, when the components of natural products are complex, DESI-MS analysis without separation may suffer from interferences of matrices as well as co-existed active components, which will affect the accuracy of identification. As a sample separation method, thin layer chromatography (TLC) has been introduced into DESI-MS for natural product analysis [62, 70], which was found to be effective to separate the active components of natural drugs from the matrices. In an early work, van Berkel et al. [71] used TLC and DESI-MS to analyze the alkaloids in *Hydrastis canadensis*. Samples were spotted on aluminum- or glass-backed TLC plates, after separation by using methanol or acetonitrile, the separated spots on the TLC including berberine, palmatine, and hydrastinine can be analyzed by both the routine fluorescence method and DESI-MS (Figure 13.2). The incorporation of DESI-MS not only offered highly selective detection of natural compounds but also found materials in overlapping bands and had the capability of identification of minor species in a mixture. Bagatela et al. [68] compared the results of the analysis of Fructus Aurantii obtained by TLC-DESI-MS with that by ESI-MS. It was found that TLC-DESI-MS can obtain characteristic mass spectra with diagnostic ions, and identify and characterize active compounds rapidly. The pigments and hormones in natural drugs can also be analyzed by the combination of TLC and DESI-MS. For example, Claude et al. [72] isolated and identified ecdysone in plants by TLC-DESI-MS. By combining the spectra of the compounds separated from the TLC with the chromatogram of the thin layer, the polarity and relative content of different compounds can be seen intuitively. In another report [73], TLC-DESI-MS was used for rapid screening of for anthraquinones in food and medicinal samples such as Chilean mushrooms.

13.3.3 DESI-Based Mass Spectrometry Imaging

Mass spectrometry imaging (MSI) is a new molecular imaging technology, which has gradually become an important research field of biomolecular imaging. In comparison with traditional imaging techniques such as optical imaging and electron microscopy imaging, MSI is characterized by the ability to image large numbers and different types of molecules without labeling with radioisotopes or fluorescent markers [74]. MSI has been well established as a useful tool to generate molecular-specific images in biomedical applications, and the introduction of ambient ionization into MSI makes it possible to image the chemical information on untreated samples under atmospheric pressure environment in a simpler manner [28]. With the rapid development of DESI-related technologies, DESI-based MSI has found broad applications in the field of biomedical analysis [75].

DESI-MSI has many advantages over traditional imaging technologies. It does not need deposition of matrix for effective ionization, which is more convenient to use in comparison with MALDI-MSI. Besides, DESI-MSI can be operated under

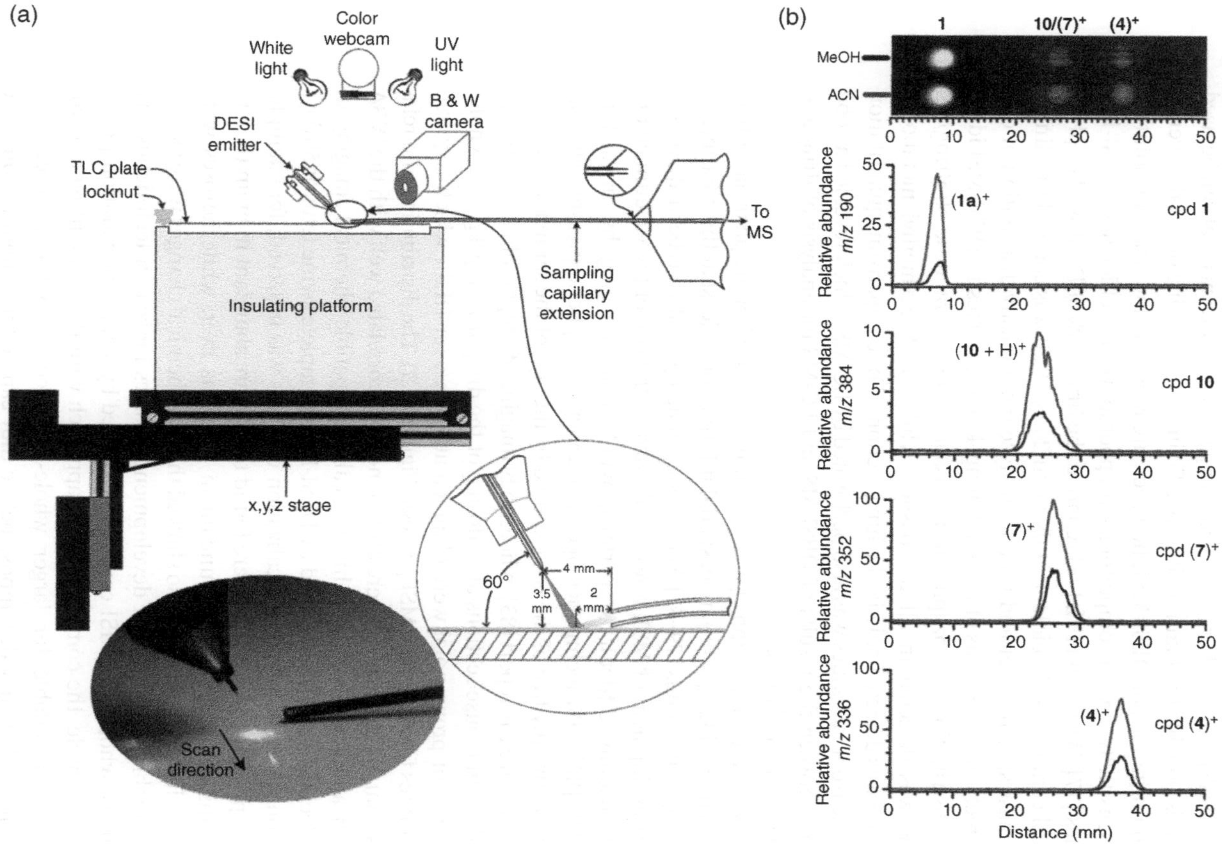

Figure 13.2 (a) Schematic of a typical TLC-DESI-MS system for natural product analysis. The photo inset shows the region of ionization, the emitter of DESI, and inlet capillary of the mass spectrometer. (b) Postanalysis photo of 5-cm-long aluminum-backed TLC plate development lane with methanol or acetonitrile. Separated components from the spotted mixture were hydrastinine (1, detected as $(1a)^+$, *m/z* 190), hydrastine ($(10 + H)^+$, *m/z* 384), palmatine ($(7)^+$, *m/z* 352), and berberine ($(4)^+$, *m/z* 336) from low to high Rf. The black line (methanol) and gray line (acetonitrile) show the extracted ion profiles obtained during the lane scans. *Source:* [71]/ reproduced with permission from American Chemical Society.

atmospheric pressure environment, overcoming the limitations of in-vacuum detection required for many MSI techniques such as secondary ion mass spectrometry (SIMS) [76]. As there are a wide range of spray solvents that can be used in DESI, it is possible to obtain highly specific imaging by using optimal solvents. The main limitations of DESI-MSI, however, are relatively low spatial resolution and low sensitivity toward compounds with poor ionization efficiency by ESI-like techniques [77]. Wiseman et al. [78] reported the first work of DESI-MSI of pharmaceutical compounds in biological tissue in 2008. DESI was coupled to a linear ion trap mass spectrometer and used to analyze the intensity distribution of clozapine directly from sections of rat brain, lung, kidney, and testis. DESI-MSI identified the distribution of clozapine in rat tissues after an oral dose of 50 mg/kg. Clozapine was detected in all the tissues, while the N-desmethyl metabolite of clozapine was detected only in the lung sections. For comparison, quantitation of clozapine from rat tissues was also performed by LC-MS, which revealed a good agreement between the results of DESI-MS and LC-MS for analysis of drug and metabolite from rat tissues.

Whole-body autoradiography (WBA) is commonly used in the pharmaceutical research as it provides quantitative, spatial information at least at the micrometer level on drug compounds [79–81]. Whole body imaging by MS has been attempted since the development of MSI, however, limited by the size of the vacuum chamber, it is usually difficult to perform whole body imaging by in-vacuum MALDI-MS [82]. MSI techniques operated under atmospheric pressure such as DESI-MS are usually free of the issue caused by the size of the sampling stages or chambers. In an early report [83], whole body imaging of propranolol was explored by DESI-MSI in mouse sections after dosing of the drug at 7.5 mg/kg. The whole-body images for propranolol were obtained at 7 mm/s for DESI, which required just 79 minutes for the whole MSI process (Figure 13.3). The distribution and relative ion signal for propranolol were also found to be correlated well with the WBA images. In another report [84], histologically compatible chemical imaging of small molecules was demonstrated by DESI-MS a complete swine fetus at 50 days of gestation. Organ-dependent localization of lipids was observed; for example, many phosphatidylserine lipids were found to be more abundant in brain than in other organs; sulfatides were abundant in the fetus liver, while hexoses were abundant in lung and heart. It also showed that the chemical changes were associated with embryonic and fetal development. Besides ionization techniques and MS methods, whole-body MSI can also be limited by sample preparation procedures. For example, the commonly used approach with thaw-mounting of cryo-sections is not suitable for larger whole-body sections with heterogeneous components as they may decompose before they can be thaw-mounted onto the sampling surface successfully. Instead of cryo-sectioning using a cryo-macrotome, Okutan et al. [85] developed an approach based on cryo-microtome. The tissue

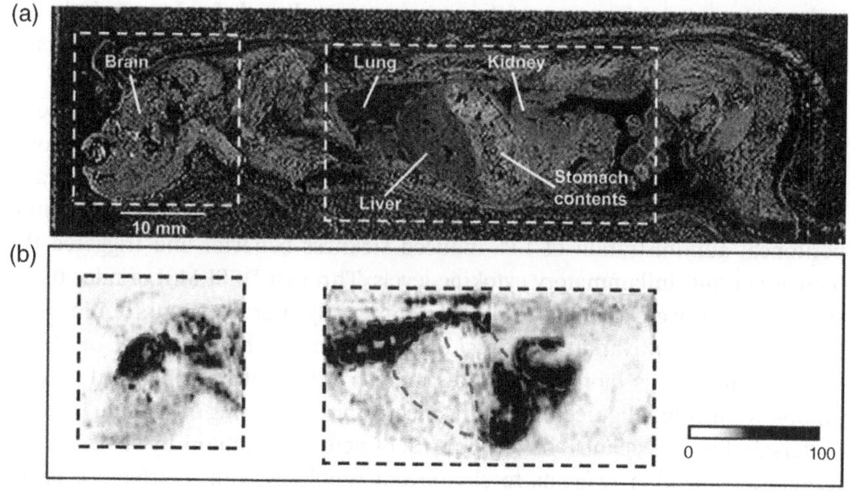

Figure 13.3 (a) Scanned optical image of a sagittal whole-body tissue section of a mouse dosed intravenously with 7.5 mg/kg propranolol. (b) Distribution of propranolol in 20 × 20 mm and 38 × 20 mm areas detected by DESI-MS/MS (*m/z* 260 → 116). *Source:* [83]/ reproduced with permission from American Chemical Society.

sections were collected on tape and analyzed directly by DESI-MS. This approach was demonstrated by mice dosed intraperitoneally with amitriptyline. By combining full-scan MS and MS/MS, the drug, its metabolites, as well lipids were simultaneously imaged. It was also relatively sensitive for imaging of drug and the metabolite in a mouse dosed with amitriptyline as low as 2.7 mg/kg, which is comparable to the normal human dose. The combination of cryo-microtome with DESI-MSI may be a simple, sensitive, and highly selective whole-body imaging for the study of drug distribution and metabolism.

Metabolites and lipids are main targets for DESI-MSI of human or animal tissue samples. Some recent clinical applications include lipid and metabolite profiling for discrimination of different types of tumor tissues [86], intraoperative analysis of metabolite 2-hydroxyglutarate (2-HG) in the glioma tissue to tell the mutation types of tumors [87], and discrimination of malignant and benign microscopic skin tumors [88]. Pirro et al. measured tumor cell percentage (TCP) based on lipid profiles obtained by DESI-MS. In the case of brain tumor diagnosis, 73 biopsies from 10 surgical resection cases show that high TCP can estimate the IDH-mutated glioma margins with 93% sensitivity and 83% specificity [89]. In another study, Banerjee et al. focused on Krebs cycle intermediates ($m/z < 200$) for DESI-MS [90]. They built up models to classify prostate cancer from benign specimens with nearly 90% accuracy per patient. They also found that the glucose/citrate ion signal ratio can predict the cancerous state if the ratio is higher 1.0.

It should be noted the purpose of these reports is not directly for drug analysis, but seeking the clinical applications of DESI-MSI through looking for biomarkers or chemical information for differentiation of disease-state tissues.

Besides biomarker discovery in disease-state samples, DESI-MSI is also useful in screening of chemical information responsible to the temporary stimulations. In a previous work, lipopolysaccharide (LPS)-induced neuroinflammation and the lipid dynamics in mice were investigated after LPS injection [91]. The intraperitoneal administration of LPS-induced sickness behavior and triggered the increase of anti-inflammatory cytokine levels. Through DESI-MSI of mice tissue slices, 14 lipids were found to be altered significantly after 3–24 hours of LPS treatment. Further multivariate statistical analyses revealed that ions of PE 38:4 and FA 22:4 may be as biomarkers to distinguish LPS-treated groups and control groups. Results highlighted the significance of the monitoring of cerebral lipids in understanding the sickness behaviors. MSI of neurotransmitters by MALDI often needs derivatization reagents to circumvent problems with interference peaks from matrix. DESI-MSI is advantageous in this application as it can directly analyze the spatial distribution of neurotransmitters and choline without derivatization. Fernandes et al. [92] reported a method for imaging of compounds including γ-aminobutyric (GABA), glutamate, aspartate, serine, acetylcholine, dopamine, and choline were simultaneously imaged using DESI-MSI. The method is potential to be used in the evaluation of imbalances in neurotransmitters levels.

The rapid development of DESI-MSI has attracted many applications in the field of biomedical science [87, 88, 90, 93–97], which in turn place higher requirements for DESI-MSI in terms of reproducibility, resolution, speed, and so on. Optimization of DESI sprayers is important to obtain high-quality MS images. The solvent systems need to be optimized to obtain sufficient signal levels, depending on the samples and target compounds analyzed. It is necessary to obtain the appropriate small spray spot by adjusting the distance between the emitter, sampling surface, and MS entrance. Besides, the spatial resolution and sensitivity should be balanced [98, 99]. Typical spatial resolution of DESI is 100–200 μm, the resolution can be improved by optimization of the spray. Tillner et al. redesigned the DESI sprayer to provide better stability of solvent capillary positioning [31]. Using a new sprayer with a 20-μm inner diameter tapered fused-silica emitter, a stable electrospray at very low flow rates was obtained, by which spatial resolution as good as 40 μm was obtained for DESI-MSI of rat cerebellum samples. However, it should be noted that increasing the spatial resolution may result in a decrease in signal intensity, which may affect the quality of the final MS images and take longer time for each analysis. The spray angle is also important for DESI-MS. A typical spray angle of ~50° provides a high signal intensity, which would be reduced if the droplet transmission efficiency reduces when the sprayer is perpendicular to the substrate [100].

The nature of the sample can also affect the imaging performance of DESI-MS, which should be considered seriously especially in the analysis of natural products [101]. Cryosections are usually used for non-flat surfaces such as organs and tissues [102]. While for plant samples such as flowers and leaves, they are soft, irregular, and very absorbent to the surface, leading to low or unstable signals during imaging process. The leaves of most plants contain hydrophobic wax cuticles that are also difficult to penetrate using regular spray solvents. An indirect sampling method was developed through imprinting of components of plant samples onto surfaces such as glass slides. To reduce ablation of metabolites deposited on the glass surface, adsorbent materials such as porous polytetrafluoroethylene or TLC plates have been introduced for sample imprinting in DESI-MSI [103–106]. These surfaces usually have good adsorption properties, retaining metabolites from plants until they are desorbed with DESI. They have been applied into imprinting and DESI-MSI of various plant organs and tissues.

As a modified DESI method, nano-DESI features with using flowing droplets for extraction of compounds from the sample surface, which can provide better ionization efficiency for some compounds and better resolution for MS imaging in some applications. In a previous report, a nano-DESI-MS method was developed for sensitive imaging of nicotine in rat brain tissue samples following single-dose nicotine administration [107]. Isotopically labeled standard was incorporated into the nano-DESI solvent, which was found to efficiently compensate for matrix effects. The increased sensitivity of this method is promising for spatially resolved detection of low-abundance drugs or metabolites in complex matrices. As to the spatial resolution, a typical nano-DESI-MS system allows for acquiring mass spectral data with a moderate resolution of ~200 μm. To improve the spatial resolution, efforts have been made to develop probes and related feedback modes. In a previous report, shear force microscopy was combined with nano-DESI-MS [108], which provided spatial resolutions as high as 10 μm for imaging of multiple lipid species in animal tissue samples.

13.3.4 Detection of Drug–Protein Interactions

Modern drug development processes demand methods that can measure binding affinities between the drug candidate with the target proteins [109]. Conventional techniques such as isothermal titration calorimetry, microscale thermophoresis, and surface plasmon resonance usually require the use of recombinant proteins, which cannot reflect the real ligand–protein interactions under biological environments. ESI-based native MS approaches have been proven to be able to retain most of biological states of a protein or protein complex and have been widely used for investigation of protein configurations, protein interactions, as well as ligand–protein complexes and their binding stoichiometry [110–113]. However, during native ESI-MS, the protein solution directly contacts with the electrode, and directly exposed with high voltages, which can affect the detection of ligand–protein interactions or drug–protein interactions [114]. In DESI-MS, the proteins are not directly exposed to high voltages or the electrode during the ionization process, making it promising for drug–protein interaction analysis.

DESI has been limited to ionization of small proteins (≤25 kDa) [100] and solid-state samples. To enable direct analysis of liquid samples by DESI, Chen and Miao [115] developed a device in which a fused silica capillary was used to introduce liquid samples continuously for DESI ionization (Figure 13.4a). In a typical procedure, a spray solvent of methanol/water/acetic acid was used to generate charged microdroplets. The sample solution was infused inside a capillary and impacted by the spray at the tip of the capillary, where the ionization of the liquid sample would happen rapidly. Chipuk and Brodbelt [117] introduced another method named transmission mode DESI (TM-DESI), which can analyze both liquid and solid samples with simplified experimental conditions (Figure 13.4b). In TM-DESI, a sampling mesh allowed the DESI spray to transmit through a 0° angle from the DESI emitter to the MS inlet. It does not require the optimization of conditions such as spray distances or angles. TM-DESI has been applied into rapid analysis of protein digests by coupling with UV photodissociation [118]. In another work, Mulligan et al. [119] developed a variant form of DESI for the detection of aqueous samples by introducing a heated surface as the sampling platform. With the thermally assisted DESI-MS, environmental contamination analytes from water sample can be sensitively detected.

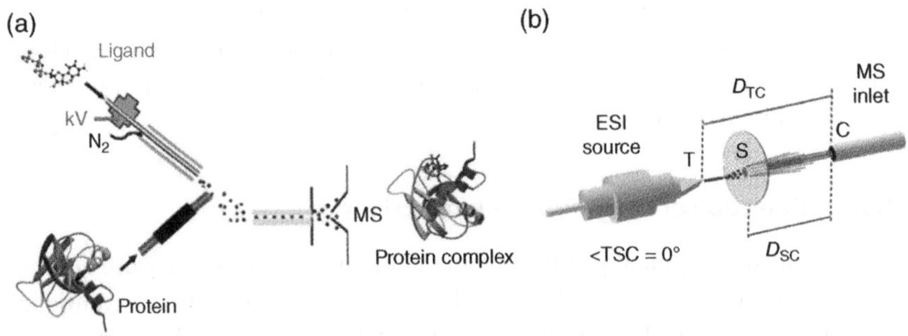

Figure 13.4 (a) Schematic of liquid sample DESI-MS for analysis protein–ligand complexes. *Source:* Liu et al. [116]/reproduced with permission of American Chemical Society. (b) Schematic of transmission mode desorption electrospray ionization (TM-DESI). The angle between the electrospray tip, sample, and capillary inlet to the mass spectrometer is 0°. D_{TC} is the distance from the electrospray tip to the capillary and D_{SC} is the distance from the sample to the capillary. *Source:* Chipuk and Brodbelt [117]/with permission of Elsevier.

Based on the liquid sample DESI, direct analysis of large proteins was also realized [120]. By doping the DESI spray solvent or protein sample with supercharging reagents, the method showed the capability of direct ionization of large noncovalent protein complexes larger than 45 kDa and proteins up to 150 kDa (immunoglobulin G). Intact protein complex ions such as superoxide dismutase, enolase, and hemoglobin were detected, indicating that the DESI can preserve weak noncovalent interactions during the ionization process. Chen and coworkers further applied the liquid sample DESI method into investigation of protein–ligand interactions [116]. Through spraying ligands toward separate protein sample solutions, it allowed the direct formation of intact protein–ligand complex ions. The method was demonstrated by binding of ribonuclease A with cytidine nucleotide ligands, and the binding of lysozyme with acetyl chitose ligands. Through reactive DESI and direct MS analysis, information such as binding stoichiometry, selectivity, and kinetics were obtained conveniently.

Ouyang and coworkers [121] developed a robust DESI-MS platform which was capable of high-throughput screening of ligands with a high affinity for proteins with anion-binding sites. It incorporated an anionic region inhibited dissociation mechanism and used the differences in detectable dissociation of protein–ligand complexes to distinguish free ligands displaced from anion-binding sites. It has been validated by α1-acid glycoprotein and amlodipine, with site-specific enantioselectivity consistent with other studies. In another work, they combined centrifugal filtration with DESI-MS for screening of drugs binding to proteins [122]. Membrane filtration was used to remove any unbound drugs. Drug–protein complexes were deposited on the DESI substrate and liberated drugs were detected by MS. Three DNA topoisomerase I inhibitors, camptothecin, hydroxycamptothecin, and 7-ethyl-10-hydroxycamptothecin, were evaluated, which may form Topo I and the DNA-Topo I complex. High-throughput screening of 21 drugs binding to human serum albumin was achieved within 1.75 minutes. The study also provided a relative quantitation method for detection of compounds binding to proteins, which was potential to be used in the process of drug design and screening.

In the analysis of ligand–protein interactions, ESI-MS and DESI-MS approaches are usually difficult in discrimination of nonspecific interactions, which can give false-positive results in the determination of ligand–protein interactions. To discriminate nonspecific bindings, a native-denatured exchange (NDX) method was proposed based on liquid sample DESI-MS [123]. The mechanism of the NDX method is that adjusting the flow rate of the DESI spraying solvents with denaturant can rapidly change the ionization environment for protein in the solution from native state to denatured state. By measuring the K_d curves of ligand–protein complexes along with shift of protein conformation from native state to denatured states, it is possible to differentiate specific interactions and nonspecific interactions. NDX-based DESI-MS has been validated by using myoglobin,

CDP-ribonuclease, and N,N′,N″-triacetylchitotriose and their binding proteins. This approach is useful to eliminate off-target effect and reduce false-positive rates during screening of ligand–protein interactions and new drug targets.

As discussed above, most DESI-based protein–ligand binding experiments were realized by mixing ligands and protein at the end of DESI emitter in a liquid sample, where the reaction took place to produce binding information. In a recent work, Ambrose et al. [124] proposed a different strategy by depositing proteins onto a solid planar for native DESI-MS analysis. Although this approach was limited in the use of volatile buffers and detergents optimized for native MS, and the requirement of high-resolution MS as the ligands were observed directly bound to proteins desorbed from the solid surface, it can bring in unique advantage for ligand–protein interaction analysis. Through adjusting different lipidic environments for membrane protein targets, this strategy can be used to screen intact membrane proteins and their complexes from a cocktail of drugs.

13.3.5 High-Throughput Experimentation

High-throughput experimentation (HTE) is useful in drug discovery process by implementing large numbers of experiments in parallel, requiring, and utilizing small amounts of raw materials and less labor in each experiment [125]. DESI-MS is an emerging tool for HTE due to its capability of acceleration of reactions and its convenience in realization of spot-to-spot analysis. Solution-phase reactions in microdroplets can be much faster than that in conventional bulk-phase settings. The first example of reaction-rate acceleration was presented by using droplets produced by DESI [126]. The transition of DESI to a system for studying reactions requires improving the ionization for both materials and reaction products. During past few years, reactive DESI approaches have been developed for investigation of chemical reactions, in which reagents are usually added in the DESI spray solvent, and products are usually formed in the secondary droplets. Acceleration of reaction rate at several orders of magnitude can be expected in the secondary droplets of reactive DESI, which may be provided by the unique environment in the droplet and/or at the droplet–surface interface. For example, the rates of acid-catalyzed reactions such as the reaction of Girard T reagent with ketosteroids was observed to increase with decreasing pH in positively charged microdroplets; while the increasing pH in negatively charged microdroplets can increase the rates of base-catalyzed Michael reactions [126]. Wleklinski et al. [127] developed a method that can analyze reaction products in droplets within about one second. The method was a combination of DESI-MS and accelerated chemical reactions in microdroplets. A continuous online process was developed, which gave 10^4 reactions per hour at area densities of about 1 spot/mm^2, while data were analyzed automatically to generate ion images of reagents and products. Using amine alkylation reactions for evaluation of the approach, reaction times less than 100 μs were achieved in microdroplets. They also used the same system to screen reaction conditions including reagents, bases, and stoichiometry for the Suzuki cross-coupling reaction [128]. The optimal conditions obtained DESI-MS screening can be transferred to reaction by the Leidenfrost technique for the purpose of accelerating metal-catalyzed cross-coupling reactions. In another work, Jaman et al. [129] employed DESI-MS to explore the reaction conditions of lomustine, which is an important agent for treatment of brain tumors and Hodgkin's lymphoma. With four commercially available reagents and a total time of nine minutes, lomustine can be successfully prepared in two telescoped flow reactors. The overall yield was improved by 63% and the production rate of 110 mg/h was achieved. To implement the DESI-MS into pharmaceutical development settings for large-scale and high-throughput experimentation, the improvement of the hardware and development of a software for automatic sampling, data collecting, and analysis are important. Sobreira described an automated system for screening of reaction conditions based on DESI [130]. A fluid-handling robot was developed, which can spot samples onto a flat polytetrafluoroethylene plate at the densities up to 6144 per plate. The custom software can control the fluid-handling robot which prepares reaction mixtures, as well as manipulates the solvent mixing for DESI spray, by which the system showed good ruggedness and repeatability.

13.3.6 High-Throughput Screening

Screening of large numbers of compounds for possible targets is required in drug discovery and development processes. Methods with capability of rapid reaction screening and optimization, which can deal with large numbers of possible drugs and are desirable for the purpose of new drug discovery. Although DESI-based high-throughput screening of organic drugs and related reactions have been widely investigated, screening of enzymatic reactions were less explored. Incorporation of enzymatic measurement capability into existing DESI-MS platform will be significant to establish a versatile system that integrates in situ analysis and high-throughput screening of both chemical and biological reactions. Recently Morato et al. described the use of DESI-MS for screening enzymatic reactions directly from the bioassay matrix [131] (Figure 13.5). The bioassay mixtures were directly deposited onto the substrate without any pretreatment or post-treatment. The substrate and product of the enzymatic reaction were monitored by MS simultaneously at the speed of 0.3 seconds per spot. The performance of this method was validated by monitoring the process of hydrolysis of acetylcholine to choline mediated by acetylcholinesterase (AChE). This work validated that DESI-MS has potential impact in high-throughput label-free enzymology and drug discovery.

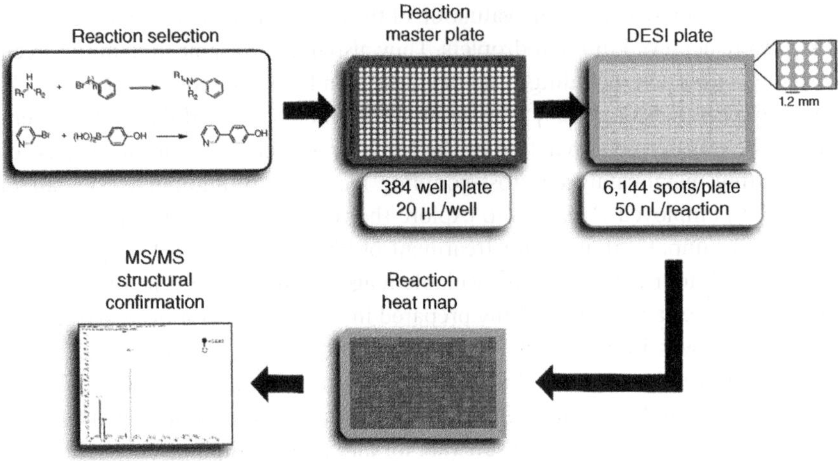

Figure 13.5 Workflow of a high-throughput reaction screening system with DESI-MS. The master plate and the sampling plate are controlled with a liquid-handling robot. The sampling plate is prepared with a 50 nL pin tool and makes up to a 6144 array. *Source:* Wleklinski et al. [127]/The Royal Society of Chemistry/CC BY-3.0.

Modified DESI methods inherit the merits of DESI and usually have capability of direct and high-throughput analysis. Very recently, Pu et al. [132] reported a novel high-throughput MS system based on infrared matrix-assisted desorption electrospray ionization (IR-MALDESI). Fast reaction monitoring was achieved, with acquisition rate of 33 spectra per second. The potential of this system for high-throughput lead discovery was validated by substrate/product types including small-molecule metabolites (isocitrate dehydrogenase 1 assay, IDH1), lipids, and short peptides (p300 histone acetyltransferase assay). Screening of about 3k compounds was achieved for the demonstration toward IDH1.

13.4 Conclusions and Future Outlook

Since the invention of DESI-MS, it has attracted tremendous attention in a wide range of fields associated with biochemical analysis. DESI has also been modified or hybrid with other techniques, which gave it new capabilities in sample clean-up, ionization, derivatization, and so on. With commercialization of DESI by Waters, we can expect to see more applications of DESI-MS outside of analytical labs. It will surely become a useful tool in drug analysis, natural product profiling, biomedical imaging, and high-throughput drug screening.

DESI-MS has been widely used in the detection of active ingredients in pharmaceutical or natural products [52–54, 62–65]. Its potential in the quality control of traditional medicines has been widely validated. In the future, it will be important to develop DESI-MS-based library of metabolites for classification of species, and further research on the synthesis pathways of plant secondary metabolism. DESI-MS is also able to connect the spatial distribution of active component in natural products, which is useful for the detection of transformation of biomolecules. Although DESI-MS has showed good ionization performances in analyzing a wide range of compounds, it still has problems such as inaccurate quantitative analysis and poor reproducibility for components in natural products with lower ionization efficiency [133]. This could be improved by incorporation of derivatization methods and other ionization techniques. It also requires the establishment of more databases and the development of software tools for data analysis [134, 135], which will be useful to transfer the DESI-MS protocols into different laboratories or application sites.

Mass spectrometry imaging is an important and growing area of biomedical analysis. The ambient nature and gentleness of DESI-MS make it more convenient to transfer into clinical settings [87, 88, 90, 93–97]. However, DESI-MSI needs further development to meet the demands of clinical analysis and drug development. Without sample clean-up and chromatographic separation, most of MSI methods suffer from ion suppression and matrix effects, resulting in low sensitivity and poor reproducibility. To resolve the issue from the matrix effect of raw biomedical samples, efforts could be paid to developing simple but integrated clean-up methods and using high-resolution MS. Besides, tandem MS approach can be incorporated into screening or quantitative analysis, which not only enhances the sensitivity of DESI-MSI but also improves molecular identification. Moreover, different imaging techniques can be combined to give complementary images of certain targets. Some reports have shown the combination of DESI imaging with SIMS and MALDI [136], DESI imaging with Fourier-transform infrared imaging for counterfeit drug analysis [137], and DESI imaging with whole-body autoradiography for drug distribution mapping [83, 84].

Although a lot of efforts have been made to transfer DESI-MS into clinical or on-site analysis applications, it would be difficult to use a commercial benchtop mass spectrometer routinely in-field or in a surgery room due to its large size and complication in operation. During the past decade, many portable or miniature mass spectrometers have been developed, such as Mini 10 [138], Mini 11 [139], backpack miniature MS [140], Mini 12 [141], brick ion size miniature mass spectrometer with sinusoidal frequency scanning [142], planar-electrode linear ion trap mass spectrometer [143], linear wire ion trap [144], miniature laser ablation time-of-flight mass spectrometer [145], and miniature dual linear ion trap mass spectrometer [146]. Many of these miniature MS systems were coupled with direct sampling methods and have been used in therapeutic drug monitoring and biomarker analysis [141, 147, 148]. However, due to the requirement of a high flow rate of nebulizing gas for producing droplets, DESI is difficult to implement into the current miniature MS system. Alterations need to be made to allow DESI-like sampling methods to be compatible with a miniature MS system [149]. In a previous report, a sampling probe was designed, which consisted of two channels for introducing the spray and transferring the charged species to the miniature mass spectrometer, respectively. In the future, we expect to see a new type of miniature MS system that can be compatible with most DESI methods, or a new type of DESI that require much lower flow rate of gases. The miniature DESI-MS system is expected to be significant to promote clinical analysis and drug discovery to a new level.

References

1 Wienkers, L.C. and Heath, T.G. (2005). Predicting in vivo drug interactions from in vitro drug discovery data. *Nat. Rev. Drug Discov.* 4 (10): 825–833.

2 Wishart, D.S. (2016). Emerging applications of metabolomics in drug discovery and precision medicine. *Nat. Rev. Drug Discov.* 15 (7): 473–484.

3 Korfmacher, W.A. (2005). Principles and applications of LC-MS in new drug discovery. *Drug Discov. Today* 10 (20): 1357–1367.

4 Kassel, D.B. (2001). Combinatorial chemistry and mass spectrometry in the 21st century drug discovery laboratory. *Chem. Rev.* 101 (2): 255–267.

5 Aksenov, A.A., da Silva, R., Knight, R. et al. (2017). Global chemical analysis of biology by mass spectrometry. *Nat. Rev. Chem.* 1: 0054.

6 Jones, H.M., Gardner, I.B., and Watson, K.J. (2009). Modelling and PBPK simulation in drug discovery. *AAPS J.* 11 (1): 155–166.

7 van der Graaf, P.H. and Benson, N. (2011). Systems pharmacology: bridging systems biology and pharmacokinetics-pharmacodynamics (PKPD) in drug discovery and development. *Pharm. Res.* 28 (7): 1460–1464.

8 Fry, D.C. (2006). Protein–protein interactions as targets for small molecule drug discovery. *Biopolymers* 84 (6): 535–552.

9 Pacholarz, K.J., Garlish, R.A., Taylor, R.J., and Barran, P.E. (2012). Mass spectrometry based tools to investigate protein–ligand interactions for drug discovery. *Chem. Soc. Rev.* 41 (11): 4335–4355.

10 Nunez, S., Venhorst, J., and Kruse, C.G. (2012). Target–drug interactions: first principles and their application to drug discovery. *Drug Discov. Today* 17 (1,2): 10–22.

11 Takáts, Z., Wiseman, J.M., Gologan, B., and Cooks, R.G. (2004). Mass spectrometry sampling under ambient conditions with desorption electrospray ionization. *Science* 306 (5695): 471–473.

12 Cooks, R.G., Ouyang, Z., Takats, Z., and Wiseman, J.M. (2006). Ambient mass spectrometry. *Science* 311 (5767): 1566–1570.

13 Takyi-Williams, J., Liu, C.-F., and Tang, K. (2015). Ambient ionization MS for bioanalysis: recent developments and challenges. *Bioanalysis* 7 (15): 1901–1923.

14 Ferreira, C.R., Yannell, K.E., Jarmusch, A.K. et al. (2016). Ambient ionization mass spectrometry for point-of-care diagnostics and other clinical measurements. *Clin. Chem.* 62 (1): 99–110.

15 Ma, X. and Ouyang, Z. (2016). Ambient ionization and miniature mass spectrometry system for chemical and biological analysis. *TrAC, Trends Anal. Chem.* 85 (Part A): 10–19.

16 Li, L.P., Feng, B.S., Yang, J.W. et al. (2013). Applications of ambient mass spectrometry in high-throughput screening. *Analyst* 138 (11): 3097–3103.

17 Petucci, C., Diffendal, J., Kaufman, D. et al. (2007). Direct analysis in real time for reaction monitoring in drug discovery. *Anal. Chem.* 79 (13): 5064–5070.

18 Crawford, E., Gordon, J., Wu, J.T. et al. (2011). Direct analysis in real time coupled with dried spot sampling for bioanalysis in a drug-discovery setting. *Bioanalysis* 3 (11): 1217–1226.

19 Shelley, J.T., Badal, S.P., Engelhard, C., and Hayen, H. (2018). Ambient desorption/ionization mass spectrometry: evolution from rapid qualitative screening to accurate quantification tool. *Anal. Bioanal. Chem.* 410 (17): 4061–4076.

20 Piehowski, P.D., Zhu, Y., Bramer, L.M. et al. (2020). Automated mass spectrometry imaging of over 2000 proteins from tissue sections at 100-μm spatial resolution. *Nat. Commun.* 11 (1): 8.

21 Morelato, M., Beavis, A., Kirkbride, P., and Roux, C. (2013). Forensic applications of desorption electrospray ionisation mass spectrometry (DESI-MS). *Forensic Sci. Int.* 226 (1–3): 10–21.

22 Beneito-Cambra, M., Gilbert-Lopez, B., Moreno-Gonzalez, D. et al. (2020). Ambient (desorption/ionization) mass spectrometry methods for pesticide testing in food: a review. *Anal. Methods* 12 (40): 4831–4852.

23 Gentili, A., Fanali, S., and Rocca, L.M. (2019). Desorption electrospray ionization mass spectrometry for food analysis. *TrAC, Trends Anal. Chem.* 115: 162–173.

24 Nielen, M.W.F., Hooijerink, H., Zomer, P., and Mol, J.G.J. (2011). Desorption electrospray ionization mass spectrometry in the analysis of chemical food contaminants in food. *TrAC, Trends Anal. Chem.* 30 (2): 165–180.

25 Cody, R.B., Laramée, J.A., and Durst, H.D. (2005). Versatile new ion source for the analysis of materials in open air under ambient conditions. *Anal. Chem.* 77 (8): 2297–2302.

26 Soparawalla, S., Salazar, G.A., Perry, R.H. et al. (2009). Pharmaceutical cleaning validation using non-proximate large-area desorption electrospray ionization mass spectrometry. *Rapid Commun. Mass Spectrom.* 23 (1): 131–137.

27 Ma, X., Zhao, M., Lin, Z. et al. (2008). Versatile platform employing desorption electrospray ionization mass spectrometry for high-throughput analysis. *Anal. Chem.* 80 (15): 6131–6136.

28 Wu, C., Dill, A.L., Eberlin, L.S. et al. (2013). Mass spectrometry imaging under ambient conditions. *Mass Spectrom. Rev.* 32 (3): 218–243.

29 Kompauer, M., Heiles, S., and Spengler, B. (2016). Atmospheric pressure MALDI mass spectrometry imaging of tissues and cells at 1.4-μm lateral resolution. *Nat. Methods* 14: 90.

30 Comi, T.J., Ryu, S.W., and Perry, R.H. (2016). Synchronized desorption electrospray ionization mass spectrometry imaging. *Anal. Chem.* 88 (2): 1169–1175.

31 Tillner, J., Wu, V., Jones, E.A. et al. (2017). Faster, more reproducible DESI-MS for biological tissue imaging. *J. Am. Soc. Mass Spectrom.* 28 (10): 2090–2098.

32 Towers, M.W., Karancsi, T., Jones, E.A. et al. (2018). Optimised desorption electrospray ionisation mass spectrometry imaging (DESI-MSI) for the analysis of proteins/peptides directly from tissue sections on a travelling wave ion mobility Q-ToF. *J. Am. Soc. Mass Spectrom.* 29 (12): 2456–2466.

33 Lin, L.-E., Chen, C.-L., Huang, Y.-C. et al. (2020). Precision biomarker discovery powered by microscopy image fusion-assisted high spatial resolution ambient ionization mass spectrometry imaging. *Anal. Chim. Acta* 1100: 75–87.

34 Roach, P.J., Laskin, J., and Laskin, A. (2010). Nanospray desorption electrospray ionization: an ambient method for liquid-extraction surface sampling in mass spectrometry. *Analyst* 135 (9): 2233–2236.

35 Lanekoff, I., Heath, B.S., Liyu, A. et al. (2012). Automated platform for high-resolution tissue imaging using nanospray desorption electrospray ionization mass spectrometry. *Anal. Chem.* 84 (19): 8351–8356.

36 Dixon, R.B., Bereman, M.S., Muddiman, D.C., and Hawkridge, A.M. (2007). Remote mass spectrometric sampling of electrospray- and desorption electrospray-generated ions using an air ejector. *J. Am. Soc. Mass Spectrom.* 18 (10): 1844–1847.

37 Dénes, J., Katona, M., Hosszú, Á. et al. (2009). Analysis of biological fluids by direct combination of solid phase extraction and desorption electrospray ionization mass spectrometry. *Anal. Chem.* 81 (4): 1669–1675.

38 Görög, S. (2007). The changing face of pharmaceutical analysis. *TrAC, Trends Anal. Chem.* 26 (1): 12–17.

39 Liu, K.-T. and Chen, C.-H. (2019). *Determination of Impurities in Pharmaceuticals: Why and How? Chapters*. London: IntechOpen.

40 Fenk, C.J., Hickman, N.M., Fincke, M.A. et al. (2010). Identification and quantitative analysis of acetaminophen, acetylsalicylic acid, and caffeine in commercial analgesic tablets by LC-MS. *J. Chem. Educ.* 87 (8): 838–841.

41 Chiang, C.-H., Lee, H.-H., Chen, B.-H. et al. (2019). Using ambient mass spectrometry and LC-MS/MS for the rapid detection and identification of multiple illicit street drugs. *J. Food Drug Anal.* 27 (2): 439–450.

42 Chen, H., Talaty, N.N., Takats, Z., and Cooks, R.G. (2005). Desorption electrospray ionization mass spectrometry for high-throughput analysis of pharmaceutical samples in the ambient environment. *Anal. Chem.* 77 (21): 6915–6927.

43 Siebenhaar, M., Kullmer, K., Fernandes, N.M. et al. (2015). Personalized monitoring of therapeutic salicylic acid in dried blood spots using a three-layer setup and desorption electrospray ionization mass spectrometry. *Anal. Bioanal. Chem.* 407 (23): 7229–7238.

44 Wiseman, J.M., Evans, C.A., Bowen, C.L., and Kennedy, J.H. (2010). Direct analysis of dried blood spots utilizing desorption electrospray ionization (DESI) mass spectrometry. *Analyst* 135 (4): 720.

45 Kennedy, J.H., Aurand, C., Shirey, R. et al. (2010). Coupling desorption electrospray ionization with solid-phase microextraction for screening and quantitative analysis of drugs in urine. *Anal. Chem.* 82 (17): 7502–7508.

46 Stojanovska, N., Tahtouh, M., Kelly, T. et al. (2015). Qualitative analysis of seized cocaine samples using desorption electrospray ionization- mass spectrometry (DESI-MS). *Drug Test Anal.* 7 (5): 393–400.

47 Abbassi-Ghadi, N., Jones, E.A., Gomez-Romero, M. et al. (2015). A comparison of DESI-MS and LC-MS for the lipidomic profiling of human cancer tissue. *J. Am. Soc. Mass Spectrom.* 27 (2): 255–264.

48 Kauppila, T.J., Talaty, N., Kuuranne, T. et al. (2007). Rapid analysis of metabolites and drugs of abuse from urine samples by desorption electrospray ionization-mass spectrometry. *Analyst* 132 (9): 868–875.

49 Atanasov, A.G., Zotchev, S.B., Dirsch, V.M. et al. (2021). Natural products in drug discovery: advances and opportunities. *Nat. Rev. Drug Discov.* 20 (3): 200–216.

50 Sorkin, B.C., Kuszak, A.J., Bloss, G. et al. (2020). Improving natural product research translation: from source to clinical trial. *FASEB J.* 34 (1): 41–65.

51 Zhang, A., Sun, H., and Wang, X. (2018). Mass spectrometry-driven drug discovery for development of herbal medicine. *Mass Spectrom. Rev.* 37 (3): 307–320.

52 Bajpai, V., Singh, A., Kumar, B., and Madhusudanan, K.P. (2018). Application of DART-MS in natural phytochemical research. In: *Direct Analysis in Real Time Mass Spectrometry: Principles and Practices of DART-MS* (ed. Y. Dong), 255–290. Wiley.

53 Patil, S.G., Patil, M.P., Maheshwari, V.L., and Patil, R.H. (2021). In situ probing of endophyte natural products with desi-imaging mass spectrometry. In: *Endophytes: Potential Source of Compounds of Commercial and Therapeutic Applications* (ed. R.H. Patil and V.L. Maheshwari), 177–193. Singapore: Springer Nature https://doi.org/10.1007/978-981-15-9371-0_9.

54 Spraker, J.E., Luu, G.T., and Sanchez, L.M. (2020). Imaging mass spectrometry for natural products discovery: a review of ionization methods. *Nat. Prod. Rep.* 37 (2): 150–162.

55 Xue, J., Bai, Y., and Liu, H. (2019). Recent advances in ambient mass spectrometry imaging. *TrAC, Trends Anal. Chem.* 120: 115659.

56 Bianchi, F., Mattarozzi, M., and Careri, M. (2020). The role of surface in desorption electrospray ionization-mass spectrometry: advances and future trends. *Anal. Bioanal. Chem.* 412 (17): 3967–3973.

57 Schwab, N.V., Ore, M.O., Eberlin, M.N. et al. (2014). Functionalized porous silicon surfaces as DESI-MS substrates for small molecules analysis. *Anal. Chem.* 86 (23): 11722–11726.

58 Penna, A., Careri, M., Spencer, N.D., and Rossi, A. (2015). Effects of tailored surface chemistry on desorption electrospray ionization mass spectrometry: a surface-analytical study by XPS and AFM. *J. Am. Soc. Mass Spectrom.* 26 (8): 1311–1319.

59 Penna, A., Elviri, L., Careri, M. et al. (2011). Investigation of novel sol–gel hydrophobic surfaces for desorption electrospray ionization-mass spectrometry analysis. *Anal. Bioanal. Chem.* 400 (5): 1515–1523.

60 Jackson, A.U., Tata, A., Wu, C. et al. (2009). Direct analysis of Stevia leaves for diterpene glycosides by desorption electrospray ionization mass spectrometry. *Analyst* 134 (5): 867–874.

61 Talaty, N., Takáts, Z., and Cooks, R.G. (2005). Rapid in situ detection of alkaloids in plant tissue under ambient conditions using desorption electrospray ionization. *Analyst* 130 (12): 1624–1633.

62 Conceição, R.S., Reis, I.M.A., Cerqueira, A.P.M. et al. (2020). Rapid structural characterisation of benzylisoquinoline and aporphine alkaloids from Ocotea spixiana acaricide extract by HPTLC-DESI-MSn. *Phytochem. Anal.* 31 (6): 711–721.

63 Suma, H.K., Kumar, V., Kumara, P.M. et al. (2017). Spatial and temporal distribution pattern of camptothecin in seeds and fruits of *Pyrenacantha volubilis* Hook. (Icacinaceae) during different fruit developmental stages. *Curr. Sci. India* 112 (5): 1034–1038.

64 Kumara, P.M., Srimany, A., Ravikanth, G. et al. (2015). Ambient ionization mass spectrometry imaging of rohitukine, a chromone anti-cancer alkaloid, during seed development in *Dysoxylum binectariferum* Hook. f (Meliaceae). *Phytochemistry* 116: 104–110.

65 Hemalatha, R.G. and Pradeep, T. (2013). Understanding the molecular signatures in leaves and flowers by desorption electrospray ionization mass spectrometry (DESI MS) imaging. *J. Agric. Food Chem.* 61 (31): 7477–7487.

66 Klejdus, B. and Kováčik, J. (2016). Quantification of phenols in cinnamon: a special focus on "total phenols" and phenolic acids including DESI-Orbitrap MS detection. *Ind. Crop. Prod.* 83: 774–780.

67 Yang, Y., Yang, Y., Qiu, H. et al. (2021). Localization of constituents for determining the age and parts of ginseng through ultraperfomance liquid chromatography quadrupole/time of flight-mass spectrometry combined with desorption electrospray ionization mass spectrometry imaging. *J. Pharm. Biomed.* 193: 113722.

68 Bagatela, B.S., Lopes, A.P., Cabral, E.C. et al. (2015). High-performance thin-layer chromatography/desorption electrospray ionization mass spectrometry imaging of the crude extract from the peels of *Citrus aurantium* L. (Rutaceae). *Rapid Commun. Mass Spectrom.* 29 (16): 1530–1534.

69 Kucharíková, A., Kimáková, K., Janfelt, C., and Čellárová, E. (2016). Interspecific variation in localization of hypericins and phloroglucinols in the genus Hypericum as revealed by desorption electrospray ionization mass spectrometry imaging. *Physiol. Plant.* 157 (1): 2–12.

70 Hu, B., Xin, G.-z., So, P.-K., and Yao, Z.-P. (2015). Thin layer chromatography coupled with electrospray ionization mass spectrometry for direct analysis of raw samples. *J. Chromatogr. A* 1415: 155–160.

71 Van Berkel, G.J., Tomkins, B.A., and Kertesz, V. (2007). Thin-layer chromatography/desorption electrospray ionization mass spectrometry: investigation of goldenseal alkaloids. *Anal. Chem.* 79 (7): 2778–2789.

72 Claude, E., Tower, M., Lafont, R. et al. (2020). High performance thin-layer chromatography of plant ecdysteroids coupled with desorption electrospray ionisation–ion mobility–time of flight high resolution mass spectrometry (HPTLC/DESI/IM/ToFMS). *Chromatographia* 83 (8): 1029–1035.

73 Laub, A., Sendatzki, A.-K., Palfner, G. et al. (2020). HPTLC-DESI-HRMS-based profiling of anthraquinones in complex mixtures – a proof-of-concept study using crude extracts of chilean mushrooms. *Foods* 9 (2): 156.

74 Prideaux, B. and Stoeckli, M. (2012). Mass spectrometry imaging for drug distribution studies. *J. Proteomics* 75 (16): 4999–5013.

75 Claude, E., Jones, E.A., and Pringle, S.D. (2017). DESI mass spectrometry imaging (MSI). *Methods Mol. Biol.* 1618: 65–75.

76 Pól, J., Strohalm, M., Havlíček, V., and Volný, M. (2010). Molecular mass spectrometry imaging in biomedical and life science research. *Histochem. Cell Biol.* 134 (5): 423–443.

77 Ferreira, C., Alfaro, C.M., Pirro, V., and Cooks, R.G. (2019). Desorption electrospray ionization mass spectrometry imaging: recent developments and perspectives. *J. Biomol. Tech.* 30 (Suppl): S46.

78 Wiseman, J.M., Ifa, D.R., Zhu, Y. et al. (2008). Desorption electrospray ionization mass spectrometry: imaging drugs and metabolites in tissues. *Proc. Natl. Acad. Sci. U.S.A.* 105 (47): 18120–18125.

79 McEwen, A. and Henson, C. (2015). Quantitative whole-body autoradiography: past, present and future. *Bioanalysis* 7 (5): 557–568.

80 Wang, L., Hong, H., and Zhang, D. (2012). Applications of quantitative whole-body autoradiography (QWBA) in drug discovery and development. In: *ADME-Enabling Technologies in Drug Design and Development* (ed. D. Zhang and S. Surapaneni), 419–434. Wiley.

81 Stumpf, W.E. (2013). Whole-body and microscopic autoradiography to determine tissue distribution of biopharmaceuticals – target discoveries with receptor micro-autoradiography engendered new concepts and therapies for vitamin D. *Adv. Drug Delivery Rev.* 65 (8): 1086–1097.

82 Reyzer, M.L., Chaurand, P., Angel, P.M., and Caprioli, R.M. (2010). Direct molecular analysis of whole-body animal tissue sections by MALDI imaging mass spectrometry. *Methods Mol. Biol.* 656: 285–301.

83 Kertesz, V., Van Berkel, G.J., Vavrek, M. et al. (2008). Comparison of drug distribution images from whole-body thin tissue sections obtained using desorption electrospray ionization tandem mass spectrometry and autoradiography. *Anal. Chem.* 80 (13): 5168–5177.

84 León, M., Ferreira, C.R., Eberlin, L.S. et al. (2019). metabolites and lipids associated with fetal swine anatomy via desorption electrospray ionization – mass spectrometry imaging. *Sci. Rep.* 9 (1): 7247.

85 Okutan, S., Hansen, H.S., and Janfelt, C. (2016). Cryo-sectioning of mice for whole-body imaging of drugs and metabolites with desorption electrospray ionization mass spectrometry imaging – a simplified approach. *Proteomics* 16 (11, 12): 1633–1641.

86 Jarmusch, A.K., Pirro, V., Baird, Z. et al. (2016). Lipid and metabolite profiles of human brain tumors by desorption electrospray ionization-MS. *Proc. Natl. Acad. Sci. U.S.A.* 113 (6): 1486–1491.

87 Santagata, S., Eberlin, L.S., Norton, I. et al. (2014). Intraoperative mass spectrometry mapping of an onco-metabolite to guide brain tumor surgery. *Proc. Natl. Acad. Sci. U.S.A.* 111 (30): 11121–11126.

88 Margulis, K., Chiou, A.S., Aasi, S.Z. et al. (2018). Distinguishing malignant from benign microscopic skin lesions using desorption electrospray ionization mass spectrometry imaging. *Proc. Natl. Acad. Sci. U.S.A.* 115 (25): 6347–6352.

89 Pirro, V., Alfaro, C.M., Jarmusch, A.K. et al. (2017). Intraoperative assessment of tumor margins during glioma resection by desorption electrospray ionization-mass spectrometry. *Proc. Natl. Acad. Sci. U.S.A.* 114 (26): 6700–6705.

90 Banerjee, S., Zare, R.N., Tibshirani, R.J. et al. (2017). Diagnosis of prostate cancer by desorption electrospray ionization mass spectrometric imaging of small metabolites and lipids. *Proc. Natl. Acad. Sci. U.S.A.* 114 (13): 3334–3339.

91 Oliveira-Lima, O.C., Carvalho-Tavares, J., Rodrigues, M.F. et al. (2019). Lipid dynamics in LPS-induced neuroinflammation by DESI-MS imaging. *Brain Behav. Immun.* 79: 186–194.

92 Fernandes, A.M.A.P., Vendramini, P.H., Galaverna, R. et al. (2016). direct visualization of neurotransmitters in rat brain slices by desorption electrospray ionization mass spectrometry imaging (DESI-MS). *J. Am. Soc. Mass Spectrom.* 27 (12): 1944–1951.

93 Eberlin, L.S., Norton, I., Dill, A.L. et al. (2012). Classifying human brain tumors by lipid imaging with mass spectrometry. *Cancer Res.* 72 (3): 645–654.

94 Calligaris, D., Caragacianu, D., Liu, X.H. et al. (2014). Application of desorption electrospray ionization mass spectrometry imaging in breast cancer margin analysis. *Proc. Natl. Acad. Sci. U.S.A.* 111 (42): 15184–15189.

95 DeHoog, R.J., Zhang, J.L., Alore, E. et al. (2019). Preoperative metabolic classification of thyroid nodules using mass spectrometry imaging of fine-needle aspiration biopsies. *Proc. Natl. Acad. Sci. U.S.A.* 116 (43): 21401–21408.

96 Eberlin, L.S., Norton, I., Orringer, D. et al. (2013). Ambient mass spectrometry for the intraoperative molecular diagnosis of human brain tumors. *Proc. Natl. Acad. Sci. U.S.A.* 110 (5): 1611–1616.

97 Eberlin, L.S., Tibshirani, R.J., Zhang, J.L. et al. (2014). Molecular assessment of surgical-resection margins of gastric cancer by mass-spectrometric imaging. *Proc. Natl. Acad. Sci. U.S.A.* 111 (7): 2436–2441.

98 Shrestha, B., Roman, G., Hattan, S. et al. (2020). High spatial resolution molecular tissue imaging by desorption electrospray ionization using novel sprayer assembly. *J. Biomol. Tech.* 31 (Suppl): S17–S18.

99 Campbell, D.I., Ferreira, C.R., Eberlin, L.S., and Cooks, R.G. (2012). Improved spatial resolution in the imaging of biological tissue using desorption electrospray ionization. *Anal. Bioanal. Chem.* 404 (2): 389–398.

100 Venter, A. and Cooks, R.G. (2007). Desorption electrospray ionization in a small pressure-tight enclosure. *Anal. Chem.* 79 (16): 6398–6403.

101 Parrot, D., Papazian, S., Foil, D., and Tasdemir, D. (2018). Imaging the unimaginable: desorption electrospray ionization – imaging mass spectrometry (DESI-IMS) in natural product research. *Planta Med.* 84 (9,10): 584–593.

102 Schwartz, S.A., Reyzer, M.L., and Caprioli, R.M. (2003). Direct tissue analysis using matrix-assisted laser desorption/ionization mass spectrometry: practical aspects of sample preparation. *J. Mass Spectrom.* 38 (7): 699–708.

103 Müller, T., Oradu, S., Ifa, D.R. et al. (2011). Direct plant tissue analysis and imprint imaging by desorption electrospray ionization mass spectrometry. *Anal. Chem.* 83 (14): 5754–5761.

104 Ifa, D.R., Manicke, N.E., Rusine, A.L., and Cooks, R.G. (2008). Quantitative analysis of small molecules by desorption electrospray ionization mass spectrometry from polytetrafluoroethylene surfaces. *Rapid Commun. Mass Spectrom.* 22 (4): 503–510.

105 Li, B., Knudsen, C., Hansen, N.K. et al. (2013). Visualizing metabolite distribution and enzymatic conversion in plant tissues by desorption electrospray ionization mass spectrometry imaging. *Plant J.* 74 (6): 1059–1071.

106 Thunig, J., Hansen, S.H., and Janfelt, C. (2011). Analysis of secondary plant metabolites by indirect desorption electrospray ionization imaging mass spectrometry. *Anal. Chem.* 83 (9): 3256–3259.

107 Lanekoff, I., Thomas, M., Carson, J.P. et al. (2013). Imaging nicotine in rat brain tissue by use of nanospray desorption electrospray ionization mass spectrometry. *Anal. Chem.* 85 (2): 882–889.

108 Nguyen, S.N., Sontag, R.L., Carson, J.P. et al. (2018). Towards high-resolution tissue imaging using nanospray desorption electrospray ionization mass spectrometry coupled to shear force microscopy. *J. Am. Soc. Mass Spectrom.* 29 (2): 316–322.

109 Wells, J.A. and McClendon, C.L. (2007). Reaching for high-hanging fruit in drug discovery at protein-protein interfaces. *Nature* 450 (7172): 1001–1009.

110 Sahasrabuddhe, A., Hsia, Y., Busch, F. et al. (2018). Confirmation of intersubunit connectivity and topology of designed protein complexes by native MS. *Proc. Natl. Acad. Sci. U.S.A.* 115 (6): 1268–1273.

111 Cong, X., Liu, Y., Liu, W. et al. (2016). Determining membrane protein–lipid binding thermodynamics using native mass spectrometry. *J. Am. Chem. Soc.* 138 (13): 4346–4349.

112 Vivat Hannah, V., Atmanene, C., Zeyer, D. et al. (2010). Native MS: an 'ESI' way to support structure-and fragment-based drug discovery. *Future Med. Chem.* 2 (1): 35–50.

113 Nguyen, G.T., Tran, T.N., Podgorski, M.N. et al. (2019). Nanoscale ion emitters in native mass spectrometry for measuring ligand–protein binding affinities. *ACS Cent. Sci.* 5 (2): 308–318.

114 Barth, M. and Schmidt, C. (2020). Native mass spectrometry – a valuable tool in structural biology. *J. Mass Spectrom.* 55 (10): e4578.

115 Miao, Z. and Chen, H. (2009). Direct analysis of liquid samples by desorption electrospray ionization-mass spectrometry (DESI-MS). *J. Am. Soc. Mass Spectrom.* 20 (1): 10–19.

116 Liu, P., Zhang, J., Ferguson, C.N. et al. (2013). Measuring protein–ligand interactions using liquid sample desorption electrospray ionization mass spectrometry. *Anal. Chem.* 85 (24): 11966–11972.

117 Chipuk, J.E. and Brodbelt, J.S. (2008). Transmission mode desorption electrospray ionization. *J. Am. Soc. Mass Spectrom.* 19 (11): 1612–1620.

118 Zheng, Q., Zhang, H., and Chen, H. (2013). Integration of online digestion and electrolytic reduction with mass spectrometry for rapid disulfide-containing protein structural analysis. *Int. J. Mass Spectrom.* 353: 84–92.

119 Campbell, I.S., Ton, A.T., and Mulligan, C.C. (2011). Direct detection of pharmaceuticals and personal care products from aqueous samples with thermally-assisted desorption electrospray ionization mass spectrometry. *J. Am. Soc. Mass Spectrom.* 22 (7): 1285–1293.

120 Ferguson, C.N., Benchaar, S.A., Miao, Z. et al. (2011). Direct ionization of large proteins and protein complexes by desorption electrospray ionization-mass spectrometry. *Anal. Chem.* 83 (17): 6468–6473.

121 Lu, X., Ning, B., He, D. et al. (2014). High-throughput screening of high-affinity ligands for proteins with anion-binding sites using desorption electrospray ionization (DESI) mass spectrometry. *J. Am. Soc. Mass Spectrom.* 25 (3): 454–463.

122 Yao, C., Na, N., Huang, L. et al. (2013). High-throughput detection of drugs binding to proteins using desorption electrospray ionization mass spectrometry. *Anal. Chim. Acta* 794: 60–66.

123 Zheng, Q., Tian, Y., Ruan, X. et al. (2018). Probing specific ligand–protein interactions by native-denatured exchange mass spectrometry. *Anal. Chim. Acta* 1036: 58–65.

124 Ambrose, S., Housden, N.G., Gupta, K. et al. (2017). Native desorption electrospray ionization liberates soluble and membrane protein complexes from surfaces. *Angew. Chem. Int. Ed.* 56 (46): 14463–14468.

125 Jaman, Z., Logsdon, D.L., Szilágyi, B. et al. (2020). High-throughput experimentation and continuous flow evaluation of nucleophilic aromatic substitution reactions. *ACS Comb. Sci.* 22 (4): 184–196.

126 Girod, M., Moyano, E., Campbell, D.I., and Cooks, R.G. (2011). Accelerated bimolecular reactions in microdroplets studied by desorption electrospray ionization mass spectrometry. *Chem. Sci.* 2 (3): 501–510.

127 Wleklinski, M., Loren, B.P., Ferreira, C.R. et al. (2018). High-throughput reaction screening using desorption electrospray ionization mass spectrometry. *Chem. Sci.* 9 (6): 1647–1653.

128 Fedick, P.W., Iyer, K., Wei, Z. et al. (2019). Screening of the suzuki cross-coupling reaction using desorption electrospray ionization in high-throughput and in leidenfrost droplet experiments. *J. Am. Soc. Mass Spectrom.* 30 (10): 2144–2151.

129 Jaman, Z., Sobreira, T.J.P., Mufti, A. et al. (2019). Rapid on-demand synthesis of lomustine under continuous flow conditions. *Org. Process Res. Dev.* 23 (3): 334–341.

130 Sobreira, T.J.P., Avramova, L., Szilagyi, B. et al. (2020). High-throughput screening of organic reactions in microdroplets using desorption electrospray ionization mass spectrometry (DESI-MS): hardware and software implementation. *Anal. Methods* 12 (28): 3654–3669.

131 Morato, N.M., Holden, D.T., and Cooks, R.G. (2020). High-throughput label-free enzymatic assays using desorption electrospray-ionization mass spectrometry. *Angew. Chem. Int. Ed.* 59 (46): 20459–20464.

132 Pu, F., Radosevich, A.J., Sawicki, J.W. et al. (2021). High-throughput label-free biochemical assays using infrared matrix-assisted desorption electrospray ionization mass spectrometry. *Anal. Chem.* 93 (17): 6792–6800.

133 Badu-Tawiah, A.K., Eberlin, L.S., Ouyang, Z., and Cooks, R.G. (2013). Chemical aspects of the extractive methods of ambient ionization mass spectrometry. *Annu. Rev. Phys. Chem.* 64: 481–505.

134 Zani, C.L. and Carroll, A.R. (2017). Database for rapid dereplication of known natural products using data from MS and fast NMR experiments. *J. Nat. Prod.* 80 (6): 1758–1766.

135 Borges, R.M., Taujale, R., de Souza, J.S. et al. (2018). Dereplication of plant phenolics using a mass-spectrometry database independent method. *Phytochem. Anal.* 29 (6): 601–612.

136 Eberlin, L.S., Liu, X., Ferreira, C.R. et al. (2011). Desorption electrospray ionization then MALDI mass spectrometry imaging of lipid and protein distributions in single tissue sections. *Anal. Chem.* 83 (22): 8366–8371.

137 Ricci, C., Nyadong, L., Fernandez, F.M. et al. (2007). Combined Fourier-transform infrared imaging and desorption electrospray-ionization linear ion-trap mass spectrometry for analysis of counterfeit antimalarial tablets. *Anal. Bioanal. Chem.* 387 (2): 551–559.

138 Gao, L., Song, Q., Patterson, G.E. et al. (2006). Handheld rectilinear ion trap mass spectrometer. *Anal. Chem.* 78 (17): 5994–6002.

139 Gao, L., Sugiarto, A., Harper, J.D. et al. (2008). Design and characterization of a multisource hand-held tandem mass spectrometer. *Anal. Chem.* 80 (19): 7198–7205.

140 Hendricks, P.I., Dalgleish, J.K., Shelley, J.T. et al. (2014). Autonomous in situ analysis and real-time chemical detection using a backpack miniature mass spectrometer: concept, instrumentation development, and performance. *Anal. Chem.* 86 (6): 2900–2908.

141 Li, L., Chen, T.-C., Ren, Y. et al. (2014). Mini 12, miniature mass spectrometer for clinical and other applications introduction and characterization. *Anal. Chem.* 86 (6): 2909–2916.

142 Jiang, T., Zhang, H., Tang, Y. et al. (2017). A "brick mass spectrometer" driven by a sinusoidal frequency scanning technique. *Anal. Chem.* 89 (10): 5578–5584.

143 Li, A., Hansen, B.J., Powell, A.T. et al. (2014). Miniaturization of a planar-electrode linear ion trap mass spectrometer. *Rapid Commun. Mass Spectrom.* 28 (12): 1338–1344.

144 Wu, Q., Li, A., Tian, Y. et al. (2016). Miniaturized linear wire ion trap mass analyzer. *Anal. Chem.* 88 (15): 7800–7806.

145 Riedo, A., Meyer, S., Heredia, B. et al. (2013). Highly accurate isotope composition measurements by a miniature laser ablation mass spectrometer designed for in situ investigations on planetary surfaces. *Planet Space Sci.* 87: 1–13.

146 Liu, X., Wang, X., Bu, J. et al. (2018). Tandem analysis by a dual-trap miniature mass spectrometer. *Anal. Chem.* 91 (2): 1391–1398.

147 Pu, F., Zhang, W., Bateman, K.P. et al. (2017). Using miniature MS system with automatic blood sampler for preclinical pharmacokinetics study. *Bioanalysis* 9 (21): 1633–1641.

148 Zou, R., Cao, W., Chong, L. et al. (2018). Point-of-care tissue analysis using miniature mass spectrometer. *Anal. Chem.* 91 (1): 1157–1163.

149 Chen, C.-H., Lin, Z., Tian, R. et al. (2015). Real-time sample analysis using a sampling probe and miniature mass spectrometer. *Anal. Chem.* 87 (17): 8867–8873.

Section 6

Conclusion

14

The Impact of HT-MS to Date and Its Potential to Shape the Future of Metrics-Based Experimentation and Analysis

Matthew D. Troutman

Hit Discovery and Optimization, Pfizer, Inc., Groton, CT, USA

High-throughput mass spectrometry (HT-MS) has had a profound impact on shaping how drug discovery, the effort to find and optimize medicines, is conducted [1, 2]. When considering the importance of MS to drug discovery, it is helpful to consider the knowledge that is provided. Mass is a fundamental molecular characteristic, and therefore using mass as an endpoint is highly desirable when possible. A measurement based on mass has the potential to be unambiguous, where the analyte of interest can be directly measured. Taken together, mass-based analysis is elegant and reliable, thus making it especially useful to drug discovery efforts where accuracy and fast-paced, informed decision making are paramount to driving success. While it is beyond the intent of this conclusion to review and capture the history and state of the art in MS instrument technology and capability, it is useful to evoke some salient aspects for sake of illustrating how these instruments play a role in driving drug discovery. The MS's track record to accurately detect a specific mass from a complex sample of interest is well established, well known, and irrefutable; indeed, it is hard to envision a technology that would supplant MS for these experimental goals in the foreseeable future [3]. Modern MS instruments are exquisite in their sensitivity, specificity, and measurement speed and can readily generate high content sample analyses routinely to support efforts such as multiplexed sample analysis, metabolite and impurity scouting, and global proteomics [4, 5]. Finally, emerging MS technology is pushing the boundaries on mass range, novel analytical experimentation in gas phase, and next-level sensitivity and specificity [6]. Analysis of macromolecules, heteromeric protein complexes, and interacting ligands is a rapidly growing area that shows high promise to deliver unique impactful data in an elegant fashion [7, 8]. In concert, throughout the chapters in this book, several ingenious creative

High-Throughput Mass Spectrometry in Drug Discovery, First Edition. Edited by Chang Liu and Hui Zhang.
© 2023 John Wiley & Sons, Inc. Published 2023 by John Wiley & Sons, Inc.

technologies and applications created to achieve a high-throughput utilization of these powerful MS instrument capabilities via rapid sampling or multiplexed sampling have been nicely described and discussed. Exquisite MS capabilities utilized in HT fashion is a synergistically impactful application that has become critical and will continue to expand in utility to driving drug discovery.

Given HT-MS is an applied approach that has emerged and grown in recent times, in this perspective thoughts will be provided to further define what is meant by HT-MS. Following this, perspective on HT-MS's unique impact on drug discovery to date and where HT-MS can continue to make impact will be discussed. The former is irrefutable, and the latter is compelling and timely.

14.1 Defining High-Throughput Mass Spectrometry (HT-MS)

As follows, some thoughts underlying what is meant by "HT" as it relates to "HT-MS" will be provided to help set the stage for understanding how HT-MS has critically enabled and influenced drug discovery to date. First, in terms of throughput, there is no codified, quantitative definition or metric that serves as a designation for HT as it relates to HT-MS; the term tends to be used descriptively and it is somewhat specific to the application and is mostly user-defined. Second, from a technology standpoint, one requirement is that the approach ultimately uses MS endpoint analysis. In that way, HT-MS is not limited to one method and can include many rather than a single technology (as is clear from the variety of approaches described in previous chapters). Furthermore, HT-MS applications themselves tend to be elegant fit for purpose approaches that ingeniously draw from a variety of sampling, sometimes separation, and ultimately, MS endpoint analysis technologies. Finally, from an applications point of view, the plasticity seen in how HT-MS is performed is an important aspect and one of the most salient attributes of HT-MS (and the scientists who perform HT-MS). Indeed, most HT-MS approaches were created to realize the potential of elegant MS solutions to produce a transformational enablement of strategies where MS endpoint analysis would be needed at a scale and pace required to be relevant and useful to drug discovery, such as driving HTS or lead optimization. Conversely, HT-MS does not typically mean high volume ad hoc analytical analysis, but rather high-volume static/consistent analytical analysis. Taken all-together, one way to define HT-MS is through impact (versus output or technologies per se): MS endpoint analysis delivered at a throughput to effectively create consistent populations and sets of data that are typically used to drive drug discovery. A practical limit to HT-MS, and HT technologies in general, is ability of the end-user to digest and prosecute the volume of data in a given timeframe. Defining HT-MS "phenotypically" with a few consistent salient criteria reflects the fluidity seen in how the approach continues to grow, along with providing some context on how it might evolve to meet future opportunities.

14.2 HT-MS: Impact to Date

To date, the clearest impacts of HT-MS have been where MS quantitative endpoint analysis was required or transformational. One of the most clarion examples of impact can be seen in the role HT-MS has played in driving druggable property optimization in design [9]. In the 1990s, it became clear that poor pharmacokinetics (PK) was a significant source of clinical candidate attrition [10]. For a compound to be useful as a drug, it needs both efficacy against the intended target and sufficient exposure (in extent and duration at site of action); PK describes exposure [11]. In response to these attrition analyses, strategies were created to eliminate poor PK as a cause of clinical failure by moving efforts to understand PK earlier into discovery and establishing sufficient PK as a selection criteria and requirement for clinical advancement [12]. Some pharma companies went even further beyond selection based on PK to establishing capabilities to allow for explicitly selecting and optimizing for favorable PK as part of drug candidate design. More specifically, the ability to determine the underlying factors that govern a compound's PK including absorption, distribution, metabolism, and elimination (ADME), in vitro, was established [13]. Large effort was invested to establish in vitro to in vivo correlations (IVIVC) for in vitro ADME endpoints to enable prospective PK predictions [14]. A key requirement for both PK and ADME is quantitative analysis of the compound (versus a more ubiquitous target-driven endpoint as is typically employed in pharmacology assays). Here, MS provides a significant experimental advantage in specificity and sensitivity to measure analyte of interest at very low levels in complex biological matrices. As these ADME science rapidly matured, considerable efforts were made to format ADME assays into high-density plate-based formats for sake of enabling strategies aimed at optimizing compound ADME properties to support high-volume, iterative, lead optimization design efforts. One of the largest hurdles to realizing this strategy was the requisite endpoint analysis; how to measure so many compounds in assays that produce multiple samples in a rapid fashion where methods for new compounds are needed each cycle to drive new molecule profiling? Meeting this challenge may be considered the genesis of HT-MS, and it drove creation of novel technologies such as high-throughput autosamplers, along with optimized analytical and electronic workflows [15, 16]. Several chapters in this book cover advanced versions of HT-MS applications that can be traced to the first days of enabling HT-ADME approaches.

Variations of HT-MS applications enabling HT-ADME have been in place now for 20 or so years and are considered critical business as usual for many pharma drug discovery efforts. It is worth reflecting on HT-MS's impact to the maturation and growth of ADME and PK sciences. First, PK-driven attrition for small-molecule therapeutics has been effectively eliminated or reduced to rare cases, and HT-ADME has played a role in driving ADME and PK in early discovery. Large, consistent data sets resulting from years of HT-ADME profiling have been amassed [17]. The HT-MS imparts high quality and specificity to data, with fast turnaround and at a large volume of profiling. With these large-scale data sets as substrates, computational scientists have been able to generate highly useful predictive in silico endpoint models that are routinely utilized during the design process. Further to this, system modelers have been successful in conducting large-scale IVIVC to generate and refine PK prediction methodologies that ultimately culminated in system-level in silico PK predictive models [18, 19].

Beyond ADME and PK, quantitative HT-MS has been used to excellent effect to enable other scientific disciplines. Specifically, HT-MS is showing great promise in for many pharmacology applications including functional (cellular and biochemical) HTS and lead optimization [20, 21]. In many cases, the attributes of MS analysis including specificity and sensitivity have allowed higher quality assays less prone to coupled read endpoints, and lower assay floors, respectively, to be achieved. Scientists in fields beyond analytical and ADME/PK are becoming aware of the potential of MS endpoint data to make significant progress beyond what is possible with conventional approaches. These robust and elegant quantitative HT-MS approaches have provided great proof of principle that will certainly lead to further utilization and approaches.

14.3 Considering How HT-MS Will Shape the Future of Drug Discovery

Modern-day drug discovery challenges are considerable [22]. The drive to deliver therapies that address unmet medical needs has led to a huge change and expansion in approaches being pursued, and how drug discovery is being performed. Addressing unmet needs oftentimes leads to exploration and evaluation of targets and pathways where the drug discovery researcher is truly on the bleeding edge of the science and the knowledge. For many of these emerging targets and pathways that have been identified as potentially useful therapeutics, evidence of potential benefit and impetus to consider as a drug target has come from "top-down" approaches including phenotypic screening, and genome-wide association studies (GWAS) [23]. The term "discovery" has never been more accurate and a necessary activity in so many areas aimed at producing potentially useful therapeutic agents. Indeed, true discovery now complements the design, execution, and optimization efforts described in the previous section capturing impact to date. Elucidating the biological mechanisms that link efficacy and disease to genetics and phenotypes, then on deeper levels to molecular and cellular biology, is challenging for precedented approaches and monumental for novel approaches. Novel therapeutic targets require exploration into their role in cellular biology for sake of understanding impacts of altering their function and to gain insight into potential ways to interact that will lead to safe, therapeutic outcomes. In response to these challenges, it is not surprising to reflect on the growth in utilization of pan-omics and counterpart computational biology disciplines to help elucidate these key links [24]. Increasingly, many of these promising new therapeutic avenues require perturbation of novel targets; many of which are in notoriously difficult to drug gene families that feature inherent disorder, multiple oligomeric states, necessary binding partners, and a multiplicity of cell functions such as enzymatic activity and scaffolding [25]. Further compounding the challenge to discovery and design of useful compounds is that many targets may require intervention in ways beyond canonical binding at a putative active site (e.g. protein–protein interfaces) and/or producing pharmacological effect beyond inhibition (e.g. agonism, activation). The ability to gain medium and high-resolution analytical information for a novel target, a potential ligand and in some case binding partner(s) is extremely critical and useful to enabling drug discovery and design [26]. Finally, for many unprecedented targets and mechanisms, safely and effectively targeting these to alter disease states in humans is a discovery effort in and of itself. Indeed, many high-confidence points of intervention play critical biological roles, where understanding the delicate balance of altering the nature of the target to produce efficacy precisely, while not completely changing overall global role, can be the difference between therapeutic utility and on-mechanism toxicity. Selectivity becomes even more important to ensure that isoforms with similar critical roles are not affected for sake of maintaining off-target safety and a wide therapeutic window. An ability to accurately measure therapeutic agents along with efficacy and safety biomarkers systemically, and at sites of action, from patient samples is critical to understanding kinetic – dynamics relationships and to subsequently balancing therapeutic and safe exposures clinically [27].

Clearly the challenges encountered in modern-day drug discovery associated with working in unknown and novel areas are daunting and high risk. However, they are not insurmountable and perhaps we now have capabilities and tools to make progress in these novel, previously unexplored areas. The key enabler and need across all of the challenges noted in the previous paragraph is generation of right content, right volume, readily actionable information delivered in timescales to enable learning, future hypothesis generation, and decision-making. Artificial intelligence and computational approaches will play a role here, but are unlikely to be entirely sufficient, particularly for unprecedented areas with little to no data [28]. There must be seed information from which to build said approaches where the current computational technologies tend to be inherently better at interpolation versus extrapolation. It is clear to see that experimental approaches will remain critical to both direct end users, and to feeding and driving next generation AI and computational approaches. In this regard, accurate, elegant, exquisite, multi-endpoint, high-throughput measurement capabilities will be a key element required for experimental success.

We are at a point where HT-MS has the potential to realize a discontinuous increase in utilization and impact to drug discovery. There are three major drivers converging to produce this special opportunity: (i) The critical need to provide new scientific information for unprecedented targets and pathways, (ii) exciting developments in MS instrumentation and applications, and (iii) critical mass of HT-MS scientist experience base in building cutting-edge HT-MS applications. The focus for the remainder of this section will be to provide a few examples of new areas of opportunity to create and utilize HT-MS for emerging and novel applications.

Our first area to consider for HT-MS application growth is rooted in emerging analytical technologies that have the potential to be used in HT-MS approaches. Current generation high-resolution mass spectrometers can provide high content analysis on the sample level rapidly. For applications seeking to understand global, system and pathway level changes in response to stimulus including injury and drug treatment, high content analysis is critical [29]. Prospectively, complex system level and phenotypic screens are created to capture more pathway and cell-level changes to drive compound optimization. High-content imaging has been useful for these sorts of analyses, but perhaps not sufficient; the technique is limited to optical endpoints [30]. It is clear to see a potential role for HRMS performed in HT to complement and enable phenotypic screens that are not best enabled via imaging, by measuring MS-amenable endpoints. Another area where high-content and high-resolution MS analysis is particularly adept and useful in reaction monitoring. For parallelized synthetic approaches generating 100s to 1000s of samples in a given experiment, the ability to measure reactant, product, and by-products is extremely useful to guide next rounds of experiments and to focus on the successful synthetic approaches [31].

Advances in sampling technologies will certainly drive novel HT-MS approaches and applications. In some cases, the sampling technologies allow for order of magnitude increases in throughput. Once mature, it is likely that acoustic [32] and MALDI-based sampling will become routinely utilized for true HTS screening [20]. In other cases, sampling technologies may allow different test articles to be profiled. For example, technologies such as FACS may allow for HT-MS single-cell analysis [33, 34].

The next area to consider for HT-MS application growth is rooted in emerging needs. Effectively prosecuting novel targets oftentimes comes with the need to find tool ligands. This is particularly difficult when natural substrates are unknown. Here, unbiased screening of large compound sets spanning drug-like chemistry space becomes the first choice for finding potential binders [35]. Especially for endpoints that require difficult to measure and/or lengthy cellular assays, affinity-based approaches with simplified cell extracts or recombinant targets may be the only plausible way to start [36]. Approaches such as affinity selection by mass spectrometry (ASMS) have shown promise for effective hit ID [37, 38]. Other approaches such as Native MS (nMS), a technique that utilizes high mass range HRMS to measure target, ligand, and target-ligand in one trace are particularly powerful in these pursuits [39]. Indeed, nMS has the promise of hit ID along with providing proof of target engagement [40, 41]. More broadly, target engagement-based approaches, rooted in stability proteomics and covalent screens, are increasingly being utilized for hit ID [42]. Finding ways to bring these powerful analytical approaches to true HT level would be very enabling.

Progressing ability to develop therapies to some diseases will require higher resolution information. More specifically, for certain disease states, there is heterogeneity in biology across components of a system and techniques such as single-cell proteomics provide the promise of delivering precise information that goes a level deeper than whole organism/tissue-level analyses. It would be powerful to gain abilities to optimize NME's using these cell (versus well) level approaches [43].

Lastly, a key driver to consider for HT-MS application growth is rooted in how data is consumed. In many cases, ability to digest and process data is the more practical limiter to throughput versus technical capability [44]. This is particularly true for systems-level and multiparametric complex data endpoints [45]. The advent of computational algorithms to process complex data will bring with it the potential to deploy higher throughput approaches. It is conceivable that we are on the cusp of an evolution from univariant experimental lead optimization to multivariant lead optimization, driven both by target and biology complexity and ability to quickly synthesize these complex endpoints; high-content HT-MS opportunities to provide this endpoint analysis are expected to increase. On a larger scale, it was noted previously that AI has promise, but a key requirement requires being fed by "big data" [28]. New "big data" sets of sufficient quality and content leading to creation of useful computational approaches will likely rely heavily on HT-MS applications to generate high-quality MS data at scale quickly.

Successfully realizing the next generation of HT-MS approaches will ultimately depend on analytical scientists to deploy their knowledge, talents, and ingenuity to drive what is possible. Additionally, it will require that analytical scientists gain a real-time understanding of emerging opportunities and a voice to contribute to how they might become critical to solving modern-day drug discovery problems. For HT-MS, the challenge is perhaps evolving from the relentless optimization of application performance to becoming keenly aware of a dynamic environment where an explosion of novel opportunities and needs, along with exquisite computational and experimental technology, provides opportunity for impact through quickly creating out of the box, fit for purpose applications that may be unprecedented and may be specific to the need in the moment. It is the opinion of this author that this emerging environment will produce synergistic advances in technologies, applications, utilization of HT-MS, and ultimately, criticality and impact of HT-MS to drug discovery.

References

1 Janiszewski, J.S., Liston, T.E., and Cole, M.J. (2008). Perspectives on bioanalytical mass spectrometry and automation in drug discovery. *Curr. Drug Metab.* 9 (9): 986–994.

2 McLaren, D.G., Shah, V., Wisniewski, T. et al. (2021). High-throughput mass spectrometry for hit identification: current landscape and future perspectives. *SLAS Discovery* 26 (2): 168–191.

3 Smith, R.D., Shen, Y., and Tang, K. (2004). Ultrasensitive and quantitative analyses from combined separations-mass spectrometry for the characterization of proteomes. *Acc. Chem. Res.* 37 (4): 269–278.

4 Meissner, F., Geddes-Mcalister, J., Mann, M., and Bantscheff, M. (2022). The emerging role of mass spectrometry-based proteomics in drug discovery. *Nat. Rev. Drug Discovery* 21 (9): 637–654.

5 Zubarev, R.A. and Makarov, A. (2013). Orbitrap mass spectrometry. *Anal. Chem.* 85 (11): 5288–5296.

6 Chen, S., Getter, T., Salom, D. et al. (2022). Capturing a rhodopsin receptor signalling cascade across a native membrane. *Nature* 604 (7905): 384–390.

7 Sahasrabuddhe, A., Hsia, Y., Busch, F. et al. (2018). Confirmation of intersubunit connectivity and topology of designed protein complexes by native MS. *Proc. Natl. Acad. Sci. U.S.A.* 115 (6): 1268–1273.

8 Gault, J., Liko, I., Landreh, M. et al. (2020). Combining native and 'omics' mass spectrometry to identify endogenous ligands bound to membrane proteins. *Nat. Methods* 17 (5): 505–508.

9 Hop, C.E., Cole, M.J., Davidson, R.E. et al. (2008). High-throughput ADME screening: practical considerations, impact on the portfolio and enabler of in silico ADME models. *Curr. Drug Metab.* 9 (9): 847–853.

10 Prentis, R.A., Lis, Y., and Walker, S.R. (1988). Pharmaceutical innovation by the seven UK-owned pharmaceutical companies (1964–1985). *Br. J. Clin. Pharmacol.* 25 (3): 387–396.

11 Morgan, P., Van Der Graaf, P.H., Arrowsmith, J. et al. (2012). Can the flow of medicines be improved? Fundamental pharmacokinetic and pharmacological principles toward improving Phase II survival. *Drug Discovery Today* 17 (9, 10): 419–424.

12 Beaumont, K. and Smith, D.A. (2009). Does human pharmacokinetic prediction add significant value to compound selection in drug discovery research? *Curr. Opin. Drug Discovery Dev.* 12 (1): 61–71.

13 Maurer, T.S., Edwards, M., Hepworth, D. et al. (2022). Designing small molecules for therapeutic success: a contemporary perspective. *Drug Discovery Today* 27 (2): 538–546.

14 Obach, R.S., Baxter, J.G., Liston, T.E. et al. (1997). The prediction of human pharmacokinetic parameters from preclinical and in vitro metabolism data. *J. Pharmacol. Exp. Ther.* 283 (1): 46–58.

15 Janiszewski, J., Schneider, R., Kapinos, B. et al. (2012). Development of a high-speed, multiplexed sample-delivery instrument for LC-MS/MS bioanalysis. *Bioanalysis* 4 (9): 1039–1056.

16 Kapinos, B., Liu, J., Piotrowski, M. et al. (2017). Development of a high-performance, enterprise-level, multimode LC-MS/MS autosampler for drug discovery. *Bioanalysis* 9 (21): 1643–1654.

17 Lombardo, F., Desai, P.V., Arimoto, R. et al. (2017). In silico absorption, distribution, metabolism, excretion, and pharmacokinetics (ADME-PK): utility and best practices. An industry perspective from the international consortium for innovation through quality in pharmaceutical development. *J. Med. Chem.* 60 (22): 9097–9113.

18 Mathew, S., Tess, D., Burchett, W. et al. (2021). Evaluation of prediction accuracy for volume of distribution in rat and human using in vitro, in vivo, PBPK and QSAR methods. *J. Pharm. Sci.* 110 (4): 1799–1823.

19 Trapa, P.E., Troutman, M.D., Lau, T.Y. et al. (2019). in vitro-in vivo extrapolation of key transporter activity at the blood-brain barrier. *Drug Metab. Dispos.* 47 (4): 405–411.

20 Haslam, C., Hellicar, J., Dunn, A. et al. (2016). The evolution of MALDI-TOF mass spectrometry toward ultrahigh-throughput screening: 1536-well format and beyond. *J. Biomol. Screening* 21 (2): 176–186.

21 Holt, T.G., Choi, B.K., Geoghagen, N.S. et al. (2009). Label-free high-throughput screening via mass spectrometry: a single cystathionine quantitative method for multiple applications. *Assay Drug Dev. Technol.* 7 (5): 495–506.

22 Moffat, J.G., Vincent, F., Lee, J.A. et al. (2017). Opportunities and challenges in phenotypic drug discovery: an industry perspective. *Nat. Rev. Drug Discovery* 16 (8): 531–543.

23 Sanseau, P., Agarwal, P., Barnes, M.R. et al. (2012). Use of genome-wide association studies for drug repositioning. *Nat. Biotechnol.* 30 (4): 317–320.

24 Berg, E.L. (2014). Systems biology in drug discovery and development. *Drug Discovery Today* 19 (2): 113–125.

25 Zhang, G., Zhang, J., Gao, Y. et al. (2022). Strategies for targeting undruggable targets. *Expert Opin. Drug Discovery* 17 (1): 55–69.

26 Schurmann, M., Janning, P., Ziegler, S., and Waldmann, H. (2016). Small-molecule target engagement in cells. *Cell Chem. Biol.* 23 (4): 435–441.

27 Kellie, J.F., Tran, J.C., Lee, J.E. et al. (2010). The emerging process of top down mass spectrometry for protein analysis: biomarkers, protein-therapeutics, and achieving high-throughput. *Mol. BioSyst.* 6 (9): 1532–1539.

28 Bender, A. and Cortes-Ciriano, I. (2021). Artificial intelligence in drug discovery: what is realistic, what are illusions? Part 2: a discussion of chemical and biological data. *Drug Discovery Today* 26 (4): 1040–1052.

29 Vincent, F., Nueda, A., Lee, J. et al. (2022). Phenotypic drug discovery: recent successes, lessons learned and new directions. *Nat. Rev. Drug Discovery.*

30 Vincent, F., Loria, P.M., Weston, A.D. et al. (2020). Hit triage and validation in phenotypic screening: considerations and strategies. *Cell Chem. Biol.* 27 (11): 1332–1346.

31 DiRico, K.J., Hua, W., Liu, C. et al. (2020). Ultrahigh-throughput acoustic droplet ejection-open port interface-mass spectrometry for parallel medicinal chemistry. *ACS Med. Chem. Lett.* 11 (6): 1101–1110.

32 Zhang, H., Liu, C., Hua, W. et al. (2021). Acoustic ejection mass spectrometry for high-throughput analysis. *Anal. Chem.* 93 (31): 10850–10861.

33 Pan, N., Rao, W., Kothapalli, N.R. et al. (2014). The single-probe: a miniaturized multifunctional device for single cell mass spectrometry analysis. *Anal. Chem.* 86 (19): 9376–9380.

34 Pan, N., Rao, W., and Yang, Z. (2020). Single-probe mass spectrometry analysis of metabolites in single cells. *Methods Mol. Biol.* 2064: 61–71.

35 Bell, A.S., Bradley, J., Everett, J.R. et al. (2016). Plate-based diversity subset screening generation 2: an improved paradigm for high-throughput screening of large compound files. *Mol. Divers.* 20 (4): 789–803.

36 Foley, T.L., Burchett, W., Chen, Q. et al. (2021). Selecting approaches for hit identification and increasing options by building the efficient discovery of actionable chemical matter from DNA-encoded libraries. *SLAS Discovery* 26 (2): 263–280.

37 Muchiri, R.N. and van Breemen, R.B. (2021). Affinity selection-mass spectrometry for the discovery of pharmacologically active compounds from combinatorial libraries and natural products. *J. Mass Spectrom.* 56 (5): e4647.

38 O'Connell, T.N., Ramsay, J., Rieth, S.F. et al. (2014). Solution-based indirect affinity selection mass spectrometry--a general tool for high-throughput screening of pharmaceutical compound libraries. *Anal. Chem.* 86 (15): 7413–7420.

39 Tamara, S., Den Boer, M.A., and Heck, A.J.R. (2022). High-resolution native mass spectrometry. *Chem. Rev.* 122 (8): 7269–7326.

40 Fenn, J.B., Mann, M., Meng, C.K. et al. (1989). Electrospray ionization for mass spectrometry of large biomolecules. *Science* 246 (4926): 64–71.

41 Khristenko, N., Amato, J., Livet, S. et al. (2019). Native ion mobility mass spectrometry: when gas-phase ion structures depend on the electrospray charging process. *J. Am. Soc. Mass Spectrom.* 30 (6): 1069–1081.

42 Masson, G.R., Jenkins, M.L., and Burke, J.E. (2017). An overview of hydrogen deuterium exchange mass spectrometry (HDX-MS) in drug discovery. *Expert Opin. Drug Discovery* 12 (10): 981–994.

43 Taylor, M.J., Lukowski, J.K., and Anderton, C.R. (2021). Spatially resolved mass spectrometry at the single cell: recent innovations in proteomics and metabolomics. *J. Am. Soc. Mass Spectrom.* 32 (4): 872–894.

44 Nickischer, D., Elkin, L., Cloutier, N. et al. (2018). Challenges and opportunities in enabling high-throughput, miniaturized high content screening. *Methods Mol. Biol.* 1683: 165–191.

45 Alvarez-Machancoses, O. and Fernandez-Martinez, J.L. (2019). Using artificial intelligence methods to speed up drug discovery. *Expert Opin. Drug Discovery* 14 (8): 769–777.

Index

Page locators in **bold** indicate tables. Page locators in *italics* indicate figures. This index uses letter-by-letter alphabetization.

High-Throughput Mass Spectrometry in Drug Discovery, First Edition. Edited by Chang Liu and Hui Zhang.
© 2023 John Wiley & Sons, Inc. Published 2023 by John Wiley & Sons, Inc.

d

u